Georg Hartwig

The Tropical World

Georg Hartwig

The Tropical World

ISBN/EAN: 9783741186332

Manufactured in Europe, USA, Canada, Australia, Japa

Cover: Foto ©Klaus-Uwe Gerhardt /pixelio.de

Manufactured and distributed by brebook publishing software (www.brebook.com)

Georg Hartwig

The Tropical World

THE TROPICAL WORLD:

A POPULAR SCIENTIFIC ACCOUNT OF THE

NATURAL HISTORY OF THE ANIMAL AND VEGETABLE KINGDOMS

IN THE EQUATORIAL REGIONS.

By DR. G. HARTWIG,

AUTHOR OF
"THE SEA AND ITS LIVING WONDERS."

WITH EIGHT CHROMOXYLOGRAPHIC PLATES AND NUMEROUS WOODCUTS.

LONDON:
LONGMAN, GREEN, LONGMAN, ROBERTS, & GREEN.
1863.

PREFACE.

ENCOURAGED by the success of "THE SEA AND ITS LIVING WONDERS," I now venture to appear before the public with a *general* survey of animal and vegetable life in the Tropical regions.

A glance over the Contents will satisfy the reader that I have condensed a great mass of information within a narrow compass; and as I have constantly endeavoured to combine the useful and the agreeable, I hope that this treatise may both please the general reader and stand the test of a severer criticism.

G. HARTWIG.

HEIDELBERG, Nov. 8, 1862.

CONTENTS.

PART I.

ASPECTS OF TROPICAL NATURE.

CHAPTER I.

THE DIVERSITY OF CLIMATES WITHIN THE TROPICS.

Causes by which it is produced — Abundance and Distribution of Rain within the Tropics — The Trade Winds — The Belt of Calms — Tropical Rains — The Monsoons — Tornados — Cyclones — Typhoons — Storms in the Pacific — Devastations caused by Hurricanes on Pitcairn Island and Rarotonga   Page 3

CHAPTER II.

THE LLANOS.

Their Aspect in the Dry Season — Vegetable Sources — Land Spouts — Effects of the Mirage — A Savannah on Fire — Opening of the Rainy Season — Miraculous Changes — Exuberance of Animal and Vegetable Life — Conflict between Horses and Electrical Eels — Beauty of the Llanos at the Termination of the Rainy Season — The Mauritia Palm   14

CHAPTER III.

THE PUNA, OR THE HIGH TABLE-LANDS OF PERU AND BOLIVIA.

Striking Contrast with the Llanos — Northern Character of their Climate — The Chuñu — The Surumpe — The Veta: its Influence upon Man, Horses, Mules, and Cats — The Vegetation of the Puna — The Maca — The Llama: its invaluable Services — The Huanacu — The Alpaca — The Vicuñas: Mode of Hunting Them — The Chacu — The Bolas — The Chinchilla — The Condor — Wild Bulls and Wild Dogs — Lovely Mountain Valleys   .   .   23

CHAPTER IV.

THE PERUVIAN SAND-COAST.

Its desolate Character — The Mule is here the "Ship of the Desert"— A Shipwreck and its Consequences — Sand-Spouts — Medanos — Summer and Winter — The Garuas — The Lomas — Change produced in their Appearance during the Season of Mists — Azara's Fox — Wild Animals — Birds — Reptiles — The Chincha or Guano Islands . . . . . Page 36

CHAPTER V.

THE AMAZONS, THE GIANT RIVER OF THE TORRID ZONE.

The Course of the Amazons and its Tributaries — The Huallaga — The Ucayale — The Içá — The Yapura, Jutay, Jurua, Teffe, Coary, and Purus — The Rio Negro — The Madeira — The Strait of Obydos — Tide Waves on the Amazon — The Tapajos — The Xingu — The Bay of the Thousand Islands — The Pororocca — Rise of the River — Inundated Forests — Lagunes — Magnificent Scenery — Fishing Scene — Different Character of the Forests beyond and within the verge of Inundation — General Character of the Banks — A Sail on the Amazons — A Night's Encampment — The "Mother of the Waters" — The Piranga — Dangers of Navigating on the Amazons — Terrific Storms — Rapids and Whirlpools — The Stream of the Future — Travels of Orellana — Madame Godin 44

CHAPTER VI.

THE KALAHARI.

Reasons why Droughts are prevalent in South Africa — Vegetation admirably suited to the Character of the Country — Number of Tuberous Roots — The Caffre Water-Melon — The Naras — The Mesembryanthemums — The Animal Life of the Kalahari — The Bushmen, a Nomadic Race of Hunters — The Bakalahari — Their Love for Agriculture — Their Ingenuity in procuring Water — Trade in Skins 61

CHAPTER VII.

THE SAHARA.

Its uncertain Limits — Caravan Routes — Ephemeral Streams — Oases — Inundations — Luxuriant Vegetation of the Oases contrasted with the surrounding Desert — The Sedentary and Vagrant Tribes — Harsh contrasts of Light and Shade — Sublimity of the Desert — The Khamsin — The Dying Slave — Sand-Spouts — Venomous Snakes — Porcupine-catching — Chase of the Gazelle — Fluctuation of Animal Life according to the Seasons 68

CHAPTER VIII.

THE PRIMITIVE FOREST.

Its peculiar Charms and Terrors — Disappointments and Difficulties of the Botanist — Variety of Trees and Plants — Character of the Primitive Forest according to its Site — Its Aspect during the Rainy Season — A Hurricane in the Forest — Beauty of the Forest after the Rainy Season — Bird Life on the rivers of Guiana — Morning Concert — Repose of Nature at Noon — Nocturnal Voices of the Forest . . . . Page 78

CHAPTER IX.

THE MEXICAN PLATEAUS, AND THE SLOPES OF SIKKIM.

Geological Formation of Mexico — The *Tierra Caliente* — The *Tierra Templada* — The *Tierra Fria*.

The Sylvan Wonders of Sikkim — Changes of the Forest on ascending — The Torrid Zone of Vegetation — The Temperate Zone — The Coniferous Belt — Limits of Arboreal Vegetation — Animal Life 86

CHAPTER X.

MANGROVE VEGETATION.

Its Peculiarities of Growth and beautiful Adaptation to its Site — Its Importance in furthering the Growth of Land — Animal Life among the Mangroves — "Jumping Johnny" — Victimised by a Grapsus — Insalubrity of the Mangrove Swamps — Uses of the Mangrove Trees — Avicennias and Sonneratias . 94

PART II.

TROPICAL PLANTS.

CHAPTER XI.

GIANT TREES AND CHARACTERISTIC FORMS OF TROPICAL VEGETATION.

General Remarks — The Baobab — Used as a Vegetable Cistern — Arborescent Euphorbias — The Dracæna of Orotava — The Sycamore — The Banyan — The sacred Bo-Tree of Anarajapoora — The Teak Tree — The Saul — The Sandal Tree — The Satinwood Tree — The Ceiba — The Mahogany Tree — The Mora — Bamboos — The Guadua — Beauty and multifarious Uses of these colossal

Grasses — Firing the Jungle — The Aloes — The Agave americana — The Bromelias — The Cactuses — The Mimosas — Bushropes — Climbing Trees — Emblems of Ingratitude — Marriage of the Fig Tree and the Palm — Epiphytes — Water-Plants — Singularly-shaped Trees — The Barrigudo — The Bottle Tree — Trees with Buttresses, fantastical Roots, and formidable Spines . . Page 101

CHAPTER XII.

PALMS.

The Cocoa-nut Tree — Its hundred Uses — Cocoa-nut Oil — Coir — Porcupine Wood — Enemies of the Cocoa Palm — The Sago Palm — The Saguer — The Gumatty — The Areca Palm — The Palmyra Palm — The Talipot — The Cocoa de Mer — Ratans — A Ratan Bridge in Ceylon — The Date Tree — The Oil Palms of Africa — The Oil Trade at Bonny — Its vast and growing Importance — American Palms — The Carnauba — The Ceroxylon andicola — The Cabbage Palm — The Gulielma speciosa — The Piaçava — Difficulties of the Botanist in ascertaining the various Species of Palms — Their wide Geographical Range — Different Physiognomy of the Palms according to their height — The Position and Form of their Fronds — Their Fruits — Their Trunk — The Yriartea Ventricosa 128

CHAPTER XIII.

THE CHIEF NUTRITIVE PLANTS OF THE TORRID ZONE.

Rice — Various Aspect of the Rice-fields at different Seasons — Ladang and Sawa Rice — The Cultivation of Rice in South Carolina of modern Date — The Rice-bird — Great Mortality among the Negroes — Arracan and Pegu — Growing Importance of the Port of Akyab — Maize — First imported from America by Columbus — Its enormous Productiveness — Its Cultivation in the United States — Its wide zone of Cultivation — Maize-beer, or Chicha — Millet, Dhourra — The Bread Fruit — Its Importance in the South Sea Islands — History of its Transplantation to the West Indies — Adventures of Bligh and Christian — Pitcairn Island — Bananas — Their ancient Cultivation — Avaca or Manilla Hemp — Humboldt's Remarks on the Banana — The Traveller's Tree of Madagascar — The Cassava Root — Tapioca — Yams — Batatas — Quinoa — Arrowroot — Taro — Tropical Fruit Trees — The Chirimoya — The Litchi — The Mangosteen — The Mango 154

CHAPTER XIV.

SUGAR.

Its commercial importance — Its original home — The progress of its cultivation throughout the Tropical Zone — The Tahitian Sugar-cane — Description of the Plant — Mode of extracting the Sugar — The enemies of the Sugar-cane — The Sugar Harvest 182

CHAPTER XV.

COFFEE.

Enarea and Kaffa — Gemaledie introduces its use into Arabia — Fanaticism endeavours to forbid it — The first Coffee-Houses in London and Marseilles — Coffee-production in Brazil — Java — Ceylon — Rapid extension of its culture in Ceylon since 1825 — The Coffee-Plant — The best situations for its growth — Its cultivation in Java described — The Musang — Expertness of the Ceylon Woodmen in preparing the Coffee Ground — Enemies of the Coffee-Plant — The Golunda Rat — The Lecanium Coffeæ . . . . Page 189

CHAPTER XVI.

CACAO AND VANILLA.

The Cacao Tree — Mode of preparing the Beans for the Market — Chocolate — The Vanilla Plant 199

CHAPTER XVII.

COCA.

Its immense Consumption in Peru and Bolivia — Mode of chewing the leaves — Its wonderfully strengthening effects — Fatal consequences of its abuse — Reasons which prompted the Spaniards to interdict its use, and finally to allow and encourage it — Its chemical analysis by Professor Wöhler of Göttingen 202

CHAPTER XVIII.

COTTON.

Amazing rise of the Cotton Manufactory, unparalleled in the Annals of Commerce — The Cotton Plants — Their culture in the Confederate States and in India — Prospects of Cotton cultivation in India — Brazilian and Egyptian Cotton — Prospects held forth by Africa — Cotton-seed Oil 207

CHAPTER XIX.

CAOUTCHOUC AND GUTTA PERCHA.

The Caoutchouc-Tree — Siphonia Elastica — Manner in which the Resin is collected — Urceola Elastica — Ficus Elastica — Has but recently become important — Mackintosh — Vulcanised Caoutchouc — Multiplicity of uses to which Caoutchouc may be applied — Marine Glue — The Gutta Percha Tree — Properties of the Resin — Its importance for Marine Telegraphy — Will the Tropical Zone be able to satisfy the growing demand ? . . . 216

CHAPTER XX.

TROPICAL SPICES.

The Cinnamon Gardens of Ceylon — Immense profits of the Dutch — Decline of the Trade — Neglected state of the Gardens — Mode of preparing the Rind — Nutmegs and Cloves — Cruel monopoly of the Dutch — A Spice Fire in Amsterdam — The Clove Tree — Beauty of an Avenue of Clove Trees — The Nutmeg Tree — Mace — The Pepper Vine — The Pimento Tree . . Page 222

CHAPTER XXI.

TROPICAL VEGETABLE DYES.

Indigo — Indigofera tinctoria — Mode of its Cultivation, and preparation of the Dye for the Market — Indigo Factories in Bengal.
Logwood — The British Logwood Cutters in Honduras — Their mode of living and disputes with the Spaniards — Brazil Wood — Red Saunders — Arnatto — Fustic — Turmeric 234

PART III.

TROPICAL ANIMALS.

CHAPTER XXII.

THE INSECT PLAGUES OF THE TROPICAL WORLD.

The Universal Dominion of Insects — Mosquitoes — Stinging Flies — *Œstrus Hominis* — The Chegoe or Jigger — The *Filaria Medinensis* — The Bête-Rouge — Blood-sucking Ticks — Garapatas — The Land-leeches in Ceylon — The Tsetsé Fly — The Tsalt-Salya — The Locust — Its dreadful Devastations — Cockroaches — The Drummer — The Cucarachas and Chilicabras . . 245

CHAPTER XXIII.

TROPICAL INSECTS DIRECTLY USEFUL TO MAN.

The Silk-worm — The Tussch and Arandi — The Cochineal Insect — The Gumlack Insect — The Locust used as Food — Other edible Insects — Insects used as Ornaments — The Diamond-beetle 259

CHAPTER XXIV.

THE ENTOMOLOGICAL WONDERS OF THE TROPICS.

Gradual Decrease of Insect-life on advancing towards the poles — Vast number of Beetles in Brazil — The Hercules Beetle — The Goliath — The Inca Beetle — Other colossal Insects — The Walking-leaf and Walking-stick Insects — The Soothsayer — Luminous Beetles — The Cocujas . . . Page 265

CHAPTER XXV.

ANTS AND TERMITES.

Vast numbers of Ants in the Tropical Zone — Excruciating pain caused by the Sting of the Ponera Clavata — The Black Fire-Ant of Guiana — The Dimiya of Ceylon — The Kaddiya — The Red Ant of Angola — Devastations of the Viviagua in the West Indian Coffee Plantations — The Atta-cephalotes, or the Umbrella Ant — Household Plagues — Difficulty of preserving Sugar from their attacks — The Ranger Ants — Wonderful construction of Tropical Ants — Slave-making Ants — Cow-keeping Ants — The Mexican Honey Ant — Devastation of the Termites — Their Services and Uses — Their marvellous Buildings — Formation of a Termite Colony — Amazing fecundity of the Termite Queen — Consequence of an Attack upon a Termite Hill — Wars between Termites and Black Ants — American Termites — Termites esteemed as a Delicacy — Marching White Ants — Mysteries of Termite Life 272

CHAPTER XXVI.

TROPICAL SPIDERS AND SCORPIONS.

Immense Webs of several Tropical Spiders — Their Means of Defence — Beautiful Colouring of the Epeiras — The Trap-door Spider — Wonderful Maternal Instincts of Spiders — Enemies of the Spiders — Their Usefulness — Mortal Combat between a Spider and a Cockroach — Scorpions — Dreadful Effects of the Sting of Tropical Scorpions — A Scorpion Battle — The Galeodes — Combat of a Galeode and a Lizard — Formidable character of the Tropical Centipedes 291

CHAPTER XXVII.

THE TROPICAL OCEAN.

Wanderings of an Iceberg — The Tropical Ocean — The Cachalot — The Frigate Bird — The Tropic Bird — The Esculent Swallow — The Flying-fish — The Bonito — The White Shark — Tropical Fishes — Crustaceans — Land Crabs — Mollusks — Jelly Fish — Coral Islands 302

CHAPTER XXVIII.

SNAKES.

First Impression of a Tropical Forest — Exaggerated Fears — Comparative rareness of Venomous Snakes — Their Habits and External Characters — Anecdote of the Prince of Neu Wied — The Bite of the Trigonocephalus — Antidotes — Fangs of the Venomous Snakes described — The Bush-Master — The Echidna Ocellata — The Rattlesnakes — Extirpated by Hogs — The Cobra de Capello — Indian Snake-Charmers — Maritime Excursions of the Cobra — The Egyptian Haje — The Cerastes — Boas and Pythons — The Jiboya — The Anaconda — Enemies of the Serpents — The Secretary — The Adjutant — The Mungoos — A Serpent swallowed by another — The Locomotion of Serpents — Anatomy of their Jaws — A Python-Meal — Serpents feeding in the Zoological Gardens — Domestication of the Rat-Snake — Water-Snakes . . . . Page 310

CHAPTER XXIX.

LIZARDS, FROGS AND TOADS.

Their Multitude within the Tropics — The Geckoes — Anatomy of their Feet — The Anolis — Their Love of Fight — The Chameleon — Its wonderful Changes of Colour — Its Habits — Peculiarities of its Organisation — The Iguana — The Teju — The Water-Lizards — Lizard Worship on the Coast of Africa — The Flying Dragon — The Basilisk — Frogs and Toads — The Pipa — The Bahia Toad — The Giant Toad — The Musical Toad — Brazilian and Surinam Tree-Frogs 328

CHAPTER XXX.

TORTOISES AND TURTLES.

The Galapagos — The Elephantine Tortoise — The Marsh-Tortoises — Mantega — River-Tortoises — Marine Turtles — On the Brazilian Coast — Their numerous Enemies — The Island of Ascension — Turtle-Catching at the Bahama and Keeling Islands — Turtle caught by means of the Sucking-Fish — The Green Turtle — The Hawksbill Turtle — Turtle-Scaling in the Feejee Islands — Barbarous mode of selling Turtle-Flesh in Ceylon — The Coriaceous Turtle — Its awful Shrieks 339

CHAPTER XXXI.

CROCODILES AND ALLIGATORS.

Their Habits — The Gavial and the Tiger — Mode of Seizing their Prey — Their Voice — Their Preference of Human Flesh — Alligator against Alligator — Wonderful Tenacity of Life — Tenderness of the Female Cayman for her Young — Enemies of the Crocodile — Torpidity of Crocodiles during the Dry Season — Their Awakening from their Lethargy with the First Rains — "Tickling a Crocodile." . . . 350

CHAPTER XXXII.

TROPICAL BIRD LIFE IN BOTH HEMISPHERES.

The Toucan—Its Quarrelsome Character—The Humming-birds—Their wide Range over the New World—Their Habits—Their Enemies—Their Courage—The Cotingas—The Campanero—The Tangaras—The Manakins—The Cock of the Rock—The Troupials—The Baltimore—The Pendulous Nests of the Cassiques—The Mocking-bird—Strange Voices of Tropical Birds—The Goat-sucker's Wail—The Organista—The Cilgero—The Flamingos—The Scarlet Ibis—The Jabiru—The Roseate Spoon-bill—The Jacana—The Calao—The Sun-birds—The Melithreptes—The Argus—The Peacock—The Lyre Bird—The Birds of Paradise—The Australian Bower-bird—The Talegalla—The Devil-bird—The Baya—The Tailor-bird—The Republican Gros-beak—The Korwé . . . . Page 358

CHAPTER XXXIII.

TROPICAL BIRDS OF PREY.

The Condor — His Marvellous Flight — His Cowardice — Various Modes of Capturing Condors — Ancient Fables circulated about them — Comparison of the Condor with the Albatross — The Carrion Vultures — The King of the Vultures — Domestication of the Urubu — Its Extraordinary Memory — The Harpy Eagle — Examples of his Ferocity — The Oricou — The Bacha — His Cruelty to the Klipdachs — The Fishing Eagle of Africa — The Musical Sparrow-hawk — The Secretary Eagle 387

CHAPTER XXXIV.

THE OSTRICH AND THE CASSOWARY.

Size of the Ostrich —His astonishing Swiftness —Ostrich Hunting — Stratagem of the Ostrich for protecting its Young — The poisoned Arrow of the Bushmen — Enemies of the Ostrich — The White Vulture — Points of Resemblance with the Camel — Voice of the Ostrich said to resemble that of the Lion — Its Voracity — Ostrich Feathers — Bechuana Parasols — Domestication of the Ostrich in Algeria — The American Rheas — The Cassowary — The Australian Emu 397

CHAPTER XXXV.

PARROTS.

Their Peculiar Manner of Climbing — Points of Resemblance with Monkeys — Their Social Habits — Their Connubial Felicity — Inseparables — Talent for Mimicry —Wonderful Powers of Speech and Memory— Their Wide Range within the Temperate Zones — Colour of Parrots Artificially Changed by the South American Indians—The Cockatoos—Cockatoo killing in Australia—The Macaw—The Parrakeets 408

CHAPTER XXXVI.

THE CAMEL.

The Ship of the Desert — Paramount Importance of the Camel on the Great Tropical Sand-wastes — His Organisation admirably adapted to his Mode of Life — Horrors and Beauties of the Desert — The Camel an Instrument of Freedom — The Robber Bedouin — Immemorial Thraldom of the Camel — His Unamiable Character — Excuses that may be urged in his behalf . Page 417

CHAPTER XXXVII.

THE GIRAFFE AND THE ZEBRA.

Beauty of the Giraffe — Its Wide Range of Vision — Use of its Horns — Giraffe Hunting — The Giraffes of the Zoological Gardens — The Quagga — The Douw — The Zebra — Its Lamentable Wailings — Its Inaccessible Retreats . . 423

CHAPTER XXXVIII.

THE HIPPOPOTAMUS.

Behemoth — Its Diminishing Numbers and Contracting Empire — Its Ugliness — A Rogue Hippopotamus or Solitaire — Dangerous Meeting — Intelligence and Memory of the Hippopotamus — Methods employed for killing the Hippopotamus — Hippopotamus Hunting on the Teoge — The Hippopotamus in Regent's Park — A Young Hippo born in Paris 432

CHAPTER XXXIX.

THE RHINOCEROS.

Brutality of the Rhinoceros — The Borelo — The Keitloa — The Monoho — The Kobaaba — Difference of Food and Disposition between the Black and the White Rhinoceros — Incarnation of Ugliness — Acute Smell and Hearing — Defective Vision — The Buphaga Africana — Paroxysms of Rage — Parental Affection — Nocturnal Habits — Rhinoceros-hunting — Adventures of the Chase — Narrow Escapes of Messrs. Oswell and Andersson — The Indian Rhinoceros — The Sumatran Rhinoceros — The Javanese Rhinoceros — Its involuntary Suicide . 440

CHAPTER XL.

THE ELEPHANT.

Love of Solitude and Pusillanimity — Miraculous Escape of an English Officer — Sagacity of the Elephant in ascending Hills — Organisation of the Stomach — The Elephant's Trunk — Use of its Tusks still Problematical — The Rogue-Elephant — Sagacity of the Elephant — The African Elephant — Tamed in Ancient Times — South African Elephant-hunting — Hair-breadth Escapes —

Abyssinian Elephant-hunters — Importance of the Ivory Trade — The Asiatic Elephant — Vast Numbers destroyed in Ceylon — Major Rogers — Elephant Catchers — Their amazing Dexterity — The Corral — Decoy Elephants — Their astonishing Sagacity — Great Mortality among the captured Elephants — Their Services . . . . Page 449

CHAPTER XLI.

THE FELIDÆ OF THE OLD WORLD.

The Lion — Conflicts with Travellers on Mount Atlas — The Lion and the Hottentot — A Lion taken in — Narrow Escapes of Andérsson and Dr. Livingstone — Lion-hunting by the Arabs of the Atlas — By the Bushmen — The Asiatic Lion — The Lion and the Dog — The Tiger — The Javanese Jungle — The Peacock — Wide Northern Range of the Tiger — Tiger-hunting in India — Miraculous Escape of an English Sportsman — Animals announcing the Tiger's Presence — Turtle-Hunting of the Tiger on the Coasts of Java — The Panther and the Leopard — The Leopard attracted by the Smell of Small-Pox — The Cheetah — The Hyæna — Fables told of these abject Animals — The Striped Hyæna — The Spotted Hyæna — The Brown Hyæna 466

CHAPTER XLII.

THE FELIDÆ OF THE NEW WORLD.

The Jaguar — Its Boldness — Jaguar Hunting — Heroic Conflict of Three Brazilian Herdsmen with a Jaguar — The Couguar, or the Puma — His Cowardice — The Ocelot 486

CHAPTER XLIII.

THE SLOTH.

Miserable Aspect of the Sloth — His Beautiful Organisation for his Peculiar Mode of Life — His Rapid Movements in the Trees — His Means of Defence — His Tenacity of Life — Fable about the Sloth refuted — The Ai — The Unau — The Mylodon Robustus 492

CHAPTER XLIV.

THE ANT-EATERS OF THE NEW AND THE OLD WORLD.

The Great Ant-Bear — Its Way of Licking up Termites — Its Formidable Weapons — A Perfect Forest-Vagabond — Its Peculiar Manner of Walking — The Smaller Ant-Eaters — The Manides — The African Orycteropi — The Porcupine Ant-Eater of Australia — The Myrmecobius Fasciatus — The Armadillo — The Glyptodon 497

CHAPTER XLV.

TROPICAL BATS.

Wonderful Organisation of the Bats — The Fox-Bat — The Vampire — Its Blood-sucking Propensities — The Vampire and the Scotchman — The Horse-Shoe Bats — The Nycteribia — The Flying Squirrel — The Galeopithecus — The Anomalurus . . . . Page 505

CHAPTER XLVI.

THE SIMIÆ OF THE OLD WORLD.

Forest-Life — Excellent Climbers — Bad Pedestrians — Imperfectly known to the Ancients — Similitude and Difference between the Human Race and the Apes — The Chimpanzee — Chim in Paris — The Gorilla — The Uran — The Gibbons — A Siamang on Board — The Proboscis Monkey — The Huniman — The Wanderoo — The Cercopitheci — The Maimon — "Happy Jerry" — The Pig-Faced Baboon — The Derryas — The Loris and Makis 514

CHAPTER XLVII.

THE SIMIÆ OF THE NEW WORLD.

Wide Difference between the Monkeys of both Hemispheres — The Prehensile Tail — The Wourali Poison — Mildness of the American Monkeys — The Stentor Monkey — The Spider Monkeys — The Sajous — The Fox-tailed Monkeys — The Saïmiris — Friendships between various kinds of Monkeys — Nocturnal Monkeys — Squirrel Monkeys — Their Lively Intelligence 531

LIST OF ILLUSTRATIONS.

CHROMOXYLOGRAPHS.

Condor Catching	.	. *Frontispiece*	Cotton Field .	. *To face page* 207
Savannah on Fire	.	*To face page* 14	Termite Hills	. ,, ,, 281
Primitive Forest	.	,, ,, 78	Flamingoes .	. ,, ,, 358
Cereus Giganteus	.	,, ,, 118	Tiger .	. ,, ,, 466

WOODCUTS.

	PAGE		PAGE
African Bushmen	61	Caravan	417
Birds:—		*Ceylonese Cocoa-nut Oil-mill	128
Adjutant . . .	322	*Coffee Estate, General Fraser's, at	
Argus Pheasant . .	376	Rangbodde, Ceylon .	198
Baltimore Bird . .	367	Coral Island . . .	302
Bird of Paradise . .	378	Fishes:—	
Campanero . . .	365	Coryphæua . . .	306
Cassowary . . .	407	Electrical Eel (Gymnotus	
Cardinal . . .	87	electricus) . .	20
Condor . . .	387	Limulus . . .	307
Emu	407	Sucking Fish . .	347
Frigate Bird . .	304	Sun Fish . . .	307
Harpy Eagle . .	393	Guano Island . . .	36
Hornbill, Rhinoceros .	374	High Table Lands of Peru .	23
Humming Bird . .	362	Insects:—	
Ibis, Egyptian . .	372	Ants and Termites .	272
Java Sparrow . .	155	Beetle, Goliath . .	263
Lory	413	Diamond . .	264
Lyre Bird . . .	377	Hercules . .	266
Macaw . . .	415	Inca . . .	267
Blue . . .	49	Buprestis gigas . .	264
Mocking-Bird . .	88	Centipede . . .	301
Ostrich . . .	398	Cicada . . .	263
-catching . .	397	Cochineal . . .	260
Parakeet, green . .	416	Cocujas . . .	270
Parrots . . .	408	Lantern Fly . .	271
Peacock, Javanese .	376	Locust . . .	254
Rice Bunting . .	158	Mantis . . .	269
Secretary Bird . .	321	Membracidæ . .	280
Talegalla . . .	380	Mosquito . . .	246
Toucan . . .	360	Mormolyce, Javanese .	270
Turkey Buzzard . .	390	Phyllium . . .	268
Vulture, King . .	390	Scorpion . . .	300
Vulture, Sociable .	394	Spider, Diadem . .	299
Woodpecker, ivory-billed	83	Termite . . .	285

LIST OF ILLUSTRATIONS

Insects—continued:
	PAGE
Termite, Soldier	286
Tsetse	258

Mammalia:—
	PAGE
Aguti	17
Alpaca	28
Ant-bear	497
Armadillo, Poyou	501
Giant	503
Axis	475
Camel, Bactrian	420
Capybara	351
Cheetah	483
Chimpanzee	517
Chinchilla	33
Coatimondi, Rufous	518
*Coffee Rat	189
Diana Monkey	526
Dromedary	421
Echidna, Porcupine	503
Elephants	449
Indian	454
*Flying Foxes	505
* at rest	513
Galeopithecus volans	512
Gibbon, or long-armed Ape	522
Giraffe, Skull of	424
Giraffes and Zebras	423
Gnu	430
Hippopotamus	432, 433
Howling Monkey	535
Ichneumon	355
Jackal	480
Jaguar	486
Koodoo	65
Lemur, slow-paced	529
handed	530
*Leopard and Cheetah	466
Lion	467
Llama	26
Malay Bear	134
Mandrill	528
*Mongoos	323
*Monkeys of the Old World	514
Musk Deer	92
Nylghau	476
Ocelot	491
Opossum	40
Orycteropus Capensis	500
Palm Squirrel	134
*Pangolin, the Indian	504
Peccary	17
Pig-faced Baboon	528
Porcupine	75
Puma	490
Quagga	428
Rhinoceros	440, 441
Indian	447
*Rhinolophus	510
Sloth	492

Mammalia—continued:
	PAGE
Spider Monkey, Black	536
Striped Hyæna	484
Tarsius Bancanus	530
Tiger	475
Uran	520
Whale, Sperm	303
Zebra	429

Plants:—
	PAGE
*Areca Palm	153
Banana and the Plantain	154
Banyan	106
*Baobab Trees at Manaar	101
*Bo-tree, the Sacred	109
Bottle-tree	124
Caoutchouc Trees—Indians incising them	216
Cocoa-nut tree	129
Coffee	193
Cinnamon	222
Clove	229
Date-tree	142
Dragon-tree	105
*Fig-tree	122
Indigo Plant	234
Mangosteen	181
Mangrove-tree	94
Mimosa	120
Nepenthes	15
Nutmeg	230
Oil Palm	145
Pepper Plant	231
*Snake-tree	125
Sugar Cane	182
Sycamore	106
Yriartea ventricosa	152

Reptiles:—
	PAGE
Alligator	351
Amblyrhyne	339
Anolis	330
Basilisk	336
Boa	319
Chameleon	331
Crocodile	350
Flying Dragon	336
Gecko	329
Iguana	332
Monitor	333
Rattlesnake	316
Toad, Bahia	337
Surinam	337
Toad and Anolis	328
Tortoise, Marsh	342
Turtle, Green	347
Loggerhead	349
*Uropeltis Philippinus	310
Water Snake	327
Tower in Agades	68
Tropical Tornado	3
Tuaryk Chieftain	77

* The illustrations marked with an asterisk are taken, by permission, from Sir James Emerson Tennent's two works on Ceylon.

PART I.

ASPECTS OF TROPICAL NATURE

Tropical Tornado.

CHAPTER I.

THE DIVERSITY OF CLIMATES WITHIN THE TROPICS.

Causes by which it is produced — Abundance and Distribution of Rain within the Tropics — The Trade Winds — The Belt of Calms — Tropical Rains — The Monsoons — Tornados — Cyclones — Typhoons — Storms in the Pacific — Devastations caused by Hurricanes on Pitcairn Island and Rarotonga.

ON surveying the various regions of the torrid zone, we find that Nature has made many wonderful provisions to mitigate the heat of the vertical sun, to endow the equatorial lands with an amazing variety of climate, and to extend the benefit of the warmth generated within the tropics to countries situated far beyond their bounds.

Thus, while the greater part of the northern temperate zone is occupied by land, the floods of ocean roll over by far the greater portion of the equatorial regions, — for both torrid America and Africa appear as mere islands in a vast expanse of sea.

The conversion of water, by evaporation, into a gaseous form is accompanied by the abstraction of heat from surrounding bodies, or, in popular language, by the production of cold; and thus over the surface of the ocean the rays of the sun have a tendency to check their own warming influence, and to impart a coolness to the atmosphere, the refreshing effects of which are felt wherever the sea breeze blows. There can, therefore, be no doubt that, if the greater part of the tropical ocean were converted into land, the heat of the torrid zone would be far more intolerable than it is.

The restless winds and currents, the perpetual migrations of the air and waters, perform a no less important part in cooling the equatorial and warming the temperate regions of the globe. Rarefied by the intense heat of a vertical sun, the equatorial air-stream ascends in perpendicular columns high above the surface of the earth, and thence flows off towards the poles; while, to fill up the void, cold air-currents come rushing from the arctic and antarctic regions.

If caloric were the sole agent on which the direction of these antagonistic air-currents depended, they would naturally flow to the north and south; but the rotation of the earth gradually diverts them to the east and west, and thus the cold air-currents, or polar streams, ultimately change into the trade winds which regularly blow over the greater part of the tropical ocean from east to west, and materially contribute, by their refreshing coolness, to the health and comfort of the navigator whom they waft over the equatorial seas.

While the polar air-currents, though gradually warming as they advance, thus mitigate the heat of the torrid zone, the opposite equatorial breezes, which reach our coasts as moist south-westerly or westerly winds, soften the cold of our winters, and clothe our fields with a lively verdure during the greater part of the year. How truly magnificent is this grand system of the winds, which, by the constant interchange of heat and cold which it produces, thus imparts to one zone the beneficial influence of another, and renders both far more fit to be inhabited by civilized man. The Greek navigators rendered homage to Æolus, but they were far from having any idea of the admirable laws which govern the unstable, ever-fluctuating domains of the " God of the Winds."

The same unequal influence of solar warmth under the line and at the poles, which sets the air in constant motion, also compels the waters of the ocean to perpetual migrations, and produces those wonderful marine currents which, like the analogous atmospheric streams, furrow in opposite directions the bosom of the sea. Thanks to this salutary interchange, the Gulf Stream, issuing from the Mexican Sea, and thence flowing to the north and east, conveys a portion of its original warmth as far as the west coast of Spitzbergen and Nowaja Semlja; while in the southern hemisphere we see the Peruvian stream impart the refrigerating influence of the antarctic waters to the eastern coast of South America.

The geographical distribution of the land within the tropics likewise tends to counterbalance or to mitigate the excessive heat of a vertical sun; for a glance over the map shows us at once that it is mostly either insular or extending its narrow length between two oceans, thus multiplying the surface over which the sea is able to exert its influence. Thus the Indian Archipelago, the peninsula of Malacca, the Antilles, and Central America, are all undoubtedly indebted to the waters which bathe their coasts for a more temperate climate than that which they would have had if grouped together in one vast continent.

The temperature of a country proportionally decreases with its elevation; and thus the high situation of many tropical lands moderates the effects of equatorial heat, and endows them with a climate similar to that of the temperate, or even of the cold regions of the globe. The Andes and the Himalaya, the most stupendous mountain-chains of the world, raise their snow-clad summits either within the tropics or immediately beyond their verge, and must be considered as colossal refrigerators, ordained by Providence to counteract the effects of the vertical sunbeams over a vast extent of land. In Western Tropical America, in Asia, and in Africa, we find immense countries rising like terraces thousands of feet above the level of the ocean, and reminding the European traveller of his distant northern home by their productions and their cooler temperature. Thus, by means of a few simple physical and geological causes acting and reacting upon each other on a magnificent scale, Nature has bestowed a wonderful variety of climate upon the tropical regions, producing a no less wonderful diversity of plants and animals.

But warmth alone is not sufficient to call forth a luxuriant vegetation: it can only exert its powers when combined with a sufficient degree of moisture; and it chiefly depends upon the presence or absence of water whether a tropical country appears as a naked waste or decked with the most gorgeous vegetation.

As the evaporation of the tropical ocean is far more considerable than that of the northern sea, the atmospherical precipitations (dew, rain) caused by the cooling of the air are far more abundant in the torrid zone than in the temperate latitudes. While the annual fall of rain within the tropics amounts, on an average, to about eight feet, it attains in Europe a height of only thirty inches; and under the clear equatorial sky the dew is often so abundant as to equal in its effects a moderate shower of rain.

But this enormous mass of moisture is most unequally distributed; for while the greater part of the Sahara and the Peruvian sand-coast are constantly arid, and South Africa and North Australia suffer from long-continued droughts, we find other tropical countries refreshed by almost daily showers, and the annual fall of rain in the West Indies and on the coast of Malabar rises to the enormous height of 274 and 283 inches. The direction of the prevailing winds, the condensing powers of high mountains and of forests, the relative position of a country, the nature of its soil, are the chief causes which produce an abundance or want of rain, and consequently determine the fertility or barrenness of the land. Of these causes, the first-mentioned is by far the most general in its effects,—so that a knowledge of the tropical winds is above all things necessary to give us an insight into the distribution of moisture over the equatorial world.

I have already mentioned the trade winds, or cool reactionary currents called forth by the ascending equatorial air-stream; but it will now be necessary to submit them to a closer examination, and follow them in their circular course throughout the tropical regions. In the Northern Atlantic, their influence, varying with the season, extends to 22° N. lat. in winter, and 39° N. lat. in summer; while in the southern hemisphere they reach no farther than 18° S. lat. in winter, and 28° or 30° S. lat. in summer.

In the Pacific, their limits vary between 21° and 31° N. lat., and between 23° and 33° S. lat.; so that, on the whole, they have here a more southern position, owing, no doubt, to the vast extent of open sea; while in the Atlantic the influence of the neighbouring continents forces them to the north, and even causes the trade winds of the southern hemisphere to ascend beyond the equatorial line. Their character is that of a continual soft breeze—strongest in the morning, remitting at noon, and again increasing in the evening. In the neighbourhood of the coasts, except over very small islands, they become weaker, and generally cease to be felt at a distance of about fifteen or twenty miles from the sea, though, of course, at greater heights they continue their course uninterruptedly over the land.

For obvious reasons the trade winds have been much more accurately investigated upon the ocean than on land, particularly in the Northern Atlantic, which is better known in its physical features than any other sea, as being a highway for numberless vessels to which the study of the winds is a matter of the greatest importance; yet, in spite of so many disturbing influences, their course, even over the continents, has been ascertained by travellers. North-easterly winds almost constantly sweep over the Sahara; and in South Africa, Dr. Livingstone informs us that north-easterly and south-easterly winds blow over the whole continent between 12° and 6° S. lat., even as far as Angola, where they unite with the sea winds.

In Brazil, the presence of the trade winds has been determined with still greater accuracy. Thus easterly breezes almost perpetually sweep over the boundless plains up to the slopes of the Andes, and even in Paraguay (25° S. lat.) a mild east wind constantly arises in summer after the setting of the sun.

As the trade winds originate in the coldest, and thence pass onwards to the warmer regions, they are, of course, constantly absorbing moisture as they advance over the seas. Saturated with vapours, they reach the islands and continents, where, meeting with various refrigerating influences (mountain-chains, forests, terrestrial radiation), their condensing vapours give rise to an abundance both of rain and dew. It is owing to their influence that in general, within the tropics, the eastern coasts, or the eastern slopes of the mountains, are better watered than the interior of the continents or lands with a western exposure.

An example on the grandest scale is afforded to us by South America, where the Andes of Peru and Bolivia so effectually drain the prevailing east winds of their moisture, that while numberless rivulets, the feeders of the gigantic Marañon, clothe their eastern gorges with a perpetual verdure, their western slopes are almost constantly arid. Such is the influence of this colossal barrier in interrupting the course of the air-current, that the trade wind only begins to be felt again on the Pacific at a distance of one hundred or even one hundred and fifty miles from the shore.

In South Africa, also, we find the eastern mountainous coastlands covered with giant timber — in striking contrast with the parched savannas or dreary wastes of the interior; and in the South Sea the difference of verdure between the east and west coasts of the Galapagos, the Sandwich Islands, the Feejees, and many other groups, never fails to arrest the attention of the mariner.

The trade winds of the northern and southern hemispheres do not, however, blow in one continuous stream over the whole breadth of the tropical ocean, but are separated from each other by a zone or belt of calms, occasioned by their mutually paralysing each other's influence on meeting from the north and the south-east, and by the attraction of the sun, which, when in the zenith, changes the easterly air-currents into an ascending stream. From this dependence on the position of the sun, it may easily be inferred that the zone of calms fluctuates, like the trade winds themselves, to the north or south, according to the seasons; and that it is far from invariably occupying the same degrees of latitude, or the same width, at all times of the year. In the Atlantic, from the causes previously mentioned, it constantly remains to the north of the line, where its breadth averages five or six degrees; in the Pacific it more generally extends, during the antarctic summer, on both sides of the equator.

Besides the intensity of its heat, the zone of calms is characterised by heavy showers, which regularly fall in the afternoon, and are caused by the cooling of the saturated air-columns in the higher regions of the air.

Daily, towards noon, dense clouds form in the sky, and dissolve in torrents of rain under fearful electrical explosions, now sooner, now later, of shorter or longer continuance, with increasing or abating violence, as the sun is more or less in his zenith. Towards evening the vapours disperse, and the sun sets in a clear, unclouded horizon. Thus towns or countries situated within the calms, such as Para, Quito, Bogota, Guayaquil, the Kingsmill Islands, may be said to have a perennial rainy season, as showers fall at every season of the year. To the north and south of the belt of calms, we find in both hemispheres a broad zone, characterised by two distinct rainy seasons, separated by two equally distinct periods of dry weather. The rainy seasons take place while the sun is crossing the zenith, and more or less paralysing the power of the trade winds. Cayenne, Honduras, Jamaica, Pernambuco, Bahia, afford us examples of these well-defined alternations. In Jamaica, for instance (18° N. lat.), the first rainy season begins in April, the second in October; the first dry season in June, and the second in December.

Towards the verge of the tropics follow the zones which are characterised by a single rainy season, but of a longer continuance, generally lasting six months, throughout the summer, or from one equinoctium to another. In these parts, the rainy season is also the warmest period of the year, since here the different height of the sun in winter and summer is already so considerable, that at the time of culmination the clouds and rain are not able to reduce the temperature below that of the clear and dry winter months; while in the zones which are situated nearer to the equator, the rainy season, in spite of the sun's culmination, is always the coolest.

To sum up the foregoing remarks in a few words: the two rainy seasons, which characterise the middle zone between each tropic and the line, have a tendency to pass into one annual rainy season on advancing towards the tropics, and to merge into a permanent rainy season on approaching the equator. As the sun goes to the north or south, he opens the sluices of heaven, and closes them as he passes to another hemisphere.

Such may be said to be the normal state which would everywhere obtain within the equatorial regions if one unbounded ocean covered their surface, and none of the disturbing influences previously mentioned interfered; but as we find the trade winds so frequently deflected from their course, we must also naturally expect the general or theoretical order of the dry and rainy seasons to be liable to great modifications.

Thus, in the Indian Ocean and in the Chinese Sea, terrestrial influences prevail during the summer which completely divert the trade wind, there called the *North-east Monsoon*, from its regular path. From the wide lands of south-eastern Asia, glowing, during the summer months, with a torrid heat, the rarefied air, as it rises into the higher regions, completely overpowers the usual course of the trade wind, and changes it into the south-western monsoon, which blows from May to September.

Hence, during the summer months, the saturated sea wind, striking against the western ghauts, brings rain to the coast of Malabar, while the opposite coast of Coromandel remains dry; but the inverse takes place when, from the sun's declining to the south, the north-east trade wind resumes its sway.

Similar deflections from the ordinary course of the trade winds occur also on the coast of Guinea (5° N. lat.), in the Mexican Gulf, and in that part of the Pacific which borders on Central America, through the influence of the heated plains of Africa, Utah, Texas, and New Mexico, and have a similar influence on the distribution of moisture. Thus the sea monsoon, which prevails during the summer months on the coast of northern Guinea, carries a rainy season over the land as far as the eighteenth or nineteenth degree of northern latitude, and fertilises a vast extent of country which, from its position on the western side of an immense continent, would otherwise have been as naked and barren as the Sahara.

As the tropical rains, though generally confined only to part of the year, and then only to a few hours of the day, fall in so much greater abundance than under our constantly drooping skies, it may naturally be supposed that the single showers must be proportionally violent. Descending in streams so close and so dense that the level ground, unable to absorb it sufficiently fast, is covered with a sheet of water, the rain rushes down the hill-sides in a volume that wears channels in the surface. For hours together the noise of the torrent, as it beats upon the trees and bursts upon the roofs, occasions an uproar that drowns the ordinary voice and renders sleep impossible. In Bombay nearly nine inches of rain have been known to fall in one day, and twelve inches in Calcutta, or nearly half the mean annual quantity of rain on the east coast of England. During one single storm which Castelnau witnessed at Pebas, on the Amazon, there fell not less than thirty inches of rain—

nearly as much as the annual supply of our west coast. The hollow trunk of an enormous tree in an exposed situation gave the French traveller the means of accurate measurement.

As in the equatorial regions the atmospherical precipitations are far more considerable than in the temperate zones, so also their storms rage with a violence unknown in our climes. In the Indian and Chinese Seas these convulsions of nature generally take place at the change of the monsoons; in the West Indies, at the beginning and at the end of the rainy seasons. The tornado which devastated the Island of Guadeloupe on the 25th July, 1846, blew down buildings constructed of solid stone, and tore the guns of a battery from their carriages; another, which raged some years back in the Mauritius, demolished a church and drove thirty-two vessels on the strand.

On the Beagle's arrival in Port Louis, after her long and arduous surveying voyage, a fleet of crippled vessels, the victims of a recent hurricane, might have been seen making their way into the harbour—some dismasted, others kept afloat with difficulty, firing guns of distress or giving other signs of their helpless condition. "On the now tranquil surface of the harbour lay a group of shattered vessels, presenting the appearance of floating wrecks. In almost all, the bulwarks, boats, and everything on deck, had been swept away; some, that were towed in, had lost all their masts; others, more or less of their spars; one had her poop and all its cabins swept away; many had four or five feet of water in the hold, and the clank of the pumps was still kept up by the weary crew." *

Such are the terrible effects of the tornados and cyclones of the Atlantic and the Indian Oceans; but the storms of the miscalled Pacific are no less furious and destructive. A hurricane, which on the 15th of April, 1845, burst over Pitcairn Island, washed all the fertile mould from the rocks, and, uprooting 300 cocoa-nut trees, cast them into the sea. Every fishing-boat on the island was destroyed, and thousands of fruit-bearing bananas were swept away. Four months of famine followed upon this terrific storm, to which the pious islanders meekly submitted as to the will of God.

The celebrated missionary, John Williams†, describes a

* Captain Stokes's "Discoveries in Australia."
† "Narrative of Missionary Enterprise in the South Sea Islands," p. 390.

similar catastrophe which befell the beautiful island of Rarotonga on the 23rd of December, 1831. The chapels, school-houses, mission-houses, and nearly all the dwellings of the natives, no less than a thousand in number, were levelled to the ground. Every particle of food on the island was destroyed. Of the thousands of banana or plantain trees which had covered and adorned the land, scarcely one was left standing, either on the plains, in the valleys, or upon the mountains. Stately trees, that had withstood the storms of ages, were laid prostrate on the ground, and thrown upon each other in the wildest confusion; while even of those that were still standing, many were left without a branch, and all perfectly leafless. So great and so general was the destruction, that no spot escaped; for the gale, veering gradually round the island, did most effectually its devastating work.

Though the tropical storms are thus frequently a scourge, they are often productive of no less signal benefits. Many a murderous epidemic has suddenly ceased after one of these natural convulsions, and myriads of insects, the destroyers of the planter's hopes, are swept away by the fierce tornado. Besides, if the equatorial hurricanes are far more furious than our storms, a more luxuriant vegetation effaces their vestiges in a shorter time. Thus Nature teaches us that a preponderance of good is frequently concealed behind the paroxysms of her apparently unbridled rage.

From these general outlines of the physical geography of the tropical world, I shall, in the next following chapters, proceed to a more detailed description of several equatorial lands distinguished by the extremes of moisture or aridity, of warmth or cold, according to their position and elevation above the level of the sea.

These "Aspects of Tropical Nature," if I thus may venture to call them, will serve to show the wonderful variety of climate which the causes above-mentioned engender under the same low latitudes, and their influence upon the development of animal and vegetable life.

Thus, while the *Sahara*, the *Kalahari*, and the *Peruvian sand-coast* afford striking examples of barrenness ensuing from the want of moisture, the vast region traversed by the mighty *Amazon* with its thousand tributary streams, and the gorgeous vegetation of the *primitive forest*, exhibit no less striking pictures of the fertility which abundant rain, together with heat, produces in the equatorial lands.

The *slopes of Sikkim* and the terraced *plateaus of Mexiço* will show us a gradual change of vegetation as we advance from the sea-borde to the higher mountainous regions,— the frigid, the temperate, and the torrid zones being, as it were, superposed one above the other; and while in the *Puna*, or high table-lands of Peru and Bolivia, we find the rigours of the north transported under the line, the influence of the alternating dry and rainy seasons appears on a magnificent scale in the vast *Llanos* of Venezuela and Guiana, to which I shall now introduce the reader.

CHAPTER II.

THE LLANOS.

Their Aspect in the Dry Season — Vegetable Sources — Land Spouts — Effects of the Mirage — A Savannah on Fire — Opening of the Rainy Season — Miraculous Changes — Exuberance of Animal and Vegetable Life — Conflict between Horses and Electrical Eels — Beauty of the Llanos at the Termination of the Rainy Season — The Mauritia Palm.

IN South America, the features of Nature are traced on a gigantic scale. Mountains, forests, rivers, plains, there appear in far more colossal dimensions than in our part of the world. Many a branch of the Marañon surpasses the Danube in size. In the boundless primitive forests of Guiana more than one Great Britain could find room. The Alps would seem but of moderate elevation if placed aside of the towering Andes; and the plains of Northern Germany and Holland are utterly insignificant when compared with the Llanos of Venezuela and New Grenada, which, stretching from the coast-chain of Caraccas to the forests of Guiana, and from the snow-crowned mountains of Merida to the Delta of the Orinoco, cover a surface of more than 250,000 square miles.

Nothing can be more remarkable than the contrast which these immeasurable plains present at various seasons of the year — now parched by a long-continued drought, and now covered with the most luxuriant vegetation. When, day after day, the sun, rising and setting in a cloudless sky, pours his vertical rays upon the thirsty Llanos, the calcined grass-plains present the monotonous aspect of an interminable waste. Like the ocean, their limits melt in the hazy distance with those of the horizon; but here the resemblance ceases, for no refreshing breeze wafts coolness over the desert, and comforts the drooping spirits of the wanderer.

In the wintry solitudes of Siberia the skin of the reindeer affords protection to man against the extreme cold; but in these sultry plains there is no refuge from the burning sun above and the heat reflected from the glowing soil, save where, at vast intervals, small clumps of the Mauritia palm afford a scanty shade. The water-pools which nourished this beneficent tree have long since disappeared; and the marks of the previous rainy season, still visible on the tall reeds that spring from the marshy ground, serve only to mock the thirst of the exhausted traveller. The long-legged jabiru and the scarlet ibis have forsaken the dried-up swamp which no longer affords them any subsistence, and only here and there a solitary Caracara falcon lingers on the spot, as if meditating on the vicissitudes of the season.

Yet even now the parched savannah has some refreshment to bestow, as Nature — which in the East Indian forests fills the pitchers of the Nepenthes with a grateful liquid, and in the waterless Kalahari causes many juicy roots to thrive under the surface of the desert — here also displays her bounty; for the globular melon-cactus, which flourishes on the driest soil, and not seldom measures a foot in diameter, conceals a juicy pulp under its tough and prickly skin. Guided by an admirable instinct, the wary mule strikes off with his fore-feet the long, sharp thorns of this remarkable plant, the emblem of good-nature under a rough exterior, and then cautiously approaches his lips to sip the refreshing juice. Yet, drinking from these living sources is not unattended with danger, and mules are often met with that have been lamed by the formidable prickles of the cactus. The wild horse and ox of the savannah, not gifted with the same sagacity, roam about a prey to hunger and burning thirst—the latter hoarsely bellowing, the former snuffing up the air with outstretched neck to discover by its moisture the neighbourhood of some pool that may have resisted the general drought.

Nepenthes.

Besides their interminable extent, the Llanos have several other points of resemblance to the sea. As here the waterspout, raised by contending air-currents, rises to the clouds and sweeps over the floods, thus also the dust of the savannah, set in motion by conflicting winds, ascends in mighty columns and glides over the desert plain. The glowing sand suspended in the air increases the sultriness of the atmosphere, and may even become dangerous to the traveller who cannot escape by a timely flight; for, seizing him with irresistible violence, it carries him along in its embrace, and then hurls him senseless to the ground.

As if "on a painted ocean," the becalmed ship lies motionless on the glassy sea. No breath of air ruffles the surface of the waters. The pennant hangs lazily from the mast; the water-casks are empty; the torments of thirst, aggravated by the heat of a vertical sun, become intolerable. But, suddenly, as if by magic, a beautiful island rises from the floods; waving palm-trees seem to welcome the mariner: he fancies he hears the purling of the brook and the splashing of the waterfall. Yet still the vessel remains immovable like a rock, and soon the fading phantom that mocked his misery leaves him the victim of increased despair.

Similar delusions of the *mirage*, produced by the refraction of the light as it passes through atmospherical strata of unequal warmth, and consequently of unequal density, likewise take place over the surface of the Llanos, which then assume the semblance of a large sea, heaving and rocking in wave-like motion. In the Lybian desert, in the dread solitudes of the polar ocean, in every zone, we meet with the same phenomenon, produced by the same cause.

As in the arctic regions the intense cold during winter retards the pulsations, or even suspends the operations of life, so in the Llanos the long continuance of drought causes a similar stagnation in animated nature. The thinly-scattered trees and shrubs do not indeed cast their foliage, but the greyish-yellow of their leaves announces that vegetation is suspended. Buried in the clay of the dried-up pools, the alligator and the water-boa lie plunged in a deep summer-sleep, like the bear of the north in his long winter slumber; and many animals which, at other times, are found roaming over the Llanos, — such as the graceful aguti, the hoggish peccary, and the timid deer of the savannah,— have left the parched plains and migrated to the forest or the river. The large maneless puma and the spotted jaguar, following their prey to less arid regions, are now no longer seen in their former hunting-grounds, and the Indian has also disappeared with the stag whom he pursued with his poisoned arrows. In the Siberian Tundras the reindeer and the migratory birds are scared away by winter; here life is banished or suspended by an intolerable heat.

Aguti.

Sometimes the ravages of fire are added to complete the image of death on the parched savannah.

"We had not yet penetrated far into the plain," says Schomburgk, "when we saw to the south-east high columns of smoke ascending to the skies, the sure signs of a savannah fire, and at the same time the Indians anxiously pressed us to speed on, as the burning torrent would most likely roll in our direction.

Peccary.

Although at first we were inclined to consider their fears as exaggerated, yet the next half-hour served to convince us of the extreme peril of our situation. In whatever direction we gazed, we nowhere saw a darker patch in the grass-plain announcing the refuge of a water-pool; we could already distinguish the flames of the advancing column, already hear the bursting and crackling of the reeds, when fortunately the sharp eye of the Indians discovered some small eminences before us, only sparingly covered with a low vegetation, and to these we now careered as if Death himself were behind us. Half a minute later, and I should never have lived to relate our adventures. With beating hearts we saw the sea of fire rolling its devouring billows towards us; the suffocating smoke, striking in our faces, forced us to turn our back upon the advancing conflagration, and to await the dreadful decision with the resignation of helpless despair.

"And now we were in the midst of the blaze. Two arms of

fire encircled the base of the little hillock on which we stood, and united before us in a waving mass, which, rolling onwards, receded farther and farther from our gaze. The flames had devoured the short grass of the hillock, but had not found sufficient nourishment for our destruction. Whole swarms of voracious vultures followed in circling flight the fiery column, like so many hungry jackals, and pounced upon the snakes and lizards which the blaze had stifled and half-calcined in its murderous embrace. When, with the rapidity of lightning, they darted upon their prey and disappeared in the clouds of smoke, it almost seemed as if they were voluntarily devoting themselves to a fiery death. Soon the deafening noise of the conflagration ceased, and the dense black clouds in the distance were the only signs that the fire was still proceeding on its devastating path over the wide wastes of the savannah."

At length, after a long drought, when all Nature seems about to expire under the want of moisture, various signs announce the approach of the rainy season. The sky, instead of its brilliant blue, assumes a leaden tint, from the vapours which are beginning to condense. The black spot of the "Southern Cross," that most beautiful of constellations, in which, as the Indians poetically fancy, the Spirit of the savannah resides, becomes more indistinct as the transparency of the atmosphere diminishes. The mild phosphoric gleam of the Magellanic clouds expires. The fixed stars, which shone with a quiet planetary light, now twinkle even in the zenith. Like distant mountain-chains, banks of clouds begin to rise over the horizon, and, forming in masses of increasing density, to ascend higher and higher, until at length the sudden lightnings flash from their dark bosom, and, with the loud crash of thunder, the first rains burst in torrents over the thirsty land. Scarcely have the showers had time to moisten the earth, when the dormant powers of vegetation begin to awaken with an almost miraculous rapidity. The dull, tawny surface of the parched savannah changes as if by magic into a carpet of the most lively green, enamelled with thousands of flowers of every colour. Stimulated by the light of early day, the mimosas expand their delicate foliage, and the fronds of the beautiful mauritias sprout forth with all the luxuriance of youthful energy.

And now, also, the animal life of the savannah awakens to

the full enjoyment of existence. The horse and the ox rejoice in the grasses, under whose covert the jaguar frequently lurks, to pounce upon them with his fatal spring. On the border of the swamps, the moist clay, slowly heaving, bursts asunder, and from the tomb in which he lay embedded rises a gigantic water-snake or a huge crocodile. The new-formed pools and lakes swarm with life, and a host of water-fowl — ibises, cranes, flamingos, mycterias — make their appearance, to regale on the prodigal banquet. A new creation of insects and other unbidden guests now seek the wretched hovels of the Indians, which are sparingly scattered over the higher parts of the savannah. Countless multitudes of ants, sandflies, and mosquitos; rattlesnakes, expelled by the cold and moisture from the lower grounds; repulsive geckos, which with incredible rapidity run along the overhanging rafters; nauseous toads, which, concealing themselves by day in the dark corners of the huts, crawl forth in the evening in quest of prey; lizards, scorpions, centipedes; in a word, worms and vermin of all names and forms, — emerge from the inundated plains, for the tropical rains have gradually converted the savannah, which erewhile exhibited a waste as dreary as that of the Sahara, into a boundless lake. The swollen rivers of the steppe — the Apure, the Arachuna, the Pajara, the Arauca — pour in mighty streams over the plains, and boats are now able to sail for miles across the land from one river-bed into another.

On the same spot where, but a short time ago, the thirsty horse anxiously snuffed the air to discover by its moisture the presence of some pool, the animal is now obliged to lead an amphibious life. The mares retreat with their foals to the higher banks, which rise like islands above the waters, and as from day to day the land contracts within narrower limits, the want of forage obliges them to swim about in quest of the grasses that raise their heads above the fermenting waters. Many foals are drowned; many are surprised by the crocodiles, that strike them down with their jagged tails, and then crush them between their jaws. Horses and oxen are not seldom met with, which, having fortunately escaped these huge saurians, bear on their limbs the marks of their sharp teeth.

"This sight," says Humboldt, "involuntarily reminds the reflecting observer of the great pliability with which nature has endowed several plants and animals. Along with the fruits of Ceres, the horse and the ox have followed man over the whole earth, from the Ganges to the La Plata, and from the coast of Africa to the mountain-plain of Antisana, which is more elevated than the lofty peak of Teneriffe. Here the northern birch-tree, there the date-palm, protects the tired ox from the heat of the mid-day sun. The same species of animal which contends in eastern Europe with bears and wolves, is attacked in another zone by the tiger and the crocodile."

But it is not the jaguar and the alligator alone which lie in wait for the South American horse, for even among the fishes he has a dangerous enemy. The rivers and marshes of the Llanos are often filled with electrical eels, whose slimy, yellow-punctured body sends forth at will from the under-part of the tail a stunning shock. These eels are from five to six feet long. They are able, when in full vigour, to kill the largest animals when they suddenly unload their electrical organs in a favourable direction. All other fishes, aware of their power, fly at the sight of the formidable gymnotes. They stun even the angler on the high river-bank, the moist line serving as a conductor for the electric fluid. The capture of these eels affords a highly entertaining and animated scene. Mules and horses are driven into the pond, which the Indians surround, until the unwonted noise and splashing of the waters rouse the fishes to an attack. Gliding along, they creep under the belly of the horses, many of whom die from the shock of their strokes; while others, with mane erect, and dilated nostrils, endeavour to flee from the electric storm which they have roused. But the Indians, armed with long poles, drive them back again into the pool.

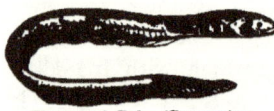

Electrical Eel. (Gymnotus electricus.)

Gradually the unequal contest subsides. Like spent thunder-clouds, the exhausted fishes disperse, for they require a long rest and plentiful food to repair the loss of their galvanic powers. Their shocks grow weaker and weaker. Terrified by the noise of the horses, they timidly approach the banks, when, wounded with harpoons, they are dragged on shore with dry and non-conducting pieces of wood; and thus the strange combat ends.

The Llanos are never more beautiful than at the end of the

rainy season, before the sun has absorbed the moisture of the soil. Then every plant is robed with the freshest green; an agreeable breeze, cooled by the evaporating waters, undulates over the sea of grasses, and at night a host of stars shines mildly upon the fragrant savannah, or the silvery moonbeam trembles on its surface. Where on the margin of the primitive forest, girt with colossal cactuses and agaves, groups of the mauritia rise majestically over the plain, the stateliest park ever planted by man must yield in beauty to the charming picture of these natural gardens, bordered here by impenetrable thickets, and there by the blue mountain-chain, behind which the fancy paints scenes of still more enchanting loveliness.

The mauritia, the chief ornament of the park-like savannah, and no less useful than the date-tree of the African oasis, provides the Indian with almost every necessary, and fully deserves the name of "tree of life"—*arbol de la vida* — bestowed upon it by the poetical fancy of the Jesuit Gumilla. Rising to the height of a hundred feet, its slender trunk is surmounted by a magnificent tuft of large, fan-shaped fronds, of a brilliant green, under whose canopy the scaly fruits, resembling pine-cones, hang in large clusters. Like the banana, they afford a food differing in taste according to the stages of ripeness in which they are plucked; and before the blossoms of the male palm have expanded, its trunk contains a nutritious pith like sago, which, dried in thin slices, forms one of the chief articles of the Indian's bill of fare. Like his brethren of the Eastern world, he also knows how to prepare an intoxicating "toddy" from the juice of the flower-spathes; the leaves serve to cover his hut; out of the fibres of their petioles he manufactures twine and cordage; and the sheaths at their base afford him material for his sandals.

At the mouth of the Orinoco the very existence of the yet unsubdued Guaranas depends on the mauritia, which gives them both food and liberty. Formerly, when this tribe was more numerous than at the present day, they raised their huts on floorings stretching from trunk to trunk, and formed of the leaf-stalks of their tutelary palm. Thus, like the monkeys and parrots of their native wilds, they lived in the trees during the inundations of the Delta in the rainy season. These platforms were partly covered with moist clay, on which fire was made for household purposes; and the flames afforded a strange sight to travellers sailing on the river at night. Even now the light-footed Guaranas owe their independence to the half-liquid soil of their territory, and to their tree life. The fruits of the mauritia, besides affording food to the Indian, are eagerly devoured by monkeys and parrots. On approaching a group of palms at the time when the fruits are ripening, the profound silence which within the tropics chiefly characterises the noon, is interrupted by a scream of warning, and soon after a numerous troop of birds wheels screeching about the grove. The green colour of their plumage seldom betrays the parrots to the eye among the equally green fronds of the palms.

When the Spaniards first settled in the beautiful mountain valleys of Caraccas and on the Orinoco, they found the Llanos, in spite of their abundant verdure, almost entirely uninhabited by man, for the Indians were unacquainted with pastoral life; and if the mauritia had not here and there tempted a few savages to settle on the open savannahs, they would have been left entirely to the animal life which from time immemorial had thriven on their herbage. But the Spaniards introduced new quadrupeds into the new world,— the ox, the horse, the ass, our faithful companions over the whole surface of the globe,— and the progeny of these domestic animals, returning to their wild state, has multiplied amazingly in the vast pastures of the Llanos. Man has followed them into their new domain; and small hamlets, often situated whole days' journeys one from another, and consisting only of a few wretched huts, though generally dignified with the name of towns, proclaim that he has at least made a beginning to establish his empire over these boundless plains.

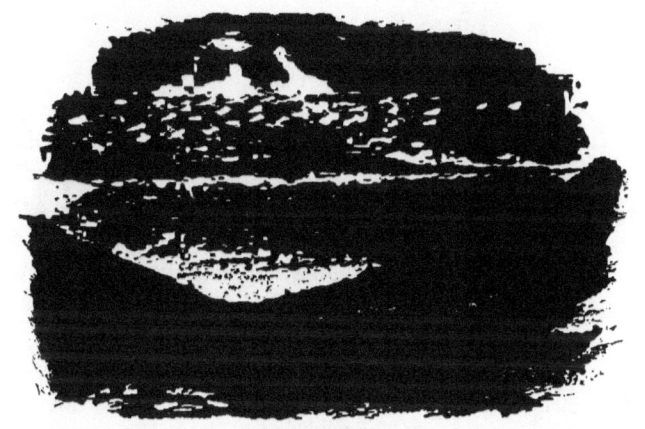

High Table-Lands of Peru.

CHAPTER III.

THE PUNA, OR THE HIGH TABLE-LANDS OF PERU AND BOLIVIA.

Striking Contrast with the Llanos — Northern Character of their Climate — The Chuñu — The Surumpe — The Veta: its Influence upon Man, Horses, Mules, and Cats — The Vegetation of the Puna — The Maca — The Llama: its invaluable Services — The Huanacu — The Alpaca — The Vicuñas: Mode of Hunting Them — The Chacu — The Bolas — The Chinchilla — The Condor — Wild Bulls and Wild Dogs — Lovely Mountain Valleys.

BETWEEN the two mighty parallel mountain-chains of the Cordillera and the Andes [*], the giant bulwarks of Western South America, we find, extending throughout the whole length of Peru and Bolivia, at a height of from ten to fourteen thousand feet above the level of the sea, vast plateaus, or table-lands, which are named, in the language of the country, the *Puna*, or "the Uninhabited." They present a striking contrast to the Llanos of Venezuela; for though situated, like these sultry plains, within the torrid zone, their great elevation paralyses the effects of a vertical sun, and transfers the rigours of the north to the very centre of the tropical world.

Their climate is hardly less bleak, unfriendly, and winterly than that of the high snow-ridges which bound them on either

[*] Though frequently confounded, even by the Peruvian Creoles, the western chain, running parallel with the coast of the Pacific, is properly the Cordillera; while the eastern chain, which generally runs in the same direction as the former, has always been named the Andes by the Indian natives.

side. Cold winds from the icy Cordillera sweep almost constantly over their surface, and during four months of the year they are daily visited by fearful storms. The suddenly darkened sky discharges, under terrific thunder and lightning, enormous masses of snow, until the sun breaks forth again. But soon the clouds obscure its brilliancy; and thus cold and warmth, winter and summer, here reign alternately, — not, as in our temperate climes, during several months, but within the short space of a single day.

In a few hours the change of temperature often amounts to forty or forty-five degrees, and the sudden fall of the thermometer is rendered still more disagreeable to the traveller by biting winds, which so violently irritate the skin of the hands and face, that it springs open and bleeds from every fissure. An intolerable burning and swelling accompany these wounds, so as to prevent the use of the hands for several days. On the lips it is also very disagreeable, as the pain increases by eating and speaking; and an incautious laugh produces deep rents, which bleed for a long time and heal with difficulty.

This evil, which is called *Chuñu* by the Peruvian Indians, is also very painful on the eyelids; but it becomes absolutely insupportable by the addition of the *Surumpe*, a very acute and violent inflammation of the eyes, caused by the sun's reflection from the snow. In consequence of the rarefied air and the biting winds, the visual organs are constantly in a state of irritation, which renders them far more sensitive to any strong light than would be the case in a more congenial atmosphere. The rapid change from a clouded sky to the brilliancy of a sunny snow-field, causes a painful stinging and burning, which increases from minute to minute to such a degree, that even the stoical Indian, when afflicted with this evil, will sit down on the road-side and utter cries of anguish and despair. Chronical ophthalmia, suppuration of the eyelids, and total blindness, are the frequent consequences of an intense Surumpe, against which the traveller over the high lands carefully guards himself by green spectacles or a dark veil.

A third plague of the wanderer in the Puna is the *Veta*, which is occasioned by the great rarefaction of the air. Its first symptoms, which generally appear at an elevation of 12,000 feet, consist in giddiness, buzzing in the ear, headache, and nausea. They frequently show themselves even in those who ride, but in a much greater degree when the traveller ascends the mountains on foot. Their intensity increases with the elevation, and is aggravated by a lassitude, which augments to such a degree as to render walking impossible, by a great difficulty of respiration, and violent palpitation of the heart. Absolute rest mitigates the symptoms; but on continuing the journey they reappear with increased violence, and are then frequently accompanied by fainting and vomiting. *The capillary vessels of the eye, nose, and lips burst, and emit drops of blood. The same phenomenon appears also in the mucous membrane of the respiratory and digestive organs; so that blood-spitting and bloody diarrhœa frequently accompany the Veta, and are sometimes so violent as to cause death.

The influence of diminished atmospheric pressure shows itself also in the horses that are unaccustomed to mountain travelling. They begin to pace more slowly, frequently stand still, tremble all over, and fall upon the ground. If not allowed to rest, they invariably die. By way of a restorative their nostrils are slit open, which seems to be of use by allowing a greater influx of air. Mules and asses are less subject to the Veta than horses, but cats are affected by it in the highest degree. These animals are unable to live at a height of 13,000 feet above the level of the sea, and die in the most violent convulsions.

Another consequence of the diminished pressure of the air in the Puna is, that water begins to boil at so low a temperature that neither eggs, potatoes, nor meat can be sufficiently boiled; and that whoever wishes to eat a warm dinner in the Puna is obliged to have it baked or roasted.

As the dry sand of the rainless coast prevents the putrefaction of animal substances which are buried in it, the power of the dry Puna-winds in a like manner arrests the progress of decomposition. Under their influence, a dead mule changes in a few days into a mummy, so that even the entrails do not exhibit the least sign of putrefaction.

It may easily be imagined that, under these peculiar atmospheric influences, vegetation can only appear in stunted proportions, and indeed the Puna presents the monotonous aspect of a northern steppe, its whole surface being covered with dun and meagre herbage, which at all times gives it an autumnal or even wintry aspect. A few arid compositæ and yellow echinocacti are quite unable to relieve the dreary landscape; and even the large-flowered calceolarias, the blue gentians, the sweet-smelling verbenas, the dwarfish cruciferæ, and many other Alpine plants, the usual ornaments of the higher mountain regions, are here almost suffocated by the dense grasses. But rarely the eye meets with a solitary queñua tree (*Polylepis racemosa*) of crippled growth, or with large spaces covered with red-brown ratania shrubs, which are carefully collected for fuel, or for roofing the wretched huts of the scanty population of these desolate highlands.

The cold climate of the Puna naturally confines agriculture to very narrow limits. The only cultivated plant which grows to maturity is the maca (a species of tropæolum), the tuberous roots of which are used like the potato, and form in many parts the chief food of the inhabitants. This plant grows best at an elevation of twelve or thirteen thousand feet, and is not planted in the lower regions, where its roots are said to be completely unpalatable. Barley is also cultivated in the Puna, but never ripens, and is cut green for forage.

The animal kingdom is more amply represented in this bleak table-land; for there is no want of food on the grass-covered plains, and wherever this exists, there is room for the development of animals appropriate to the climate.

Thus the *Llama* and its near relations, the *Alpaca*, the *Huanacu*, and the *Vicuña*, the largest four-footed animals which

The Llama.

Peru possessed before the Spaniards introduced the horse and the ox, are all natives of the Puna. Long before the invasion of Pizarro, the llama was used by the ancient Peruvians as a beast of burthen, and was not less serviceable to them than the camel to the Arabs of the desert. The wool served for the fabrication of a coarse cloth; the milk and flesh, as food; the skin, as a warm covering or mantle; and without the assistance of the llama, it would have been impossible for the Indians to transport goods or provisions over the high tablelands of the Andes, or for the Incas to have founded and maintained their vast empire. The llama is also historically remarkable as being the only animal domesticated by the aboriginal Americans. The reindeer of the north* and the bison of the prairies enjoyed then, as they do now, their savage independence: the llama alone was obliged to submit to the yoke of man. But the llama reminds us of the dromedary not only by a similar destiny and similar services, but also by a strong resemblance in form and structure, so as to be classed by naturalists in the same family. The unsightly hump is wanting, but the llama possesses the same callosities on the breast and on the knees, the same divided hoof, the same formation of the toes and stomach; and the microscope teaches us that the resemblance extends even to the globules of the blood, which are elliptical only in the camelides and some species of deer, but circular in all other quadrupeds. Thus Nature has formed in the llama a species of mountain camel, admirably adapted to the exigencies of a totally different soil and climate; and surely it is not one of the least wonders of creation to see animals so similar in many respects emerge, without any connecting links, at the opposite extremities of the globe.

The size of the llama is about that of the stag; the neck is very long and habitually upright, the eyes large and brilliant, the lips thick, the ears long and movable. Its general colour is a light brown, the under parts being whitish, but it is also frequently dappled — seldom quite white or black.

The ordinary load of the llama is about one hundred pounds, and its rate of travelling with this burthen over rugged mountain passes is from twelve to fifteen miles a-day. When overloaded it lies down, and will not rise until relieved of part of its burthen. In spite of their weakness, the llamas are invaluable in the silver mines; for they are often obliged to transport the ore along precipices so abrupt, that even the hoof of the mule would find no support. Yet their price does not exceed three or four dollars, as the introduction of the stronger solipedes, the horse, and the mule, have very generally superseded them as beasts of burthen.

"The Indians," says Tschudi, "often travel with large herds of llamas to the coast to fetch salt. Their journeys are very small, rarely more than three or four leagues; for the llamas never feed after sunset, and are thus obliged to graze while

---

* It is only in the Old World that the reindeer has ever been domesticated.

journeying, or to rest for several hours. While reposing, they utter a peculiar low tone, which at a distance, and when the herd is large, resembles the sound of several Æolian harps. A loaded herd of llamas traversing the high table-lands affords an interesting spectacle. Slowly and stately they proceed, casting inquisitive glances on every side. On seeing any strange object which excites their fears, they immediately scatter in every direction, and their poor drivers have great difficulty to re-unite the herd."

The Indians, who are very fond of these animals, decorate their ears with ribbons, hang little bells about their necks, and always caress them before placing the burthen on their back. When one of them drops from fatigue, they kneel at its side and strive to encourage it for further exertion by a profusion of flattering epithets and gentle warnings. Yet, in spite of good treatment, a number of llamas perish on the way to the coast or to the forests, as they cannot stand the hot climate.

The *Huanacu* is of a greater size than the llama, and resembles it so much that it was supposed to be the wild variety until Tschudi, in his "Fauna Peruana," pointed out the specific differences between both. The huanacu is of a larger size; its fleece is shorter and less fine; its colour is brown, the under parts being whitish — but varieties of colour are never observed, as in the llama; the face is blackish grey, lighter and almost white about the lips. The huanacus generally live in small troops of from five to seven. They are very shy, but when caught young are easily tamed, though they always remain spiteful, and can hardly ever be trained to carry burthens. They are frequently met with in the European menageries, where they pass under the name of llamas.

The *Alpaca* is smaller than the llama, and resembles the

The Alpaca.

sheep; but its neck is longer, and it has a more elegantly formed head. The wool is very long, soft, fine, and of a silky lustre — sometimes quite white or black, but often also variegated. On account of its admirable qualities it fetches a high price, and is extensively used in England; so that from 1835 to 1839, 98,808 bales, weighing near 7,000,000 lbs., were imported into Liverpool alone. It is particularly valuable from its being able to be woven with common wool, silk, or cotton, into a variety of equally beautiful and durable fabrics; so that, in spite of a continual increase of the importation, its price has trebled since 1840.

It is not wonderful that frequent attempts have been made to acclimatise the alpaca in other countries, though unfortunately they have almost always failed, from want of knowledge or care. That the alpaca might be successfully transplanted to other parts of the world, is sufficiently proved by its thriving condition in the Zoological Gardens of Antwerp; so that there can be no doubt that, with proper attention, it would come on as well, or even better, in the mountainous regions of central Europe. A herd of alpacas is said to have been introduced into Australia in 1859. Should this attempt succeed, it would be of great importance to the English manufacturers, as the eternal civil wars of Peru and Bolivia prevent all pastoral or agricultural progress in those unhappy countries.

The alpacas are kept in large herds, and graze all the year round on the high table-lands. They are only driven to the huts to be shorn, and are therefore extremely shy. There is, perhaps, no more obstinate animal in existence. When one of them is separated from the herd, it throws itself upon the ground, and neither good words nor blows, nor even the greatest tortures, are able to force it from the spot.

Shy, like the chamois or the steinbock, the *Vicuña* inhabits the most solitary mountain-valleys of the Andes. It is of a more elegant shape than the alpaca, with a longer and more graceful neck, and a much shorter and more curly wool, of such extreme fineness as to be worth about twenty-seven shillings the pound. The upper parts of the body are of a peculiar reddish-yellow colour (*color de vicuña*), the under side of the neck and the limbs light ochre, the long wool of the breast and the abdomen white. During the rainy season, the vicuñas retire to the crests of the Cordillera, where vegetation is reduced to the scantiest limits; but they never venture on the bare summits, as their hoof, accustomed to tread only on the turf, is very tender and sensitive. When pursued, they never fly to the ice-fields, but only along the grass-grown slopes. In the dry season, when vegetation withers on the heights, they descend to seek their food along the sources and swampy grounds. From six to fifteen she-vicuñas live under the protection and guidance of a single male, who always remains a few paces apart from his harem, and keeps watch with the most attentive care. At the least approach of danger he immediately gives the alarm by a shrill cry, and rapidly steps forward. The herd, immediately assembling, turns inquisitively towards the side whence danger is apprehended, advances a few paces, and then, suddenly wheeling, flies, at first slowly, and constantly looking back, but soon with unrivalled swiftness. The male covers the retreat, frequently standing still and watching the enemy. The females reward this faithful care of their leader with an equally rare attachment; for when he is wounded or killed, they will keep running round him with shrill notes of sorrow, and rather be shot than flee. But when a bullet strikes a female, the whole troop continues its flight with increased velocity.

The cry of the vicuña is a peculiar sharp piping or whistling, which, though greatly resembling the shrill neighing of the llama and the other American camelides, may easily be distinguished by a practised ear, when it suddenly pierces the thin air of the Puna, even from a distance where the sharpest eye is no longer able to distinguish the form of the animal. Like their native congeners and the camels of the old world, the vicuñas have the habit of ejecting a quantity of saliva and half-digested food upon those that come within their reach. The llamas and huanacus, however, only do so when angry or attacked; while the alpacas and vicuñas spitefully bespatter the harmless passer-by, and generally aim at the face, which they seldom miss. The ejected mass, which has a disgusting smell, imparts a deep green colour to the skin, which can only be cleaned with difficulty.

The hunting of the vicuñas, which is very singular and interesting, takes place in April or May. Each family in the Puna villages is obliged to furnish the contingent of one of its members at least; and the widows accompany the hunters, to serve as cooks. The whole troop, frequently consisting of seventy or eighty persons, and carrying bundles of poles and large quantities of cordage, sets out for the more elevated plateaus of the Puna, where the vicuñas are grazing. In an appropriate plain the poles are fixed into the earth, at intervals of twelve or fifteen paces, and united by the cordage, about two feet from the ground. In this manner a circular space, called *Chacu*, of about half a league in circumference, is enclosed, leaving on one side an entrance several hundred paces wide. The women attach to the cordage coloured rags, which are moved to and fro by the wind. As soon as the Chacu is ready, the men, who are partly mounted, disperse, and, forming a ring many miles in circumference, drive all the intervening vicuña herds through the entrance into the circle, which is closed as soon as a sufficient number has been collected. The shy animals do not venture to spring over the cord and its fluttering rags, and are thus easily killed by the bolas of the Indians. These bolas consist of three balls of lead or stone, two of which are heavy, and one lighter, each ball being attached to a long leather thong. The thongs are knotted together at their free extremity. When used, the lighter ball is taken in the hand, and the two others swung in a wide circle over the head. At a certain distance from the mark, about fifteen or twenty paces, the hand-ball is let loose, and then all three fly in hissing circles towards the object which they are intended to strike, and encompass it in their formidable embrace. The Esquimaux use a similar device for catching the birds of passage which a short summer attracts to their inhospitable shores.

The hind-legs of the vicuñas are generally aimed at. It is no easy matter to throw the bolas adroitly, particularly when on horseback; for the novice often wounds either himself or his horse mortally, by not giving the balls the proper swing, or letting them escape too soon from his hand.

The flesh of the vicuñas is divided in equal portions among the hunters. When dried in the air, and then pounded and mixed with Spanish pepper, its taste is not unpleasing. The Church, however, manages to get the best part of the animal, for the priest generally appropriates the skin. In 1827 Bolivar issued a decree, ordering the vicuñas, when caught, to be merely shorn and then let loose again; but their uncommon wildness rendered the execution of this well-meant law impossible. If a huanacu comes into the chacu, it bounds at one leap over the cord, or breaks through it, and is then followed by all the vicuñas, so that great care is taken not to chase any of these animals into the circle.

As soon as all the entrapped vicuñas are killed, the chacu is taken to pieces, and put up again ten or twelve miles further off. The whole chase lasts a week, and the number of the animals slaughtered frequently amounts to several hundreds. During a chacu-chase in the Altos of Huayhuay, at which Tschudi assisted, 122 vicuñas were caught, and the produce of their skins served for the building of a new altar in the village church.

In the times of the Incas, the Puna chases were conducted on a much grander scale. Annually from 25,000 to 30,000 Indians assembled, who were obliged to drive all the wild animals from a circuit of more than a hundred miles into an enormous chacu. As the circle narrowed, the ranks of the Indians were doubled and trebled, so that no animal could escape. The pernicious quadrupeds, such as bears, cuguars, and foxes, were all killed, but only a limited number of stags, deer, vicuñas, and huanacus: for the provident Incas did not lose sight of the wants of futurity, and were more economical of the lives of animals than their brutal successors, the christian Spaniards, were of the lives of men.*

In spite of the persecutions to which they are subject, not only from hunters but from the ravenous condor, who frequently robs them of their young, the vicuñas do not seem to diminish, and, particularly in the more remote Puna regions, they are often seen roaming about in large numbers,—the inaccessible wilds to which they are able to retreat amply securing them against extermination.

Besides these four remarkable Camelides, we find among the animals peculiar to the Puna the stag-like Tarush (*Cervus antisiensis*), whose horns consist but of two branches; the timid deer, who also descends from the high mountain-plains into the coast-valleys and the forest region; the Viscachas and the Chinchillas.

The Peruvian Viscachas (*Lagidium peruanum* and *pallipes*), which must not be confounded with the Viscachas of the Pampas (*Lagostomus viscacha*), live at an elevation of from 10,000 to 12,000 feet, between 33° and 18° S. lat., and resemble the rabbit in form and colour, but have shorter ears and a long

rough tail. Their fur is soft, but not nearly so fine as that of the near-related Chinchilla (*Chinchilla lanigera*). This little creature, which is somewhat larger than our squirrel, has large and brilliant eyes, an erect tail, strong bristles on the upper lip, and almost naked, rounded ears. It lives in burrows, feeding chiefly upon roots, and is found in such numbers in the Chilian Andes that its holes considerably increase the difficulty of travelling.
Chinchilla.

The fur, which is of a remarkably close texture, is too well known to require any further description.

Where ruminants and rodents abound it may easily be imagined that beasts of prey will not be wanting. The cunning fox (*Canis Azaræ*) waylays both the chinchillas and the waterbirds; and, impelled by hunger, the Puma, or American lion, ascends even to the borders of eternal snow in quest of the vicuña and the deer. The veta, therefore, which is so destructive to the domestic cat, seems to have no influence upon him. But the monarch of the Puna is, unquestionably, the mighty condor, who, soaring over the highest peaks of the Andes, sees on one side the Pacific rolling its heavy breakers against the coast, and on the other the Marañon, or Maragnon, disappearing in the hazy distance of the primitive forest. No created being embraces with one glance so vast an horizon,— a scenery of such unparalleled grandeur; but, indifferent to the beauties of Nature, the piercing eye of the lordly vulture sweeps over the Puna, only to espy the mule that has sunk under its load, or the llama or sheep on the point of giving birth to its young; then, descending with the rapidity of an arrow, he tears the entrails of the new-born creature, indifferent to the cries of the defenceless mother.

The frequent showers and snow-falls of the Puna naturally give rise to numerous swamps and lagunes, which afford nourishment to an abundance of birds, — such as the beautiful snow-white Huachua goose (*Chloëphaga melanoptera*), with dark-green wings of a metallic lustre; the licli, a species of plover; the ibis; the long-legged flamingo; the Quiulla gull (*Larus serranus*); and the gigantic coot (*Fulica gigantea*), which, unable to fly, dives in the cold waters, and builds its nest on the solitary stones which rise above the surface.

To the aboriginal animals of the Puna man has added the horse, the ox, the dog, and the sheep. In the more sheltered Puna valleys there are estates or haciendas possessing from 60,000 to 80,000 sheep, and from 400 to 500 cows. During the wet season these flocks are driven into the Altos or highest regions, often to a height of 15,000 feet; but when the cold, frosty nights of the dry period of the year parch the grass, they are obliged to descend to the swampy valleys, where they have much to suffer from hunger. The wool serves mostly for home consumption, and is partly exported to Europe, where, however, it fetches a much lower price than the produce of the South African or Australian flocks, which most likely would not be the case if the high mountain-plains of the Andes were in the possession of Germans or Anglo-Saxons.

The herds of oxen generally graze in the most distant Altos. As they seldom see man, they are so savage that sometimes even the shepherds do not venture to catch them. These wild bulls render travelling in many parts of the Puna very dangerous, as they will sometimes rush upon man without any provocation or previous notice, though they generally announce their approach by a hoarse bellowing. But even then it is almost impossible to escape them in the open plain, and more than once Tschudi was only able, by a well-aimed shot, to save himself from the attack of one of these formidable animals.

Though not so dangerous, the half-wild Puna Dogs (*Canis Ingæ*, Tschudi) are extremely troublesome to the traveller,— false, spiteful animals, which ferociously attack enemies far stronger than themselves; and, like the bull-dog, will rather suffer themselves to be cut to pieces than retreat. They have a particular antipathy to the white race, and it is rather a bold undertaking for the European traveller to approach the hut of an Indian that is guarded by these animals.

The frosts of winter and an eternal spring are nowhere found in closer proximity than in the Peruvian highlands, for deep valleys cleave or furrow the windy Puna; and when the traveller, benumbed by the cold blasts of the mountain-plains, descends into these sheltered gorges, he almost suddenly finds himself transported from a northern climate to a terrestrial paradise.

Situated at a height where the enervating power of the tropical sun is not felt, and where at the same time the air is not too rarefied, these pleasant mountain vales, protected by their rocky walls against the gusts of the Puna, enjoy all the advantages of a mild and genial sky. Here the astonished European sees himself surrounded by the rich corn-fields, the green lucerne meadows, and the well-known fruit trees of his distant home, so that he might almost fancy that some friendly enchanter had transported him to his native country, if the cactuses and the agaves on the mountain-slopes by day, and the constellations of another hemisphere by night, did not remind him of the vast distance which separates him from the land of his birth.

There are regions in this remarkable country where the traveller may in the morning leave the snow-decked Puna hut, and before sunset pluck pine-apples and bananas on the cultivated margin of the primeval forest; where in the morning the stunted grasses and arid lichens of the naked plain remind him of the arctic regions, and where he may repose at night under the fronds of gigantic palms.

Guano Island.

CHAPTER IV.

THE PERUVIAN SAND-COAST.

Its desolate Character — The Mule is here the "Ship of the Desert"— A Shipwreck and its Consequences — Sand-Spouts — Medanos — Summer and Winter — The Garuas — The Lomas — Change produced in their Appearance during the Season of Mists — Azara's Fox — Wild Animals — Birds — Reptiles — The Chincha or Guano Islands.

BETWEEN the Cordilleras to the east and the Pacific to the west extends, from 3° to 21° S. lat., 540 leagues long and from 3 to 20 leagues broad, a desert coast, the picture of death and desolation. Traversed by spurs of the mighty mountainchain, which either gradually sink into the plain, or form steep promontories washed by the ocean, it rises and falls in alternate heights and valleys, where the eye seldom sees anything but fine drift-sand or sterile heaps of stone.

Only where, at considerable intervals, some rivulet, fed by the melting glaciers or by the small mountain lakes, issues from the ravines of the Andes to lose itself after a short course in the Pacific, green belts, like the oases of the African desert, break the general monotony, and appear more charming from the contrast with the nakedness of the surrounding waste. The planter carefully husbands the last drop of water from those scanty streams to moisten his stony fields; for, as the tribes of the Sahara can only, by dint of constant industry, preserve their date-palm islands against the waves of the surrounding sand-sea, thus also the inhabitant of the Peruvian coast can only by perpetual irrigation protect his plantations and gardens from the encroachments of the neighbouring desert! But the fruits which he reaps and garners are very different from those which are produced by the African oasis; for, while none of the plants of the Peruvian sand-coast has ever found its way to the Sahara, the sycamores and tamarinds of the latter are equally unknown on the eastern shores of the Pacific. Cotton and sugar, maize and batatas, manioc and bananas, here take the place of the date-palm of the Arab, and thrive only so far as the limits of irrigation extend.

In the surrounding wastes, where for miles and miles the traveller meets no traces of vegetation, and finds not one drop of water, the mule performs the part of the African camel; for, satisfied with a scantier food than the horse, it more easily supports the fatigues of a prolonged journey through the sand, and in Peru is fully entitled to be called the *ship of the desert*. The horse cannot support hunger and thirst longer than forty-eight hours without becoming so weak as hardly to be able to carry its rider; and if the latter is imprudent enough to urge it on to a more rapid pace, it falls a victim to his obstinacy, as it will obey the spur until it sinks never to rise again. Not so the mule, which, on feeling itself unable to advance, stands still, and will not move an inch until it has rested for a time; after which it willingly continues its journey. Yet, in spite of these excellent qualities, many mules succumb to the fatigues and privations of the desert; and as in the Sahara the caravan-routes are marked by camel-skeletons, so here long rows of mule-skulls and bones point out the road along the Peruvian sand-coast. Woe to him whom a shipwreck casts on these shores; for he is almost inevitably doomed to destruction!

In the year 1823 a transport, with 320 dragoons on board, under the command of Colonel Lavalle, was stranded in the neighbourhood of Pisco. The soldiers saved themselves on the land. They had only been thirty-six hours in the desert, when they were found by a regiment of cavalry which had been sent to their assistance from Pisco with provisions and water, yet in this short space of time 116 of the poor wretches had expired of fatigue and thirst, and above fifty more died on the following day from their previous exhaustion. In general, a healthy man can withstand hunger and thirst during four or five days, but only in a temperate climate and when the body is at rest; while in the burning deserts of Peru, the want of water during forty-eight hours, combined with the fatigue of wading through the deep sands, can only end in death. Thirst can, undoubtedly, be supported ten times longer in the moist sea-air than in the thoroughly desiccated atmosphere of a tropical waste. The dangers of these solitudes are increased by the great mobility of the soil. When a strong wind blows, huge sand-columns, rising like water-spouts to a height of eighty or a hundred feet, advance whirling through the desert, and suddenly encompass the traveller, who can only save himself by a rapid flight. Such is the instability of the soil, that in a few hours a plain will be covered with hillocks or *Medanos*, and recover after a few days its former level. The most experienced muleteers are thus constantly deceived in their knowledge of the road, and are the first to give way to despair, while seeking to extricate themselves from a labyrinth of newly-formed medanos. These constant transformations and shiftings in the desert, which Tschudi graphically calls "a life in death," take place more particularly in the hot season, when the least pressure of the atmosphere suffices to disturb the dried-up sands, whose weight increases during the winter by the absorption of moisture. The single grains then unite to larger masses, and more easily withstand the pressure of the wind.

The summer, or dry season, begins in November. The rays of the vertical sun strike upon the light-coloured sands, and are reflected with suffocating power. No plant, except the cactuses and tillandsias, which manage to thrive where nothing else exists, takes root in the glowing soil: no animal finds food on the lifeless plain; no bird, no insect, hovers or buzzes in the stifling atmosphere. Only in the highest regions the condor, the monarch of the air, is seen sailing along in lonely majesty.

In May, which in these southern latitudes corresponds to our October, the scene changes. A thin, misty veil extends over the sea and the coast, and, increasing in density during the following months, only begins to diminish in October. At the beginning and the end of this damp season the mist generally ascends between nine and ten in the morning, and falls again at about three in the afternoon; but in August and September, when it is most dense, it rests for weeks immovably over the earth, never dissolving in rain, but merely descending in a fine, penetrating drizzle, which is called "garua" by the inhabitants. In many parts of the Peruvian coast rain has not been known to fall for centuries, except only after very severe earthquakes, and even then the phenomenon is not of constant occurrence. The mist seldom ascends to a vertical height of more than 1200 feet, when it is replaced by violent showers of rain; and, remarkably enough, the limits between both can be determined with almost mathematical precision, as there are plantations, one half of whose surface is invariably moistened by garuas and the other by rain.

When the mists appear, the Lomas, or chains of hills which bound the sand-coast towards the east, begin to assume a new character; and, as if by magic, a garden is seen where but a few days before a desert extended its dreary nakedness. Soon also, animal life begins to animate the scene, as the Lomeros drive their cattle and horses to these newly-formed pasture-grounds, where for several months they find an abundance of juicy food, but no water. This, however, they do not require, as they always leave the Lomas in the best condition.

In some of the northern coast-districts, situated near the equatorial line, where the garuas seldom appear, the fertility of the land depends wholly upon the streams which issue from the mountains. The dew, which along the coasts of central and south Peru hardly moistens the soil to the depth of half an inch, is there so completely wanting, that a piece of paper exposed to the air during the night shows no sign of moisture in the morning; and so thoroughly does the dryness of the soil prevent putrefaction, that after 300 years the mummified corpses are still found unaltered, which the ancient Peruvians buried in a sitting posture.

Thus the aridity of a great part of the Sahara repeats itself in these American deserts, and is in some measure owing to the same cause, though their geographical position to the west of the Andes, whose eastern slopes absorb all the moisture of the prevailing trade-winds, chiefly accounts for their nakedness. Rain is wanting, as there is no vegetation of any great extent to condense the passing vapours; and, on the other hand, the want of moisture prevents plants from rooting on the unstable soil.

A glance at the animal world of the Peruvian coast shows us the same poverty of species as in the great African desert. A fox (*Canis Azaræ*) seems here to play the part of the hyæna and the jackall; and is found as well in the cotton-plantations along the streams, as in the Lomas, where he is destructive to the young lambs.

The large American felidæ, the puma, and the jaguar, seldom appear on the coast, where they attain a more considerable size than in the mountains. The cowardly puma is afraid of man; while the bloodthirsty jaguar penetrates into the plantations, where he lies in wait for the oxen and horses, and avoids, with remarkable sagacity, the manifold traps and pitfalls that are laid for him by the slaves of the hacienderos. It is almost superfluous to remark that these beasts of prey do not stray about in the sand-deserts; but, descending from the mountains, and following the course of the rivers, disturb the inhabitants of the oases, or prey upon the herds on the Lomas in the moist season.

In the cultivated districts some species of Opossums are likewise found among the low bushes, in deserted dwellings, or in the store-rooms of the plantations; and armadillos (*Dasypus tatuay*) are sometimes shot in the fields.

Opossum.

There are several species of indigenous *Rodents*; but the cosmopolite rat, which seems to have been but recently imported, is as yet not very common.

Wild hogs of an enormous size are sometimes met with in the thickets near some of the plantations in the valley of Lima.

Instead of the antelope and the gazelle of the African deserts, the Venado, a species of deer, makes its appearance on the Peruvian coast. It chiefly lives in the low thickets, which are scattered here and there, and after sunset visits the plantations, where it causes considerable damage.

Besides the numerous sea- and strandbirds, the carrion vultures and the condor, often found in large numbers feasting upon the marine animals that have been cast ashore, are the most conspicuous among the feathered tribes of the coast. A small falcon (*Falco sparverius*) is likewise often seen, and a small burrowing owl (*Athene cunicularia*) haunts almost every ruin on the coast. The pearl-owl, performing the useful services of our own barn-owl (*Strix perlata*), is protected and encouraged in many plantations, as it thins the ranks of the mice. Swallows are scarce; nor do they build their nests on the houses, but on solitary walls, far from the habitations of man.

Among the singing birds, the beautiful crowned fly-catcher (*Myoarchus coronatus*) is one of the most remarkable. Its head, breast, and belly are of a burning red; its wings and back blackish brown. It always sits upon the highest top of the bushes, flies vertically upwards, whirls about a short time singing in the air, and then again descends in a straight line upon its former resting-place. Some tanagras and parrots, and two starling-like birds, the red-breasted picho and the lustrous black chivillo, that are frequently kept in cages on account of their agreeable song, are found in the coast-valleys; and various pigeons, among others the neat little turtuli and the more stately cuculi, frequent the neighbourhood of the plantations. The latter has a monotonous but very melodious song, which lasts from early morning to the forenoon, and is again resumed towards sunset. It consists in repeating three times the word "cŭcŭli," and resuming the same notes after a shorter or longer pause. Some of the birds repeat more frequently their cŭcŭli, and their value increases in the same proportion. In Cocachacra Tschudi saw a bird that reiterated its note fourteen times consecutively, and was so highly prized by its possessor that he would not sell it for less than two ounces of gold, or thirty-four dollars.

Among the lizard tribes large and brilliantly green iguanas are found on the southern coast; but much more frequently dull and sombre agamas lurk among the rocks and stones. Some geckos creep about the houses and walls in the inhabited valleys; but it would be vain to seek in the Peruvian oasis for the chameleon, which frequents every desert island of the Sahara.

Snakes, both venomous and harmless, are in general tolerably rare, and occur both in the fruitful lands and the sand-plains.

The animated sea-shore forms a striking contrast to the death-like solitude of the interior. Troops of carrion vultures gather about the large marine animals cast ashore by the surf; numerous strand-birds are greedily on the look-out for the shell-fish left by the retreating tide, or for the crabs and sea-spiders that everywhere draw their furrows about the beach; and sea-otters and seals sun themselves on the cliffs along the whole coast, except in the neighbourhood of the seaports, where they have been extirpated, or driven away by incessant persecutions. Several promontories and islands (Punta de Lobos, Isla de Lobos) owe their names to the numbers of Phocæ which frequent their shores.

To the north of Chancay, steep sand-hills rise to the height of 300 or 400 feet, abruptly verging to the sea. The way, leading along the side of these hills, would be extremely dangerous but for the unstable nature of the soil. For though at each false step the mule slides with his rider towards the sea, it is very easy for him to regain his footing on the yielding sand.

A large stone on one of these hills bears a striking resemblance to a sleeping sea-lion, and almost perpendicularly beneath it lies a little cove, inhabited by a number of seals. At night the bark of these animals, mixing with the hollow roar of the breakers, fills the traveller with a kind of involuntary terror.

Myriads of sea-birds breed on the small islands along the coast or swarm about the bays, where the fish supply them with abundant food. The number of these birds, a matter formerly of only local interest, is now a subject of general importance, as to them are owing the deep Guano beds which, richer and more useful than the silver mines of Peru, increase the harvests of the English husbandman. These beds have now been worked for many years with a constantly increasing energy, so that the actual annual exportation of the Chincha Islands[*] amounts to no less than half a million of tons; yet so enormous are the deposits, that, according to Castelnau, the excavations hitherto made in the mass appear but as small quarries on the slopes of a chain of hills. How many centuries — what numbers of birds — how many legions of fishes must have been required for the accumulation of these huge mounds!

[*] For a more detailed account of the Peruvian Guano Islands, see "The Sea and its Living Wonders." Second Edition, pp. 144, 147.

The want of rain, which renders the greatest part of the Peruvian coast so utterly barren, is of the utmost advantage for the production of the guano; for if the Chincha Islands, like the Orkneys or the Hebrides, had been exposed to frequent storms, or washed by unceasing showers, they would have been mere naked rocks, instead of affording the richest deposits of manure which the world can boast of.

CHAPTER V.

THE AMAZONS, THE GIANT RIVER OF THE TORRID ZONE.

The Course of the Amazons and its Tributaries — The Huallaga — The Ucayale — The Ica — The Yapura, Jutay, Jurua, Teffe, Coary, and Purus — The Rio Negro — The Madeira — The Strait of Obydos — Tide Waves on the Amazon — The Tapajos — The Xingu — The Bay of the Thousand Islands — The Pororocca — Rise of the River — Inundated Forests — Lagunes — Magnificent Scenery — Fishing Scene — Different Character of the Forests beyond and within the verge of Inundation — General Character of the Banks — A Sail on the Amazons — A Night's Encampment — The "Mother of the Waters" — The Piranga — Dangers of Navigating on the Amazons — Terrific Storms — Rapids and Whirlpools — The Stream of the Future — Travels of Orellana — Madame Godin.

THE Amazons, the giant stream of the tropical world, is of no less magnificent proportions than the Andes, where it takes its source. From the small Peruvian mountain-lake of Lauricocha, 12,500 feet above the sea, the Tunguragua, which is generally considered as the chief branch, rushes down the valleys. At Tomependa, in the province of Juan de Bracamoros, rafts first begin to burden its free waters; but, as if impatient of the yoke, it still throws many an obstacle in the navigator's way; for twenty-seven rapids and cataracts follow each other as far as the Pongo de Manseriche, where, at the height of 1164 feet above the level of the sea, it for ever bids adieu to the romance of mountain scenery.

Its width, which at Tomependa exceeds that of the Thames at Westminster bridge, narrows to 150 feet in the defile of the Pongo, which in some places is obscured by overhanging rocks and trees, and where huge masses of drift-wood, torn from the slopes by the mountain torrents, are crushed and disappear in the vortex.

From the Pongo to the ocean, a distance of more than 2000 miles, no rocky barrier impedes the further course of the monarch of streams; and according to Herndon (Exploration of the Valley of the Amazons, 1851-1853), its depth constantly remains above eighteen feet, so that it is navigable for large ships all the way from Para to the foot of the Andes! No other river runs in so deep a channel at so great a distance from its mouth, and the tropical rains, spreading over a territory nearly equal in extent to one-half of Europe, are alone able to feed a stream of such colossal dimensions!

The first considerable tributary of the Amazons is the Huallaga, which rises near the famous silver-mines of Cerro de Pasco, 8600 feet above the level of the sea, and is 2500 paces broad at the point where the rivers meet. Lower down at Nauta, the Ucayale, descending from the distant mountains of Cuzco, adds his waters to the growing stream, after a course nearly 400 miles longer than that of the Tunguragua itself. Where these mighty rivers meet, Lieutenant Lister Maw found a depth of thirty-five fathoms.

From the Brazilian frontier, where it still flows at an elevation of 630 feet above the sea, to the influx of the Rio Negro, the Amazons is called the Solimoens, as if one name were not sufficient for its grandeur. During its progress between these two points it receives on the left, or from the north, the Içа and the Yapura, on the right, or from the south, the Xavari, the Jutay, the Jurua, the Teffe, the Coary, and the Purus, streams which in Europe would only be surpassed by the Danube, but are here merely the obscure branches of a giant trunk.

No traveller has ever yet visited the banks of the Içа. The many-armed Yapura, which during the rainy season inundates the left bank of the chief stream, has been navigated by Von Martius, the distinguished naturalist, to whom botany owes the best monograph on palms, but science knows next to nothing of the banks of the Jutay, the Jurua, the Teffe, and the Purus. At the conflux of the Yapura, Herndon estimates the breadth of the Solimoens at four or five English miles, and yet it still rolls its waters deep in the heart of a vast continent, and has not yet been joined by its chief tributaries.

Of these the gigantic Rio Negro is its most considerable northern vassal. This stream, which owes its name to the black colour of its waters, rises in the Sierra Tunuhy, an isolated mountain-group in the Llanos, and conveys part of the waters of the Orinoco to the Amazons, as if the latter were not already sufficiently great. After a course of 1500 miles it throws itself into the vast stream, 3600 paces broad and 19 fathoms deep. Brigs of war have already ascended the Amazons as far as the Rio Negro, and frigates would find no obstacle in their way.

The Madeira, the next great tributary of the regal stream, has thus been named from the vast quantities of drift-wood floating on its waters. Three considerable rivers (two of which, the Mamore and the Beni, issue from the Andes, while the third, the Guapore, arises in the Campos de Parecis, but fifteen miles from the sources of the Paraguay), form the enormous Madeira, which rolls its waters slowly through the plains, as if, conscious of being among the largest of rivers, it unwillingly stooped to minister to the greatness of a still mightier stream.

After the influx of the Madeira, the Amazons frequently swells to a breadth of seven miles, and hollows its bed to an average depth of twenty-four fathoms. Farther on, after having with a side-arm embraced the island of Tupinambaranas, which almost equals Yorkshire in extent, the Amazons now reaches the famous strait of Obydos, where it narrows to 2126 paces, and rolls along between low banks in a bed whose depth as yet no plummet hath sounded. The mass of waters which, during the rainy season, rushes in one second through the strait, is estimated by Von Martius at 500,000 cubic feet,—enough to fill all the streams of Europe with an exuberant current.

The tides extend as far as Obydos, though still 400 miles from the sea; and, according to La Condamine, they are even perceptible as far as the confluence of the Madeira. But so slow is their progress upwards, that seven floods, with their intervening ebbs, roll simultaneously along upon the giant stream; and thus, four days after the tide-wave was first raised in the wide deserts of the South Sea, its last undulations expire in the solitudes of Brazil.*

The next considerable vassal of the Amazons is the shallow Tapajos, which, however, is preferred to the Madeira for water-communication between the provinces of Matto-Grosso and

---

* "The Sea and its Living Wonders." Second Edition, p. 41.

Para, on account of the frequent rapids and cataracts which interrupt the navigation of that wood-drifting river.

Fancy six streams, like the Thames, strung successively together, and you have the length of the Tapajos; take the Rhine twice from its source in the glacier of Mount Adula to the sands of Katwyck, and you have the measure of the Xingu. Before the confluence of this last of its great tributaries, — for the Tocantines, though considered by some geographers as a vassal, is in reality an independent stream, — the breadth of the Amazons appeared to Von Martius equal to that of the Lake of Constance; but soon even this enormous bed becomes too narrow for the vast volume of its waters, for below Gurupa it widens to an enormous gulf, which might justly be called the "Bay of the Thousand Isles." Nobody has ever counted their numbers; no map gives us an idea of this labyrinth: as the Brazilian government, in its wretched jealousy and ignorance, scorns to lay out any money on hydrography, and will not allow any other power to undertake the task. If we reckon the island of Marajo, which equals Sicily in size, to the delta of the Amazons, its extreme width on reaching the ocean is not inferior to that of the Baltic in its greatest breadth.

Dangerous sand-banks guard the giant's threshold; and no less perilous to the navigator is the famous Pororocca, or the rapid rising of the spring-tide at the shallow mouths of the chief stream and of some of its embranchments, — a phenomenon which, though taking place at the mouth of many other rivers, such as the Hooghly, the Indus, the Dordogne, and the Seine[*], nowhere assumes such dimensions as here, where the colossal wave frequently rises suddenly along the whole width of the stream to a height of twelve or fifteen feet, and then collapses with a roar so dreadful that it is heard at the distance of more than six miles. Then the advancing flood-wave glides almost imperceptibly over the deeper parts of the river-bed, but again rises angrily as soon as a more shallow bottom arrests its triumphant career.

The territory drained by the Amazons is so vast that, at the sources of its northern and southern tributaries, the rainy season

---

[*] "The Sea and its Living Wonders." Second Edition, p. 40.

takes place at opposite times of the year. So wonderful is the length of the stream that, while at the foot of the Andes it begins to rise early in January, the Solimoens swells only in February; and below the Rio Negro the Amazons does not attain its full height before the end of March.

The swelling of the river is colossal as itself. In the Solimoens and farther westwards the water rises above forty feet; and Von Martius even saw trees whose trunks bore marks of the previous inundation fifty feet above the height of the stream during the dry season.

Then for miles and miles the swelling giant inundates his low banks, and, majestic at all times, becomes terrible in his grandeur when rolling his angry torrents through the wilderness. The largest forest-trees tremble under the pressure of the waters, and trunks, uprooted and carried away by the stream, bear witness to its power. Fishes and alligators now swim where a short while ago the jaguar lay in wait for the tapir, and only a few birds, perching on the highest tree-tops, remain to witness the tumult which disturbs the silence of the woods.

Meanwhile the waters stimulate vegetation; numberless blossoms break forth from the luxuriant foliage; and while the turbid waters still play round the trunks of the submerged trees, the gayest flowers enamel their green crowns, and convert the inundated forest into an enchanted garden. When at length the river retires within its usual limits, new islands have been formed in its bed, while others have been swept away; and in many places the banks, undermined by the floods, threaten to crush the passing boat by their fall, — a misfortune which not seldom happens, particularly when high trees come falling headlong down with the banks into the river.

The periodical overflowing of rivers so broad as the Amazons and its chief tributaries, and the astonishing rapidity with which their waters rise above their usual bounds, implies a quantity of rain superior to all European ideas. If, in general, the atmospherical precipitations of the tropical zone are much greater than those of the temperate latitudes, Brazil again surpasses in this respect most other equatorial lands. I have already mentioned in the first chapter that, during a single thunderstorm witnessed by Castelnau at Pebas, no less than thirty inches of rain fell; and when we consider that the average fall in the equatorial plains of the new world is estimated at nine feet, we must cease to wonder at their rivers exhibiting a rapid rise on a scale unknown to our smaller and less prodigally nourished streams.

Countless lagunes stretch along the course of the Marañon and his tributaries. Possibly they may owe their existence to the waters remaining after the inundations have subsided; perhaps also, as Von Martius believes, to the numberless streams which, gushing out of the earth, more or less distant from the rivers, either extend as standing waters, or flow as brooks into the larger rivers. Most of these lagunes communicate with the larger currents by channels, which, however, are generally dried up before the next rainy season sets in.

The magical beauty of tropical vegetation reveals itself in all its glory to the traveller who steers his boat through the solitudes of these aquatic mazes. Here the forest forms a canopy over his head; there it opens, allowing the sunshine to disclose the secrets of the wilderness; while on either side the eye penetrates through beautiful vistas into the depths of the woods. Sometimes, on a higher spot of ground, a clump of trees forms an island worthy of Eden. A chaos of bushropes and creepers flings its garlands of gay flowers over the forest, and fills the air with the sweetest odour. Numerous birds, partly rivalling in beauty of colour the passifloras and bignonias of these hanging gardens, animate the banks of the lagune, while gaudy macaws perch on the loftiest trees; and, as if to remind one that death is not banished from this scene of paradise, a dark-robed vulture screeches through the woods, or an alligator rests, like a black log of wood or a sombre rock, on the tranquil waters. Well he knows that food will not be wanting; for river tortoises and large fish are fond of retiring to these lagunes.

Blue Macaw.

In one of these shallow lakes Castelnau witnessed fish-catching on a grand scale. On the previous evening a quantity of branches of the Barbasco (*Jacquinia armillaris*), after having been beaten with clubs and divided among the canoes that were

to take part in the sport, had been steeped in water, and then
flung along with the infusion into the lagune. At least five
hundred Indians stood on the banks among the high rushes, or
on the trunks of trees, armed with arrows, harpoons, and clubs.
At first only small fishes appeared upon the surface, and, as if
stunned and then suddenly awakening, sought to leap upon the
bank. Then the larger species were seen to float on the waters,
or to make similar efforts to escape from the poisoned element.
The whole day long the canoes of the Indians were passing on
the lagune, and the same bustle reigned along the banks. The
whistling of the arrows was incessantly heard, along with the
beating of the clubs upon the water, while on land no less
activity was displayed in cutting up, smoking, and salting the
fish. Castelnau counted thirty-five different species,—among
others the famous electrical eel, — and estimated the number
caught at 50,000 or 60,000, many of whom measured a foot
or more in length. Although the lagune was thus poisoned,
the Indians drank the water with impunity, and the river tortoises and alligators seemed to be equally untouched by the
Barbasco juice, which proved so fatal to the fishes.

The inundations of the Amazon, which often extend many
miles inland, essentially modify the character of the bordering
forest; for it is only beyond their verge that the enormous fig
and laurel trees, the Lecythas and the Bertholletias, appear
in all their grandeur. As here the underwood is less dense
and more dwarfish, it is easy to measure the colossal trunks,
and to admire their proportions, often towering to a height of
120 feet, and measuring fifteen feet in diameter above the projecting roots. Enormous mushrooms spring from the decayed
leaves, and numberless parasites rest upon the trunks and
branches. The littoral forest, on the contrary, is of more
humble growth. The trunks, branchless in their lower part,
clothed with a thinner and a smoother bark, and covered with
a coat of mud according to the height of the previous inundation, stand close together, and form above a mass of interlacing
branches. These are the sites of the cacao-tree and of the
prickly sarsaparilla, which is here gathered in large quantities
for the druggists of Europe. Leafless bushropes wind in grotesque festoons among the trees, between whose trunks a dense
underwood shoots up, to perish by the next overflowing of the

stream. Instead of the larger parasites, mosses and jungermannias weave their carpets over the drooping branches. But few animals besides the numerous water-birds inhabit this damp forest zone, in which, as it is almost superfluous to add, no plantation has been formed by man.

The many windings of the water channels which traverse the littoral woods are so overgrown with bushes, that the boat can only with difficulty be pushed onwards through these retreats, whose silence is only broken by the splashing of a fish or the snorting of a crocodile.

The many islands of the delta of the Amazon are everywhere encircled by mangroves; but sailing stream upwards, the monotonous green of these monarchs of the shore is gradually replaced by flowers and foliage, which, in every variety of form and colour, for hundreds and hundreds of miles characterise the banks of the river.

During the dry season prickly astricarias, large musaceæ, enormous bamboo-like grasses, white plumed ingas, and scarlet poivreas, are most frequently seen among the numberless plants growing along the bank of the stream, or projecting over its margin; while above the shrubbery of the littoral forest numberless palms tower, like stately columns, to the height of a hundred feet; others of a lower stature are remarkable for the size of their trunks, on which the foot-stalks of the fallen leaves serve as supports for ferns and other parasites.

On the trees which often lie floating on the river, though still attached by their roots to the bank on which they had flourished, petrels or scarlet ibises frequently perch; and as a boat approaches, hideous bats, disturbed in their holes, fly out of the mouldering trunks.

It stands to reason that in a length of more than 3000 miles the species of plants must frequently change; yet the low banks of the Amazon, and of its vassals, as soon as they have emerged from the mountains where they rise, have everywhere a similar character.

On sailing down the river for hundreds of miles, the eye may at length grow weary of the uniformity of a landscape, which remains constantly the same; but the interest increases as the mind becomes more and more impressed by the grandeur of its dimensions. A broad stream, now dividing into numerous arms, and now looking like a lake; a dark forest-border, which on so flat a ground seems at a distance like an artificial but colossal hedge: these are the only elements of which the landscape is composed. No busy towns rise upon the banks, and it is only at vast intervals that one finds a few wretched huts, which are soon again lost in the forest; but a sky so brilliant spreads over the whole scene, and the rays of the sun beam upon a nature of such luxuriance, that the traveller, far from feeling the voyage monotonous, proceeds on his journey with increasing interest, and every morn salutes with new joy the wilderness, reposing in the stillness of its early grandeur.

The boat floats along, borne by the current of the river, which, in the dry season, generally flows at the rate of four English miles in an hour. Even during the night the journey is usually continued, when no special danger claims a greater caution, and a landing only takes place when the desire becomes general to enjoy a perfectly quiet night's rest, or when a broad sandy bank happens to be invitingly near: the raft is then attached to the bank, and preparations are made for camping in the wilderness. Generally an island is selected, as affording both greater security from beasts of prey and a clearer ground: The Indians are not obliged to fetch fire-wood from a distance, for trees, drifted by the floods, are constantly found at the upper end of the river-islands, where they remain until the next inundation once more raises them; and thus many of them, though born at the foot of the Andes, ultimately find their way to the ocean, and by means of the Equatorial and Gulf Streams,* perhaps even to the desert shores of Lapland, Spitzbergen, and Nowaya Semlya. The Indians, who are as fond of sights as most other people, sometimes set fire to the whole pile, and then the fire, taking an unexpected direction, may force the company to flee as fast as possible to the raft, and to settle in a safer place, while the flames continue to blaze over the forest, or to cast a lurid light over the waters.

Fires are frequently lighted for a more useful purpose on the banks of the stream, as they never fail to attract a number of large fishes, which the dexterous Indians know how to strike with their harpoons. While some are thus engaged, others are

---

* "The Sea and its Living Wonders," p. 46.

lurking for the tortoises that pay their nightly visits to the bank, anxious to bury their numerous eggs in the sand. But the jaguar prowls about, intent upon the same prey, and on this account the Indians never go to any distance from the watch-fires, either alone or unarmed.

Thus almost every landing on one of these river-islands furnishes fresh provisions for the continuance of the journey; for the captured tortoises are bound to the raft, where, in the enjoyment of water and shade, they continue to live for a long time.

As soon as the supper is finished, the Indians invariably splash about in the water; and having thrown an additional log upon the watch-fire, they all stretch themselves on the ground, under their dark-coloured toldos, or mosquito covers, which on the white sand have the appearance of as many coffins. Their tranquil breathing soon tells that they are enjoying the deep repose peculiar to their race; but sleep forsakes the European amid scenes so novel and so grand. The soul is struck with impressions which compel it to reflection. The ripple breaks lightly on the bank; no noise, save the crackling of the fire, breaks the stillness of the night. Only from time to time the splashing of a fish is heard in the distant centre of the stream. The same stillness reigns in the skies; for not the slightest cloud dims the brightness of the stars. But suddenly the waters begin to rustle at a distance, as if wave were rolling after wave; and as the strange sound draws nigh, an unusual agitation becomes apparent in the water. The awakening Indians whisper anxiously, for they imagine an enormous reptile to be the cause of the phenomenon. They also believe the lagunes of the great stream to be the seat of a prodigious serpent, equal in size and power to the fabulous sea-snake; for the yacu-mama, or "mother of the waters," as this imaginary monster is called, attracts by a single inspiration every living creature — man, quadruped, or bird — that passes within a hundred feet of its jaws. As the maelstrom sucks down the helpless boat that comes within its vortex, thus the mighty air-current forces its prey into the wide mouth of the monster lurking in the thicket. For this reason an Indian will never venture to enter an unknown lagune without blowing his horn, as the yacu-mama is said to answer, and thus to give him time for a speedy flight. The "mother of the waters" is said to be at least fifty paces long, and to measure ten or twelve yards in circumference. Thus fancy is as busy in creating imaginary terrors in the lagunes of the Marañon as on the rocky shores of Scandinavia.

Infinitely more dangerous than this fabulous serpent, more dreadful even than the cayman or the anaconda, are the pirangas, a small species of salmon, which in many places attack the unfortunate swimmer with their sharp teeth, and taint the waters with his blood. Castelnau saw how a stag, which threw itself into the river to avoid the hunters' pursuit, was soon killed by the pirangas. The Roman knight that cast his slaves to the murænas,[*] would, no doubt, have been rejoiced to people his ponds with fish like these; and how delighted Tiberius would have been to have possessed them at Capræa!

The pirangas frequently lacerate the tails of the alligators, but no animal in the world is without its enemies, and the pirangas in their turn suffer much from large crustaceous parasites. The flesh of the pirangas is delicate, and their voracity facilitates their capture

A night encampment in the Amazon is, however, not always so pleasant as the foregoing description might lead one to suppose; for many islands are so infested with mosquitos that they are quite intolerable, and the growl of a jaguar or the sight of a crocodile (for this animal is by no means afraid of fire) not unfrequently disturbs the company. Complete security from these persecutions and visits is only to be found in the centre of the stream; for here a cayman is seldom seen, and the wings of the insects are too weak to carry them to such a distance from the shore.

In more than one respect the Amazon reminds one of the ocean, from whose bosom its waters originally arose. Like the sea, it forms a barrier between various species of animals; for the monkeys on its northern bank are different from those of the forests on its southern side, and many an insect—nay, even many a bird,—finds an impassable barrier in the enormous width of the river. Like the sea, it has a peculiar species of dolphin, and hundreds of miles up the stream, sea-mews and petrels, deceived by its grandeur, screech or shoot in arrowy

---

[*] "The Sea and its Living Wonders," p. 195.

flight over its fish-teeming waters. As over the ocean, or in the desert, the illusions of the mirage are also produced over the surface of the Marañon. The distant banks, not always clearly defined even in the morning, disappear wholly at noon, and the rays of the sun are then so refracted that the long rows of palms appear in an inverted position.

The dreadful storms which burst suddenly over the Amazon, likewise recal to memory the tornados of the ocean. The howlings of the monkeys, the shrill tones of the mews, and the visible terror of all animals, first announce the approaching conflict of the elements. The crowns of the palms rustle and bend, while as yet no breeze is perceptible on the surface of the stream; but, like a warning voice, a hollow murmur in the air precedes the black clouds ascending from the horizon, like grim warriors ready for battle. And now the old forest groans under the shock of the hurricane; a night-like darkness veils the face of nature; and, while torrents of rain descend amid uninterrupted sheets of lightning and terrific peals of thunder, the river rises and falls in waves of a dangerous height. Then it requires a skilful hand to preserve the boat from sinking; but the Indian pilots steer with so masterly a hand, and understand so well the first symptoms of the storm, that it seldom takes them by surprise, or renders them victims of its fury.

Among the dangers of the Amazon, the rapids must not be forgotten that frequently arise where large tracts of the bank, undermined by the floods, have been cast into the river. The boat is almost unavoidably lost when carried by the current among the branches of the trees, which, though submerged, still remain attached to the ground, and sweep furiously through the eddy, overturning or smashing all that comes within their reach.

If the Nile — so remarkable for its historical recollections, which carry us far back into the by-gone ages — and the Thames, unparalleled by the greatness of a commerce which far eclipses that of ancient Carthage or Tyre — may justly be called *the* rivers of the *past* and the *present,* the Amazon has equal claims to be called the stream of the *future;* for a more splendid field nowhere lies open to the enterprise of man.

All the gifts of Nature are scattered in profusion over the

vast territory drained by the river. The mountains, where it rises, teem with mineral treasures, and the very ideal of fertility is realised in those well-watered plains, where the equatorial sun developes life in boundless luxuriance. The most useful and costly productions of the tropical world, — sugar, cotton, coffee, indigo, tobacco, maize, rice; quinquina in the higher regions of the Marañon, where wheat and the vine find a congenial climate; cacao and vanilla, sarsaparilla and caoutchouc, various palms of the most manifold uses; trees and shrubs, some rivalling our oaks in the solidity of their timber, others fit by the beauty of their grain to adorn palaces; dyes, resins, gums, spices, drugs,— all, in one word, that is capable of satisfying the wants of the frugal or the fancies of the rich, might there be raised in profusion over a space surpassing England at least forty times in extent. The whole actual population of the globe could easily live in content and plenty in the almost uninhabited valleys of the Marañon and its tributary streams.

And where has Nature better provided for the facility of communications so necessary for unsealing and multiplying the resources of a country? For the Amazon and its great tributaries are not only all of them navigable for hundreds of miles by larger vessels to the very foot of the mountains, where they rise, but to the north we find the Rio Negro communicating, through the Cassiquiare, with the Orinoco, and through the Rio Branco and the Parima with the Essequibo; while to the south it would but require an insignificant canal to unite the Guapore and the Paraguay, both navigable for smaller ships nearly up to their sources, which are only fifteen miles distant from each other. What a magnificent prospect for future steam-boat intercourse! where, without unloading, the same ship will be able to traverse the Rio Negro, the Marañon, and, far to the south, the La Plata.

With these splendid prospects the present forms a melancholy contrast. Here and there some small town or wretched village rises on the banks of the mighty stream; and a few Indians roam over the forests, through which it rolls along, or enjoy the produce of its prolific waters. The vast province of Para, the garden of Brazil, the paradise of unborn millions, has scarcely four inhabitants on a geographical square mile; while even the northern province of Archangel, the land of the stunted fir and the mossy tundra, has a population four times as large. Weak governments, fearful of foreign intrusion,— which would alone be able to carry the living spirit of progress into the wilderness,—have hitherto, with all the power of inertness, weighed upon the valley of the Amazon; but in a century, when the most exclusive nations of Asia have been made to open their harbours to commerce, and when even Central Africa rushes into the current of trade, jealous imbecility can hardly close much longer a territory of such importance and promise to the growing spirit of enterprise.

Eight years after Columbus had revealed the existence of a new world, Vincent Yañez Pinson, the companion of his first voyage, sailed with four ships from the port of Palos (13th January, 1500), steered boldly towards the south, crossed the line, and discovered the mouth of the Amazon. Forty years later Gonzalo Pizarro, governor of Quito, left his capital with 340 Spaniards and 4000 Indian carriers to conquer the unknown countries to the east of the Andes. The march over the Puna and the high mountain ridges proved fatal to the greater part of their wretched attendants; and even the Spaniards — accustomed to brave every climate and hardship wherever gold held forth its glittering promise — had much to suffer from the excess of cold and fatigue. But when they descended into the low country their distress increased. During two months it rained incessantly, without any interval of fair weather long enough to enable them to dry their clothes. They could not advance a step, unless they cut a road through woods, or made it through marshes. The land, either altogether without inhabitants, or occupied by the rudest and least industrious tribes in the New World, yielded little food. Such incessant toil and continual scarcity were enough to shake the most stedfast hearts; but the heroism and perseverance of the Spaniards of the sixteenth century surmounted obstacles which to all others would have seemed insuperable. Allured by false accounts of rich countries before them, they struggled on, until they reached the banks of the Napo, one of the rivers whose waters add to the greatness of the Marañon. There, with infinite labour, they built a bark, which they expected would prove of great use in conveying them over rivers, in procuring provisions, and in exploring the country. This was manned with fifty soldiers, under the command of Francis Orellana, the officer next in rank to Pizarro.

The stream carried them down so quickly that they were soon far ahead of their countrymen, who followed slowly and with difficulty by land. At first Orellana may have had no intention to betray the trust bestowed upon him by his commander; but on reaching the Marañon, the aspect of the stream rolling majestically to the east proved a temptation too strong for his ambition; and, forgetting his duty to his fellow soldiers, he resolved to follow the course of the river, which seemed to beckon him onwards to riches and renown. But one among his followers, *Sanchez de Vargas*, whose name well deserves a record, had the courage to remonstrate against this breach of faith, for which he was landed as a criminal, without food or help of any kind. After a dangerous and romantic navigation of seven months, whose real adventures he afterwards embellished with fabulous tales of El Dorados and warlike Amazons, Orellana at length reached the mouth of the stream. Drifted by the current, he thence safely steered for the Spanish settlement in the island of Cubagua, and soon after embarked for Spain. The magnificence of his discovery threw a veil over his guilt; and, having been appointed governor of the territory whose grandeur he had been the first to reveal, he once more crossed the ocean. But he was not destined to reach the scene where his ambition dreamt of exploits worthy to eclipse the fame of Cortez or Pizarro; a mortal disease befel him on the passage, and in the sea he found a nameless grave.

But what had meanwhile become of the leader whom he had so basely abandoned in the wilderness? The consternation of Pizarro on not finding the bark at the confluence of the Napo and the Marañon, where he had ordered Orellana to wait for him, may well be imagined. But, imputing his absence to some unknown accident, he advanced above fifty leagues along the banks of the river, expecting every moment to see the bark appear with abundant provisions and joyful tidings. At length he met with the faithful Vargas, and now no doubt remained about the treachery of his lieutenant and his own desperate situation. The spirit of the stoutest-hearted veteran sank within him; all demanded to be led back instantly, and Pizarro, though he assumed an appearance of tranquillity, did not oppose their inclination. But they were now 1200 miles from Quito, and a march of many months had to be made without the hopes which had soothed their previous sufferings. Hunger compelled them to sacrifice all their dogs and horses, to devour the most loathsome reptiles, to gnaw the leather of their saddles and sword-belts. All the Indians and 210 Spaniards perished in this wild expedition, which lasted nearly two years. When at length the survivors arrived at Quito, they were naked like savages, and so worn out with famine and fatigue, that they looked more like spectres than men.

Two hundred years after the adventures of Pizarro and Orellana, the French naturalist, La Condamine, performed his celebrated voyage from Bracamoros to Para. He was accompanied by the learned M. Godin des Odonnais, who, leaving his wife on the eastern slope of the Andes, returned alone to Europe in the year 1749. After a separation of twenty years, Madame Godin undertook to descend the Amazons to Para, where her husband was waiting for her. She embarked with her two brothers, a doctor, three female servants, and some Indians, in a large open boat. At the very first opportunity the doctor abandoned the party, and was soon followed by the Indians, who had been paid beforehand. The unskilled travellers vainly attempted to steer their boat; it foundered on the bank, and Madame Godin with difficulty saved her life. They then made a raft, which met with the same misfortune. Undaunted by these repeated disasters, but completely inexperienced, they now resolved to proceed on foot through the forest; but hunger and fatigue soon drove them to despair, and they all perished, except Madame Godin, who, though physically the weakest, was morally the strongest of the party. Who could describe her sufferings after the last of her companions died! Tattered, emaciated, exhausted, she at length met some Indians who treated her with the greatest kindness. The long struggle for her life, amid dangers and hardships without number, had bleached her hair, and stamped her with the marks of extreme old age. The good-natured

Indians guided her to the next European settlement, whence she continued her journey to Para without any further adventures. But the dreadful scenes she had witnessed, and the loss of the dear relations and faithful companions, who one after the other had dropped from her side, had too severely shocked her nerves; and, though she escaped death in the wilderness, it was only to fall a prey to hopeless insanity.

African Bushmen.

CHAPTER VI.

THE KALAHARI.

Reasons why Droughts are prevalent in South Africa — Vegetation admirably suited to the Character of the Country — Number of Tuberous Roots — The Caffre Water-Melon — The Naras — The Mesembryanthemums — The Animal Life of the Kalahari — The Bushmen, a Nomadic Race of Hunters — The Bakalahari — Their Love for Agriculture — Their Ingenuity in procuring Water — Trade in Skins.

A GEOGRAPHICAL position, not unlike that which condemns the plains along the western foot of the Peruvian and Bolivian Andes to perpetual aridity, renders also the greater part of tropical and sub-tropical Southern Africa subject to severe droughts, and in general to great scarcity of rain. For the emanations of the Indian Ocean, which the easterly winds carry towards that continent, and which, if equally distributed over the whole surface, would render it capable of bearing the richest productions of the torrid zone, are mostly deposited on the eastern slopes of the mountain-chains, which, under various denominations, traverse eastern South Africa from north to south; and when the moving mass of air, having crossed their highest elevations, reaches the great heated inland plains, the ascending warmth of that hot dry surface gives it greater power of retaining its remaining moisture, and few showers can be given to the central and western lands. Thus, while the sea-borde gorges of the eastern zone are clad with gigantic forests, and an annual supply of rain there keeps a large number of streams perpetually flowing, Damara Land, the Namaqua country, and the Kalahari, are almost constantly deprived of moving water.

From these general remarks it might be imagined that regions so scantily supplied with one of the prime necessaries of life could be nothing but a dead and naked waste; yet, strange to say, even the great Kalahari, extending from the Orange river in the south, lat. 29°, to Lake Ngami in the north, lat. 21°, and from about 24° E. long. to near the west coast, has been called a desert, simply because it contains no flowing streams and very little water in wells; as, far from being destitute of vegetable or animal life, it is covered with grass and a great variety of creeping plants, interspersed with large patches of bushes and even trees. In general, the soil is a light-coloured, soft sand; but the beds of the ancient rivers contain much alluvial soil, and, as that is baked hard by the burning sun, rain-water stands in pools in some of them for several months in the year.

The abundance of vegetation on so unpromising a soil may partly be explained by the geological formation of the country; for as the basin-shape prevails over large tracts, and as the strata on the slopes where most of the rain falls dip in towards the centre, they probably guide water beneath the plains, which are but ill-supplied with moisture from the clouds.

Another cause, which serves to counteract the want or scarcity of rain, is the admirable foresight of Nature in providing these arid lands with plants suited to their peculiar climate. Thus creepers abound which, having their roots buried far beneath the soil, feel but little the effects of the scorching sun. The number of these which have tuberous roots is very great,— a structure evidently intended to supply nutriment and moisture when, during the long droughts, they can be obtained nowhere else.

One of these blessings to the inhabitants of the desert is a small plant named *Leroshúa*, with linear leaves, and a stalk not thicker than a crow's quill; but on digging down a foot or eighteen inches beneath, the root enlarges to a tuber, often as big as the head of a young child, which, on the rind being removed, is found to be a mass of cellular tissue, filled with fluid much like that in a young turnip. Owing to the depth beneath the surface at which it is found, it is generally deliciously cool and refreshing. Another kind, named *mokuri*, is seen in other parts of the country, where long-continued heat parches the soil. This plant is an herbaceous creeper, and deposits under ground a number of tubers, some as large as a man's head, often in a circle, a yard or more horizontally from the stem. The natives strike the ground on the circumference of the circle with stones, till, by hearing a difference of sound, they know the water-bearing tuber to be beneath. They then dig down a foot or so and find it.

But the most wonderful plant of the desert is the Kengwe, or Kēme (*Cucumis Caffer*), the water-melon of the Caffres. In years when more than the usual quantity of rain falls, vast tracts of the country are literally covered with these juicy gourds, and then animals of every sort and name, including man, rejoice in the rich supply. Crossing the desert from Kolobeng to Lake Ngami, Mr. J. Macabe found them in such profusion that his cattle lived on the fluid contained in them for not less than twenty-one days; and when at last they reached a supply of water, they did not seem to care much about it.

On the west coast, along the banks of the Kiusep,—a river which, *when it has water*, flows into Walfish Bay,—Anderson found almost every little sand-hillock covered with a creeper, which produced a kind of prickly gourd of the most delicious flavour. The *naras*, as it is called by the natives, is about the size of an ordinary turnip, and when ripe has a greenish exterior, with a tinge of lemon; while the interior, which is of a deep orange colour, presents a most cooling and inviting appearance, and for three or four months in the year constitutes the chief food both of man and beast.

The *naras* contains a great number of seeds, not unlike a peeled almond in appearance and taste, which, being easily separated from the fleshy parts, are carefully collected, dried in the sun, and then stored away in little skin bags. When the fruit fails the natives have recourse to these seeds, which are equally nutritious, and perhaps even more wholesome.

Thus even in the desert the bounty of the Almighty raises sustenance for man and all His creatures; for, in this barren and poverty-stricken country, food is so scarce, that without the naras the land would be all but uninhabitable. The creeping plants of the desert serve, moreover, a double purpose; for, besides their use as food, they fix, by means of their extensive ramifications, the constantly shifting sands,—thus rendering similar services to those of the sand-reed (*Ammophila arundinacea*) on the dunes along the sandy coasts of the North Sea.

The Mesembryanthemums are another family of plants admirably adapted to the desert, as their seed-vessels remain firmly shut while the soil is hot and dry, and thus preserve the vegetative power intact during the highest heat of the torrid sun; but when rain falls the seed-vessel opens and sheds its contents, just when there is the greatest likelihood of their vegetating. This is the more wonderful, as in other plants heat and drought cause the seed-vessels to burst and shed their charge.

One of this family is edible (*M. edule*); another possesses a tuberous root, which may be eaten raw; and all are furnished with thick, fleshy leaves, with pores capable of imbibing and retaining moisture from a very dry atmosphere and soil; so that if a leaf is broken during the greatest drought it shows abundant circulating sap. The oblong tubers, with which some of the mesembryanthemums are furnished, serve them as an additional means for withstanding the want of rain; for, being buried deep enough beneath the soil for complete protection from the sun, they serve as reservoirs of sap during those rainless periods which recur very frequently in even the most favoured spots of these parts of Africa.

Many useful plants have been conveyed from one country to another; but I doubt whether any attempts have as yet been made to acclimatise the tuberous cucurbitaceæ and the mesembryanthemums of Southern Africa in the northern deserts of that continent, or on the arid coast of Peru, where perhaps some of them might thrive and prove of great use. The propagation of all that is beautiful or useful in nature to every other part of the world adapted to its growth, will be a work for future generations.

The peculiar and comparatively abundant vegetation of the arid plains of South Africa explains how these wastes are peopled by herds of herbivorous animals, which in their turn are preyed upon by the lion, the panther, or the python. Hundreds of elands (*Boselaphus oreas*), gemsbucks, koodoos (*Strepsiceros capensis*), or duikers (*Cephalopus mergens*), may often be seen thirty or forty miles from the nearest water. These, having sharp-pointed hoofs well adapted for digging, are able to subsist without water for many months at a time, by living on moist bulbs and tubers; while the presence of the rhinoceros, of the buffalo and gnu (*Catoblepas Gnu*), of the giraffe, the zebra, and pallah (*Antilope melampus*), is always a certain indication of water being within a distance of seven or eight miles.

Koodoo.

The tribes of the Kalahari consist of bushmen, probably the aborigines of the southern part of the continent, and of Bakalahari, the remnants of an ancient Bechuana emigration. The nomadic bushmen are a nation of hunters, and so thoroughly acquainted with the habits of the game, that they follow them in their migrations, and prey upon them from place to place; thus proving as complete a check upon their inordinate increase as the other carnivora. Supplying by cunning or invention their lack of strength, like the Indian tribes of Guiana, they make use of poisoned arrows, fatal even to the lion; but though their chief food is the flesh of animals, they eke it out by what the women collect of roots and beans and fruits of the desert. Their thin wiry forms are capable of great exertion, and well able to bear the severe privations to which the inhabitants of such a country must submit.

The Bakalahari are traditionally reported to be the oldest of the Bechuana tribes driven into the desert by a fresh migration of their own nation. Though living ever since on the same plains with the bushmen, under the same influences of climate, enduring the same thirst, and living on the same food for centuries, they still retain in undying vigour the Bechuana love for agriculture and domestic animals, hoeing their gardens annually, though often all that they can hope for is a supply of melons and pumpkins, and carefully rearing small herds of goats, although to provide them with water is a task of no small difficulty, since the dread of hostile visits from the adjacent Bechuana tribes makes them choose their abode far from the nearest spring or pool, and leads them not unfrequently to hide their supplies by filling the pits with sand and making a fire over the spot. When they wish to draw water for use, the women come with twenty or thirty of their water vessels in a bag or net on their backs. These water vessels consist of ostrich egg shells, with a hole in the end of each, such as would admit one's finger. The women tie a bunch of grass to one end of a reed about two feet long, which they insert in a hole dug as deep as the arm will reach, and then ram down the wet sand firmly round it. Applying the mouth to the free end of the reed, they form a vacuum in the grass beneath, in which the water collects, and in a short time rises into the mouth. An egg-shell is placed on the ground alongside the reed, some inches below the mouth of the sucker. A straw guides the water into the hole of the vessel as she draws mouthful after mouthful from below; and thus the whole stock of water passes through her mouth as a pump, and when taken home is carefully buried to prevent its loss by evaporation. A short stay among the thirsty Bakalaharis might teach us better to appreciate the blessings of an abundant supply of water.

These poor people generally attach themselves to influential men in the different Bechuana tribes near to their desert home, in order to obtain supplies of spears, knives, tobacco, and dogs, in exchange for the skins of animals which they kill. These are small carnivora of the feline race, including two species of jackal, the dark and the golden, the former of which has the warmest fur the country yields, while the latter is very handsome when made into the skin-mantle called *kaross*. Next in value follow the small ocelot, the lynx, the wild and the spotted cat. Great numbers of duiker and steinbuck skins are also obtained, besides those of lions, leopards, panthers, and hyænas.

During Dr. Livingstone's stay in the Bechuana country, between twenty and thirty thousand skins were made up into *karosses*; part of them were worn by the inhabitants, and part sold to traders, many ultimately finding their way to China. The Bechuanas buy tobacco from the eastern tribes, then, purchasing skins with it from the Bakalahari, tan and sew them into karosses; and these go south to purchase heifer-calves, —

cows being the highest form of riches known,—so that the missionary was frequently asked by the natives, "if Queen Victoria had many cows."

Thus the Kalahari desert, besides supporting multitudes of animals, both small and large, adds its mite to the market of the world, and has afforded a refuge to many a fugitive tribe, which, though contending with many privations in its new abodes, found there at least the blessing of freedom from an alien yoke.

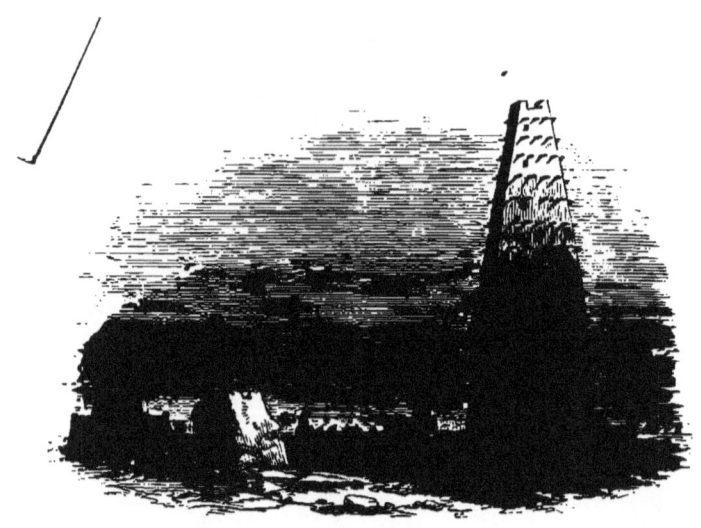

Tower in Agades.

CHAPTER VII.

THE SAHARA.

Its uncertain Limits — Caravan Routes — Ephemeral Streams — Oases — Inundations — Luxuriant Vegetation of the Oases contrasted with the surrounding Desert — The Sedentary and Vagrant Tribes — Harsh contrasts of Light and Shade — Sublimity of the Desert — The Khamsin — The Dying Slave — Sand-Spouts — Venomous Snakes — Porcupine-catching — Chase of the Gazelle — Fluctuation of Animal Life according to the Seasons.

FROM the Nile to the Senegal, and from the vicinity of Agades or of Timbuctoo to the southern slopes of the Atlas, extends the desert, which above all others has been named the Great.

Surpassing the neighbouring Mediterranean at least three times in extent, and partly situated within the tropical zone, partly bordering on its confines, its limits are in many places as undetermined as the depths of its hidden solitudes. For from the mountain chains which separate it in the north from the fertile coast-lands of Barbary and intercept the winter rains, the steppes covered with sedgy pale green Alfa-grass, and dotted here and there with grey wormwood and rosemary shrubs or dark-leaved pistacias, only gradually merge into the naked wilderness; and in the south no geographer is able as yet to draw the line between the rainless Sahara and the well-watered lands of Nigritia. No European traveller has ever followed the southern limits of the desert from east to west, nor is its interior known except only along a few roads, traced for many a century by the wandering caravans. From Tafilet to Timbuctoo, or from Murzuk to Bornu, the long train traverses the desert to exchange cotton-goods, silk, iron, glass, pearls, and other articles of northern industry, for the ivory, gold-dust, camels, slaves, ostrich-feathers, and tanned hides of the wealthy Soudan; and annually from the oases of the Touat, situated to the south of Algeria, a stream of pilgrims flows to the east, and growing as it advances through the Fezzan, Augila, and Siwah, at length reaches Kosseir on the Red Sea, where it finds vessels waiting to transport it to Djedda, situated on the opposite shore, in the vicinity of Mecca the Holy.

In general the desert may be said to extend in breadth from the thirty-ninth to the seventieth degree of northern latitude; but while in many parts it passes these bounds, in others fruitful districts penetrate far into its bosom, like large peninsulas or promontories jutting into the sea.

Until within the last years, it was supposed to be a low plain, partly situated even below the level of the ocean; but the journeys of Barth, Overweg, and Vogel have proved it, on the contrary, to be a high table-land, rising 1000 or 2000 feet above the sea. Nor is it the uniform sand-plain which former descriptions led one to imagine; for it is frequently traversed by chains of hills, as desolate and wild as the expanse from which they emerge. But the plains also have a different character in various parts: sometimes over a vast extent of country the ground is strewed with blocks of stone or small boulders, no less fatiguing to the traveller than the loose drift sand, which, particularly in its western part (most likely in consequence of the prevailing east winds), covers the dreary waste of the Sahara. Often also the plain is rent by deep chasms, or hollowed into vast basins. In the former, particularly on the northern limits of the desert, the rain descending from the gulleys of the Atlas, sometimes forms streams, which are soon swallowed up by the thirsty sands, or dried by the burning sunbeams. In spite of this short duration, the sudden appearance of these streams is not unfrequently the cause of serious distress to the oases which border the northern limits of the desert.

For this reason, as soon as the Atlas veils itself with clouds, horsemen from the oases of the Beni Mzab are sent at full speed into the mountains. They form a chain as they proceed, and announce by the firing of their rifles, the approach of the waters. The inhabitants of the oases instantly hurry to their gardens to convey their agricultural implements to a place of safety. A rushing sound is heard; in a short time the ground is inundated; and the little village seems suddenly as if by magic transported to the banks of a lake, from which the green tufts of the palm-trees emerge like islands. But this singular spectacle soon passes away like the fantastic visions of the mirage.

The deeper basins of the Sahara are frequently of great extent, and sometimes contain valuable deposits of salt. Wherever perennial springs rise from the earth, or wherever it has been possible to collect water in artificial wells, green oases, often many a day's journey apart from each other, break the monotony of the desert. They might be compared with the charming islands that stud the vast solitudes of the South Sea; but they do not appear, like them, as elevations over surrounding plains of sea, but as depressions, where animals and plants find a sufficient supply of water, and a protection, not less necessary, against the terrific blasts of the desert.

A wonderful luxuriance of vegetation characterises these oases of the wilderness. Under and between the date-palms, that are planted about six paces apart, grow apricot and peach trees, pomegranates and oranges, the henneh, so indispensable to oriental beauty; and even the apple-tree, the pride of European orchards. The vine twines from one date-palm to another, and every spot susceptible of culture bears corn, particularly dourrah or barley, and also clover and tobacco. With prudent economy the villages are built on the borders of the oases on the unfruitful soil, so that not a foot of ground susceptible of culture may be lost.

Sedentary Berber tribes inhabit the oases, and chiefly live upon the fruits of their date-trees; while the nomadic Tuaryks and Tibbos wander, with their camels and sheep, over the desert in quest of scanty forage and thorny shrubbery. In spite of their mutual hatred, the bonds of a common interest connect the vagrant and the agricultural tribes. Condemned to perpetual migrations, the nomade is forced to confide all the property which he is unable to carry about with him to the inhabitant of the oasis; he may even possess a small piece of land, the cultivation or care of which he intrusts to the latter, who, on his part, as soon as he has saved something, buys a sheep or a goat, which he gives in charge to the nomade.

An unmitigated hatred, on the contrary, exists between the various erratic tribes, as here no mediating self-interest softens the antipathies which are almost universally found to exist between neighbouring barbarians; and their robber-expeditions not merely attack the richly laden caravan, but also the oasis which may be connected by the bonds of intercourse with their hereditary enemies.

The vast tracts of sterile sand, where not even the smallest plant takes root, and which might be called the "desert of the desert," present the greatest conceivable contrast to its green oases. With the vegetable world the animal kingdom likewise disappears, and for days the traveller pursues his journey without meeting with a single quadruped, bird, or insect. Nowhere are the transitions of light and shade more abrupt than in the desert, for nowhere is the atmosphere more thoroughly free of all vapours. The sun pours a dazzling light on the ground, so that every object stands forth with wonderful clearness, while all that remains in the shade is sharply defined, and appears like a dark spot in the surrounding glare.

These harsh contrasts between light and shade deprive the landscape of all grace and harmony; but this want is amply compensated by its singular grandeur. The boundless horizon and the silence which reigns over the whole scene, have a powerful effect upon every one who for the first time enters upon so strange a world. The ocean and the ice-fields of the polar regions make a similar impression.

Thus, in spite of all he may have endured, the traveller that has once crossed the desert will ever after remember it with pleasure, and long for the renewal of its deep emotions. For the life of the Sahara resembles that of the ocean. During a continuance of bad weather or a calm, the mariner may vow to forsake the sea for ever; but he has scarcely landed when his affection revives, and he longs for the sea again. We may grow fatigued with the life of the town, but we cannot tire of the majestic uniformity of the ocean or of the awful sublimity of the desert.

The stillness of these wastes is sometimes awfully interrupted by the loud voice of the khamsin or simoom. The crystal transparency of the sky is veiled with a hazy dimness. The wind rises and blows in intermittent gusts, like the laborious breathing of a feverish patient. Gradually the convulsions of the storm grow more violent and frequent; and although the sun is unable to pierce the thick dust-clouds, and the shadow of the traveller is scarcely visible on the ground, yet so suffocating is the heat, that it seems to him as if the fiercest rays of the sun were scorching his brain. The dun atmosphere gradually changes to a leaden blackness; the wind becomes constant; and even the camels stretch themselves upon the ground and turn their backs to the whirling sand-storm. At night the darkness is complete: no light or fire burns in the tents, which are hardly able to resist the gusts of the simoom. Silence reigns throughout the whole caravan, yet no one sleeps; the bark of the jackal or the howl of the hyæna alone sounds dismally from time to time through the loud roaring of the storm.

Modern travellers (Russegger, D'Escayrac de Lauture) assure us that the often repeated tale of whole caravans having been buried in the sand by one of these tornados is very much exaggerated; but there can be no doubt that the simoom causes the death of many a poor slave. Walking at the side of the camels, beaten, miserably fed, and but niggardly supplied with brackish water, without hope—the last support of the wretched —he sinks under an apoplectic stroke; or else, worn out by fatigue and want, is unable any longer to move his paralysed limbs. The caravan abandons him to his fate; for every one thinks but of himself; and who, but his owner, cares about a slave? In a short time the dry atmosphere changes the corpse into a natural mummy, which, "grinning horribly a ghastly smile," seems to defy the desert. But often the crows, wheeling with dismal cries over the dying wretch, hack out his eyes before death relieves him of all pain. These corsairs of the air accompany the caravan, as sharks accompany a vessel, reckoning, like the tyrants of the seas, upon the tribute of the journey.

The sultry breath of the desert is felt far beyond its bounds; for the Sahara is great, and all greatness has a wide extended influence. It blows over Italy, where it is known as the sirocco, and crosses even the Alps, where, under the name of the Fönwind, it rapidly melts the snow of the higher valleys, and causes dangerous inundations. The dust of the desert, whirled high into the air, frequently falls upon the decks of vessels crossing the Atlantic, far from the coast of Africa, and flies in clouds over the Red Sea—a greeting from Nubia to Arabia.

Russegger gives us a graphic account of a khamsin storm which he witnessed at Scheibun, on the Nile (11° 13′ N. lat.): "At three in the afternoon storm-clouds arose, that rapidly approached with an increasing intensity of colour: a dark, and yet burningly vivid, brown-red passing into deep black and grey. A thundering roar was heard: it seemed as if a vast city like London was in flames, and as if peals of heavy artillery were continually booming out of the fiery mass. Menscherah, with its palms and mimosas, and the broad river in the foreground, had an indescribably magnificent effect; for never did a more vivid green stand forth from a more dark and lurid background. But suddenly the whole prospect was veiled, and a strange and ghastly twilight cast its funereal glare over the scene. Sheets of lightning, instantly followed by terrific claps of thunder, flamed uninterruptedly across the sky; the groaning trees crashed and fell; and the river rose in large waves, as if it had been a wide lake agitated by a storm."

The air, filled with dust and sand to suffocation, had a temperature of 102°. When the khamsin was most violent, Russegger was obliged to sit down and bury his face in his mantle, his breast heaving with oppression, his brain reeling as if he were intoxicated. The indescribably grand phenomenon terminated with floods of rain, which at first literally showered down torrents of liquid mud.

As the conflicting air-currents of the ocean occasion waterspouts, the terror of the mariner, who sees them approach in rapid whirls; so also sand-spouts, or trombs, arise in rotatory eddies from the midst of the desert, menacing with instant destruction all living things that come in their way. "At eleven o'clock," says Mr. Bruce, "while we contemplated with great pleasure the rugged top of Chiggre, to which we were fast approaching, and where we were to solace ourselves with plenty of good water, Idris our guide cried out with a loud voice, 'Fall upon your faces, for here is the simoom!' I saw from the south-east a haze come, in colour like the purple part of the rainbow, but not so compressed or thick. It did not occupy twenty yards in breadth, and was about twelve feet high from the ground. It was a kind of blush upon the air, and it moved very rapidly; for I scarce could turn to fall upon the ground with my head to the northward, when I felt the heat of its current plainly upon my face. We all lay flat on the ground, as if dead, till Idris told us it was blown over. The meteor, or purple haze, which I saw was indeed past, but the light air which still blew was of heat to threaten suffocation. For my part, I found distinctly in my breast that I had imbibed a part of it; nor was I free of an asthmatic sensation till I had been some months in Italy at the baths of Poretta, near two years afterwards."

In this case the severity of the blast seems to have past almost instantaneously; yet the hot wind continued to blow about six hours longer with gradually decreasing violence, so as to leave the traveller in a state of depression such as is occasioned, though in a far less degree, by the blowing of the sirocco.

When we consider the scanty vegetation of the Sahara, we cannot wonder that animal life is but sparingly scattered over it. The lion, whom our poets so frequently name the "king of the desert," only shows himself on its borders; and on asking the nomades of the interior whether it is ever seen in their parts, they gravely answer that in Europe lions may perhaps feed on shrubs or drink the air, but that in Africa they cannot exist without flesh and water, and therefore avoid the sandy desert. In fact, they never leave the wooded mountains of the Atlas, or the fruitful plains of the Soudan, to wander far away into the Sahara, where snakes and scorpions are the only dangerous animals to be met with. The snakes, which belong to the genus *Cerastes*, which is distinguished by two small horns upon the head, have a deadly bite, and are remarkable for their almost total abstinence from water. When a caravan, on first entering the desert, meets with one of these venomous reptiles, it is not killed, "for it is of good omen to leave evil behind;" but farther on the snakes are mercilessly destroyed wherever they are seen.

Among the animals which inhabit those parts of the desert, which are covered with prickly shrubs, we find hares and rabbits, hyænas and jackals, the hedgehog and the porcupine.

Well-beaten paths, and here and there a scattered quill, lead to the hole which this proverbially fretful animal burrows in the sand. The hunters widen the entrance with their poniards or swords, until a hoarse, prolonged growl, and the peculiar noise which the enraged porcupine

Porcupine.

makes on raising his quills, warn them to be on their guard. Suddenly the creature rushes from its burrow to cast itself into the thicket; but the well-aimed blow of a poniard stretches it upon the sand. A fire is then kindled, and the animal buried under the embers; the quills then easily separate from the roasted and excellently-flavoured meat.

The ostrich, which is proverbially said to drink only every five days when there is water, and to be able to endure thirst for a much longer period when there is none, and the gazelle, which even the greyhound finds it difficult to catch, venture deeper into the desert. The chase of the gazelle is a favourite amusement of the Saharians. On seeing a herd at a distance they approach as cautiously as possible; and when about a mile distant, they unleash their greyhounds, who dart off with the rapidity of arrows, and are excited, by loud cries, to their utmost speed. Yet they only reach the flying herd after a long chase; and now the scene acquires the interest of a drama. The best greyhound selects the finest gazelle for his prey, which uses all its cunning to avoid its pursuer, springing to the right, to the left, now forwards, then backwards, sometimes even right over the greyhound's head; but all these zig-zag evolutions fail to save it from its indefatigable enemy. When seized it utters a piteous scream, the signal of the greyhound's triumph, who kills it with one bite in the neck.

Several lizards inhabit the desert; among others, a large grey monitor, and a small white skink, with very short legs, called Zelgague by the Arabs. Its movements are so rapid that it seems to swim on the sand like a fish in the water, and when one fancies one has caught it, it suddenly dives under the surface. Its traces, however, betray its retreat, and it is easily extracted from its hole,—a trouble which, in spite of the meagre booty, is not considered too great when provisions are scarce.

According to the seasons animal life fluctuates in the Sahara from north to south. In winter and spring, when heavy rains, falling on its northern borders, provide wide districts, thoroughly parched by the summer heat, with the water and pasturage needed for the herds, the nomadic tribes wander farther into the desert with their camels, horses, sheep, and goats, and retreat again to the coast-lands as the sun gains power. At this time of the year the wild animals—the lion, the gazelle, and the antelope—also wander farther to the south, which at that time provides them, each according to its taste, with the nourishment which the dry summer is unable to bestow; while the ostrich, who during the summer ranged farther to the north, then retreats to the south; for hot and sandy plains are the paradise in which this singular bird delights to roam.

In the southern part of the Sahara the tropical rains, whose limits extend to 19° N. lat., and in some parts still farther to the north, produce similar periodical changes in the character of the desert. Under their influence the sandy plains are soon covered with grasses and shrubs. In the dry season, on the contrary, the green carpet disappears, and the country then changes into a dry waste, covered with stubbles and tufts of mimosas. This beneficial change, however, does not take place every year; for the tropical rains frequently fail to appear on their northern boundaries, and thus disappoint the hopes of the thirsty desert.

Tuaryk Chieftain.

CHAPTER VIII.

THE PRIMITIVE FOREST.

Its peculiar Charms and Terrors—Disappointments and Difficulties of the Botanist — Variety of Trees and Plants — Character of the Primitive Forest according to its Site — Its Aspect during the Rainy Season — A Hurricane in the Forest — Beauty of the Forest after the Rainy Season — Bird Life on the rivers of Guiana — Morning Concert — Repose of Nature at Noon — Nocturnal Voices of the Forest.

THE peculiar charms of the tropical primitive forest are enhanced by the mystery of its impenetrable thickets; for however lovely its lofty vaults and ever-changing forms of leaf or blossom may be, fancy paints scenes still more beautiful beyond, where the eye cannot penetrate, and where, as yet, no wanderer has ever strayed. But imagination also peoples the forest with peculiar terrors; for man feels himself here surrounded by an alien, or even hostile, nature: the solitude and silence of the woods weigh heavily on his mind; in every rustling of the fallen leaves a venomous snake seems ready to dart forth; and who knows what ravenous beast may not be lurking in the dense underwood that skirts the tangled path. In Europe there is no room for such feelings; for in our part of the world there are no woods that may not be visited, even in their deepest recesses: no thorny bushropes stretch their intricate cordage before the wanderer; no masses of matted shrubbery block up his way.

But it is very different in the boundless forests of tropical America, through which roll the Orinoco or the Amazon, with their numberless tributaries. Here the jaguar sometimes loses himself in such impenetrable thickets that, unable to hunt upon the ground, he lives for a long time on the trees, a terror to the

monkeys; here the *padres* of the mission-stations, which are not many miles apart in a direct line, often require more than a day's navigation to visit each other, following the windings of small rivulets in their courses, as the forest renders communication by land impossible.

Even the more open parts of the forest are full of mysteries. In our woods the summits of the highest trees are accessible; there is no blossom that we are not able to pluck,—no plant that we are not able to examine, from its root to its topmost branches; but in the Brazilian forest, where the matted bush-ropes, climbing along the trunks and branches, extend like the rigging of a ship from one tree to another, and blossom at a giddy height, it is frequently as impossible to reach these flowers, as it is to distinguish to which of the many interlacing stems they may belong.

If any one should be inclined to tax this description with exaggeration, let him try to pluck the flowers of the lianas, or to ascend by climbing their flexible cordage. The tiger-cat and the monkey, perhaps also the agile Indian, may be able to accomplish the feat; but it would be utterly hopeless for the European to undertake it. Nor is it possible to drag down one of these inaccessible creepers; for, owing to their strength and toughness, it would be easier to pull down the tree to which it attaches itself than to force the liana from its hold.

No botanist ever entered a primitive forest without envying the bird to whom no blossom is inaccessible, who, high above the loftiest trees, looks down upon the sea of verdure, and enjoys prospects whose beauty can hardly be imagined by man.

A majestic uniformity is the character of our woods, which often consist but of one species of tree, while in the tropical forests an immense variety of families strive for existence, and even in a small space one neighbour scarcely ever resembles the other. Even at a distance this difference becomes apparent in the irregular outlines of the forest, as here an airy dome-shaped crown, there a pointed pyramid, rises above the broad flat masses of green, in ever-varying succession. On approaching, the differences of colour are added to the irregularities of form; for while our forests are deprived of the ornament of flowers, many tropical trees have large blossoms, mixing in thick bunches with the leaves, and often entirely overpowering the verdure of the foliage by their gaudy tints. Thus splendid white, yellow, or red coloured crowns are mingled with those of darker or more humble hue. At length when, on entering the forest, the single leaves become distinguishable, even the last traces of harmony disappear. Here they are delicately feathered, there lobed, — here narrow, there broad, — here pointed, there obtuse, — here lustrous and fleshy, as if in the full luxuriance of youth, there dark and arid, as if decayed with age. In many the inferior surface is covered with hair; and as the wind plays with the foliage, it appears now silvery, now dark green, — now of a lively, now of a melancholy, hue. Thus the foliage exhibits an endless variety of form and colour; and where plants of the same species unite in a small group, they are mostly shoots from the roots of an old stem. This is chiefly the case with the palms; but the species of the larger trees are generally so isolated in the wood, that one rarely sees two alike from the same spot. Each is surrounded by strangers that begrudge it the necessary space and air; and where so many thousand forms of equal pretensions vie for the possession of the soil, none is able to expand its crown or extend its branches at full liberty. Hence there is a universal tendency upwards; for it is only by overtopping its neighbours that each tree can hope to attain the region of freedom and of light; and hence also the crowns borne aloft on those high columnar trunks are comparatively small.

As the tropical primitive forest occupies sites of a very different character, — here extending along the low banks of rivers there climbing the slopes of gigantic mountains, — here under the equator, there on the verge of the tropics, where many of the trees, annually casting their foliage, remind one of the winter of the temperate zone, — it is of course quite impossible to embrace all these varieties of form and aspect in one general description.

On descending from the heights of the Andes to the plains of the Marañon, the eye is attracted, in the more elevated forests (the region of the Quinquina trees), by a variety of fantastically flowering orchids, — and of arborescent ferns, with their lace-like giant leaves, — by large dendritic urticeas, — by wonderful bignonias, banisterias, passifloras, and many other inextricably tangled bushropes and creepers. Farther downwards, though the lianas still appear in large numbers, the eye delights in palms of every variety of form, in terebinthinaceas, in leguminosas, whose sap is rich with many a costly balsam; in laurels, bearing an abundance of aromatic fruit; or it admires the broad-leaved heliconias, the large blossoms of the solaneas, and thousands of other flowers, remarkable for the beauty of their colour, the strangeness of their form, or their exquisite aroma.

In the deep lowlands the forest assumes a severe and gloomy character: dense crowns of foliage form lofty vaults almost impenetrable to the light of day; no underwood thrives on the swampy ground; no parasite puts forth its delicate blossoms where the mighty trees stand in interminable confusion; and only mushrooms sprout abundantly from the humid soil.

Nothing can equal the gloom of these forests during the rainy season. Thick fogs obscure the damp and sultry air, and clouds of mosquitos whirl about in the mist. The trees are dripping with moisture; the flowers expand their petals only during the few dry hours of the day, and every animal seeks shelter in the thicket. No bird, no butterfly comes forth; the snorting of the capybaras, and the monotonous croaking of frogs and toads, are the only sounds that break the dull silence. Night darkens with increasing sadness over these dismal solitudes; no star is visible; the moon disappears behind thick clouds; and the roar of the jaguar, or the howlings of the stentor-monkey, issue like notes of distress from the depth of the melancholy woods.

A hurricane bursting over the primitive forest is one of the most terrific scenes of nature. A hollow uproar in the higher regions of the air, as if the wild huntsman of the German legends were sweeping along with his whole pack of phantom hounds, precedes the explosion of the storm, while the lower atmosphere still lies in deep repose. The roaring and rushing descends lower and lower; the higher branches of the trees strike wildly against each other; the forked lightning flashes through the night-like darkness; the thunder, repeated by a hundred echoes, rolls through the trembling thicket; and trees, uprooted by the fury of the storm, fall with a loud crash, crushing every stem of minor growth in their sweeping ruin. The howlings and wailings of terrified animals accompany the wild sounds of the tempest.

After the wet season the woods appear in their full beauty. Before the first showers, the long-continued drought had withered their leaves, and dried up many of the more tender parasites; during its continuance the torrents of rain despoiled them of all ornament; but when the clouds disperse and the animals come forth from their retreats to stretch their stiffened limbs in the warm sunshine, then also the vegetable world awakens to new life; and where, a few days before, the eye met only with green in every variety of shade, it now revels in the luxuriance of beautiful flowers, which embalm the air with exquisite fragrance.

At this time of the year the banks of the rivers of Guiana winding through the primitive woods are of magical beauty, and sailing on their placid waters one might fancy oneself gliding along on the fairy streams of the lost gardens of Eden.

Through the underwood, which often overhangs wide spaces of the stream, the large white blossoms of the inga shine forth, along with the scarlet brushes of the magnificent combretias. Elegant palms, armed with a panoply of thorns, and bearing a profusion of red fruit, rise above this lovely foreground; and farther on, noble forest trees are seen festooned with creepers and parasites, which are covered with flowers.

With every stroke of the oar new scenes reveal themselves; so that in the variety of forms it is impossible to determine which plant or flower, most deserves the prize of beauty.

These fairy bowers are enlivened by birds of splendid plumage, particularly in the early morning, when the luscious green of the high palm-fronds or the burning yellow of the lofty leopoldinias, touched by the first rays of the sun, suddenly shines forth. Then hundreds of gaudy parrots fly across the river; numberless colibris dart like winged gems through the air; whole herds of cotingas flutter among the blossoms; ducks of brilliant plumage cackle on the branches of submerged trees; on the highest tree-tops the toucan yelps his loud pia-po-ko! while, peeping from his nest, the oriole endeavours to imitate the sound; and the scarlet ibis flies in troops to the coast, while the white egrette flutters along before the boat, rests, and then again rises for a new career.

In general the morning hours are the loudest in the primitive forest; for the animals that delight in daylight, though not more numerous than the nocturnal species, have generally a louder voice. Their full concert, however, does not begin immediately after sunrise; for they are mostly so chilled by the colder night, that they need to be warmed for some time before awakening to the complete use of their faculties. First, single tones ring from the high tree-crown, and gradually thousands of voices join in various modulation,—now approaching, now melting into distance. Pre-eminent in loudness is the roar of the howling monkeys, though without being able fully to stifle the discordant cries and chattering of the noisy parrots. But the sun rapidly ascends towards the zenith, and one musician after the other grows mute and seeks the cool forest shade, until finally the whole morning concert ceases to be heard. Where the rays of light break through the foliage and play upon the underwood, or on the damp ground, gaudy butterflies flutter about, beetles of metallic brilliancy warm themselves, and richly robed or dark-vested snakes creep forth; for these indolent creatures are also fond of basking in the sun.

As the heat grows more intense, the stillness of the forest is only interrupted at intervals by single animal voices. Sometimes it is the note of the ivory-billed woodpecker, resounding like the distant axe of the forester, or the wail of the sloth breaking forth from the dense thicket. Sometimes human voices seem to issue from the depth of the forest, and the astonished huntsman fancies himself close to his comrades of the chase, or in the more dangerous neighbourhood of a wild tribe of Indians. With deep attention he listens to the sounds, until

Ivory-billed Woodpecker

he discovers them to be the melancholy cries of the wood-pigeon.

The deepest silence reigns at noon, when the sun becomes too powerful even for the children of the torrid zone; and many creatures, particularly the birds, sink into a profound sleep. Then all the warm-blooded animals seek the shade, and only the cold reptiles,—alligators, lizards, salamanders,—stretch themselves upon the glowing rocks in the bed of the forest-streams, or on sunny slopes, and, with raised head and distended jaws, seem to inhale with delight the sultry air.

As evening approaches, the noise of the morning begins to re-awaken. With loud cries the parrots return from their distant feeding-grounds to the trees on which they are accustomed to rest at night; and, as the monkeys saluted the rising sun, so chattering or howling, they watch him sinking in the west.

With twilight a new world of animals,—which, as long as the day lasted, remained concealed in the recesses of the forest, —awakens from its mid-day torpor, and prepares to enjoy its nightly revels. Then bats of hideous size wing their noiseless flight through the wood, chasing the giant hawk-moths and beetles, which have also waited for the evening hour, while the felidæ quit their lairs, ready to spring on the red stag near some solitary pool, or on the unwieldy tapir, who, having slept during the heat of the day, seeks, as soon as evening approaches, the low-banked river, where he loves to wallow in the mud. Then also the shy opossum quits his nest in hollow trees, or under some arch-like vaulted root, to search for insects or fruits, and the cautious agouti sallies from the bush.

In our forests scarcely a single tone is heard after sunset; but in the tropical zone many loud voices celebrate the night, where, for hours after the sun has disappeared, the cicadas, toads, frogs, owls, and goatsuckers chirrup, cry, croak, howl, and wail. The quietest hours are from midnight until about three in the morning. Complete silence, however, occurs only during very short intervals; for there is always some cause or other that prompts some animal to break the stillness. Sometimes the din grows so loud, that one might fancy a legion of evil spirits were celebrating their orgies in the darkness of the forest. The howling of the aluates, the whine of the little sapajous, the snarl of the duruculi, the roaring of the jaguar, the grunt of the pecari, the cry of the sloth, and the shrill voices of birds, join in dreadful discord. Humboldt supposes the first cause of these tumults to be a conflict among animals, which, arising by chance, gradually swells to larger dimensions. The jaguar pursues a herd of pecaris or tapirs which break wildly through the bushes.

Terrified by the noise, the monkeys howl, awakening parrots and toucans from their slumber; and thus the din spreads through the wood. A long time passes before the forest returns to its stillness. Towards the approach of day the owls, the goatsuckers, the toads, the frogs, howl, groan, and croak for the last time; and as soon as the first beams of morning purple the sky, the shrill notes of the cicadas mix with their expiring cries.

CHAPTER IX.

THE MEXICAN PLATEAUS, AND THE SLOPES OF SIKKIM.

Geological Formation of Mexico — The *Tierra Caliente* — The *Tierra Templada* — The *Tierra Fria*.

The Sylvan Wonders of Sikkim — Changes of the Forest on ascending — The Torrid Zone of Vegetation — The Temperate Zone — The Coniferous Belt — Limits of Arboreal Vegetation — Animal Life.

THE prodigious height attained in the torrid zone, not only by single mountains, but by vast tracts of land, and the diminution of temperature, which is the necessary consequence of their elevation above the level of the sea, enable the inhabitants of many tropical countries, without leaving their native land, to view the vegetable forms of every zone, and to pluck nearly every fruit that is found between the equator and the arctic circle. In Asia, Africa, and America, in the islands of the Indian Archipelago, and in the Hawaian group, where the Mauna Loa towers to the height of Mont Blanc, and girdles his foot with palms, while snow rests for a great part of the year upon his summit, we find numerous examples of a rapid transition from the torrid to the temperate or frigid zone, often within the range of a single day's journey.

It would far exceed the limits of a popular work were I to attempt to follow all these gradations of climate throughout the wide extent of the tropics; but a short description of the Mexican plateaus, and of the slopes of Sikkim, which I have selected as remarkable instances of the wonderful change of vegetation resulting from the progressive elevation of the land, will, I hope, prove not uninteresting to the reader.

After traversing South America and the Isthmus of Darien,

the giant chain of the Andes spreads out, as it enters Mexico, into a vast sheet of table-land, which maintains an elevation of from 6000 to 9000 feet for the distance of 200 leagues, until it gradually declines in the higher latitudes of the north, or descends in successive stages to the sea-borde of the Atlantic. To this remarkable geological formation the land, though warmed during part of the year by the rays of a vertical sun, owes that astonishing variety of climate and productions which would make it the envy of the earth, if its wretched inhabitants, the victims of ignorance, superstition, and anarchy, did not brutally trample on the gifts of Nature, and render Mexico the scoff and by-word of the civilised world.

All along the Mexican gulf stretches a broad zone of lowlands, called the *tierra caliente,* or hot region, which has the usual high temperature of the tropics. Parched and sandy plains, dotted with mimosas and prickly opuntias, are intermingled with savannahs, overshadowed by groves of palms and woodlands of exuberant fertility, and glowing with all the splendour of equinoctial vegetation. The branches of the stately forest trees are festooned with clustering vines of the dark purple grape, convolvuli, and other flowering parasites of the most brilliant dyes. The undergrowth of prickly aloe, matted with wild rose and honeysuckle, makes in many places an almost impervious thicket. In this wilderness of sweet-smelling buds and blossoms flutter birds of the parrot tribe, and clouds of butterflies, whose colours, nowhere so gorgeous as here, rival those of the vegetable world; while birds of exquisite song, — the scarlet cardinal, and the mocking-bird that comprehends in his own notes the whole music of a forest,— fill the air with melody.

Cardinal.

But, like the genius of evil, the malaria engendered by the decomposition of rank vegetable substances in the hot and humid soil, poisons these enchanting retreats, and from the spring to the autumnal equinox renders them dangerous or fatal to man.

Hastening to escape from its influence, the traveller, after passing some twenty leagues across the dreaded region of the yellow-fever, finds himself rising into a purer atmosphere. His limbs recover their elasticity. He breathes more freely, for his senses are not now oppressed by the sultry heats and intoxicating perfume of the lowlands. The aspect of Nature, too, has changed, and his eye no longer rests on the gay variety of colours with which the landscape was painted there. The vanilla, the indigo, the chocolate-tree, disappear as he advances, but the sugar-cane and the glossy-leaved banana still remain ; and when he has ascended about four thousand feet, he sees, in the unchanging green and the rich foliage of the liquidambar-tree, that he has reached the height where clouds and mists settle in their passage from the Mexican gulf, and keep up a perpetual moisture.

Mocking-Bird.

He is now beyond the influence of the deadly *vomito* on the confines of the *tierra templada*, or temperate region, where evergreen oaks begin to remind him of the forests of central Europe. The features of the scenery become grand, and even terrible. His road sweeps along the base of mighty mountains, once gleaming with volcanic fires, and still glistening in their mantles of snow, which serve as beacons to the mariner for many a league at sea. All along he beholds traces of their ancient combustion as his road passes over vast tracts of lava, bristling in the fantastic forms into which the fiery torrent has been thrown by obstacles in its career. Perhaps at the same moment, as he casts his eyes down one of those unfathomable ravines or barrancas, which often, to a depth of more than 1200 feet, rend the mountain-side, he sees its sheltered and sultry recesses glowing with the rich vegetation of the tropics : as if these wonderful regions were anxious to exhibit, at one glance, the boundless variety of their flora. Cactuses, euphorbias, and dracænæ, with a multitude of minor plants, cling to the rocky walls ; while in the depth of the gorge stand huge laurels, fig-trees, and bombaceæ, whose blossoms exhale almost overpowering odours, and whose trunks are covered with magnificent creepers, expanding their gay petals in the torpid air. Still pressing upwards, he mounts into regions favourable to other kinds of cultivation. He has traced the yellow maize growing from the lowest level ; but he now first sees fields of wheat and the other European cereals, brought into the country by the

Spanish conquerors, and with these, plantations of the American agave, which, among other uses, provides the Mexican with his favourite beverage. The oaks acquire a sturdier growth; and at an elevation of about eight thousand feet, the dark forests of pine announce that he has entered the *tierra fria*, or cold region,—the third and last of the great natural terraces into which the country is divided.

Loaded with vapours, the prevailing southerly sea-winds, after crossing the dead level occupied by the delta of the Ganges and Burrampooter, strike against the mountain-spurs of Sikkim, the dampest region of that stupendous chain, and, expending their moisture on their flanks, clothe them with a thick mantle of verdure to an enormous height. The giant peaks of Donkiah, Kinchinghow, and Kinchinginga, the third great mountain of the world (28,178 feet), equalling in height more than two Mont Blancs piled one above the other, and as far as we know, only surpassed in altitude by the Korakorum (28,278 feet) and Mount Everest (29,002 feet), form the culminating points of this magnificently wooded region, and look down upon the dense forests which, varying as they rise, extend between the plains of Bengal and their own perpetual snows.

The railway from Calcutta to Radjmahal on the Ganges has placed the wonders of Sikkim within easy range of the traveller; for as the foot of the Himalaya is not much more than a hundred miles from the capital of our Indian empire, a few days' travelling suffices to reach the Sanatorium of Dorjiling, which, at an elevation of 7000 feet in the Sikkim mountains, greets the European with the bracing coolness of his native land.

The distance from the Bengal plain to Dorjiling is hardly more than eighty miles, and about as much from Dorjiling to the regions of perpetual snow; so that in a very short time the botanist is here able to wander through every zone of vegetation, and through every variety of climate, from the heat of the tropical plain to the confines of perpetual winter.

Dark green forests, of an exclusively tropical character, cover the valleys and declivities to a height of from 4000 to 5000 feet. They chiefly consist of duabanga, terminalis, cedrela, and Gordonia Wallichii, the commonest tree in Sikkim, and much prized for ploughshares and other purposes requiring a hard wood. Mighty palms rise above the mass of the forest, while innumerable shrubs cover the ground. The prevalent timber is gigantic, and scaled by climbing leguminosas, bauhinias, and robinias, which sometimes sheath the trunks, or span the forest with huge cables, joining tree to tree. Large bamboos rather crest the hills than court the deeper shade, of which there is abundance, for the torrents cut a straight and steep course down the hill-flanks. The gulleys which they traverse are choked by vegetation, and bridged by fallen trees, whose trunks are clothed with epiphytical orchids, pendulous lycopodia, ferns, pothos, peppers, vines, bignonias, and similar types of the hottest and dampest climates. The beauty of the drapery of the pothos leaves is pre-eminent, whether for the graceful folds of the foliage or for the liveliness of its colour. Of the more conspicuous smaller trees the wild banana is the most abundant, its crown of very beautiful foliage contrasting with the smaller-leaved plants amongst which it nestles; next comes a screw pine, with a straight stem and a tuft of leaves, each eight or ten feet long, waving on all sides.

At an elevation of about four thousand feet many plants of the temperate zone, increasing in numbers as the traveller ascends, begin to mingle with the tropical vegetation, and to impart new charms to the forest; oaks and walnuts are here seen thriving near palms and arborescent ferns; mighty rhododendrons expand over thickets of tropical herbage; parasitical orchids adorn the trunks of the oaks, while thalictrons and geraniums blossom underneath.

At a height of about seven thousand feet the forest assuming a decidedly temperate physiognomy, is chiefly composed of oaks, magnolias, chestnuts, laurels, and walnuts. In many parts arborescent rhododendrons prevail, and ferns are generally very abundant.

About ten thousand feet above the level of the sea begins a zone or belt of coniferæ, chiefly characterised by the silver fir (*Abies Webbiana*) and the *Abies Brunoniana*, a beautiful species, which forms a stately blunt pyramid, with branches spreading like the cedar, but not so stiff, and drooping gracefully on all sides. Only at intervals other trees, such as willows, magnolias, ashes, birches, poplars, apple and cherry trees, appear among the thick pine-woods. The shrubbery and herbaceous plants of this zone are representatives of the whole temperate flora of Europe and America, intermixed with many Chinese, Japanese, and Malayan plants in the richest variety. Several epiphytic orchids grow to an elevation of 10,000 feet, and large spaces are frequently occupied by rhododendrons, which either ascend from the temperate zone into the coniferous belt, or first appear in the latter. But very few trees such as the willows, birches, maples, and ashes, rise above the coniferous forest, which reaches an upper limit of about thirteen thousand feet. Most arboreal plants now appear only in a dwarfed condition; but the willows still rise in powerful growth over the many Alpine shrubs,—juniperus, rosa, lonicera, potentilla, rhododendron,—which cover the ground; and single specimens, though low and stunted, are even found at a height of 16,000 feet.

The whole zone between the extreme limits of arboreal vegetation and the upper boundary of shrubs, generally occupies an elevation of from 13,500 to 16,000 feet, and may justly be called the region of the Alpine rhododendrons: these plants are here by far the most numerous, and frequently belt the mountains with a girdle of richly coloured blossoms, even to the verge of the perennial snows.

A large number of herbs, cruciferæ, compositæ, ranunculaceæ, grasses, sedges, green and bloom beyond the limits of the shrubs, frequently forming luxuriant pastures, on which numerous herds of yacks, or *grunting*-oxen, graze during the summer. Many plants are even exclusively confined to these enormous heights; such as the Rhododendron nivale, the most Alpine of woody plants, which Dr. Hooker found at 17,000 feet elevation, Delphinium glaciale, and Arenaria rupifraga, a curious species, forming great hemispherical balls, and altogether resembling in habit the curious balsam-bog of the Falkland Islands, which thrives in similar scenes. While on the summits of the Swiss Alps, lichens but sparely cover the rocks, wherever they are denuded of snow, the wanderer in Sikkim enjoys the sight of many a gay-coloured flower in regions 3000 or 4000 feet higher than the summit of Mont Blanc.

A zone of mosses and lichens does not exist on the Sikkim Himalaya, for they nowhere occur in large quantities, and scarcely ascend beyond the average limits of the Phanerogamous plants. It is only in very favourable stations that they considerably transgress these bounds; but this is just as often the case with some of the latter. If, for instance, the Tripe de roche (*Gyrophora*) of the arctic voyagers, and Lichen geographicus, which under 52° N. lat. and 50° S. lat. grows at the level of the sea, are found in the Himalaya at an elevation of 18,000 feet,—if Dr. Hooker was surprised, and no doubt delighted, when, at the Donkiah pass, at a height of 18,460 feet, he discovered Lecanora miniata, an old acquaintance of Cockburn Island, in the Antarctic Ocean,—the Delphinium glaciale and other flowering plants attain a similar altitude.

While thus in Sikkim a wonderful variety of vegetation rises in successive zones from the foot of the mountains to heights unparalleled in any other part of the world, animal life abounds only in its lowest classes; for the higher orders appear only in few species, and in very scanty numbers. On ascending from the foot of the Himalaya, one is astonished at the silence of the woods, broken at intervals only by the voice of a bird, or the chirping of a cicada. The solitude increases on penetrating into the interior of Sikkim, and is but rarely enlivened by a few monkeys in the valleys, some musk-deer on the spare grass of the mountains, in heights of from 8000 to 13,000 feet, or a few larks, sparrows, finches, pigeons, swallows, falcons, and other birds, some of which ascend to a surprising height. Thus Dr. Hooker found the Himalaya partridge 16,080 feet above the sea, and crows and ravens at 16,500. The Khaidge pheasants never descend below 12,000 feet, and high over the Kintschinghow (22,756) flocks of wild-geese are seen to wing their flight to unknown regions.

Musk-Deer.

The insects, however, and other invertebrata, make up by their numbers for the absence of warm-blooded animals, and are often insupportable plagues to the wanderer. Beautiful tropical butterflies sometimes ascend to heights of 10,000 feet, along with the less agreeable mosquitos and ticks; and in all the streams, up to an elevation of 7000 feet, hill-leeches occur in such multitudes that bathing is almost impossible.

Mangrove.

CHAPTER X.

MANGROVE VEGETATION.

Its Peculiarities of Growth and beautiful Adaptation to its Site — Its Importance in furthering the Growth of Land — Animal Life among the Mangroves — "Jumping Johnny" — Victimised by a Grapsus — Insalubrity of the Mangrove Swamps — Uses of the Mangrove Trees — Avicennias and Sonneratias.

IN the tropical zone, wherever the reflux of the tide exposes a broad belt of alluvial soil, the shores of the sea, particularly along the estuaries of rivers or in the shallow lagoons, are generally found fringed with a dense vegetation of mangroves. For no plants are more admirably adapted for securing a footing on the unstable brink of the ocean,—none are better formed to lead an amphibious life.

The growth of these salt-water loving trees (*Rhizophora gymnorrhiza, R. Mangle*) is equally peculiar and picturesque. The seeds germinate on the branches, and, increasing to a considerable length, finally fall down into the mud, where they stick, with their sharp point buried, and soon take root. The fruits of many plants are furnished with wings, that the winds may carry them far away and propagate them from land to land; others, enveloped in hard, waterproof shells, float on the surface of the sea, and are wafted by the currents to distant coasts; but here we have a plant, the seeds of which were destined to remain fixed on an uncertain soil, close to the parent-plant, and surely this end could not have been attained in a more beautiful manner!

As the young mangrove grows upwards, pendulous roots issue from the trunk and low branches, and ultimately strike into the muddy ground, where they increase to the thickness of a man's leg; so that the whole has the appearance of a complicated series of loops and arches, from five to ten feet high, supporting the body of the tree like so many artificial stakes.

It may easily be imagined what dense and inextricable thickets, what incomparable breakwaters, plants like these — through whose mazes even the light-footed Indian can only penetrate by stepping from root to root — are capable of forming.

Their influence in promoting the growth of land is very great, and in course of time they advance over the shallow borders of the ocean. Their matted roots stem the flow of the waters, and, retaining the earthy particles that sink to the bottom between them, gradually raise the level of the soil. As the new formation progresses, thousands of seeds begin to germinate upon its muddy foundation, thousands of cables descend, still farther to consolidate it; and thus foot by foot, year after year, the mangroves extend their empire and encroach upon the maritime domains.

The enormous deltas of many tropical rivers partly owe their immense development to the unceasing expansion of these littoral woods; and their influence should by no means be overlooked by the geologist when describing the ancient and eternal strife between the ocean and the land.

When the waters retire from under the tangled arcades of the mangroves, the black mud, which forms the congenial soil of these plants, appears teeming with a boundless variety of life. It absolutely swarms with the lower marine animals, with myriads of holothurias, annelides, sea-urchins, entomostraca, paguri, and crabs, whose often brilliantly coloured carapaces form a strong contrast to the black ooze in which they are seen to crawl about. Life clings even to the roots and branches bathed by the rising floods; for they are found covered with muscles, barnacles, and oysters, which thus have the appearance of growing upon trees, and pass one half of their existence under water, the other in the sultry atmosphere of a tropical shore.

The close-eyed Gudgeon (*Periophthalmus*), or "Jumping Johnny," as he is more familiarly named by the sailors, plays a conspicuous part in the animal world of the mangrove swamps, where the uncouth form of this strange amphibious fish may be seen jumping about in the mud like a frog, or sliding awkwardly along on its belly with a gliding motion. By means of its pectoral fins, it is even enabled to climb with great facility among the roots of the mangroves, where it finds a goodly harvest of minute crustaceas. It must, however, not be supposed that "Johnny" has all the swamp to himself; for though he manages to swallow many a victim, he is not seldom doomed to become the prey of creatures more wily or stronger than himself. A large and powerful crab of the Grapsus family may often be observed stealing, with an almost imperceptible motion, and in a cautious, sidelong manner, towards a Periophthalmus basking on the shore, and, before the fish has time to plunge into the sea, the pincer of the crab secures it in a vice-like gripe, from which it is perfectly hopeless to escape. While watching the evolutions of this lively and sagacious crustacean, one cannot help comparing it to an enormous jumping spider, which in a somewhat similar manner creeps towards the flies on which it preys, and suddenly surprises them by leaping on their backs, and then feasting on their blood.

This vast multitude of marine animals naturally attracts a great number of strand, lacustrine, and sea birds; for it would be strange, indeed, if guests were wanting where the table is so prodigally supplied. The red ibis, the snow-white egrette, the rosy spoonbill, the tall flamingo, and an abundance of herons and other water-fowl, love to frequent the mangrove thickets, enhancing by their magnificent plumage the beauty of the scene. For, however repulsive may be the swampy ground on which these strange trees delight, yet their bright green foliage, growing in radiated tufts at the ends of the branches, and frequently bespangled with large gaily-coloured flowers, affords a most pleasing spectacle. A whole world of interesting discoveries would here, no doubt, reward the naturalist's attention; but the mangroves know well how to guard their secrets, and to repel the curiosity of man. Should he attempt to invade their domains, clouds of bloodthirsty insects would instantly make him repent of his temerity; for the plague of the mosquitos is nowhere more dreadful than in the thickets of the semi-aquatic Rhizophoræ. And supposing his scientific zeal intense enough to bid defiance to the torture of their stings, and to scorn the attacks of every other visible foe — insect or serpent, crocodile or beast of prey — that may be lurking among the mangroves, yet the reflection may well bid him pause, that poisonous vapours, pregnant with cholera or yellow fever, are constantly rising from that muddy soil. Even in the temperate regions of Europe the emanations from marshy grounds are pregnant with disease, but the malaria ascending from the sultry morasses of the torrid zone is absolutely deadly.

Thus there cannot possibly be a better natural bulwark for a land than to be belted with mangroves; and if Borneo, Madagascar, Celebes, and many other tropical islands and coasts, have to the present day remained free from the European yoke, they are principally indebted for their independence to the miasms and tangles of a Rhizophora girdle, bidding defiance alike to the sharp edge of the axe or the destructive agency of fire.

As the mangroves are found in places suited to their growth throughout the whole torrid zone, it is not surprising that there are many species, some rising to the height of stately trees, while others are content with a shrub-like growth. Some are peculiar to America, others to the Old World; some grow near the sea, others prefer a brackish water and the low swampy banks of rivers.

Their uses are various; the bark and roots of the R. Mangle and gymnorrhiza serve for tanning leather, and as a black dye; and the crooked branches of the *Mangium celsum*, which seems to be the loftiest of the family, are employed by the Chinese as anchors and rudders for their clumsy junks; while tribes of the eastern Archipelago, in those parts which produce neither sago nor rice, are fain to live upon the pith of the seeds of this tree, which they boil with fish, or the milk of the cocoa-nut. The leaves are also eaten as a vegetable, and the wood forms excellent palisades, particularly on a swampy soil, as the worm soon fastens upon it when out of the water.

Next to the mangroves, the bruguieras, the avicennias, the sonneratias, and various species of palms, such as the Nipa fruticans and the Phœnix paludosa, a dwarf date-tree, which literally covers the islands of the Sunderbunds, at the delta of the Ganges, form conspicuous features in the marsh-forests of the torrid zone.

The magnificent Avicennia tomentosa, which, with a more majestic growth than the rhizophora raises its crown to the height of seventy feet, and is said to flourish throughout the whole range of the tropics as far as the flood extends, mixes with the mangroves, standing like them on overarching roots.

The sonneratias (*acida, alba*) grows along the marshy banks of the large rivers of India, the Moluccas, and New Guinea; their roots spread far and wide through the soft mud, and at various distances send up, like the avicennias, extraordinarily long spindle-shaped excrescences four or five feet above the surface. These curious formations spring very narrow from the root, expand as they rise, and then become gradually attenuated, occasionally forking, but never throwing out shoots or leaves, or in any way resembling the parent root. For lining insect-boxes and making setting-boards they are unequalled, as the finest pin passes in easily and smoothly, and is held so firmly and tightly, that there is no risk of the insects becoming disengaged. In fact Nature, while forming them, seems to have had the entomologist in view, and to have studied how to gratify his wishes.

PART II.

TROPICAL PLANTS

Baobab Trees at Manaar.

CHAPTER XI.

GIANT TREES AND CHARACTERISTIC FORMS OF TROPICAL VEGETATION.

General Remarks—The Baobab—Used as a Vegetable Cistern—Arborescent Euphorbias—The Dracæna of Orotava—The Sycamore—The Banyan—The sacred Bo-Tree of Anarajapoora—The Teak Tree—The Saul—The Sandal Tree—The Satinwood Tree—The Ceiba—The Mahogany Tree—The Mora—Bamboos—The Guadua—Beauty and multifarious Uses of these colossal Grasses—Firing the Jungle—The Aloes—The Agave americana—The Bromelias—The Cactuses—The Mimosas—Bushropes—Climbing Trees—Emblems of Ingratitude—Marriage of the Fig Tree and the Palm—Epiphytes—Water-Plants—Singularly-shaped Trees—The Barrigudo—The Bottle Tree—Trees with Buttresses, fantastical Roots, and formidable Spines.

WHEREVER in the tropical regions periodical rains saturate the earth, vegetable life expands in a wonderful variety of forms. In the higher latitudes of the frozen north, a rapidly evanescent summer produces but few and rare flowers in sheltered situations, soon again to disappear under the winter's snow; in the temperate zones, the number, beauty, and variety of plants increase with the warmth of a genial sky; but it is only where the vertical rays of an equatorial sun awaken and foster life on humid grounds that ever-youthful Flora appears in the full exuberance of her creative power. It is only there we find the majestic palms, the elegant mimosas, the large-leafed bananas, and so many other beautiful forms of vegetation alien to our cold and variable clime. While our trees are but sparingly clad with scanty lichens and mosses, they are there covered with stately bromelias and wondrous orchids. Sweet-smelling vanillas and passifloras wind round the giants of the forest, and large flowers break forth from their rough bark, or even from their very roots.

"The tropical trees," says Humboldt, "are endowed with richer juices, ornamented with a fresher green, and decked with larger and more lustrous leaves than those of the more northerly regions. Social plants, which render European vegetation so monotonous, are but rarely found within the tropics. Trees, nearly twice as high as our oaks, there glow with blossoms large and magnificent as those of our lilies. On the shady banks of the Magdalena river, in South America, grows a climbing Aristolochia, whose flower, of a circumference of four feet, the children, while playing, sometimes wear as a helmet; and in the Indian Archipelago the blossom of the Rafflesia measures three feet in diameter, and weighs more than fourteen pounds."

The number of known plants is estimated at about 200,000, and the greater part of this vast multitude of species belongs to the torrid zone. But if we consider how very imperfectly these sunny regions have as yet been explored,—that in South America enormous forest lands and river basins have never yet been visited by a naturalist,—that the vegetation of the greater part of Central Africa is still completely hidden in mystery,—that no botanist has ever yet penetrated into the interior of Madagascar, Borneo, New Guinea, South-Western China, and Ultra-Gangetic India,— and that, moreover, many of the countries visited by travellers have been but very superficially and hastily examined,—we may well doubt whether even one fourth part of the tropical plants is actually known to science. What a vast field for future naturalists! What prospects for the trade and industry of future generations!

After these general remarks on the variety and exuberance of tropical vegetation, I shall now briefly review those plants which, by their enormous size, their singularity of form, or their frequency in the landscape, chiefly characterise the various regions of the torrid zone in different parts of the globe.

The African Baobab, or monkey-bread tree (*Adansonia*

digitata), may justly be called the elephant of the vegetable world. Near the village Gumer, in Fassokl, Russegger saw a baobab thirty feet in diameter and ninety-five in circumference; the horizontally outstretched branches were so large that the negroes could comfortably sleep upon them. The Venetian traveller Cadamosto (1454) found, near the mouths of the Senegal, baobabs measuring more than a hundred feet in circumference. As these vegetable giants are generally hollow, like our ancient willows, they are frequently made use of as dwellings or stables; and Dr. Livingstone mentions one in which twenty or thirty men could lie down and sleep, as in a hut. In the village of Grand Galarques, in Senegambia, the negroes have decorated the entrance into the cavity of a monstrous baobab with rude sculptures cut into the living wood, and make use of the interior as a kind of assembly room, where they meet to deliberate on the interests of their small community, "reminding one," says Humboldt, "of the celebrated plantain in Lycia, in whose hollow trunk the Roman consul, Lucinius Mutianus, once dined with a party of twenty-one." As the baobab begins to decay in the part where the trunk divides into the larger branches, and the process of destruction thence continues downwards, the hollow space fills, during the rainy season, with water, which keeps a long time, from its being protected against the rays of the sun. The baobab thus forms a *vegetable cistern*, whose water the neighbouring villagers sell to travellers. In Kordofan the Arabs climb upon the tree, fill the water in leathern buckets, and let it down from above; but the people in Congo more ingeniously bore a hole in the trunk, which they stop, after having tapped as much as they require. *

The height of the baobab does not correspond to its amazing bulk, as it seldom exceeds sixty feet. As it is of very rapid growth, it acquires a diameter of three or four feet and its full altitude in about thirty years, and then continues to grow in circumference. The larger beam-like branches, almost as thick at their extremity as at their origin, are abruptly rounded, and then send forth smaller branches, with large, light green, palmated leaves. The bark is smooth and greyish. The oval fruits,

---

* D'Escayrac, "Le Desert et le Soudan."

which are of the size of large cucumbers, and brownish yellow when ripe, hang from long twisted spongy stalks, and contain a white farinaceous substance, of an agreeable acidulated taste, enveloping the dark brown seeds. They are a favourite food of the monkeys, whence the tree has derived one of its names.

From the depth of the incrustations formed on the marks which the Portuguese navigators of the fifteenth century used to cut in the large baobabs which they found growing on the African coast, and by comparing the relative dimensions of several trunks of a known age, Adanson concluded that a baobab of thirty feet in diameter must have lived at least 5000 years; but a more careful investigation of the rapid growth of the spongy wood has reduced the age of the giant tree to more moderate limits, and proved that, even in comparative youth, it attains the hoary aspect of extreme senility.

The baobab belongs to the same family as the mallow or the hollyhock, and is, like them, emollient and mucilaginous in all its parts. The dried and powdered leaves constitute the *lalo*, which the Africans mix daily with their food for diminishing excessive perspiration. The natives, likewise, make a strong cord from the fibres contained in the pounded bark. The whole of the trunk, as high as they can reach, is consequently often quite denuded of its covering, which in the case of almost any other tree would cause its death; but this has no effect on the baobab, except to make it throw out a new bark, and, as the stripping is repeated frequently, it is common to see the lower five or six feet an inch or two less in diameter than the parts above. This monstrous tree ranges over a wide extent of Africa, particularly in parts where the summer rains fall in abundance, as in Senegambia, in Soudan, and in Nubia. Dr. Livingstone admired its colossal proportions on the banks of the Zouga and the Zambesi; and William Peters found it on the eastern coast, near 26° S. lat. It forms a conspicuous feature in the landscape at Manaar in Ceylon, where it has most likely been introduced by early mariners, perhaps even by the Phœnicians, as the prodigious dimensions of the trees are altogether inconsistent with the popular conjecture of a Portuguese origin. Sir Emerson Tennent found one of the largest, measuring upwards of thirty feet in circumference; and another at Putten, since destroyed by the digging of a well under part of its roots, which, though but seventy feet high, was forty-six feet in girth.

Another tree very characteristic of Africa, and frequently seen along with the baobab, is the large arborescent Euphorbia (*E. arborescens*), surmounted at the top with stiff leaves, branching out like the arms of a huge candelabra. It adds greatly to the strange wildness of the landscape, and seems quite in character with the aspect of the unwieldy rhinoceros and the long-necked giraffe.

Dracænas, or dragon-trees, are found growing on the west coast of Africa and in the Cape Colony, in Bourbon and in China; but it is only in the Canary Islands, in Madeira, and Porto Santo, that they attain such gigantic dimensions as to entitle them to rank among the vegetable wonders of the world.

Dragon Tree.

Near Orotava, in Teneriffe, still flourishes the venerable dragon-tree, which was already reverenced for its age by the extirpated nation of the Guanches, the aboriginal inhabitants of the island, and which the adventurous Bethencourts, the conquerors of the Canaries, found hardly less colossal and cavernous in 1402 than Humboldt, who visited it in 1799. Above the roots, the illustrious traveller measured a circumference of forty-five feet; and according to Sir George Staunton, the trunk has still a diameter of four yards, at an elevation of ten feet above the ground. The whole height of the tree is not much above sixty-five feet. The trunk divides in numerous upright branches, terminating in tufts of evergreen leaves, resembling those of the pine-apple.

Next to the baobab and the dracæna, the Sycamore (*Ficus Sycomorus*) holds a conspicuous rank among the giant trees of Africa. It attains a height of only forty or fifty feet, but in the course of many centuries its trunk swells to a colossal size, and its vast crown covers a large space of ground with an impenetrable shade. Its leaves are about four inches long and as many broad, and its figs have an excellent flavour. In Egypt it is almost the only grove-forming tree; and most of the mummy coffins are made of its incorruptible wood.

Sycamore.

No baobab rears its monstrous trunk on the banks of the Ganges; no dragon-tree of patriarchal age here reminds the wanderer of centuries long past; but the beautiful and stately

Banyan.

Banyan (*Ficus indica*) gives him but little reason to regret their absence. Each tree is in itself a grove, and some of them are of an astonishing size, as they are continually increasing, and, contrary to most other animal and vegetable productions, seem to be exempted from decay; for every branch from the main body throws out its own roots, at first in small tender fibres, several yards from the ground, which continually grow thicker, until, by a gradual descent, they reach its surface, where, striking in, they increase to a large trunk and become a parent-tree, throwing out new branches from the top. These in time suspend their roots, and, receiving nourishment from the earth, swell into trunks and send forth other branches, thus continuing in a state of progression so long as the first parent of them all supplies her sustenance.

> "The bended twigs take root, and daughters grow
> About the mother-tree; a pillar'd shade
> High overarch'd, and echoing walks between.
> There oft the Indian herdsman, shunning heat,
> Shelters in cool, and tends his pasturing herds
> At loopholes cut through thickest shade."

These beautiful lines of Milton's are by no means overdrawn; as a banyan tree, with many trunks, forms the most beautiful walks and cool recesses that can be imagined. The leaves are large, soft, and of a lively green; the fruit is a small fig (when ripe of a bright scarlet), affording sustenance to monkeys, squirrels, peacocks, and birds of various kinds, which dwell among the branches.

The Hindoos are peculiarly fond of this tree; they consider its long duration, its outstretching arms and overshadowing beneficence, as emblems of the Deity; they plant it near their dewals or temples; and in those villages where there is no structure for public worship they place an image under a banyan, and there perform a morning and evening sacrifice.

Many of these beautiful trees have acquired an historic celebrity; and the famous Cubbeer-burr, on the banks of the Nerbuddah, thus called by the Hindoos in memory of a favourite saint, is supposed to be the same as that described by Nearchus, the admiral of Alexander the Great, as being able to shelter an army under its far-spreading shade. "High floods have at various times swept away a considerable part of this extraordinary tree, but what still remains is near 2000 feet in circumference, measured round the principal stems; the overhanging branches not yet struck down cover a much larger space; and under it grow a number of custard-apple and other fruit trees. The large trunks of this single colossus amount to a greater number than the days of the year, and the smaller ones exceed 3000, each constantly sending forth branches and hanging roots, to form other trunks and become the parents of a future progeny.

"About a century ago a neighbouring rajah, who was extremely fond of field diversions, used to encamp under it in a magnificent style, having a saloon, drawingroom, dining room, bedchamber, bath, kitchen, and every other accommodation, all in separate tents; yet the noble tree not only covered the whole, together with his carriages, horses, camels, guards, and attendants, but also afforded with its spreading branches shady spots for the tents of his friends, with their servants and cattle. And in the march of an army it has been known to shelter 7000 men." *

Such is the banyan — more wonderful, and infinitely more beautiful and majestic, than all the temples and palaces which the pride of the Moguls has ever reared!

The nearly related Pippul of India, or Bo-tree (*Ficus religiosa*), which differs from the banyan (*F. indica*) by sending down no roots from its branches, is reverenced by the Buddhists as the sacred plant, under whose shade Gotama, the founder of their religion, reclined when he underwent his divine transfiguration. Its heart-shaped leaves, which, like those of the aspen, appear in the profoundest calm to be ever in motion, are supposed to tremble in recollection of the mysterious scene of which they were the witnesses.

The sacred Pippul at Anarajapoora, the fallen capital of the ancient kings of Ceylon, is probably the oldest *historical tree* in the world; as it was planted 288 years before Christ, and hence is now 2150 years old. The enormous age of the baobabs of Senegal, and of the wondrous wellingtonias of California, can only be conjectured; but the antiquity of the Bo-tree is matter of record, as its preservation has been an object of solicitude to successive dynasties; and the story of its fortunes has been preserved in a series of continuous chronicles amongst the most authentic that have been handed down by mankind.

* Forbes's "Oriental Memoirs."

"Compared with it, the Oak of Ellerslie is but a sapling, and the Conqueror's Oak in Windsor Forest barely numbers half its years. The yew trees of Fountains Abbey are believed to have flourished there 1200 years ago; the olives in the garden of Gethsemane were full-grown when the Saracens were expelled

The Sacred Bo-Tree.

from Jerusalem, and the cypress of Somma in Lombardy is said to have been a tree in the time of Julius Cæsar. Yet the Bo-tree is older than the oldest of these by a century, and would almost seem to verify the prophecy pronounced when it was planted, that it would 'flourish and be green for ever.'

"The degree of sanctity with which this extraordinary tree has been invested in the imagination of the Buddhists, may be compared to the feeling of veneration with which Christians would regard the attested wood of the cross. To it kings have even dedicated their dominions, in testimony of their belief that it is a branch of the identical fig-tree under which Gotama Buddha reclined at Uruwelaya when he underwent his apotheosis.

"When the king of Magadha, in compliance with the request of the sovereign of Ceylon, was willing to send him a portion of that sanctified tree to be planted at Anarajapoora, he was deterred by the reflection that it 'cannot be meet to lop it with any weapon;' but under the instruction of the high priest, using vermilion in a golden pencil he made a streak on the branch, which, 'severing itself, hovered over the mouth of a vase filled with scented soil,' into which it struck its roots and descended. Taking the legend as a sacred law, the Buddhist priests to the present day object religiously to 'lop it with any weapon,' and are contented to collect any leaves, which, 'severing themselves,' may chance to fall to the ground. These are regarded as treasures by the pilgrims, who carry them away to the remotest parts of the island. . . . .

"At the present day the aspect of the tree suggests the idea of extreme antiquity: the branches which have rambled at their will far beyond the outline of its inclosure, the rude pillars of masonry that have been carried out to support them, the retaining walls which shore up the parent-stem, the time-worn steps by which the place is approached, and the grotesque carvings that decorate the stone-work and friezes, all impart the conviction that the tree which they encompass has been watched over with abiding solicitude, and regarded with an excess of veneration that could never attach to an object of dubious authenticity."*

Although far inferior to these wonders of the vegetable world in amplitude of growth, yet the Teak tree, or Indian oak (*Tectona grandis*), far surpasses them in value, as the ship-worm in the water, and the termite on land, equally refrain from attacking its close-grained strongly scented wood; and no timber equals it for ship-building purposes.

* Tennent's "Ceylon," vol. ii. pp. 614, 618.

It grows wild over a great part of British India; in the mountainous districts along the Malabar coast, in Guzerat, the valley of the Nerbuddah, in Tenasserim and Pegu; which, for this reason alone, was well worth being wrenched from his gold-footed majesty of Birmah and annexed to our Indian empire. Unlike the oak and fir forests of Europe, where large spaces of ground are covered by a single species, the teak forests of India are composed of a great variety of trees, among which the teak itself does not even predominate. After a long neglect, which, in some parts, had almost caused its total extirpation, government has at length taken steps for its more effectual protection, and appointed experienced foresters to watch over this invaluable tree. Since 1843, hundreds of thousands of young plants have been raised from seeds; and in 1848, above 500,000 had been already set in the teak districts on the Malabar coast. Unfortunately the teak is of as slow a growth as our oak, and many years will still be necessary to repair the ruinous improvidence of the past.

In Java also the teak forests, both those of natural growth and those that have been planted by the Dutch, are carefully administered. This tree, which requires a century to attain its full diameter of four feet, loses its leaves in the dry season, when the grass and undergrowth of shrubbery is burnt, as the heat which is developed does the trees no injury. The ashes afford an excellent manure, and the fire makes crevices and rents in the soil, through which the fertilising rain can afterwards more easily penetrate to the roots. In Java the teak tree attains only a height of eighty feet, inferior to its loftier Hindostanic stature.

Next to the teak tree, the Saul (*Shorea robusta*) is renowned in India for its excellent timber. It grows in the forests at the foot of the Himalayan chain, on the dry grounds immediately rising above the swampy Terai, and ranges over an enormous though narrow belt from Kumaon to Assam.

Among the numerous timber-trees of Ceylon, the Satinwood (*Chloroxylon Swietenia*) is by far the first, in point of size and durability. All the forests around Batticaloa and Trincomalee, and as far north as Jaffna, are thickly set with this valuable tree, under whose ample shade the traveller rides for days together. It grows to the height of a hundred feet, with a rugged grey bark, small white flowers, and polished leaves, with a somewhat unpleasant odour. Owing to the difficulty of carrying its heavy beams, the natives cut it only near the banks of the rivers, down which it is floated to the coast, whence large quantities are exported to every part of the colony. The richly coloured and feathery pieces are used for cabinet-work, and the more ordinary logs for building purposes, every house in the eastern province being floored and timbered with satinwood.

The Sandal tree, which furnishes the sweet-scented, fine-grained wood, so highly prized by the Chinese, and so much used in small cabinets, escritoires, and similar articles, because no insect can exist within its influence, also deserves to be noticed as one of the most valuable productions of the Malabar coast. It chiefly grows on rocky hills, and, if permitted, would attain a tolerable size, but, from its great value, is generally cut down at an early stage. On low land and a richer soil it degenerates, and is in all respects less esteemed. A variety of the same tree, but furnishing a wood of inferior quality, grows on many of the South Sea islands,— Hawaii, Feejee, the New Hebrides; but in many parts the excessive avidity of the traders has almost caused its total extirpation. The sandal is a beautiful tree; the branches regular and tapering; the leaf like that of the willow, but shorter and delicately soft. The blossoms hang in bunches of small flowers, either red or white, according to the colour of the wood.

On turning our attention to America we find that Nature, delighting in infinite varieties of developement, and disdaining a servile copy of what she had elsewhere formed, covers the earth with new and no less remarkable forms of vegetation. Thus, while in Africa the baobab attracts the traveller's attention by its colossal size and peculiarity of growth, the gigantic Ceiba (*Bombax Ceiba*), belonging to the same family of plants, raises his astonishment in the forests of Yucatan. Like the baobab, this noble tree rises only to a moderate height of sixty feet, but its trunk swells to such dimensions that fifteen men are hardly able to span it, while a thousand may easily screen themselves under its canopy from the scorching sun. The leaves fall off in January; and then at the end of every branch bunches of large, glossy, purple-red flowers make their appearance, affording, as one may well imagine, a magnificent sight. In Guiana the savages take refuge upon the ceiba trees during the inundations. The seeds have an agreeable taste, and are frequently eaten, as well as the young and mucilaginous leaves.

In British Honduras, in the neighbourhood of Balize, and along the Motagua river, the Mahogany tree (*Swietenia Mahagoni*) is found scattered in the forests, attracting the woodman's attention from a distance by its light-coloured foliage. Such are its dimensions, and such is the value of peculiarly fine specimens, that in October 1823 a tree was felled which weighed more than seven tons, and cost, when landed at Liverpool, above 375*l.*; here it was sold for 525*l.*, and the expense of sawing amounted to 750*l.* more: so that the wood of this single tree, before passing into the hands of the cabinet-maker, was worth as much as a moderately sized farm. The African mahogany wood is furnished by the near-related *Khaya senegalensis*, which likewise towers to the height of a hundred feet, and has been transplanted to the Antilles.

"Heedless and bankrupt in all curiosity must he be," says Waterton[*], "who can journey through the forests of Guiana without stopping to take a view of the towering Mora (*Mora excelsa*). Its topmost branch, when naked with age, or dried by accident, is the favourite resort of the toucan. Many a time has this singular bird felt the shot faintly strike him from the gun of the fowler beneath, and owed his life to the distance betwixt them. The wild fig tree, as large as a common English apple tree, often rears itself from one of the thick branches at the top of the mora; and when its fruit is ripe, to it the birds resort for nourishment. It was to an indigested seed passing through the body of this bird, which had perched on the mora, that the fig tree first owed its elevated station there. The sap of the mora raised it into full bearing; but now, in its turn, it is doomed to contribute a portion of its own sap and juices towards the growth of different species of vines, the seeds of which also the birds deposited on its branches. These soon vegetate and bear fruit in great quantities; so what with their usurpation of the resources of the fig tree, and the fig tree of the mora, the mora, unable to support a charge which Nature never intended it should, languishes and dies under its burden;

---

[*] "Wanderings," p. 5.

and then the fig tree and its usurping progeny of vines, receiving no more succour from their late foster-parent, droop and perish in their turn."

Our stateliest oaks would look like pygmies near this "chieftain of the forests," who raises his dark green cupola over all the neighbouring trees, and deceives the traveller, who fancies that a verdant hill is rising before him. Its wood is much firmer than that of the fir, and is, or will be, of great importance to the British navy. On the Upper Barima alone, a river of Guiana hardly even known by name in Europe, Schomburgk found the giant tree growing in such profusion that it could easily afford sufficient timber for the proudest fleet that ever rode the ocean.

The graceful tapering form of the *Gramineæ*, or grasses, belongs to every zone; but it is only in the warmer regions of the globe that we find the colossal *Bambusaceæ*, rivalling in grandeur the loftiest trees of the primeval forest. Such is the rapidity of their growth, that in the Royal Botanical Garden in Edinburgh, a bamboo was observed to increase six inches a day in a temperature of from 65° to 70°. The *Bambusa gigantea* of Burmah has been known to grow eighteen inches in twenty-four hours; and as the *Bambusa Tulda* in Bengal attains its full height of seventy feet in a single month, its average increase cannot be less than an inch per hour!

In New Grenada and Quito the Guadua, one of these giant grasses, ranks next to the sugar-cane and maize as the plant most indispensable to man. It forms dense jungles, not only in the lower regions of the country, but in the valleys of the Andes, 5000 feet above the level of the sea. The culms attain a thickness of six inches, the single joints are twenty inches long, and the leaves are of indescribable beauty. A whole hut can be built and thatched with the guadua, while the single joints are extensively used as water-vessels and drinking-cups.

India, South China, and the Eastern Archipelago are the seats of the real bamboos, which grow in a variety of genera and species, as well on the banks of lakes and rivers in low marshy grounds, as in the more elevated mountainous regions. They chiefly form the impenetrable jungles, the seat of the tiger and the python. Sometimes a hundred culms spring from a single root, not seldom as thick as a man, and towering to a height of eighty or a hundred feet. Fancy the grace of our meadow grasses, united with the lordly growth of the Italian poplar, and you will have a faint idea of the beauty of a clump of bamboos.

The variety of purposes to which these colossal reeds can be applied almost rivals the multifarious uses of the cocoa-nut palm itself. Splitting the culm in its whole length into very thin pieces, the industrious Chinese then twist them together into strong ropes, for tracking their vessels on their numerous rivers and canals. The sails of their junks, as well as their cables and rigging, are made of bamboo; and in the southern province of Sechuen, not only nearly every house is built solely of this strong cane, but almost every article of furniture which it contains — mats, screens, chairs, tables, bedsteads, bedding — is of the same material. From the young shoots they also fabricate their fine writing-paper, which is so superior to the produce of our own manufactories. Although the bamboo grows spontaneously and most profusely in nearly all the southern portion of their vast empire, they do not entirely rely on the beneficence of Nature, but cultivate it with the greatest care. They have treatises and whole volumes devoted solely to this subject, laying down rules derived from experience, and showing the proper soils, the best kinds of water, and the seasons for planting and transplanting the bamboos, whose use is scarce less extensive throughout the whole East Indian world. In Mysore and Orizza the seeds of several species are eaten with honey; and in Sikkim the grain of the Praong, a small bamboo, is boiled, or made into cakes, or into beer. In Java, the prickly bamboo, whose wood is of such flinty hardness that sparks are emitted on its being struck with an axe, and whose formidable thorns project from every node, is used to form impenetrable hedges.

At one season of the year the bamboos are easily destroyed by fire; and as the great stem-joints burst from the expansion of the air confined within, the report almost rivals the roar of cannon. In Sikkim firing the jungle is a frequent practice, and Dr. Hooker, who often witnessed the spectacle, describes the effect by night as exceedingly grand. "Heavy clouds canopy the mountains above, and, stretching across the valleys, shut out the sky; the air is a dead calm, as usual in the deep gorges, and the fires, invisible by day, are seen raging all around, appearing to an inexperienced eye in all but dangerous proximity. The voices of birds and insects being hushed, nothing is audible but the harsh roar of the rivers, and occasionally rising far above it, that of the forest fires. At night we were literally surrounded by them; some smouldering like the shale-heaps at a colliery, others fitfully bursting forth, whilst others again stalked along with a steadily increasing and enlarging flame, shooting out great tongues of fire, which spared nothing as they advanced with irresistible might. At Dorjiling the blaze is visible, and the deadened reports of the bamboos bursting is heard throughout the night; but in the valley, and within a mile of the scene of destruction, the effect is the most grand, being heightened by the glare reflected from the masses of mist which hover above." *

The aloes form the strongest contrast to the airy lightness of the grasses, by the stately repose and strength of their thick, fleshy, and inflexible leaves. They generally stand solitary in the parched plains, and impart a peculiarly austere or melancholy character to the landscape. The real aloes are chiefly African, but the American yuccas and agaves have a similar physiognomical character. The *Agave americana*, the usual ornament of our hot-houses, bears on a short and massive stem a tuft of fleshy leaves, sometimes no less than ten feet long, fifteen inches wide, and eight inches thick! After many years a flower-stalk twenty feet high shoots forth in a few weeks from the heart of the plant, expanding like a rich candelabrum, and clustered with several thousands of greenish yellow aromatic flowers. But a rapid decline succeeds this brilliant efflorescence, for it is soon followed by the death of the exhausted plant.

In Mexico, where the agave is indigenous, and whence it has found its way to Spain and Italy, it is reckoned one of the most valuable productions of Nature. At the time when the flower-stalk is beginning to sprout, the heart of the plant is cut out, and the juice, which otherwise would have nourished the blossom, collects in the hollow. About three pounds exude

* "Himalayan Journals," vol. i. p. 146.

daily, during a period of two or three months. Thus a single agave, or maguey, gives about a hundred and fifty bottles; and though the plant perishes in consequence, and produces but once after ten or fifteen years, its culture must be profitable, as extensive fields of it cover the Mexican plains at an elevation of 7000 feet above the sea. After standing for a short time, the sweet juice undergoes a vinous fermentation, and the stranger, when once accustomed to its disagreeable odour, prefers the *pulque* to all other wines, and joins in the enthusiastic praises of the Mexican. The consumption is enormous; and before the revolution, which severed the country from the yoke of Spain, the revenue derived from a very small municipal duty exacted on the pulque at the gates of Mexico and La Puebla averaged 600,000 dollars.*

But the use of the agave is not confined to the production of a vinous liquor, as the tough fibres of its leaves furnish an excellent material for the strongest ropes, or the formation of coarse cloth. Long before the conquest of the country by Cortez, the aborigines applied the agave to a great variety of purposes. From it they made their paper (pieces of which of various thickness are still found covered with curious hieroglyphic writing), their threads, their needles (from its sharp points), and many articles of clothing and cordage.

The American bromelias likewise resemble the aloes of torrid Africa by the form and arrangement of their leaves. To this useful family belongs the pine-apple (*Bromelia Ananas*), which grows best and largest in Brazil, where it is so common that the pigs fatten on the fruit. Formerly confined in our country to the tables of the wealthier classes as long as it was only supplied by our hot-houses, it can now be enjoyed at a very moderate expense, since thousands are imported by every West Indian steamer.

The leaves of several species of bromelia furnish excellent twine for ropes. The inhabitants of the banks of the river San Francisco, in Brazil, weave their fishing-nets with the fibres of the Caroa (*B. variegata*), and the filaments of the Crauata de rede (*B. sagenaria*) furnish a cordage of amazing strength and durability. Koster mentions in his "Travels in Brazil" (1816),

* Ward's "Mexico," 1827.

that a rope thus composed had been in constant use during many years upon the wharf of the city of Paraiba, where it was employed for embarking merchandise, and it retained its strength unimpaired. On one occasion the heavy anchors belonging to a line-of-battle ship were hoisted on board a vessel with this same old rope, after hempen cables of a larger diameter had been found inefficient for the purpose. A plant like this, though hitherto neglected by the routine of commerce, seems destined to a great future importance in the markets of the world.

The foliage of the screw pines, so widely extended over the East Indian and South Sea Isles, where they form a prominent feature in the landscape, closely resembles that of the bromelias, while the stem (round which the serrated leaves ascend in spiral convolutions, till they terminate in a pendulous crown), the aërial roots, and the fruit of the strong plants, remind one of the palms, the mangroves, and the coniferæ.

The *Pandanus odoratissimus*, or sweet-smelling screw pine, whose fruits, when perfectly mature, resemble large rich-coloured pine-apples, plays an important part in the household economy of the coral-islanders of the South Sea. The inhabitants of the Mulgrave Archipelago, where the cocoa-nut is rare, live almost exclusively on the juicy pulp, and the pleasant kernels of the fruit. The dried leaves serve to thatch their cottages, or are made use of as a material for mats and raiment. The wood is hard and durable. They string together the beautiful red and yellow coloured nuts for ornaments, and wear the flowers as garlands. When the tree is in full blossom, the air around is impregnated with delicious aromas.

The grotesque forms of the Cactuses possess the stiff rigidity of the aloes. Their fleshy stems, covered with a gray green coriaceous rind, generally exhibit bunches of hair and thorns instead of leaves. The angular columns of the Cerei, or torch-cactuses, rise to the height of sixty feet,—generally branchless, sometimes strangely ramified, as candelabras, while others creep like ropes upon the ground, or hang, snake-like, from the trees, on which they are parasitically rooted. The opuntias are unsymmetrically constructed of thick flat joints springing one from the other, while the melon-shaped Echinocacti and Mammillariæ, longitudinally ribbed or covered with warts, remain

attached to the soil. The dimensions of these monstrous plants are exceedingly variable. One of the Mexican echino-cacti (*E. Visnaga*) measures four feet in height, three in diameter, and weighs about two hundred pounds; while the dwarf-cactus (*E. nana*) is so small that, loosely rooted in the sand, it frequently remains sticking between the toes of the dogs that pass over it. The splendid purple flowers of the cactuses form a strange contrast to the deformity of their stems, and the spectator stands astonished at the glowing life that springs forth from so unpromising a stock. These strange compounds of ugliness and beauty are in many respects useful to man. The pulp of the melocacti, which remains juicy during the driest season of the year, is one of the vegetable sources of the wilderness, and refreshes the traveller after he has carefully removed the thorns. Almost all of them bear an agreeable acid fruit, which, under the name of the Indian fig, is consumed in large quantities in the West Indies and Mexico. The light and incorruptible wood is admirably adapted for the construction of oars and many other implements. The farmer fences his garden with the prickly opuntias; but the services which they render, as the plants on which the valuable cochenille insect feeds and multiplies, are far more important.

The cactuses prefer the most arid situation, naked plains, or slopes, where they are fully exposed to the burning rays of the sun, and impart a peculiar physiognomy to a great part of tropical America.

None of the plants belonging to this family existed in the Old World previously to the discovery of America; but some species have since then rapidly spread over the warmer regions of our hemisphere. The Nopal (*Cactus Opuntia*) skirts the Mediterranean along with the American agave, and from the coasts has even penetrated far into the interior of Africa, everywhere maintaining its ground, and conspicuously figuring among the primitive vegetation of the land.

Although chiefly tropical, the cactuses have a perpendicular range, which but few other families enjoy. From the low sandcoasts of Peru and Bolivia they ascend through vales and ravines to the highest ridges of the Andes.

Magnificent dark-brown Peireskias (the only cactus genus bearing leaves instead of prickles) bloom on the banks of the

Lake of Titicaca, 12,700 feet above the level of the sea; and in the bleak Puna*, even at the very limits of vegetation, the traveller is astonished at meeting with low bushes of cactuses thickly beset with yellow prickles.

What a contrast between these deformities and the delicately feathered mimosas, unrivalled among the loveliest children of flora in the matchless elegance of their foliage! Our acacias give but a faint idea of the beauty which these plants attain under the fostering rays of a tropical sun. In most species the branches extend horizontally, or umbrella-shaped, somewhat like those of the Italian pine, and the deep-blue sky shining through the light green foliage, whose delicacy rivals the finest embroidery, has an extremely picturesque effect. Endowed with a wonderful sensibility, many of the mimosas seem, as it were, to have outstepped the bounds of vegetable life, and to rival in acuteness of feeling the coral polyps and the sea-anemones of the submarine gardens.

Mimosa.

The Porliera hygrometrica foretells serene or rainy weather by the opening or closing of its leaves. Large tracts of country in Brazil are almost entirely covered with sensitive plants. The tramp of a horse sets the nearest ones in motion, and, as if by magic, the contraction of the small grey green leaflets spreads in quivering circles over the field, making one almost believe, with Darwin and Dutrochet, that plants have feeling, or tempting one to exclaim with Wordsworth —

> "It is my faith, that every flower
> Enjoys the air it breathes."

Among the most remarkable forms of tropical vegetation, the creeping plants, bushropes, or lianas (cissus, bauhinia, bignonia, banisteria, passiflora), that contribute so largely to the impenetrability of the forests, hold a conspicuous rank. Often three or four bushropes, like strands in a cable, join tree to tree, and

---

* See Chapter III.

branch to branch; others, descending from on high, take root
as soon as their extremity touches the ground, and appear like
shrouds and stays supporting the mainmast of a line-of-battle
ship; while others send out parallel, oblique, horizontal, and
perpendicular shoots in all directions. Frequently trees above
a hundred feet high, uprooted by the storm, are stopped in
their fall by these amazing cables of Nature, and are thus
enabled to send forth vigorous shoots, though far from their
perpendicular, with their trunks inclined to every degree from
the meridian to the horizon.

Their heads remain firmly supported by the bushropes;
many of their roots soon refix themselves in the earth, and frequently a strong shoot will sprout out perpendicularly from near
the root of the reclined trunk, and in time become a stately
tree.

No European is able to penetrate the intricate network of a
forest thus matted together: astonished and despairing he
stands before the dense cordage that impedes his path, and,
should he attempt to force his way through the maze, the
strong thorns and hooks with which the tropical creepers are
generally armed would soon make him repent of his boldness. Even the Brazilian planter never thinks of entering
the forest without a large knife, or without being accompanied
by slaves, who, with heavy scythe-like axes attached to long
poles, clear the way by severing the otherwise impenetrable
cordage.

But if the naturalist vents dire imprecations on the creepers,
which render forest exploration so difficult, they are the delight
of the monkeys and tiger-cats that climb them with astonishing
celerity. The rapid ascension of the highest trees, or even the
passage over streams, is rendered easy by their means to whole
herds. No less pliable than tough, the lianas of the western
hemisphere are used by the Brazilians as cordage to fasten
the rafters of their houses, in the same manner as the equally
flexible ratans are employed throughout the East Indian
world.

The enormous climbing trees, that stifle the life of the mightiest giants of the forest, offer a still more wonderful spectacle.
At first, these emblems of ingratitude grow straight upwards like
any feeble shrub, but as soon as they have found a support in

other trees, they begin to extend over their surface; for, while the stems of other plants generally assume a cylindrical form, these climbers have the peculiarity of divesting themselves of their rind when brought into contact with an extraneous body, and of spreading over it, until they at length enclose it in a tubular mass. When, during this process, the powers of the original root are weakened, the trunk sends forth new props to restore the equilibrium; and thus this tough and hardy race continually acquires fresh strength for the ruin of its neighbours.

Several species of the fig trees are peculiarly remarkable for this distinctive property, and from the facility with which their

Fig Tree.

seeds take root where there is a sufficiency of moisture to permit of germination, are formidable assailants of ancient monuments. Sir Emerson Tennent mentions one which had fixed itself on the walls of a ruined edifice at Polanarrua, and formed one of the most remarkable objects of the place, its roots streaming downwards over the walls as if their wood had once been fluid, and following every sinuosity of the building and terraces till they reached the earth.

On the borders of the Rio Guama, the celebrated botanist, Von Martius, saw whole groups of Macauba palms encased by fig trees that formed thick tubes round the shafts of the palms, whose noble crowns rose high above them; and a similar spectacle occurs in India and Ceylon, when the Tamils look with increased veneration on their sacred pippul thus united in marriage with the palmyra. After the incarcerated trunk has been stifled and destroyed, the grotesque form of the parasite, tubular, cork-screw like, or otherwise fantastically contorted, and frequently admitting the light through interstices like loopholes in a turret, continues to maintain an independent existence among the straight-stemmed trees of the forest, — the image of an eccentric genius in the midst of a group of steady citizens.

Like the mosses and lichens of our woods, epiphytes of endless variety and almost inconceivable size and luxuriance (ferns, bromelias, tillandsias, orchids, and pothos) cover in the tropical zone the trunks and branches of the forest trees, forming hanging gardens, far more splendid than those of ancient Babylon. While the orchids are distinguished by the eccentric forms and splendid colouring of their flowers, sometimes resembling winged insects or birds, the pothos family (caladium, calla, arum, dracontium, pothos) attracts attention by the beauty of their large, thick-veined, generally arrow-shaped, digitated, or elongated leaves, and form a beautiful contrast to the stiff bromelias or the hairy tillandsias that conjointly adorn the knotty stems and branches of the ancient trees.

In size of leaf, the Pothos family is surpassed by the large tropical water-plants, the Nymphæas and Nelumbias, among which the Victoria regia, discovered in 1837 by Robert Schomburgk in the river Berbice, enjoys the greatest celebrity. The round light-green leaves of this queen of water-plants measure no less than six feet in diameter, and are surrounded by an elevated rim several inches high, and exhibiting the pale, carmine red of the under surface. The odorous white blossoms, deepening into roseate hues, are composed of several hundred petals; and, measuring no less than fourteen inches in diameter, rival the colossal proportions of the leaves.

The trunk of several tropical trees offers the remarkable peculiarity of bulging out in the middle like a barrel. In the Brazilian forests, the Pao Barrigudo (*Chorisia ventricosa*) arrests the attention of every traveller by its odd ventricose shape, nearly half as broad in the centre as long, and gradually tapering towards the bottom and the top, whence spring a few thin and scanty branches.

Bottle-Tree.

The Delabechea, or bottle-tree, discovered by Mr. Mitchell in tropical Australia, has the same lumpish mode of growth. Its wood is of so loose a texture, that when boiling water is poured over its shavings a clear jelly is formed and becomes a thick viscid mass.

In other trees which, struggling upwards to air and light, attain a prodigious altitude, or from their enormous girth and the colossal expansion of their branches require steadying from beneath, we find buttresses projecting like rays from all sides of the trunk. They are frequently from six to twelve inches thick, and project from five to fifteen feet, and, as they ascend, gradually sink into the bole and disappear at the height of from ten to twenty feet from the ground. By the firm resistance which they offer below, the trees are effectually protected from the leverage of the crown, by which they would otherwise be uprooted. Some of these buttresses are so smooth and flat as almost to resemble sawn planks; as, for instance, in the Bombax Ceiba, one of the most remarkable examples of this wonderful device of Nature.

In other cases we find the roots fantastically spreading and revelling in a variety of grotesque shapes, such as we nowhere find in the less exuberant vegetation of Europe. Thus, in the india rubber tree (*Ficus elastica*), masses of the roots appear above ground, extending on all sides from the base, and writhing over the surface in serpentine undulations, so that the Indian villagers give it the name of the snake-tree. Sir Emerson Tennent mentions an avenue of these trees leading to the botanical garden of Peradenia, in Ceylon, the roots of which meet from either side of the road, and have so covered the surface, as to form a wooden framework, the interstices of which retain the materials that form the roadway.

Snake Tree.

These tangled roots sometimes trail to such an extent, that they have been found upwards of 140 feet in length, whilst the tree itself was not thirty feet high.

The mangroves, the iiarteas, the screw-pines, are equally singular in the formation of their roots; but those of the lum, a large tree which Kittlitz found growing on the island of Ualan, are perhaps without a parallel in the vegetable world. Each of the roots, running above ground for a considerable distance, is surmounted by a perfectly vertical crest, gradually diminishing in size as the root recedes from the trunk, but often three, or even four, feet high near its base. These crests, which are very thin but perfectly smooth, regularly follow all the sinuosities of the root, and thus form, to a considerable distance round the tree, a labyrinth of the strangest appearance. Large spaces of swampy ground are often covered with their windings, and it is no easy matter to walk on the sharp edges of these vertical bands, whose interstices are generally filled with deep mud. On being struck, the larger crests emit a deep sonorous sound, like that of a kettledrum.

The thorns and spines with which many European plants are armed, give but a faint idea of the size which these defensive weapons attain in the tropical zone. The cactuses, the acacias, and many of the palm trees, bristle with sharp-pointed shafts, affording ample protection against the attacks of hungry animals, and might appropriately be called vegetable hedge-hogs, or porcupines. The Toddalia aculeata, a climbing plant, very common in the hill-jungles of Ceylon, is thickly studded with knobs, about half an inch high, and from the extremity of each a thorn protrudes, as large and sharp as the bill of a sparrow-hawk.

The black twigs of the buffalo-thorn (*Acacia latronum*) a low shrub, abounding in northern Ceylon, are beset at every joint by a pair of thorns set opposite each other, like the horns of an ox, as sharp as a needle, from two to three inches in length, and thicker at the base than the stem they grow on; and the Acacia tomentosa, another member of the same numerous genus, has thorns so large as to be called the jungle-nail by Europeans, and the elephant-thorn by the natives. In some of these thorny plants, the spines grow, not singly, but in branching clusters, each point presenting a spike as sharp as a lancet; and where these shrubs abound, they render the forest absolutely impassable, even to animals of the greatest size and strength.

The formidable thorny plants of the torrid zone, which are often made use of by man to protect his fields and plantations against wild beasts and robbers, have sometimes even been made to serve as a bulwark against hostile invasions. Thus Sir Emerson Tennent informs us, that during the existence of the Kandyan kingdom, before its conquest by the British, the frontier forests were so thickened and defended by dense plantations of thorny plants, as to form a natural fortification impregnable to the feeble tribes on the other side; and at each pass which led to the level country, movable gates, formed of the same thorny beams, were suspended as an ample security against the incursions of the naked and timid lowlanders.

A Ceylonese Cocoa-nut Oil-Mill.

CHAPTER XII.

PALMS.

The Cocoa-nut Tree — Its hundred Uses — Cocoa-nut Oil — Coir — Porcupine Wood — Enemies of the Cocoa Palm — The Sago Palm — The Saguer — The Gumatty — The Areca Palm — The Palmyra Palm — The Talipot — The Cocoa de Mer — Ratans — A Ratan Bridge in Ceylon — The Date Tree — The Oil Palms of Africa — The Oil Trade at Bonny — Its vast and growing Importance — American Palms — The Carnauba — The Ceroxylon andicola — The Cabbage Palm — The Gulielma speciosa — The Piacava — Difficulties of the Botanist in ascertaining the various Species of Palms — Their wide Geographical Range — Different Physiognomy of the Palms according to their height — The Position and Form of their Fronds — Their Fruits — Their Trunk — The Yriartea ventricosa.

THE graceful acanthus gave the imaginative Greeks the first idea of the Corinthian capital; but the shady canopy of the cocoa-nut tree would no doubt form a still more beautiful ornament of architecture, were it possible for art to imitate its feathery fronds and carve their delicate tracery in stone.

Essentially littoral, this noble palm requires an atmosphere damp with the spray and moisture of the sea to acquire its full stateliness of growth, and while along the bleak shores of the Northern Ocean the trees are generally bent landward by the rough sea breeze, and send forth no branches to face its violence, the cocoa, on the contrary, loves to bend over the rolling surf, and to drop its fruits into the tidal wave. Wafted by the winds and currents over the sea, the nuts float along without losing their germinating power, like other seeds which migrate through the air; and thus, during the lapse of centuries, the cocoa-palm has spread its wide domain from coast to coast throughout the whole extent of the tropical zone. It waves its graceful fronds over the emerald islands of the Pacific, fringes the West Indian shores, and from the Philippines to Madagascar crowns the atolls, or girds the sea-borde of the Indian Ocean.

But nowhere is it met with in such abundance as on the coasts of Ceylon, where for miles and miles one continuous grove of palms, preeminent for beauty, encircles the " Eden of the eastern wave." Multiplied

Cocoa-nut Tree.

by plantations and fostered with assiduous care, the total number in the island cannot be less than twenty millions of full-grown trees; and such is its luxuriance in those favoured districts, where it meets with a rare combination of every advantage essential to its growth, a sandy and pervious soil, a free and genial air, unobstructed solar heat, and abundance of water, that, when in full bearing, it will annually yield as much as a ton's weight of nuts — an example of fruitfulness almost unrivalled even in the torrid zone.

No other tree in the world, no other plant cultivated by man, contributes in *so many ways* to his wants and comforts as this inestimable palm; and it is a curious illustration of its innumerable uses, that some years ago a ship from the Maldive Islands touched at Galle, which was entirely built, rigged, provisioned, and laden with the produce of the cocoa-tree. Besides furnishing their chief food to many tribes on the coast within the torrid zone, the nut contains a valuable oil, which burns without smoke or smell, and serves, when fresh, for culinary

K

purposes. Consisting of a mixture of solid (*stearine*) and fluid (*elain*) fat, it congeals at a temperature of 72°; but both its component substances acquire additional value after having been separated by means of the hydraulic press; for while the liquid part furnishes an excellent lamp-oil, the solid fat is manufactured into candles rivalling wax, and at the same time not much dearer than tallow.

This important product first became known in the European markets at the beginning of the present century, and is now a considerable article of commerce, which has drawn the attention of speculators even to the remotest islands of the tropical ocean. Thus Mr. Darwin mentions an Englishman of the name of Hare, who, with a number of Malays, had settled on the uninhabited Keeling Islands for the purpose of collecting the produce of their cocoa-nut groves; and Skogman (Circumnavigation of the Swedish frigate "Eugenie") tells us of a steam oil-mill, erected by a Mr. Brodin on Foa, one of the Tonga Islands.

The Kingsmill group, Puynipet, the loftiest and most beautiful of the Carolinas, and many other Polynesian atolls, have, in a similar manner, been drawn within the vortex of commerce; and in 1835 the Society Islands furnished 170 tons of cocoa-nut oil, worth about 13,000 dollars.

But the whole produce of the Pacific is utterly insignificant when compared to the vast export of Ceylon, which in 1857 amounted to no less than 1,767,413 gallons, valued at 212,184*l*. Reckoning forty nuts to a gallon, and taking into account that at least as much oil is consumed in the island as is sent out of it, no less than 140 millions of nuts were necessary for producing this enormous supply.*

It may easily be imagined that the crushing of the fruit for the expression of the oil gives employment to a great number of people, and constitutes a flourishing branch of trade. For this purpose the natives generally erect their creaking mills under the shade of the palm groves near their houses. These primitive engines consist of the trunk of a tree hollowed

* I find stated in the "Economist" that during the first eight months of 1861 188,053 cwts. of cocoa-nut oil were imported into Great Britain, worth 50*l*. per ton; so that the annual importation probably amounts to no less than 600,000*l*. During the same period the importations of African palm-oil amounted to 347,486 cwts.

into a mortar, in which a heavy upright pestle is worked round by a bullock yoked to a transverse beam.

No wonder that, to meet the constantly increasing demand, new plantations are continually forming on the coast, wherever the *Cocos nucifera* can be advantageously cultivated; and that the Dutch, ever anxious to add new sources of wealth to the produce of Java, have within the last twenty years planted more than twenty millions of cocoa-nut trees along the shores of their splendid colony—the gem and envy of the Indian world.

The fibrous rind or husk of the nut furnishes the coir of commerce, a scarce less important article of trade than the oil itself. It is prepared by being soaked for some months in water, for the purpose of decomposing the interstitial pith, after which it is beaten to pieces until the fibres have completely separated, and ultimately dried in the sun. Ropes made of coir, though not so neat in appearance as hempen cords, are superior in lightness, and exceed them in durability, particularly if wetted frequently by salt water. From their elasticity and strength they are exceedingly valuable for cables. Thus Mr. Bennett ("Whaling Voyage round the Globe") mentions having once been on board a ship during a severe gale, when hemp and chain gave way, and the vessel at last most unexpectedly rode out the gale with a coir-rope. In the year 1854, 43,957 cwt. of this valuable fibre were exported from Ceylon to England, and its consumption increases rapidly. Besides cordage of every calibre, beds, cushions, carpets, brushes, and nets are manufactured from the filaments of the cocoa-nut husk, while the hard shell is fashioned into drinking-cups, spoons, beads, bottles, and knife-handles. From the spathes of the unopened flowers a delicious "toddy" is drawn, which, drunk at sunrise before fermentation has taken place, acts as a cooling gentle aperient, but in a few hours changes into an intoxicating wine, and may either be distilled into arrack—the only pernicious purpose to which the gifts of the bounteous tree are perverted—or soured into vinegar, or inspissated by boiling into sugar. So vast is the consumption of "toddy" in Ceylon for these various purposes, that while it has been computed that only 3,500,000 of cocoa-nut trees are destined for the production of oil, and 11,500,000 for the food of the islanders, no less than 5,000,000 are said to be devoted to the drawing of toddy.

The strong tough foot-stalks of the fronds, which attain a length of from eighteen to twenty feet, are used for fences, for yokes, for carrying burthens on the shoulders, for fishing-rods; the leaflets serve for roofing, for mats, for baskets, for cattle-fodder; and their midribs form good brooms for the decks of ships. Cooked or stewed, the cabbage or cluster of unexpanded leaves is an excellent vegetable, though rarely used, as it necessarily involves the destruction of the tree; and even the tough web or network, which sustains the foot-stalks of the leaves, may be stripped off in large pieces and used for straining.

After the cocoa-nut tree has ceased to bear, its wood serves for many valuable purposes — for the building of ships, bungalows, and huts, for furniture and farming implements of every description; and, as it admits of a fine polish, and its reddish ground colour is beautifully veined with dark lines, it is frequently imported into England under the name of porcupine-wood.

When we consider the many benefits conferred upon mankind by this inestimable tree, we cannot wonder at the animation with which the islander of the Indian Ocean recounts its "hundred uses," or at the superstition which makes him believe that by some mysterious sympathy it pines when beyond the reach of the human voice.

Yet, strange to say, as Sir Emerson Tennent informs us, one preeminent use of the cocoa-nut tree is omitted in all the popular enumerations of its virtues: it acts as a conductor in protecting the houses from lightning; thus shielding the life and property of those whom it nourishes with its fruit and enriches with its manifold productions.

As experience has shown that the cocoa-nut palm requires both sea-air and plentiful irrigation for its vigorous growth, those portions of the Ceylon coast are selected for new plantations which are flanked by estuaries and intersected by inland lakes, where wells can be sunk at a small expense and water can be most conveniently carried.

The ripe nuts are first planted in a nursery, where they are covered an inch deep with sand and sea-weed, or soft mud from the beach, and watered daily till they germinate. After two or three months, a white shoot, in which the foliaceous rudiments are distinctly to be perceived, rises from one of the three holes of the nut, while the radicles emerge from the other two orifices in a direction opposite to the shoot, and penetrate the ground. The nuts put down in April are sufficiently grown to be planted out before the rains of September, when they are set out in holes, three feet deep, and twenty to thirty feet apart. During the first years of their growth the young plants require incessant care. They must be assiduously watered, and protected from the glare of the sun under shades made of the plaited fronds of the cocoa-nut palm, or the fan-like leaves of the palmyra; and as their tender shoots are especial favourites with wild hogs, rats, elephants, and porcupines, constant attention is required to ward off the attacks of these animals. As the stem ascends it has to encounter the most formidable enemy of all — the Cooroominya beetle (*Batocera rubus*), which penetrates the trunk near the ground, and deposits its eggs in the cavity, the grubs when hatched eating their way upwards through the centre of the tree to the top, where they pierce the young leaf-buds, and do such damage, as frequently to leave not a single young tree untouched in large plantations.

Notwithstanding the repulsive aspect of the large pulpy grubs of these pernicious insects, they are esteemed as great a delicacy by the natives as the termites, which often prove equally destructive. All the injuries from these united causes involve the loss of about one-fourth of the plants put down, and constant renewal is required in order to replace those which have been destroyed. After the second year irrigation becomes unnecessary, and all that is then required is to keep the ground clean, and in each alternate year to dress the young palms with sea-weed and salt manure. In six or seven years the trees usually commence to bear fruit, and yield an abundance for sixty years, when the influence of age begins to show itself in decreasing production. The cultivation of the cocoa-palm, which was formerly confined to the native Singalese, has latterly been undertaken with growing spirit by European planters, who in the year 1853 had already covered 21,412 acres with this valuable tree; and, as Professor Schmarda ("Voyage round the World") informs us that there is yet room in Ceylon for a tenfold increase of its palm-groves, a vast field still lies open here to British capital and enterprise.

Wherever the cocoa grows, its sweet and nutritious nuts are

eagerly sought for by many animals. The small, black, long-clawed cocoa-nut bear (*Ursus malayanus*), which inhabits

Malay Bear.

Sumatra and Borneo, and surpasses all other members of the Ursine family by its surprising agility in climbing, though far from despising other fruit, yet shows by its name to which side its inclinations chiefly lean. The East Indian Palm-martin (*Paradoxurus typus* or *Pougouni*) and the sprightly Palm-squirrel (*Sciurus palmarum*) likewise climb the cocoa-palms, and, per-

Palm Squirrel.

forating the soft and unripe nuts, eagerly sip their juice. The ubiquitous Rat, the most disgusting of the whole mammalian race, bites holes into the cocoa-nuts close to their stalk, taking good care not to gnaw the shell, where the juice would run out and defraud it of its meal. Its devastations in the island of Mauritius, where cocoa-planting is likewise conducted on a great scale, are so great, that a price is set upon its head, and the negro who succeeds in bringing the planter a dozen rat-tails is rewarded with a glass of rum!

Even the birds diminish the produce of the cocoa-nut grove. The Noddy (*Sterna stolida*) builds his nest between the foot-stalks, and picks so busily at the blossom, when stormy weather prevents him making any long excursions, that on many islands he is considered as a chief cause of the sterility of numerous palms.

But of all the animals that infest the cocoa-nut tree, and live upon its fruits, there is none more remarkable than the famous East Indian cocoa-nut crab (*Birgus latro*), a kind of intermediate link between the short and long-tailed crabs. "It is said to climb the palm trees for the sake of detaching the heavy nuts; but Mr. Darwin, who attentively observed the animal on the Keeling Islands, tells us that it merely lives upon those that spontaneously fall from the tree. To extract its nourishment from the hard case, it shows an ingenuity which is one of the most wonderful instances of animal instinct. It must first of all be remarked that its front pair of legs are terminated by very strong and heavy pincers, the last pair by others narrow and weak. After having selected a nut fit for its dinner, the crab begins its operations by tearing the husk, fibre by fibre, from that end under which the three eye-holes are situated; it then hammers upon one of them with its heavy claws, until an opening is made; hereupon it turns round, and, by the aid of its posterior pincers, extracts the whole albuminous substance. It inhabits deep burrows, where it accumulates surprising quantities of picked fibres of cocoa-nut husks, on which it rests as on a bed. Its habits are diurnal, but every night it is said to pay a visit to the sea, no doubt for the purpose of moistening its branchiæ. It is very good to eat, living, as it does, on choice vegetable substances, and the great mass of fat accumulated under the tail of the larger ones sometimes yields, when melted, as much as a quart of limpid oil. Thus, our taking possession of the Keeling Islands, as a coaling station for the steamers from Australia to Ceylon, bodes no good to the Birgus." *

In every zone we find nations in a low degree of civilisation living almost exclusively upon a single animal or plant. The Laplander has his reindeer, the Eskimo his seal, the Sandwich Islander his taro-root; and thus also we find the natives of a great part of the Indian Archipelago living almost exclusively upon the pith of the sago palm (*Sagus fariniferus*). This tree, which is of such great importance to the indolent Malay, as it almost entirely relieves him of the necessity of labour, grows at first very slowly, and is covered with thorns. As soon, however, as the stem is once formed, it shoots upwards with such rapidity, that it speedily attains its full height of ten yards, with a girth of five or six feet, losing in this stage its thorny accompaniments. The crown is larger and thicker than that of the cocoa-nut tree; the efflorescence colossal, forming an immense bunch, the branches of which spread out like the arms of a gigantic candelabrum. The tree must, however, be felled before the fruit begins to form, as otherwise the farina would be exhausted, which man destines for his food. When the trunk has been cut and split into convenient pieces, the pith is scooped out, kneaded with water, and strained, to separate the meal from the fibres. One tree will produce from

---

* "The Sea and its Living Wonders," second edition, p. 211.

two to four hundredweight of flour, which is mostly consumed on the spot, and serves to feed several millions of men; but a great quantity is also exported to Europe. Thus, in Singapore, there are upwards of thirty refineries, which convert the sago imported in an impure state from Borneo into a beautiful white flour, which is extensively used by English manufacturers for stiffening calicoes.

The sago palm forms large forests, particularly on swampy ground in Borneo and Sumatra, in the Moluccas and New Guinea. Mushrooms of an excellent flavour frequently cover the mouldering trunks, and in the pith, the fat and whitish grubs of the *Cossus saguarius*, a large lamellicorn beetle, are found, which the natives consider a great delicacy when roasted.

The Saguer or Gomuti (*Gomutus vulgaris*), the ugliest of palms, but almost rivalling the cocoa-nut tree by the multiplicity of its uses, is likewise a native of the Indian Archipelago. On seeing its rough and swarthy rind, and the dull dark-green colour of its fronds, the stranger wonders how it is allowed to stand, but when he has tasted its delicious wine he is astonished not to see it cultivated in greater numbers. Although the outer covering of the fruits has venomous qualities, and is used by the Malays to poison springs, the nuts have a delicate flavour, and the wounded spathe yields an excellent toddy, which, like that of the cocoa-nut and the palmyra palm, changes by fermentation into an intoxicating wine, and on being thickened by boiling furnishes a kind of black sugar, much used by the natives of Java and the adjacent isles. The reticulum or fibrous net at the base of the petioles of the leaves constitutes the gumatty, a substance admirably adapted to the manufacture of cables, and extensively used for cordage of every description. The gumatty is black as jet, the hairs extremely strong, and resembling coir, except that they are longer and finer. The small hard twigs found mixed up with this material are employed as pens, besides forming the shafts of the sumpits, or little poisoned arrows of the Malays, and underneath the reticulum is a soft silky material, used as tinder by the Chinese, and applied as oakum in caulking the seams of ships, while from the interior of the trunk a kind of sago is prepared.

Throughout the whole of the Indian Archipelago, and

extending over part of Polynesia and Australia, we find a
custom almost as filthy as tobacco-smoking, and causing an
almost equal waste of time and money. As many a countryman of ours is hardly ever seen without a cigar in his mouth,
thus you will rarely meet with a Malay — man or woman,
old or young — that does not indulge from morning till night
in the luxury of chewing the astringent nut of the areca palm,
mixed with lime and the leaf of the betel-pepper. This combination has the disgusting property of colouring the saliva of
so deep a red, that the lips and teeth appear as if covered
with blood; it spoils the teeth, and sometimes even produces
a peculiar kind of cancer in the cheek; but, as excuses are
never wanting to justify bad habits, it is said to have tonic
effects and to promote digestion.

The Areca palm (*Areca Catechu*) bears a great resemblance to the cocoa-nut tree, but is of a still more graceful
form, rising to the height of forty or fifty feet, without any
inequality on its thin polished stem, which is dark-green
towards the top, and sustains a crown of feathery foliage, in the
midst of which are clustered the astringent nuts, for whose
sake it is carefully tended. In the gardens of Ceylon the
areca palm is invariably planted near the wells and watercourses, and the betel plant, which immemorial custom has
associated to its use, is frequently seen twining round its
trunk.

The Palmyra palm (*Borassus flabelliformis*), the sacred Talgaha of the Brahminical Tamils of Ceylon, extends from the
confines of Arabia to the Moluccas, and is found in every
region of Hindostan from the Indus to Siam, the cocoa and
the date tree being probably the only palms that enjoy a still
wider geographical range. In northern Ceylon, and particularly
in the peninsula of Jaffna, it forms extensive forests; and such
is its importance in the Southern Dekkan and along the
Coromandel coast, that its fruits afford a compensating resource
to seven millions of Hindoos on every occasion of famine or
failure of the rice crop. Unlike the cocoa-nut palm, which
gracefully bends under its ponderous crown, the palmyra rises
vertically to its full height of seventy or eighty feet, and
presents a truly majestic sight when laden with its huge
clusters of fruits, each the size of an ostrich's egg, and of a rich

brown tint, fading into bright golden at its base. It is not till the tree has attained a mature age that its broad fan-like leaves begin to detach themselves from the stem; they climb from the ground to its summit in spiral convolutions, forming a dense cover for many animals — ichneumons, squirrels, and monkeys, that resort to it for concealment. In these hiding-places the latter might easily defy the sportsman; but they frequently fall victims to a silly curiosity, for when he is accompanied by his dog, they cannot resist the temptation of watching the animal's movements, and, coming forth to peep, expose themselves to a fatal shot.

The stalks of the decayed leaves remain partly attached to the trunk, affording supports to a profusion of climbing and epiphytic plants, which hide the stem under a brilliant tapestry of flower and verdure.

The palmyra rivals the cocoa-nut and the gomuti by its many uses, and Hindoo poets celebrate the numerous blessings it confers upon mankind.

When the spathes of the fruit-bearing trees exhibit themselves, the toddy-drawer forthwith commences his operations, climbing by the assistance of a loop of flexible jungle-vine, sufficiently wide to admit both his ancles and leave a space between them, thus enabling him to grasp the trunk of the tree with his feet and support himself as he ascends. Having pruned off the stalks of fallen leaves, and cleansed the crown from old fruit-stalks and other superfluous matter, he binds the spathes tightly with thongs to prevent them from farther expansion, and descends, after having thoroughly bruised the embryo flowers within to facilitate the exit of the juice. For several succeeding mornings the operation of crushing is repeated, and each day a thin slice is taken off the end of the racemes, to facilitate the exit of the sap and prevent its bursting the spathe. About the eighth morning the sap begins to exude, an event which is notified by the immediate appearance of birds, especially of the "toddy bird," a species of shrike, (*Artamus fuscus*), attracted by the flies and other insects which come to feed on the luscious juice of the palm. The crows, ever on the alert when any unusual movement is in progress, keep up a constant chattering and wrangling; and about this time the palmyra becomes the resort of the palm-martin and the graceful genet, which frequent the trees, and especially the crown of the cocoa-nuts, in quest of birds. On ascertaining that the first flow of the sap has taken place, the toddy-drawer again trims the wounded spathe, and inserts its extremity in an earthen chatty to collect the juice. Morning and evening these vessels are emptied, and for four or five months the palmyra will continue to pour forth its sap at the rate of three quarts a day. But once in every three years the operation is omitted, and the fruit is allowed to form, without which the natives assert that the tree would pine and die.*

Most of the "palmyra toddy" drawn in Ceylon is made into sugar or *jaggery*, which sells in the bazaars for about three farthings a pound. Of the produce of Jaffna alone about 500 tons are annually exported to the opposite coast of India, where it undergoes the process of refining. The granulation is said to surpass that of the sugar-cane, and large quantities of palmyra sugar are even sent to Europe from Cuddalore and Madras. The produce of jaggery might be greatly increased, but, by a strange anomaly in these free-trade times, a high duty is levied on its export by the Colonial Government.† The hard and durable wood of the palmyra, which, consisting like the other palms of straight horny fibres, can easily be split into lengths, is said to resist the attacks of the termites, and is used universally in Ceylon and India for roofing and similar purposes. The exports of Jaffna alone consume annually between 7,000 and 8,000 palms, each worth from three to six shillings. The tough and polished shell of the fruit contains three intensely hard seeds imbedded in a farinaceous orange pulp, which has a sweet and oily taste, and is eaten in various ways.

The leaves, finally, are employed for roofs, fences, mats, baskets, fans, and paper. According to a common fallacy, "the two nuts of India, the cocoa-nut and the palmyra, cherish such secret envy and hatred towards each other, that they will not grow in the same field, nor in one and the same region; which, however, must be attributed to the great wisdom of the Creator, who is unwilling that these trees, so productive and so

---

* Tennent's "Ceylon," vol. ii. p. 523.
† Schmarda, "Reise um die Erde, von 1853 bis 1857," vol. i. p. 283.

necessary to the human race, should grow in the same locality;"[*] but, unfortunately for the palmyra, the erroneousness of this belief has of late years been conclusively demonstrated, as, even in its ancient head-quarters at Jaffna, it is now almost outnumbered by the recent plantations of cocoa-nuts, and has in many instances been felled to make room for its more fortunate and still more valuable rival.

The Talpot or Talipot of the Singalese (*Corypha umbraculifera*) rises to the height of one hundred feet, and expands into a crown of enormous fan-like leaves, each of which when laid upon the ground will form a semicircle of sixteen feet in diameter, and cover an area of nearly two hundred superficial feet. These gigantic foliaceous expansions are employed by the Singalese for many purposes. They form excellent fans, umbrellas, or portable tents, one leaf being sufficient to shelter seven or eight persons; but their most interesting use is for the manufacture of a kind of paper, so durable as to resist for many ages the ravages of time. The leaves are taken whilst still tender, cut into strips, boiled in spring water, dried, and finally smoothed and polished, so as to enable them to be written on with a style, the furrow made by the pressure of the sharp point being rendered visible by the application of charcoal ground with a fragrant oil. The leaves of the palmyra similarly prepared are used for ordinary purposes; but the most valuable books and documents are written to-day, as they have been for ages past, on *olas* or strips of the talipot.

The currents of the sea sometimes drift to the shores of the Maldives, and even to the south and west coasts of Java and Sumatra, a nut, exceeding the ordinary cocoa many times in size, with the additional peculiarity of presenting a double, or sometimes even a triple form, as if two separate fruits had grown together. These mysterious nuts were formerly believed to be of submarine origin, whence they derived their name of *coco de mer*, and to have the wonderful power of neutralising poisons.

On the Maldive Islands these wonderful drift-nuts were the exclusive property of the king, who either sold them at an exorbitant price, or made presents of them to other potentates. At length, about a hundred years ago, the French traveller

---

[*] Rumphius, "Herbarium Amboinense."

Sonnerat discovered in the Seychelles the home of the *Lodoicea Sechellarum*, which, like the cocoa, grows on the strand of that small and secluded group, and drops its large nuts into the sea, which then carries them along to the east. The trunk of the lodoicea rises to the height of forty or fifty feet, and bears a crown of immense fan-like leaves, upwards of twenty feet long and fifteen broad, with foot-stalks seven feet long. As soon as the origin of the nuts became known, they of course immediately fell in price, so that a *coco de mer* which could formerly not have been had for 400 rupees is now hardly worth so much as ten. The lodoicea has been introduced into Ceylon, but Sir E. Tennent is not aware that it has yet fruited there.

The Ratans, a most singular genus of creeping palms, luxuriate in the forests of tropical Asia. Sometimes their slender stems, armed with dreadful spines at every joint, climb to the summit of the highest tree; sometimes they run along the ground; and while it is impossible to find out their roots among the intricate tangles of the matted underwood, their palm-like topes expand in the sunshine, the emblems of successful parasitism. They frequently render the forest so impervious, that the distinguished naturalist Junghuhn, while exploring the woods of Java, was obliged to be accompanied by a vanguard of eight men, one half of whom were busy cutting the ratans with their hatchets, while the others removed the stems. These rope-like plants frequently grow to the incredible length of four or even six hundred feet, often consisting of a couple of hundred joints two or three feet long, and bearing at every knot a feathery leaf, armed with thorns on its lower surface. Sir. E. Tennent mentions having seen a specimen two hundred and fifty feet long and an inch in diameter, without a single irregularity, and no appearance of foliage other than the bunch of feathery leaves at the extremity.

Though often extremely disagreeable to the traveller, yet the ratans are far from being useless. The natives of Java and the other islands of the Eastern Archipelago cut the cane into fine slips, which they plait into beautiful mats, manufacture into strong and neat baskets, or twist into cordage; and they are also extensively exported to Europe, where they are chiefly employed for the making of chair bottoms.

So great is the strength of these slender and seemingly

fragile plants, that they are frequently used with success in the formation of bridges across the watercourses and ravines. Sir E. Tennent mentions one which crossed the Falls of the Mahawelli Ganga, in Ceylon, and was constructed with all the precision of scientific engineering.

It was entirely composed of a species of ratan, "the extremities of which were fastened to living trees on the opposite sides of the ravine, through which a furious and otherwise impassable mountain torrent thundered and fell from rock to rock, with a descent of nearly a hundred feet. The flooring of this aerial bridge consisted of short splints of wood laid transversely, and bound in their place, by thin strips of the ratan itself. The whole structure vibrated and swayed with fearful ease, but the coolies traversed it, though heavily laden, and the European, between whose estate and the high road it lay, rode over it daily without dismounting."

Date Tree.

On turning from Asia to the adjoining continent of Africa, we find a new world of palms, several of which are no less valuable than the cocoa-nut or the palmyra, either as affording food, or enriching by their produce the commerce of the world.

The date-tree, sung from time immemorial by the poets of the East, is as indispensable as the camel to the inhabitants of the wastes of North Africa and Arabia, and, next to the "ship of the desert," the devout Mussulman esteems it the chief gift of Allah. Few palms have a wider range, for it extends from the Persian Gulf to the borders of the Atlantic, and flourishes from the twelfth to the thirty-seventh degree of northern latitude.

Groves of dates adorn the coasts of Valencia in Spain; near Genoa its plantations afford leaves for the celebration of Palm Sunday; and in the gardens of southern France a date tree sometimes mixes among the oranges and olives. But it never bears fruit on these northern limits of its empire, and thrives best in the oases on the borders of the sandy desert. Here it is cultivated with the greatest care, and irrigated every morning; for though it will grow on an arid soil, it absolutely requires water to be fruitful.

The date-palm is propagated by shoots, and the female tree bears its first fruits after four or five years. It is said to attain to an age of two centuries, but is rarely left standing longer than eighty years, when the trunk is tapped in spring, producing a kind of toddy, which is consumed in great quantities in "Biledulgerid," or the long line of oases to the south of the Atlas, which has been preeminently called the "land of dates." Like the cocoa-nut tree, the *Phœnix dactylifera* bears a crown of feathery fronds, though without equalling the "littoral palm" in beauty. It blooms in March or April, and its fecundity is such that it annually bears as much as two hundredweight of dates, which are plucked in October or November. The fresh fruits (*Tamr*) are the best, but they are mostly dried in the sun (*Bela*), and either eaten without any other preparation, or mixed with flour and baked into a kind of bread, which is very nutritious and of an agreeable taste. D'Escayrac de Lauture tells us that a fresh date, the kernel of which has been extracted and replaced by an almond or a piece of butter, is delicious.

There are at least sixty varieties of dates, and in the oases of Tozer and Nefta the chieftains told D'Escayrac the names of thirty-five different kinds all cultivated there. The most esteemed varieties are the *Monakhir*, which is very rare, and exclusively reserved for the table of the Bey of Tunis, and the *Degleh*, which grows to the height of eighty feet and produces from eight to ten clusters or large bunches of fruit, weighing each from twelve to twenty pounds. The common people hardly ever taste this date, which is either exported, or consumed only by the wealthier classes, and must content themselves with the common *Halig*, which grows in abundance wherever the Phœnix can flourish. It is not to be wondered at that the tribes of the desert so highly value a tree which, by enabling a family to live on the produce of a small spot of ground, extends as it were the bounds of the green islands of the desert, and rarely disappoints the industry that has been bestowed on its culture. It is considered criminal to fell it while still in its vigour, and both the Bible and the Koran forbid the warriors of the true God to apply the axe to the date trees of an enemy.

In Nubia, between Wadi Halfa and Khartum, there are at least a million of date-palms, which yield the extortionary government of the Pasha of Egypt an annual revenue of as many piastres. The date trees of Egypt and Nubia not only amply provide for home consumption, but furnish above 60,000 hundredweight for exportation to Syria and Turkey, besides a large quantity which the Nubian merchants sell in Sennaar, Kordofan, and Darfur. Here the Phœnix disappears, while the Doum (*Hyphæne thebaica*), distinguished from most other palms by its branching trunk, each branch being surmounted by a tuft of large stiff flabelliform leaves, assumes a conspicuous place in the landscape. Its fruits, which are of the size of a small apple and covered with a tough yellow lustrous rind, have a sugary taste, and serve for the preparation of sherbet. The old leaf-stalks with their thorns and sheathes remain attached to the trunk, increase its dimensions in an extraordinary degree, and render the task of climbing it next to impossible. The chief seat of this beautiful palm are the banks of the Nile, in the region of the cataracts. In Kordofan the Delebl palms form large clumps, with tamarinds, cassias, adansonias, and various mimosas. Straight as an arrow and perfectly smooth-rinded, this magnificent tree rises to the height of a hundred feet, bearing large fan-like leaves, attached to foot-stalks ten feet long, and armed with mighty thorns. From ten to twenty large bunches of nuts, as big as a man's head, hang beneath the fronds, but unfortunately these fine-looking fruits disappoint the taste.

Thus various forms of palms flourish along the banks of the Nile, but in general Africa has a less number of these trees to boast of than either Asia or America. On the other hand neither India nor Brazil have palms of such vast commercial importance as the Cocos butyracea, and the Elæis guineensis, the oil-teeming fruit trees of tropical West Africa.

The productiveness of the Elæis may be inferred from its bearing clusters of from 600 to 800 nuts, larger than a pigeon's egg, and so full of oil that it may be pressed out with the fingers. As long as the slave trade reigned along the coast of Guinea, these vegetable treasures remained unnoticed; but since England began to raise her voice against this infamous traffic, they have become the object of a great and constantly increasing commerce.

Liverpool, which was principally indebted to the slave trade for its rapid increase during the last century, now almost mo-

Oil Palm.

nopolises the palm-oil market; as, with the exception of a few vessels from Bristol, and now and then a solitary sail from London, almost every ship under the British flag that makes its appearance at Bonny, Benin, or Old and New Calabar, belongs to that great port.

Several Liverpool houses have eight or ten ships regularly trading with Bonny, and, by this constant intercourse, are able to provide such of their vessels as are busy loading with fresh provisions or fresh hands, to replace the losses caused by the destructive climate — an advantage which, of course, does not belong to those firms which carry on the trade on a less extensive scale.

The stay of a ship at Bonny depends in a great measure on the activity of the captain, the assortment of goods, the number of vessels that may be in port at the same time, the size of the ship, and the arrival of supplies from the interior; on an average, however, a vessel of 400 tons requires about four months for the completion of its cargo of 800 or 850 puncheons. On account of the murderous insalubrity of the climate, the crews are more numerous than is ordinarily the case in trading vessels; the pay is also considerably higher, as of course no sailor would think of confronting the Bight of Benin without an adequate reward. For the same reason each larger ship is provided with a surgeon, whose salary is increased at each repetition of the voyage. On their first trip, these disciples of the healing art confine themselves to their medical duties; on the second, as they are now supposed to be acquainted with the manner in which the trade is carried on, they act as supercargoes; on the third, they command the vessel, and receive their percentage of the profits, while a sailing master, or an upper steersman, placed under their orders, conducts the manœuvres of the ship. Thus, on the coast of Guinea, many captains are found that were originally the votaries of Esculapius, but now render a more profitable homage to Mercury.

With the exception of glass beads from Germany, and corals from France, all the other articles in request at Bonny are the produce of English manufactories.

As soon as a ship has anchored, it is immediately visited by a number of canoes, but trading is not allowed to begin before certain duties have been paid to the king. The first care is to purchase mats for roofing the deck against the sun and the tropical showers; and, this having been accomplished, not only the negro traders, but also their chief attendants, are presented with muskets, rum, powder, tobacco, strings of beads, calicoes, and cotton or woollen caps. Every black begs for "dash," so that there is no escaping their importunity, and though they return the gift with a "dash" of their own, yet they take good care not to lose by the exchange.

The cabin and the deck now swarm with negroes from early morning till noon. The dealers breakfast with the captain, in the chief cabin, and, when they come too late, ask for refreshments without ceremony, frequently raising their pretensions to champagne, though ultimately satisfied with a bottle of brown stout. Generally these black gentry bring one of their little boys or slaves with them, leaving him on board to learn English. All preliminaries having been adjusted, the trade begins by making advances to the dealers, who go into the interior, to purchase the oil with the goods they have obtained on credit.

A cask-house is erected on shore to put together the puncheons, the staves of which, along with the iron hoops, are brought ready-made from England.

The palm-oil, or *pulla*, is of a rich orange colour, and of the consistence of honey, at the ordinary temperature of the air. When eaten fresh, it is a delicate and wholesome article of diet, differing as much from the palm-oil imported into England as fresh from rancid butter.

Bonny itself produces no oil; most of it comes from New Calabar, Ibo, and the Brass country, and, owing to the difference of preparation, is of different quality, according to the place of its origin.

The trade is constantly on the increase, and cannot fail to introduce some civilisation among the barbarians of the coast; though, to judge by their present state, it needs must be a work of time.

In the year 1830, only 800 tons of palm-oil were imported into England; while, in 1854, Bonny alone exported 15,124 tons. The exportation from the Brass River amounted in 1856 to 3,280 tons; from Old Calabar, to 4,090; and from Camaroons, to 2,110.

The various ports in the Bight of Benin (Benin, Lagos, Talma, Badagry, Porto Novo, Ahguay) exported in the same year 17,480 tons, so that the whole coast of Guinea probably furnishes no less than fifty thousand tons of oil, worth at least 2,000,000*l*.

The American palms are pre-eminent in beauty, and many of them bid fair to rank highly in the future commerce of the world.

The leaves of the Carnauba (*Corypha cerifera*) furnish an abundance of wax. The lowlands of Guiana, between 3° and 7° N. lat., are frequently covered with this social fan-palm, whose full-grown fronds, when cut and dried in the shade, cover themselves with light-coloured scales. These melt in a warmth of 206° F., and then form a straw-coloured liquid, which again concretes on cooling. It burns with as clear and bright a flame as the best bees'-wax, and will no doubt become a considerable article of trade, when once the spirit of industry awakens in those rich but thinly-populated regions. Like many other palms, the Carnauba does not confine her gifts to one single product. The boiled fruit is edible, and the pith of the young stems affords a nutritious fecula. Roofs thatched with its leaves resist for many years the effects of the weather, and its wood may be used for a variety of purposes.

A kind of wax, exuding from the rings of its trunk, is also produced by the beautiful *Ceroxylon andicola*, which grows on the slopes of the Andes, up to an elevation of eight thousand feet. Even the lofty vault of the Crystal Palace would be unable to span this majestic palm, which, according to Humboldt's accurate measurement, towers one hundred and eighty feet above the ground, and bears a tuft of fronds each twenty-four feet long.

The cabbage-palm of the Antilles (*Oreodoxa oleracea*) almost rivals the mountain Ceroxylon in magnificence of growth, as its stem, which near to its base is about seven feet in circumference, ascends straight and tapering to the height of 130 feet. Its lofty fronds, moved by the gentlest breeze, are an object of beauty which can hardly be conceived by those who are unused to the magnificent vegetation of a tropical sun. Within the leaves which surround the top of the trunk, the cabbage, composed of longitudinal flakes, like ribands, but so compact as to form a crisp and solid body, lies concealed. It is white, about two or three feet long, as thick as a man's arm, and perfectly cylindrical. When eaten raw, it resembles the almond in flavour, but is more tender and delicious. It is usually cut into pieces, boiled, and served as an auxiliary vegetable with meat. To obtain this small portion, borne on the pinnacle of the tree, and hidden from the eye of man, the axe is applied to the stately trunk, and its towering pride laid low.

Besides its cabbage, the Oreodoxa furnishes another great delicacy to the table. After the removal of the heart, a kind of black beetle deposits its eggs in the cavity, from which fat grubs are developed, growing to the size and thickness of a man's thumb. These, though disgusting in appearance, when fried in a pan, with a very little butter and salt, have a taste which savours of all the spices of India.

Both the Oreodoxa and the Ceroxylon are far surpassed in height by the Californian firs and the Eucalypti of Australia, but no other trees rise so proudly in the air on shafts comparatively so slender. While the enormous trunks of the Sequoias and Wellingtonias remind one of the massy pillars of our old gothic churches, or of the ruins of Thebes, the graceful palms recall to our memory the slender Ionic or Corinthian columns which adorn the masterpieces of Grecian architecture.

The oil of the Corozo (*Elæis oleifera*) is usually burnt in the houses and churches of Carthagena and New Granada; and the *Oenocarpus disticha* is cultivated in Brazil, as it furnishes an excellent oil for culinary purposes.

The Pirijao (*Gulielma speciosa*) is planted round the huts of the Indians, and replaces in some districts the Mauritia,* as the tree of life.

The Piaçava (*Attalia funifera*), whose stone-hard darkbrown nuts are manufactured into rosaries by the inhabitants of Villa Nova de Olivenza, is far more important, on account of its fibres, which, unknown a few years ago, are now imported into England in large quantities, where they serve for making brooms; and the amazingly hard nuts of the Cabeza di Negro (*Phytelephas*), rivalling ivory in whiteness, solidity, and beauty, are extensively used by our turners for similar purposes.

To these the names of many other useful palms might be added, for there is scarce a member of this widely-extended family which has not some valuable quality; but not to fatigue

* See Chapter II.

the reader's patience, I will now conclude the chapter with a few general remarks.

Until Linnæus's death, but fifteen species of palms were known; while at present more than 440 have been methodically described, and every new voyage of discovery in the tropical zone reveals new forms to the botanist.

The difficulty of collecting palm-blossoms for the purpose of ascertaining the species is far greater than any one, judging merely from the vegetation of Europe, would imagine. Most palms bloom but once a year — in January and February — in the central equatorial regions. Many remain in flower only a few days, and frequently over spaces of many thousand square miles there are but three or four species, so that the botanist would have to be ubiquitous in order to become acquainted with the palms of a continent. Add to this the difficulty of gathering blossoms growing in dense forests or on swampy banks, and often hanging from the top of shafts sixty feet high, and armed with formidable thorns! In the missions of Guiana, the Indians are so barbarous, so stoically indifferent, and their wants so few, that no inducement can tempt them to move a foot from their path, to the great vexation of the naturalist, who sees these same people climb the bush-ropes, and ascend the highest trees, with cat-like agility, whenever the caprice of the moment prompts them to do so.

During Humboldt's stay in the Havana, the Palma Real (*Oreodoxa regia*) was in full bloom about the town, and, though the great naturalist offered two dollars for a single bunch of the snow-white flowers, he was for several days unable to obtain a single specimen; for in a tropical country man never works, unless compelled by absolute necessity.

A knowledge of these difficulties serves to show in a brighter light the indefatigable perseverance of the naturalists who have carried the torch of science into the dark shades of the primeval forest.

Though no trees are more characteristic of the tropics than the palms, yet specimens are found far within the temperate regions.

Along with the date-tree the *Chamærops humilis* graces the environs of Nizza, while in America the *Chamærops palmette* reaches 34° N. lat. In Australia, the *Corypha australis* grows under 34° S. lat. and, thanks to its mild insular climate, New Zealand boasts of the *Areca sapida* in the still higher latitude of 38°. In Africa, the *Hyphæne coriacea* penetrates to the south as far as Port Natal (30° S. lat.), while in America the palms extend to 35° S. lat., both in the Pampas of Buenos Ayres, and westward of the colossal barrier of the Andes, where the Choco, the only Chilian palm, indicates the extreme limits of the family.

As these palms of the temperate regions are able to exist under a mean annual temperature of 58°, they might possibly be made to adorn the gardens of Penzance. Most palms, however, require a mean temperature of from 70° to 72°, and on advancing towards the Equator they are found to increase in beauty, stateliness of growth, and variety of form. Their chief seats are the lower regions of the torrid zone; but as some species range far to the North or South, thus others ascend the mountain-slopes, almost to the limits of perpetual snow.

In South America, the *Ceroxylon andicola*, the palmetto of Azufral, and the *Kunthia montana* are found growing on heights of 6,000 and 9,000 feet, where the thermometer often falls during the night to 43° and 47°; and in the Paramo de Guanacos, Humboldt even saw palms growing at an elevation of 13,000 feet above the level of the sea.

Besides the height of the shaft, the position of the leaves serves chiefly to impart a more or less majestic character to the palms: those with drooping leaves being far less stately than those whose fronds shoot more or less upwards to the skies. Nothing can exceed the elegance of the Jagua palm, which along with the splendid Cucurito adorns the granite-rocks in the cataracts of the Orinoco at Atures and Maypures. The fronds, which are but few in number, rise almost perpendicularly sixteen feet high, from the top of the lofty columnar shaft, and their feathery leaflets of a thin and grass-like texture play lightly round the tall leaf-stalks, slowly bending in the breeze. The physiognomy of the palms depends also upon the various character of their efflorescence. The spathe is seldom vertical, with erect fruits; generally it hangs downwards, sometimes smooth, frequently armed with large thorns.

In the palms with a feathery foliage, the leaf-stalks rise either immediately from the brown rugged ligneous trunk (cocoa-nut, date), or, as in the beautiful Palma Real of the Havana, from a smooth, slender, and grass-green shaft, placed like an additional column upon the dark-coloured trunk. In the fan-palms, the crown frequently rests upon a layer of dried leaves, which impart a severe character to the tree.

The form of the trunk also varies greatly, sometimes almost entirely disappearing, as in *Chamærops humilis*; sometimes, as in the Calami, assuming a bush-rope appearance, smooth or rugged, unarmed or bristling with spines.

In the American Yriarteas, the trunk, as in the mangroves, and many of the screw-pines, rests upon a number of roots rising above the ground. Thus the *Y. exorrhiza*, which grows on the banks of the Amazon to the height of a hundred feet, frequently stands upon a dozen or more supports, embracing a circumference of twenty feet, and the trunk begins only six or eight feet from the ground. The *Yriartea ventricosa* is still more curious, as the spindle-shaped trunk, which at the top and at the bottom is scarce a foot thick, swells in the middle to a threefold diameter, and, from its convenient form, is frequently used by the Indians for the construction of their canoes.

Yriartea ventricosa.

The form and colour of the fruits is also extremely various. What a difference between the large coco de mer and the date—between the egg-shaped fruits of the Mauritia, whose scaly dark rind gives them the appearance of fir-cones, and the gold and purple peaches of the Pirijao, hanging in colossal clusters of sixty or eighty from the summit of the majestic trunk.

Notwithstanding the fecundity of the palms, generally but few individuals of each species are found growing wild, partly in consequence of the frequent abortive developement of the fruits, but chiefly on account of the large number of animals — from the grub to the monkey — that are constantly feeding upon them.

When we consider the enormous range of territory over which the palm-trees extend, and how very few of their many hundred species have hitherto been multiplied and improved by cultivation, we cannot doubt that many benefits are yet to be expected from this noble family of plants.

Many a palm possesses virtues hitherto unknown to man, and, though now neglected, is destined to hold an important rank in the future commercial annals of the world.

Areca Palm.

The Banana and the Plantain.

CHAPTER XIII.

THE CHIEF NUTRITIVE PLANTS OF THE TORRID ZONE.

Rice — Various Aspect of the Rice-fields at different Seasons — Ladang and Sawa Rice — The Cultivation of Rice in South Carolina of modern Date — The Rice-bird — Great Mortality among the Negroes — Arracan and Pegu — Growing Importance of the Port of Akyab — Maize — First imported from America by Columbus — Its enormous Productiveness — Its Cultivation in the United States — Its wide zone of Cultivation — Maize-beer, or Chicha — Millet, Dhourra — The Bread Fruit — Its Importance in the South Sea Islands — History of its Transplantation to the West Indies — Adventures of Bligh and Christian — Pitcairn Island — Bananas — Their ancient Cultivation — Avaca or Manilla Hemp — Humboldt's Remarks on the Banana — The Traveller's Tree of Madagascar — The Cassava Root — Tapioca — Yams — Batatas — Quinoa — Arrowroot — Taro — Tropical Fruit Trees -- The Chirimoya — The Litchi — The Mangosteen — The Mango.

OF all the cereals there is none that affords food to so vast a multitude as the rice-plant (*Oryza sativa*), on whose grains from time immemorial the countless millions of south-eastern Asia chiefly subsist.

From its primitive seat, on the Ganges or the Sikiang, its cultivation has gradually spread not only over the whole tropical zone, but even far beyond its bounds, as it thrives both in the swamps of South Carolina and in the rich alluvial plains of the Danube and the Po.

Along the low river banks, in the delta-lands which the rains of the tropics annually change into a boundless lake, or where, by artificial embankments, the waters of the mountain streams have been collected into tanks for irrigation, the rice-plant attains its utmost luxuriance of growth, and but rarely deceives the hopes of the husbandman.

The aspect of the lowland rice-fields of India and its isles is very different at various seasons of the year. Where, in Java, for instance, you see to-day long-legged herons gravely stalking over the inundated plain partitioned by small dykes, or a yoke of indolent buffaloes slowly wading through the mud, you will three or four months later be charmed by the view of a gracefully undulating corn-field, bearing a great resemblance to our indigenous barley. Cords, to which scare-crows are attached, traverse the field in every direction, and converge to a small watch-house, erected on high poles. Here the attentive villager sits, like a spider in the centre of its web, and by pulling the cords, puts them from time to time into motion, whenever the wind is unwilling to undertake the office. Then the grotesque and noisy figures begin to rustle and to caper, and whole flocks of the neat little rice-bird or Java sparrow (*Loxia oryzivora*), rise on the wing, and hurry off with all the haste of guilty fright. After another month has elapsed, and the waters have

Java Sparrow.

long since evaporated or been withdrawn, the harvest takes place, and the rice-fields are enlivened by a motley crowd, for all the villagers, old and young, are busy reaping the golden ears.

The rice-fields offer a peculiarly charming picture when, as in the mountain valleys of Ceylon, they rise in terraces along the slopes. "Selecting an angular recess where two hills converge, the Kandyans construct a series of terraces, raised stage above stage, and retiring as they ascend along the slope of the acclivity, up which they are carried as high as the soil extends. Each terrace is furnished with a low ledge in front, behind which the requisite depth of water is retained during the germination of the seed, and what is superfluous is permitted to trickle down to the one below it. In order to carry on this peculiar cultivation the streams are led along the level of the hills, often from a distance of many miles, with a skill and perseverance for which the natives of these mountains have attained a great renown."

The prodigious embankments constructed in ancient times for the purposes of irrigation may be reckoned among the wonders of the island.

"Many of the tanks, though partially in ruins, cover an area from ten to fifteen miles in circumference. They are now generally broken and decayed, the waters which would fertilise a province are allowed to waste themselves in the sands, and hundreds of square miles, capable of furnishing food for all the inhabitants of Ceylon, are abandoned to solitude and malaria, whilst rice for the support of the non-agricultural population is annually imported from the opposite coast of India." *

It will be the duty and glory of England to restore these monuments of ancient greatness, and by raising one of her fairest colonies to her former affluence, contribute at the same time to her own wealth.

Rice does not invariably require the marsh or the irrigated terrace for its growth, as there is a variety which thrives on the slopes of hills, where it is not continuously watered. In the mountain regions of Sumatra, rains fall at almost every season of the year, though dry weather is more frequent from April to July. In August, the rainy days are as three to one, and this is the time generally chosen for the sowing of the *Ladang*, or mountain rice. After the harvest, the field is sown a second time with maize; it then lies fallow for a few years, and is soon covered with a thick vegetation of wild shrubbery, generally with glagah, a species of grass which attains a height of twelve feet.

When the field is again to be cultivated, fire is resorted to to destroy the dense jungle, in which the tiger has made his lair, or where the rhinoceros grazes. At night, these fires, ascending the slopes of the mountains, present a fine sight;

---

* Sir Emerson Tennent's Ceylon, vol. i. p. 26–27.

during the day time, they cover the land with a dun mist, and veil the prospect, like the *Höhrauch*, which (caused by the firing of the heaths of Oldenburg and Hanover) so often spoils the beauty of the spring and early summer in northern Germany.

The rapidity with which the dry culms of the glagah take fire is not seldom dangerous to the traveller when his path leads him across the slope of a hill at whose foot the grass-field begins to burn, for the rustling fire-columns ascend with the swiftness of the wind, and soon wrap the side of the mountain in a sheet of flame.

The ashes of the glagah afford the richest manure, so that these fields are only surpassed in fertility by the virgin soil of the cleared forest, a laborious work, which is seldom undertaken in this thinly-populated country.

Sawa is the general Malay name for artificially-irrigated rice-fields. In the Indian Archipelago, the Sawa, or marsh-rice, is at first thickly sown in small beds, and transplanted after a fortnight into the fields, the soil of which has been softened by water. As the plant grows, copious irrigations supply it with the necessary moisture, but as maturity approaches the field is laid dry, and about two months later the ears assume the rich golden colour so pleasing to the husbandman. Each field could easily be made to produce two annual harvests, but when not compelled to labour, the tropical peasant never thinks of taxing his industry beyond the supply of his immediate wants.

The swamps of South Carolina, both those which are occasioned by the periodical visits of the tides, and those which are caused by the overflowing of the rivers, are admirably adapted to the production of rice, yet the culture of the valuable cereal on this congenial soil is of comparatively modern date. About the beginning of the last century, a brigantine from the island of Madagascar happened to put in at Carolina, having a little seed-rice left, which the captain gave to a gentleman of the name of Woodward. From part of this, the latter had a very good crop, but was ignorant for some years how to clean it. It was soon dispersed over the province, and, by frequent experiments and observations, the planters ultimately raised the culture to its present perfection. By the introduction of this water-loving cereal, various swamps which previously had only afforded food to frogs and water-birds, have been changed into the most fruitful fields, so that South Carolina, before her ports were blockaded, not merely supplied the whole of the United States with all the rice they require, but also annually exported more than a hundred thousand large casks to the various markets of Europe.

Unfortunately, the cultivation of the rice-plant is very unhealthy, and the swamp-fevers, engendered by the alternate flooding and drying of the land, annually carry away so many of the slaves engaged in the labour, that the natural increase of the population does not suffice to cover the loss, which can only be repaired by continually importing fresh negroes from the more northern slave-states — a scandalous traffic, the disgrace and shame of what was once the Union.

Besides the devastations which the atmosphere of the rice-fields causes among his labourers, the planter frequently suffers heavy losses in consequence of the depredations of the rice-bunting (*Dolichonyx oryzivorus*), a species of ortolan, known familiarly in the country by the name of Bob Lincoln. This bird is about six or seven inches long; its head and the under part of its body are black, the upper part is a mixture of black, white, and yellow, and the legs are red.

Rice-bunting.

It migrates over the continent of America from Labrador to Mexico, and over the Great Antilles, appearing in the southern extremity of the States about the end of March. During the three weeks to which its unwelcome visit to the rice-fields is usually limited, it grows so fat upon the milky grains of its favourite cereal, that its flesh becomes little inferior in flavour to that of the European ortolan. As long as the female is sitting, the song of the male continues with little interruption: it is singular and pleasant, consisting of a jingling medley of short, variable notes, confused, rapid, and continuous.

Besides Carolina — Brazil, Java, Bengal, and, more recently, Arracan and Pegu, chiefly supply the European market with rice. As long as the two last-named provinces groaned under the yoke of his gold-footed majesty of Birmah, they were totally unknown to the commerce of the world; but scarce had they been annexed to the British empire, and security of life and property restored to their inhabitants, when they began to flourish with a rapidity like that of their own luxuriant vegetation.

Who ever heard, a quarter of a century ago, of the magnificent port of Akyab, which now harbours whole fleets as they arrive from all parts of the world—from the east and from the west, from China and Europe—to gather the inexhaustible supplies of the Arracan rice-fields, and convey them to their distant homes?

The exportation of Akyab rice amounted in

1847 to	70,537 tons
1853 ,,	99,487 ,,
1854 ,,	103,120 ,,
1855 ,,	165,047 ,,

and is still continually increasing.

More than half of this enormous mass finds its way to England, while the rest is divided, in a diminishing ratio, between Holland, Bremen, Hamburg, Sweden and Norway, America, and Belgium, which imports directly only about 4,000 tons, but receives indirectly a much larger quantity from London. Most of the Arracan rice is exported in the unshelled state, or as *paddy*, and cleaned in Europe, where the operation can be more effectually and cheaply performed than in the country of production. The loss by waste is also found to be less on the transport of *paddy* than of shelled rice.

Thus, new branches of industry are constantly shooting forth, through the growing intercourse of nations; and every new victory of civilisation over barbarism, even at the very extremity of the globe, reacts beneficially upon the trading interests of Europe.

Maize is no less important to the rapidly-growing nations of America than the rice-plant to the followers of Buddh or of Brama; and when hereafter the banks of the Mississippi, of the Amazon, and of the Orinoco, shall be covered by as dense a population as the plains of Bengal, the number of maize-eaters will probably be greater than that of the consumers of any other species of grain.

The time when the cereals of the old world — wheat, rye, barley — were first transplanted from their unknown Asiatic homes to our part of the world is, and ever will be, hidden in legendary obscurity; but the epoch when maize was for the first time seen and tasted by Europeans lies before us in the broad daylight of authentic history. For, when Columbus discovered Cuba, in the year 1492, he found maize cultivated by the Indians, and was equally pleased with the taste of the roasted grains and astonished at their size. In the following year, when he made his triumphant entry into Barcelona, and presented his royal patrons — Ferdinand and Isabella — with specimens of the various productions of the New World, the maize-spikes he laid down before their throne, though but little noticed, were in reality of far greater importance than the heaps of gold which were so falsely deemed to be the richest prizes of his grand discovery.

In this manner maize, which is found growing wild from the Rocky Mountains to Paraguay, and had been cultivated from time immemorial, as well in the Antilles as in the dominions of the Mexican Aztecs and of the Peruvian Incas, was first conveyed from the New World to Spain, whence its cultivation gradually extended over the tropical and temperate zone of the eastern hemisphere. Round the whole basin of the Mediterranean, maize has found a new home, and its grain now nourishes the Lombard and the Hungarian, as it does the Egyptian fellah or the Syrian peasant.

While our northern cereals only produce a pleasing effect when covering extensive fields, but are individually too insignificant to claim attention, the maize-plant almost reminds the spectator of the lofty Bambusaceæ of the tropical world. Even in our gardens it rises above man's height, and in warmer countries not seldom attains the gigantic stature of fourteen feet. Ensiform, dark green, lustrous leaves, somewhat resembling those of the large Oarweeds, or Laminariæ, of the northern seas, spring alternately from every joint of this cereal, streaming like pennants in the wind. The top produces a bunch of male flowers of various colours, which is called the *tassel*. Each plant likewise bears one or more spikes or ears, the usual number being three, though as many as seven have been seen occasionally on one stalk. These ears proceed from the stem, at various distances from the ground, and are closely enveloped by several thin leaves, forming a sheath, or *husk*. They consist of a cylindrical substance of the nature of pith, which is called the *cobb*, and over the entire surface of which the seeds are ranged and fixed, in eight or more straight rows. Each of these has generally as many as thirty or more seeds, and each seed weighs at least as much as five or six grains of wheat or barley. Surely a cereal like this deserves beyond all others to symbolise abundance, and, had it been known to the Greeks, it would beyond all doubt have figured conspicuously in the teeming horn of Amalthea.

While the British farmer is satisfied with an increase of twenty for one, the productiveness of maize, under the circumstances most favourable to its growth, is such as almost to surpass belief. In the low and sultry districts of Mexico, it is quite a common thing, in situations where artificial irrigation is practised, to gather from 350 to 400 measures of grain for every one measure that has been sown; and some particularly favoured spots have even been known to yield the incredible increase of 800. In other situations, where reliance is placed only on the natural supply of moisture to the soil from the periodical rains, such an abundant supply is not expected; but even then, and in the least fertile spots, it is rare for the cultivator to realise less than from forty to sixty bushels for each one sown.

When I add that in some of the warm and humid regions of Mexico three such harvests might annually be raised, though it is not usual to take more than one, an idea may be formed of the amazing capabilities and of the wide prospects of a land which now lies prostrate under the triple scourge of anarchy, ignorance, and fanaticism.

The productiveness of maize diminishes in the more temperate climate of the now no longer United States; but even in Pennsylvania, almost on the northern limits of its zone, it still yields double the increase of wheat; and such is the quantity annually grown in the once great republic, that, in spite of its low price, the value of the maize-harvest more than twice surpasses that of all the other cereals.*

---

* In the year 1853 the United States produced —
    6,500,000 bushels of barley, valued at 4,815,000 dollars
    14,000,000 „ rye „ 12,600,000 „
    110,000,000 „ wheat „ 100,000,000 „
    600,000,000 „ maize „ 240,000,000 „

Another great advantage attending the cultivation of maize is, that of all the cereals it is the least subject to disease. Blight, mildew, or rust are unknown to it. . It is never liable to be beaten down by rain, or by the most violent storms of wind, and in climates and seasons which are favourable to its growth, the only enemies which the maize-farmer has to dread are insects in the early stages, and birds in the later periods, of its cultivation. In mountainous countries, and the farther it advances beyond the tropics, maize — a child of the sun — naturally suffers from the ungenial influence of a cold and wet summer, which not only prevents the ripening of the grain, but also developes a poisonous ergot in its ears, similar to that which an inclement sky is apt to engender in our rye.

Through the length and breadth of the field which he destines for the raising of a maize-crop, the American farmer draws his furrows three or four feet apart, and thus divides it in small regular squares. At every period of intersection, he sows three or four grains and covers them about three inches deep with earth. A few weeks after the tender light-green shoots have sprouted forth, and once more at a later period, he drives a small plough between their rows, for the purpose of loosening the soil around the roots and of cutting up the weeds. At the same time he takes care to remove the suckers that spring from the bottom of the plant, not only as they draw away part of the nourishment which should go to support the main stalk, but because the ears, which the suckers bear, ripen at later periods than the others, and the harvest could not all be simultaneously secured in the fittest state of maturity. In order to admit the sun as much as possible to the plant, and to afford more nutriment to the grain, he also usually removes the blades, together with the top and tassel, as soon as its office of dropping its fecundating farina upon the ears has been fully accomplished, which is known by its putting on a withered appearance. In the northern states maize is not sown before May, to prevent the blighting influence of the night-frosts, and the harvest takes place in October, a few weeks later than that of wheat—a great advantage to the farmer, who, in a country where wages are so high, would otherwise find it difficult and costly to garner the abundance of his fields. The stripped stalks are allowed to stand some time longer, after which they are cut off close to the ground, and carefully piled up under a shed, as, from their saccharine quality, they are an excellent food for horned cattle. Thus the field, which in summer emulated the luxuriance of tropical vegetation, now lies shorn of all its beauty, the northern blasts soon begin to howl over the naked waste, and a thick cover of snow bears testimony to the advance of winter.

In the southern states the periods of sowing and harvesting are naturally earlier, on advancing farther to the tropic, the former taking place in Texas in February, the latter in June.

The ears are preserved in bins or cribs, either with or without the husk, and it is not considered good farming to shell the corn before it is required to be sent to market. This operation is very easily performed. The only implement required for the purpose is a blunt piece of iron, like a sword-blade; when this has been fixed across the top of a tub, the ear is taken in both hands and scraped lengthwise across its edge until all the grains are removed. In this manner an industrious man will shell from twenty to twenty-five bushels of corn in the course of the day. The cobb which remains makes a very tolerable quick-burning fuel, and thus no part of the plant proves altogether without use.

The grain forms one-half the measure of the ear; and so correct is this estimate found to be, that in the markets of the United States, where Indian corn is sold both shelled and with the cobb, two bushels of the latter are taken without question by the purchasers as being equal to one bushel of shelled grain.

When we consider that the zone of cultivation of the maize-plant extends from 49° N. lat. to 40 S. lat., and that it grows as well in the low countries of the equatorial regions as on the islands of Lake Titicaca, 12,000 feet above the level of the sea, it is not to be wondered at that there are numerous varieties, from the gigantic tlaouili of the Mexicans, which absolutely requires a hot sun, and bears ears ten inches in length and five or six inches in circumference, and the small variety with both yellow and white seeds, and ears four or five inches long, which in ordinary seasons will ripen its grain, even under the variable and weeping sky of England!

The various uses to which the maize-plant and grain may be applied cannot be better enumerated than in the words of the celebrated Dr. Franklin.

"It is remarked in North America that the English farmers, when they first arrived there, finding a soil and climate proper for the husbandry they have been accustomed to, and particularly suitable for raising wheat, they despise and neglect the culture of maize or Indian corn; but, observing the advantage it affords their neighbours, the older inhabitants, they by degrees get more and more into the practice of raising it, and the face of the country shows from time to time that the culture of that grain goes on visibly augmenting.

"The inducements are the many different ways in which it may be prepared so as to afford a wholesome and pleasing nourishment to men and other animals. First, the family can begin to make use of it before the time of full harvest: for the tender green ears, stripped of their leaves and roasted by a quick fire till the grain is brown, and eaten with a little salt or butter, are a delicacy. Secondly, when the grain is riper and harder, the ears boiled in their leaves and eaten with butter are also good and agreeable food. The tender green grain dried may be kept all the year, and, mixed with green kidney beans, also dried, make at any time a pleasing dish, being first soaked some hours in water and then boiled. When the grain is ripe and hard there are also several ways of using it. One is to soak it all night in a *lessive* or lye, and then pound it in a large wooden mortar with a wooden pestle; the skin of each grain is by that means skinned off, and the farinaceous part left whole, which, being boiled, swells into a white soft pulp, and, eaten with milk or with butter and sugar, is delicious. The dry grain is also sometimes ground loosely so as to be broken into pieces of the size of rice, and, being winnowed to separate the bran, it is then boiled and eaten with turkies or other fowls as rice. Ground into a finer meal, they make of it by boiling a hasty-pudding or bouilli, to be eaten with milk or with butter and sugar, that resembles what the Italians call polenta. They make of the same meal with water and salt a hasty-cake, which, being stuck against a hoe or other flat-iron, is placed erect before the fire, and so baked, to be used as bread. They also parch it in this manner. An iron pot is filled with sand, and set on the fire till the sand is very hot. Two or three pounds of the grain are then thrown in, and well mixed with the sand by stirring. Each grain bursts and throws out a white substance of twice its bigness. The sand is separated by a wire sieve, and returned into the pot to be again heated, and repeat the operation with fresh grain. That which is parched is pounded to a powder in a mortar. This being sifted will keep long for use. An Indian will travel far and subsist long on a small bag of it, taking only six or eight ounces of it per day mixed with water. The flour of maize mixed with that of wheat makes excellent bread, sweeter and more agreeable than that of wheat alone. To feed horses it is good to soak the grain twelve hours: they mash it easier with their teeth, and it yields them more nourishment. The leaves stripped off the stalks after the grain is ripe, tied up in bundles when dry, are excellent forage for horses, cows, &c. The stalks, pressed like sugar-cane, yield a sweet juice, which, being fermented and distilled, yields an excellent spirit; boiled without fermentation it affords an excellent syrup. In Mexico fields are sown with it thick that multitudes of small stalks may arise, which, being cut from time to time, like asparagus, are served in desserts, their thin sweet juice being extracted in the mouth by chewing them. The meal wetted is excellent food for young chickens and the old grain for grown fowls."

In addition to the many uses enumerated by Franklin, the Indians in Peru and Mexico prepare a kind of beer, by first moistening the grains of maize to promote germination, then drying them in the oven, pounding them, and finally boiling them in water and causing the decoction to ferment. The somewhat sharp and bitter taste of the chicha or maize-beer is generally displeasing to strangers, but the Peruvians are all of them determined chicheros, and assemble every evening in their filthy taverns for the enjoyment of this detestable beverage. On dirty plates, that generally remain unwashed from the time they first were used, their favourite *picantes* or stews, well seasoned with Cayenne, are served to these unsophisticated guests. Instead of bread, a portion of roasted maize is either set before them in small pumpkin-bowls, or cast without ceremony upon the greasy table. The chicha itself is presented in large glasses, on whose unclean rim each consumer is able to see the marks of his predecessor's lips.

As maize contains but very little gluten, some theorists have been inclined to set but a small value on its nutritive power, but this is an error which is sufficiently refuted by facts. Domestic animals which are fed with maize very speedily fatten, their flesh being at the same time remarkably firm, and horses which consume this corn are enabled to perform their full portion of labour. The Indian miners are perhaps the most hard-working people in existence, carrying large weights of ore on steep ladders several hundred feet high, and yet they hardly taste anything else but maize. Chemistry, moreover, teaches us that this species of grain contains nearly as much albumen as the best wheat, and is consequently by no means deficient in nourishing azote.

Dr. Moritz Wagner* is inclined to ascribe the great mortality among the children in Costa-Rica, and the frequent dyspeptic complaints, as well as the moral apathy of the adult population, to their almost exclusively living on maize; but these evils may far more justly be ascribed to the indolence of that wretched people, who could so easily raise cattle with their abundance of maize, and thus procure the variety of food which, as experience teaches, is most conducive to the health and vigour of man.

In light sandy soils, under the scorching rays of the sun, and in situations where sufficient moisture cannot be obtained for the production of rice, numerous varieties of millet (*Sorghum vulgare*) are successfully cultivated in many tropical countries—in India, Arabia, the West Indies, in Central Africa, and in Nubia, where it is grown almost to the exclusion of every other esculent plant. Though the seeds are by much the smallest of any of the cereal plants, the number borne upon each stalk is so great as to counterbalance this disadvantage, and to render the cultivation of millet as productive as that of any other grain.

As the Peruvians make a kind of beer from maize, thus the Nubians prepare a fermented liquor from millet or *dhourra*, which they consider very wholesome and nutritious, though intemperance will render it as hurtful as any other intoxicating drink.

The bread-fruit tree is the great gift of Providence to the fairest isles of Polynesia. No fruit or forest tree in the north of Europe, with the exception of the oak or linden, is its equal

---

* Travels in Central America.

in regularity of growth and comeliness of shape : it far surpasses the wild chestnut, which somewhat resembles it in appearance. Its large oblong leaves, frequently a foot and a half long, are deeply lobed like those of the fig tree, which they resemble not only in colour and consistence, but also in exuding a milky juice when broken. About the time when the sun, advancing towards the Tropic of Capricorn, announces to the Tahitians that summer is approaching, it begins to produce new leaves and young fruits, which commence ripening in October, and may be plucked about eight months long in luxuriant succession. The fruit is about the size and shape of a new-born infant's head; and the surface is reticulated, not much unlike a truffle; it is covered with a thin skin, and has a core about as big as the handle of a small knife. The eatable part lies between the skin and the core; it is as white as snow, and somewhat of the consistence of new bread; it must be roasted before it is eaten, being first divided into three or four parts; its taste is insipid, with a slight sweetness, somewhat resembling that of the crumb of wheaten bread mixed with boiled and mealy potatoes.

When the season draws to an end, the last fruits are gathered just before they are perfectly ripe, and, being laid in heaps, are closely covered with leaves. In this state they undergo a fermentation and become disagreeably sweet; the core is then taken out entire, which is done by gently pulling out the stalk, and the rest of the fruit is thrown into a hole, which is dug for that purpose, generally in the house, and neatly lined in the bottom and sides with grass; the whole is then covered with leaves, and heavy stones laid upon them; in this state it undergoes a second fermentation, and becomes sour, after which it will suffer no change for many months.

It is taken out of the hole as it is wanted for use, and, being made into balls, it is wrapped up in leaves and baked: after it is dressed it will keep five or six weeks. It is eaten both cold and hot, and the natives seldom make a meal without it, though to Europeans the taste is as disagreeable as that of a pickled olive generally is the first time it is eaten. The fruit itself is in season eight months in the year, and the mahei or sour paste formed in the manner above described fills up the remaining cycle of the year.

To procure this principal article of their food costs the fortunate

South Sea Islanders no more trouble than plucking and preparing it in the manner above described; for, though the tree which produces it does not grow spontaneously, yet, if a man plants but ten of them in his lifetime, which he may do in about an hour, he will, as Cook remarks, "as completely fulfil his duty to his own and future generations, as the native of our less genial climate by ploughing in the cold of winter and reaping in the summer's heat as often as the seasons return."

The bread-fruit tree (*Artocarpus incisa*) is useful to the South Sea Islanders in several other respects, as they make a kind of cloth of its bark or inner rind, and its soft light wood is particularly serviceable for the building of their canoes.

Though it has a far extended range over the islands and coasts of the Indian and Pacific Ocean, yet its importance as an article of food is chiefly confined to the Tahitian, Friendly, Samoan, Fiji, and Marquesan groups, while in the Indian Archipelago it is either neglected or only used for fuel.

The celebrated navigator Dampier (1688) is the first English writer that mentions the bread-fruit tree, which he found growing in the Ladrones, and a few years later Lord Anson enjoyed its fruits at Tinian, where they contributed to save the lives of his emaciated and scurvy-stricken followers. It continued, however, to remain unnoticed in Europe, until the voyages of Wallis and Cook attracted the attention of the whole civilised world to the fortunate islands, whose inhabitants, instead of gaining their bread by the sweat of their brow, plucked it ready formed from the teeming branches of their groves.

What could be more natural than the desire of an enlightened government to transfer a similar boon to the tropical countries subject to its rule; and thus Lieutenant Bligh, who had accompanied Cook in his last voyage, was sent in the " Bounty "— an excellent name for a vessel destined for such a mission — to transport a number of young bread-fruit trees from Tahiti to the West Indies.

On October 26, 1788, Bligh landed at Tahiti, and was cordially welcomed by the inhabitants, to whom he brought the citron and the orange as a return for their valuable artocarpus.

Between the roomy decks, where the open port-holes admitted a constant current of fresh air, more than 1000 bread-tree saplings were carefully planted in tubs, and in the following April Bligh left the island with his floating garden, perhaps the most remarkable which the world had ever seen.

Impelled by the mild trade-wind, the "Bounty" sped her way through the tropical ocean, but the peace which reigned over the face of nature dwelt not in the hearts of her ill-fated crew. On April 18 a mutinous band, headed by Christian, one of the mates, burst into the captain's cabin, seized him, bound his hands with a cord behind his back, and forced him, along with eighteen faithful adherents, into a boat prepared for their reception. Thirty-two pounds of bacon, 150 pounds of bread, an eight and twenty gallon cask of water, four cutlasses, a small quantity of rum and wine, a quadrant, a compass, and some canvas and cordage, were handed down to them, and thus they were cast adrift, exposed without a cover to the scorching rays of the sun and all the changes of a stormy ocean.

This strange scene took place about thirty miles from Tofoa, one of the islands which Cook had named the "Friendly," but which now did but little honour to their name; for on Bligh's attempting to land for the purpose of obtaining a supply of bread-fruit and water, one of his men was killed by the treacherous savages, and the rest, having no firearms for their defence, with difficulty escaped.

The last hope of the unfortunate outcasts now consisted in reaching the Dutch colony of Timor, which, although at a distance of 3600 miles, was still the nearest European settlement — a terrible journey in a small heavily-laden boat, whose gunwale rose scarcely six inches above the surface of the sea. Fortunately obedience and discipline reigned in that little crew, and after each man had solemnly promised the captain to be satisfied with an ounce of bread and a quarter of a pint of water daily — for their scanty provisions were to last them at least six weeks — they began to steer towards the west.

The boat was so crowded, that while one half of the crew was sitting and performing its duty, the other half was obliged to lie down on the bottom, where, unable to stretch their limbs, now scorched by the sun, and now exposed in wet clothes to the cold night wind, they were soon tormented with pinching cramps and rheumatic pains.

On the fifth day of their journey they saw two large canoes

of Fiji Islanders paddling in all haste towards them. Without arms for their defence, they must have fallen into the hands of these merciless cannibals, who had already approached within two miles, when suddenly from some inexplicable caprice the savages abandoned the pursuit, and were soon lost in the distance.

After May 10 the weather became stormy, so that every moment the waves struck into the boat and forced the exhausted crew continually to bale out the water. To increase their distress, the brine spoiled their small supply of bread, so that the rations had to be reduced to one twenty-fifth part of a pound, morning and noon. On May 25 a couple of sea-birds of the size of pigeons approached so near that they were caught with the hand. Each bird was divided into eighteen parts, and eaten raw with all the keen relish of voracious hunger. On the morning of May 29 they found themselves close to a coral-bank, against which the breakers were striking with furious uproar; but fortunately they discovered a passage in the reef, through which they safely reached the quiet waters of a lagune. They landed on a small island, and we can easily imagine the delight with which, after more than five weeks of cramped confinement, they once more stretched out their stiffened limbs on the earth. No fruit trees graced the island, but they found sweet water and oysters, more delicious than they had ever tasted. Fortunately, also, they had a copper kettle on board; a fire was lighted, and a warm soup made of bread, oysters, and bacon, revived their exhausted bodies.

They would willingly have remained some time longer; but, seeing at a distance a troop of savages approaching with loud cries, and not relying much on their hospitality, they immediately embarked without putting their intentions to the test. They were now between the east coast of New Holland and the large barrier-reef, and, steering through the broad channel towards the north, they passed one small island after another. Wherever they saw groups of savages watching on the strand they sailed on as fast as they could, but where the screw-pine or the cocoa-nut raised its green canopy over a desert shore, they frequently landed and gathered oysters, or a small bean which Nelson, the botanist of the expedition, recommended as nutritious and healthy. The man of science had often been derided by his seafaring companions, but they now found that the knowledge of plants was not so useless as they had imagined.

On the evening of June 3 the small boat passed Endeavour Straits and once more entered the open sea, which greeted it with high winds and pelting showers.

On the morning of the 10th, after a sleepless night, the health of the crew exhibited alarming symptoms of exhaustion. An almost total prostration of strength, swollen legs, hollow and spectral features, a great inclination to sleep, and mental apathy, proclaimed that the vital powers were ready to sink under so many hardships and privations. A few teaspoonfuls of wine, reserved for the hour of utmost need, and still more the spiritual cordial of hope — for they knew that they were now not far distant from the goal — served to reanimate their energies, and on the morning of June 12, Timor, the longed-for island, lay before them.

What words could express their delight, their intense gratitude to Providence, which had safely guided them through such unnumbered perils, and so mercifully protected them that, with the single exception of the sailor killed by the Friendly Islanders, not one of them had hitherto succumbed to the hardships of that unparalleled voyage!

The way-worn mariners were most hospitably received by the Dutch, and Bligh returned safely to England by way of Batavia, carrying along with him the boat endeared to him by the remembrance of the most trying passage of his life.

As captain of the "Providence" he was once more sent out to realise the benevolent plan which Christian's mutiny had frustrated, reached Tahiti on April 9, 1792, took 1281 young bread-fruit trees on board, and in January 1793 safely arrived at St. Vincent, where he landed 333 trees, and thence sailed with the remainder to Jamaica.

In this manner, after having given rise to so many romantic adventures, the bread-fruit tree found its way from the distant islands of the South Sea to the tropical lands of the Atlantic, but without fulfilling the sanguine expectations that had been raised upon it; for it rapidly degenerated in its new home, and the negro spurns the fruit which in Tonga and Tahiti forms the chief sustenance of life.

But what had meanwhile been the lot of the mutineers?

how had they fared on the distant ocean, the scene of their crimes? It may easily be imagined that government could not allow so gross an outrage to discipline and law to remain unpunished, and thus the "Pandora" was immediately sent out to the South Sea in quest of Christian and his comrades. Fourteen of the rebel band were found in Tahiti, and brought in chains to England, where some of them were reprieved and some hung; others had already died of illness or a violent death; but of Christian, the leader of the mutiny, who, with eight of his followers and a number of male and female natives, had long since sailed away in the "Bounty," no trace was to be discovered, and many a year had still to pass before the secret of his fate was revealed.

In lonely majesty Pitcairn Island rises 1046 feet above the unfathomable depths of the South Sea. A furious surf encircles it with a girdle of white foam, which forms a beautiful contrast to the surrounding deep blue waters. There is but one single landing-place, and even this is only practicable in the calmest weather. The northern side of the island is extremely picturesque, rising from the sea as a steep amphitheatre, wooded to the very summit with palms, plantains, bread-fruit trees, and majestic banians, and bounded on either side by precipitous cliffs and naked rocks.

This was the place, still uninhabited by man, which Christian and his comrades chose for their abode: it was here they hoped to realise the dreams which had prompted them to crime. But in their own bosoms they carried along with them the fiend who made a hell out of this paradise. At first Christian succeeded in maintaining some authority over his associates; but, tormented by the constant fear of detection, he spent many a lonely hour on the summit of a cliff, casting anxious glances over the sea, above whose horizon the pennant of England could at any time arise for his destruction!

Soon after the arrival of the mutineers at Pitcairn the "Bounty" was committed to the flames, lest she should betray their residence; and thus they condemned themselves to a lasting exile on a lonely rock, 10,000 miles from their home, their families, or their friends. Soon a bloody feud arose between the mutineers and the Tahitians, who had embarked to share their fortunes, and before a year had passed Christian and four of his comrades were murdered by the latter, who in their turn were killed by the remaining four Europeans. One of these men discovered the art of distilling an intoxicating liquor from a root, and thus introduced a new element of discord among the remnants of the unhappy band. He died the victim of his own invention, by precipitating himself in a fit of delirium tremens into the sea; another of the four was shot in self-defence by Young and Adams, the only two that were destined to die of a natural death; and thus in this case also we see the ancient truth confirmed, that none but bitter fruits proceed from crime.

The wonderful luxuriance of tropical vegetation is perhaps nowhere more conspicuous and surprising than in the magnificent Musaceæ, the banana (*Musa sapientum*), and the plantain (*Musa paradisiaca*), whose fruits most probably nourished mankind long before the gifts of Ceres became known. A succulent shaft or stem, rising to the height of fifteen or twenty feet, and frequently two feet in diameter, is formed of the sheath-like leaf-stalks rolled one over the other, and terminating in enormous light-green and glossy blades, ten feet long and two feet broad, of so delicate a tissue that the slightest wind suffices to tear them transversely as far as the middle rib. A stout foot-stalk, arising from the centre of the leaves, and reclining over one side of the trunk, supports numerous clusters of flowers, and subsequently a great weight of several hundred fruits about the size and shape of full-grown cucumbers. On seeing the stately plant, one might suppose that many years had been required for its growth; and yet only eight or ten months were necessary for its full developement.

Each shaft produces its fruit but once, when it withers and dies; but new shoots spring forth from the root, and before the year has elapsed unfold themselves with the same luxuriance. Thus, without any other labour than now and then weeding the field, fruit follows upon fruit and harvest upon harvest. A single bunch of bananas often weighs from sixty to seventy pounds, and Humboldt has calculated that thirty-three pounds of wheat and ninety-nine pounds of potatoes require the same space of ground to grow upon as will produce 4000 pounds of bananas.

This prodigality of nature, seemingly so favourable to the human race, is however attended with great disadvantages; for where the life of man is rendered too easy, his best powers remain dormant, and he almost sinks to the level of the plant which affords him subsistence without labour. Exertion awakens our faculties as it increases our enjoyments, and well may we rejoice that wheat and not the banana ripens in our fields.

The esculent musaceæ are cultivated throughout the whole tropical zone in many varieties. The Hindoo and the Malay, the Negro and the South Sea Islander, enjoy their fruits, and the cool shade of their colossal leaves. In the Society Islands, the Marquesas, and the Sandwich group, a species of pisang, with upright fruit-stalks, grows wild in the mountains. In the deep ravines, protected from the wind, it forms thick groves, whose succulent foliage forms a picturesque contrast to the wild character of the surrounding highland vegetation.

As the seeds of the cultivated plantain and banana never or very rarely ripen, they can only be propagated by suckers. "In both hemispheres," says Humboldt, "as far as tradition or history reaches, we find plantains cultivated in the tropical zone. It is as certain that African slaves have introduced, in the course of centuries, varieties of the banana into America, as that before the discovery of Columbus the pisang was cultivated by the aboriginal Indians.

"These plants are the ornaments of humid countries. Like the farinaceous cereals of the north, they accompany man from the first infancy of his civilisation. Semitical traditions place their original home on the banks of the Euphrates; others, with greater probability, at the foot of the Himalaya. According to the Greek mythology, the plains of Enna were the fortunate birthplace of the cereals; but while the monotonous fields of the latter add but little to the beauty of the northern regions, the tropical husbandman multiplies by his pisang plantations one of the most magnificent forms of vegetable life."

The fruits of the musaceæ are used in various ways—boiled, roasted, raw, dried and converted into meal—so that the multiplicity of their preparations vies with that of rice or maize.

These beneficial plants are not only useful to man by their mealy, wholesome, and agreeable fruits, but also by the fibres of their long leaf-stalks. Some species furnish filaments for the finest musselin, and the coarse fibres of the *Musa textilis*, which are known in trade under the name of Avaca or Manilla hemp, serve for the preparation of very durable cordage.

To the same family belongs also the traveller-tree of Madagascar (*Ravenala speciosa*), one of those wonderful sources of refreshment which nature has provided for the thirsty wanderer in the wilderness. The foot-stalks of the elliptical, alternate leaves embrace the trunk with broad sheathes, in which the dew trickling from their surface is collected. The hollow baobab, the pitcher-plant, the juicy cactuses, the nara, all answer a similar purpose, and it is impossible to say which of them is most to be admired, or which most evinces the goodness and wisdom of the Creator.

Life and death are strangely blended in the Cassava or Mandioca root (*Jatropha Manihot*); the juice a rapidly destructive poison, the meal a nutritious and agreeable food, which, in tropical America, and chiefly in Brazil, forms a great part of the people's sustenance. The height to which the cassava attains varies from four to six feet: it rises by a slender, woody, knotted stalk, furnished with alternate palmated leaves, and springs from a tough branched woody root, the slender collateral fibres of which swell into those farinaceous parsnip-like masses, for which alone the plant is cultivated. It requires a dry soil, and is not found at a greater elevation than 2000 feet above the level of the sea. It is propagated by cuttings, which very quickly take root, and in about eight months from the time of their being planted, the tubers will generally be in a fit state to be collected; they may, however, be left in the ground for many months without sustaining any injury. The usual mode of preparing the cassava is to grind the roots after pealing off the dark-coloured rind, to draw out the poisonous juice, and finally to bake the meal into thin cakes on a hot iron hearth. Fortunately the deleterious principle is so volatile as to be entirely dissipated by exposure to heat; for when the root has been cut into small pieces, and exposed during some hours to the direct rays of the sun, cattle may be fed on it with perfect safety. If the recently extracted juice be drunk by cattle or poultry, the animals soon die in convulsions, but if this same liquid is boiled with meat and seasoned, it forms a wholesome and nutritious soup. The *Jatropha Janipha,* or Sweet Cassava, though very similar to the Manihot or bitter variety, and wholly innocuous, is far less extensively cultivated.

A palatable and wholesome bread is made of both kinds; and although its taste may be thought somewhat harsh by persons accustomed to soft fermented bread made from wheaten flour, yet those who have been accustomed to its use are so fond of it, that Creole families who have gone to live in Europe frequently have it sent to them from the West Indies.

The kind of starch so well known under the name of tapioca is prepared from the farina of cassava roots. A large quantity is exported from Brazil to Europe, and may well be considered as a more useful production than all the diamonds of Minas Geraes.

The yam-roots, which are so frequently mentioned in narratives of travel through the tropical regions, are the produce of two climbing plants — the dioscorea sativa and alata — with tender stems of from eighteen to twenty feet in length, and smooth sharp-pointed leaves on long foot-stalks, from the base of which arise spikes of small flowers. The roots of the dioscorea sativa are flat and palmated, about a foot in breadth, white within and externally of a dark brown colour, almost approaching to black, those of the D. alata, are still larger, being frequently about three feet long, and weighing about thirty pounds. Both kinds are cultivated like the common potato, which they resemble in taste, though of a closer texture. When dug out of the earth, the roots are placed in the sun to dry, and are then put into sand or casks, where, if guarded from moisture they may be preserved for a long time without being in any way injured in their quality.

The Dioscoreæ are natives of South Asia, and are supposed to have been thence transplanted to the West Indies, as they have never been found growing wild in any part of America, while in the island of Ceylon, and on the coast of Malabar, they flourish in the woods with spontaneous and luxuriant growth. They are now very extensively cultivated in Africa, Asia, and America, as their large and nutritious roots amply reward the labour of the husbandman.

The Spanish or Sweet Potato—Convolvulus Batatas, commonly

cultivated in the tropical climates both of the eastern and the western hemispheres, is an herbaceous perennial, which sends out many trailing stalks, extending six or eight feet every way, and putting forth at each joint roots which in a genial climate grow to be very large tubers, so that from a single plant forty or fifty large roots are produced. The leaves are angular and stand on long petioles, the flowers are purple. The batata is propagated by laying down the young shoots in the spring; indeed in its native climate it multiplies almost spontaneously, for if the branches of roots that have been pulled up are suffered to remain on the ground, and a shower of rain falls soon after, their vegetation will recommence. From its abundant growth, it is surprising that in Brazil the mandioc should be cultivated in preference as food for the negro, the batata being raised more as a luxury for the planter's table.

The batata was introduced into England by Sir Francis Drake and Sir John Hawkins in the middle of the sixteenth century, and known long before the common potato, which, as recently as the beginning of last century, was of less note than the horseradish.

At first, attempts were made to naturalise the batata, but it was found too tender to thrive in the open air through an English winter. The roots were formerly imported in considerable quantities from Spain and the Canaries, and used as a confection, but a more abundant supply of indigenous fruit has caused the sweet potato gradually to decline in favour, and for many years it has altogether ceased to be an article of consumption.

The Quinoa (*Chenopodium Quinoa*), a species of goose-foot, has been cultivated from time immemorial in the Peruvian and Bolivian Andes, and is said to be both nutritious and agreeable. The leaves, before the plant has attained its full maturity, are eaten as spinage, but much more rarely than the seeds, which are used in various ways, either boiled with milk or prepared with cheese and Spanish pepper.

Arrowroot is chiefly obtained from two different plants — the Marantha arundinacea and the Tacca pinnatifida. The former, a native of South America, is an herbaceous perennial and is propagated by parting the roots. It rises to the height of two or three feet, has broad pointed leaves, and is crowned by a spike of small white flowers. It is much cultivated, both for domestic use and for exportation in the West Indies, and in some parts of Hindostan. The arrowroot is obtained by first pounding the long stalky roots in a large wooden mortar, and pouring a quantity of water over them. After the whole has been agitated for some time, the starch, separated from the fibres, collects at the bottom of the vessel, and having been cleansed by repeated washings is dried in the sun.

The Tacca pinnatifida, likewise an herbaceous plant with pinnated leaves, an umbelliform blossom, and large potato-like roots, is scattered over most of the South Sea Islands. It is not cultivated in the Hawaiian group, but found growing wild in abundance in the more elevated districts, where it is satisfied with the most meagre soil, and sprouts forth among the lava blocks of those volcanic islands. Arrowroot is prepared from it in the same manner as from the West Indian Marantha, but, as the improvident Polynesians only think of digging it out of the earth, and never give themselves the trouble of replanting the small and useless tubers, its quantity has very much diminished.

The Caladium esculentum, an aquatic plant, furnishes the large Taro roots which form the chief food of the Sandwich Islanders, and are extensively cultivated in many other groups of the South Seas. It grows like rice on a marshy ground, the large sagittated leaves rise on high foot-stalks, immediately springing from the root, and are likewise very agreeable to the taste, but are more seldom eaten, as they are used for propagation. Severed from the root, they merely require to be planted in the mud to produce after six months a new harvest of roots. The growth is so abundant that 1,500 persons can live upon the produce of a single square mile, so that supposing the United Kingdom to be one vast taro-field, its surface would be able to nourish about two thousand millions of souls!

The Sandwich Islanders boil the root to a thick paste (poé), which they eat with their hands, for spoons, knives, and forks are as yet but rare articles among the subjects of His Majesty Kaméhaméha the fourth.

As there is a mountain-rice which thrives without artificial irrigation, there is also a mountain-taro (*Caladium cristatum*), which resembles the former in general appearance, but prefers a more dry and elevated soil. Although the plant grows wild both in the Society and Marquesas Islands, yet Pitcairn was the only spot where Mr. Bennett saw it cultivated.

The possession of the esculent Caladium, which furnishes so much food with so little labour, can hardly be considered as a benefit for the Sandwich Islanders, whose natural indolence is too much encouraged by the abundance it creates. The Hawaiian constantly sees before his eyes the coffee-groves and sugar-plantations, the cotton and indigo fields, which, cultivated by Chinese coolies, amply reward the enterprise of the European and American settlers in his native land, and yet he saunters by, too indolent even to stretch out his hand and gather the berries from the trees.

It may easily be imagined that the tropical sun, which distils so many costly juices and fiery spices in indescribable multiplicity and abundance, must also produce a variety of fruits. But man has as yet done but little to improve by care and art these gifts of Nature, and, with rare exceptions, the delicious flavour for which our native fruits are indebted to centuries of cultivation, is found wanting in those of the torrid zone. Even the pine-apple acquires in our hot-houses a size and perfection which it does not attain in its original home, and shows us how much it might be improved in its native seats, if cultivated with the same industry and knowledge. In our gardens Pomona appears in the refined garb of civilisation, while in the tropics she still shows herself as a savage beauty, requiring the aid of culture for the full developement of her attractions.

Yet there are exceptions to the rule, and among others the Peruvian Chirimoya (*Anona tripetala*) is vaunted by travellers in such terms of admiration that it can hardly be inferior to, and probably surpasses, the most exquisite fruits of European growth. Hänke calls it in one of his letters a masterpiece of nature, and Tschudi says that its taste is quite incomparable. It grows to perfection at Huanuco, where it attains a weight of from fourteen to sixteen pounds. The fruit is generally heart-shaped, with the broad base attached to the branch. The rind is green, covered with small tubercles and scales, and encloses a snow-white, juicy pulp, with many black kernels. Both the fruit and the blossoms exhale a delightful odour. The tree is about twenty feet high, and has a broad dull green crown.

In the eastern hemisphere, the litchi, the mangosteen, and the mango enjoy the highest reputation.

The Litchi (*Nephelium Litchi*), a small insignificant tree, with lanceolate leaves, and small greenish-white flowers, is a native of China and Cochin-China, but its cultivation has spread over the East and the West Indies. The plum-like scarlet fruit is generally eaten by the Chinese to their tea, but it is also dried in ovens and exported. In order to obtain the fruit in perfection, for the use of the Imperial Court, the trees, as soon as they blossom, are conveyed from Canton to Pekin on rafts, at a very great trouble and expense, so that the plum may just be ripe on their arrival in the northern capital. It is, however, to be feared that the rebels have within the last few years deprived the Emperor of his accustomed delicacy, unless the Kings of Nankin are urbane enough to allow the Litchi-rafts to pass — a piece of politeness which, from all we hear of them, is hardly to be expected.

The beautiful Mangosteen (*Garcinia Mangostana*), a native of the Moluccas, and thence transplanted to Java, Siam, the Philippines, and Ceylon, resembles at a distance the citron-tree, and bears large flowers like roses. The dark brown capsular fruit, about the size of a small apple, is described as of unequalled flavour — juicy and aromatic, like a mixture of strawberries, raspberries, grapes, and oranges. It is said that the patient who has lost an appetite for everything else still relishes the mangosteen, and that the case is perfectly hopeless when he refuses even this.

The stately Mango (*Mangifera indica*) is frequently represented on the silk tissues of the Hindoos, who venerate, under the ugly form of the ape Huniman (*Semnopithecus Entellus*), the transformed hero who first robbed the gardens of a Ceylonese giant of its sweet fruit, and presented their forefathers with this inestimable gift. The mango bears beautiful girandoles of flowers, followed by large plum-like fruits, of which, however, but four or five ripen on each branch. More than forty varieties are grown at Kew, the finest sorts of which are reserved for the Queen's table. From Ceylon and the East Indies, the mango has not only been transplanted to Persia, the Moluccas, and Mauritius, but even to the Antilles and South America.

Mangosteen.

The Sugar Cane.

CHAPTER XIV.

SUGAR.

Its commercial importance — Its original home — The progress of its cultivation throughout the Tropical Zone — The Tahitian Sugar-cane — Description of the Plant — Mode of extracting the Sugar — The enemies of the Sugar-cane — The Sugar Harvest.

SUGAR is undoubtedly one of the most valuable products of the vegetable world, and may be said with truth to be only surpassed in importance by the nourishing meal of the cereals, or the textile fibres of the cotton-plant.

Our garden fruit owes its agreeable taste to the sugar which the ripening sun developes in its juices. The sap of many a plant — the palm, the birch, the maple, the American agave — is rendered useful to man by the sugar it contains. It is this substance which imparts sweetness to the honey gathered by bees from flowers, and, after undergoing fermentation, changes the juice of the grape into delicious wine.

But although sugar is of almost universal occurrence throughout the vegetable world, yet few plants contain it in such abundance as to render its extraction profitable; and even the beet-root requires high protective duties to be able to compete with the tropical sugar-cane, a member of the extensive family of the grasses, or the Gramineæ, which probably spreads more than twenty thousand species over the surface of the globe.

The original home of this plant — for which, doubtless, the lively fancy of the ancient Greeks, had they been better acquainted with it, would have invented a peculiar god, as for the vine or the cereals — is most likely to be sought for in South-Eastern Asia, where the Chinese seem to have been the first people that learnt the art to multiply it by culture, and to extract the sugar from its juice.

From China its cultivation spread westwards to India and Arabia, at a time unknown to history, and the conquests of Alexander the Great first made Europe acquainted with the sweet-juiced cane, while sugar itself had long before been imported by the Phœnicians as a rare production of the Eastern world.

At a later period, both the plant and its produce are mentioned by several classical authors. Pliny speaks of the Arabian and Indian *saccharum*, which then was only used as a medicine; and a passage in Lucan's " Pharsalia " —

" Quique bibunt tenerâ dulces ab arundine succos "

— evidently alludes to the sweet juice of the sugar-cane.

During the dark ages which followed the fall of the Roman Empire, all previous knowledge of the Oriental sugar-plant became lost, until the Crusades, and, still more, the revival of commerce in Venice and Genoa reopened the ancient intercourse between the Eastern and the Western world. From Egypt, where the cultivation of the sugar-cane had meanwhile been introduced, it now extended to the Morea, to Rhodes, and Malta; and at the beginning of the twelfth century we find it growing in Italy, on the sultry plains at the foot of Mount Etna.

After the discovery of Madeira by the Portuguese, in the year 1419, the first colonists added the vine of Cyprus and the Sicilian sugar-cane to the indigenous productions of that lovely island; and both succeeded so well, as to become after a few years the objects of a lively trade with the mother country.

Yet, in spite of this extension of its culture, the importance of sugar as an article of international trade continued to be very limited, until the discovery of tropical America* by Columbus opened a new world to commerce. As early as the year 1506 the sugar-cane was transplanted from the Canary Islands to Hispaniola, where its culture, favoured by the fertility of a virgin soil and the heat of a tropical sun, was soon found to be so profitable, that it became the chief occupation of the European settlers, and the principal source of their wealth.

The Portuguese, in their turn, conveyed the cane to Brazil; from Hispaniola it spread over the other West Indian Islands; thence wandered to the Spanish main, and followed Pedrarias and Pizarro to the shores of the Pacific. Unfortunately, a dark shade obscures its triumphal march, as its cultivation was the chief cause which entailed the curse of negro slavery on some of the fairest regions of the globe.

Towards the middle of the last century, the Chinese or Oriental sugar-cane had thus multiplied to an amazing extent over both hemispheres, when the introduction of the Tahitian variety, which was found to attain a statelier growth, to contain more sugar, and to ripen in a shorter time, began to dispossess it of its old domains. This new and superior plant is now universally cultivated in all the sugar-growing European colonies; and if Cook's voyages had produced no other benefit than making the world acquainted with the Tahitian sugar-cane, they would for this alone deserve to be reckoned by the political economist among the most successful and important ever performed by man.

The sugar-cane bears a great resemblance to the common

---

* The northern part of the New Continent had been visited and colonized centuries before by the mariners of Iceland. For an account of this discovery, see "The Sea and its Living Wonders," second edition, p. 362.

reed, but the blossom is different. It has a knotty stalk, like most grasses, frequently rising to the height of fourteen feet, and produces at each joint a long, pointed, and sharply serrated leaf or blade. The joints in one stalk are from forty to sixty in number, and the stalks rising from one root are sometimes very numerous. As the plant grows up, the lower leaves fall off. A field of canes, when agitated by a light breeze, affords one of the most pleasing sights, particularly when, towards the period of their maturity, the golden plants appear crowned with plumes of silvery feathers, delicately fringed with a lilac dye.

As the cane is a rank succulent plant, it requires a strong deep soil to bring it to perfection, and generally grows best in a low moist situation. On the eastern well-watered slopes of the Andes, however, it still thrives at a height of 6,000 feet above the level of the sea.

In preparing a field for planting with the cuttings of cane — for the cultivator nowhere resorts to the sowing of seed, which in America, at least, has never been known to vegetate — the ground is marked out in rows, three or four feet apart, and in these lines holes are dug, from eight to twelve inches deep, and with an interval of two feet between the holes. In these the cuttings are inserted, which invariably consist of the top joints of the plant, because they are less rich in saccharine juice than the lower parts of the cane, while their power of vegetation is equally strong. While the shoots are growing and progressing to ripeness, great care must be taken to irrigate and weed the field. The canes annually yield fresh shoots, or rattoons, but as they have a tendency to deteriorate — at least in size — it is customary in all well-managed estates to renew every year one sixth part of the plantation.

The manufacture of sugar, as far as it is conducted in the colonies (its refining being an object of European industry) has been greatly improved by the introduction of steam-power, which thoroughly presses out all the juice of the canes on their being passed but once between the three iron rollers which the crushing-machine sets in motion. The sap is collected in a cistern, and must be immediately heated, to prevent its becoming acid — an effect which frequently commences in less than an hour from the time of its being expressed. A certain quantity of lime is added to promote the separation of the feculent matters contained in the juice, and these being removed, the cane liquor is then subjected to a very rapid boiling, to evaporate the watery particles and bring the syrup to such a consistency that it will granulate on cooling.

In order to separate the granulated or crystallised sugar from the molasses, which are incapable of crystallisation and even attract the moisture of the air, it is placed in a large square iron and air-tight case, divided into two compartments by a sieve-like bottom of wire with narrow meshes. The sugar is placed in the upper compartment, and the lower one communicates with two air-pumps, which are set in motion by the same engine which crushes the canes. On the air being exhausted in this lower compartment, the liquid molasses come pouring in to fill up the void, while the crystallised mass remains almost thoroughly purified at the top. This used formerly to be a very tedious operation: the sugar was placed in large casks whose bottoms were pierced with holes, and though left to drain for at least eight days, it still retained a quantity of molasses, while by the new process the cleansing is most effectually performed in a couple of hours, and the sugar, which has of course a much better appearance, can immediately be packed in hogsheads and cases ready for shipment.

Though the steam-engine has been generally introduced in the large sugar-plantations in the West Indies, Brazil, and Mauritius, it has not yet penetrated into all sugar-growing countries. In 1842, Tschudi found in Peru the art of cane-crushing and sugar-cleaning still in its infancy, oxen or horses imperfectly performing the part of steam, and in India both poverty and prejudice combine to prevent the *ryot* from adopting any improved method of manufacture.

The sugar-cane is liable to be destroyed by many enemies. Sometimes herds of monkeys come down from the mountains by night, and having posted sentinels to give the alarm if anything approaches, destroy incredible quantities of the cane by their gambols as well as their greediness. It is in vain to set traps for these creatures, however baited; and the only way to protect a plantation and destroy them, is to set a numerous watch, well armed with fowling-pieces, and furnished with dogs. Fortunately the negroes perform this service cheerfully, for they consider a roasted monkey as a great delicacy. The ubiquitous rat, which the extension of commerce has gradually spread over the world, is still more destructive to the sugar-cane, and great pains are taken to keep it in check by poison, or by its arch-enemy the cat.

The sugar-cane is also subject to the *blast*—a disease which no foresight can obviate, and for which human wisdom has hitherto in vain attempted to find a remedy. When this happens, the fine broad green blades become sickly, dry, and withered; soon after they appear stained in spots, and if these are carefully examined, they will be found to contain countless eggs of an insect like a bug, which are soon quickened, and cover the plants with the vermin; the juice of the canes thus affected becomes sour, and no future shoot issues from the joints. The ravages of the ants, which I shall have occasion to mention more amply in another chapter, concur with those of the bugs in ruining the prospects of many a sugar-field, and often a long continued drought or the fury of the tornado will destroy the hopes of the planter.

The land crabs are frequently very injurious to the sugar-fields, some of the species being particularly fond of the cane, the juice of which they suck and chiefly subsist on. They are of course narrowly watched, and no opportunity of catching them is lost sight of; but such is their activity in running, or rather darting in any direction, or with any part of their bodies foremost, that they are almost always enabled to escape. It is seldom, however, that they go far from their burrows in day-time; and their watchfulness is such that they regain them in a moment, and disappear as soon as a man or dog comes near enough to be seen.

The labour in the hot sugar-fields is most irksome, and it may well be imagined that it is no trifling exertion to hoe the hard ground under the rays of a tropical sun, or to cut the thick canes and transport them to the mill. Yet this hard work is not without its pleasures, and harvest-time in the sugar-plantation is no less a season of gladness than in the corn-fields of England or the vineyards of France.

So palatable, wholesome, and nourishing is the juice of the cane, that every animal drinking freely of it, derives health and vigour from its use. The meagre and sickly among the negroes exhibit a surprising alteration in a few weeks after the mill is set in action. The labouring oxen, horses, and mules, though almost constantly at work during this season, yet being indulged with plenty of the green tops and some of the scummings from the boiling-house, improve more than at any other period of the year. Even the pigs and poultry fatten on the refuse, and enjoy their share of the banquet. The wholesome effects of the juice of the sugar-cane has not escaped the attention of English physicians, and many a weak-breasted patient, instead of coughing and freezing at home over what is ironically termed a comfortable fireside, now spends his winter in the West Indian Islands, chewing the sweet cane and enjoying in January a genial warmth of seventy-two degrees in the shade.

Coffee Rat.

CHAPTER XV.

COFFEE.

Enarea and Kaffa — Gemaledie introduces its use into Arabia — Fanaticism endeavours to forbid it — The first Coffee-Houses in London and Marseilles — Coffee-production in Brazil — Java — Ceylon — Rapid extension of its culture in Ceylon since 1825 — The Coffee-Plant — The best situations for its growth — Its cultivation in Java described — The Musang — Expertness of the Ceylon Woodmen in preparing the Coffee Ground — Enemies of the Coffee-Plant — The Golunda Rat — The Lecanium Coffeæ.

THE mountain regions of Enarea and Caffa, which the reader, on consulting a map of Africa, will find situated to the south of Abyssinia (and not Arabia-Felix as Linnæus erroneously supposed), are most probably the countries where the coffee-tree was first planted by nature, as it has here not only been cultivated from time immemorial, but is everywhere found growing wild in the forests.

Here also the art of preparing a beverage from its berries seems to have been first discovered. Arabic authors inform us that about four hundred years ago Gemaledie, a learned mufti of Aden, having become acquainted with its virtues on a journey to the opposite shore of Africa, recommended it on his return to the dervises of his convent as an excellent means for keeping awake during their devotional exercises. The example of these holy men brought coffee into vogue, and its use spreading from tribe to tribe, and from town to town, finally reached Mecca about the end of the fifteenth century. There fanaticism endeavoured to oppose its progress, and in 1511 a council of theologians condemned it as being contrary to the law of Mahomet, on account of its intoxicating like wine, and sentenced the culprit who should be found indulging in his cup of coffee to be led about the town on the back of an ass. The sultan of Egypt, however, who happened to be a great coffee-drinker himself, convoked a new assembly of the learned, who declared its use to be not only innocent, but healthy; and thus coffee advanced rapidly from the Red Sea and the Nile to Syria, and from Asia Minor to Constantinople, where the first coffee-house was opened in 1554, and soon called forth a number of rival establishments. But here also the zealots began to murmur at the mosques being neglected for the attractions of the ungodly coffee divans, and declaimed against it from the Koran, which positively says that *coal* is not of the number of things created by God for good. Accordingly the mufti ordered the coffee-houses to be closed; but his successor declaring coffee not to be *coal*, unless when over-roasted, they were allowed to reopen, and ever since the most pious mussulman drinks his coffee without any scruples of conscience. The commercial intercourse with the Levant could not fail to make Europe acquainted with this new source of enjoyment. In 1652, Pasqua, a Greek, opened the first coffee-house in London, and twenty years later the first French cafés were established in Paris and Marseilles.

As the demand for coffee continually increased, the small province of Yemen, the only country which at that time supplied the market, could no longer produce a sufficient quantity, and the high price of the article naturally prompted the European governments to introduce the cultivation of so valuable a plant into their colonies. The islands of Mauritius and Bourbon took the lead in 1718, and Batavia followed in 1723. Some years before, a few plants had been sent to Amsterdam, one of which found its way to Marly, where it was multiplied by seeds. Captain Descleux, a French naval officer, took some of these young coffee-plants with him to Martinique, desirous of adding a new source of wealth to the resources of the colony. The passage was very tedious and stormy; water began to fail, and all the gods seemed to conspire against the introduction of the coffee-tree into the new world. But Descleux patiently endured the extremity of thirst that his tender shoots might not droop for want of water, and succeeded in safely bringing over one single plant, the parent stock whence all the vast coffee-plantations of the West Indies and Brazil are said to have derived their origin. The names of such men should not be forgotten, and deserve, in my opinion, to be recorded before those of the many worthless monarchs which uselessly fill the pages of history.

On examining the present state of coffee-production throughout the world, we find that it has undergone great revolutions within the last twenty years, as some of the countries that were formerly prominent in this respect now occupy but an inferior rank, while in others the cultivation of coffee has rapidly attained gigantic proportions.

Thus Brazil, which at the beginning of the century was hardly known in the coffee market, now furnishes nearly as much as all the rest of the world besides. Its exportation, which in 1820 amounted to 97,500 sacks, rose to a million in 1840 and attained in 1855 the enormous quantity of 2,392,100 sacks, or more than 350 millions of pounds!

Java ranks next to Brazil among the coffee-producing countries, for though slavery does not exist in this splendid colony, yet the Dutch have introduced a system which answers the purpose fully as well. Every Javanese peasant is obliged to work sixty-six days out of the year for government; and the residents or administrators of the various districts distribute this compulsory labour among the several plantations, which are all in the hands of private individuals. Thus the latter are provided with the necessary hands at a very cheap rate; but on the other hand, they are compelled to sell their whole produce to the *Handels Maatschappy*, or Dutch East India Company, at a price fixed by the government, which of course takes care to secure the lion's share of the profit.

When we consider that Java has more than nine millions of

inhabitants, and possesses a soil of inexhaustible fertility, the masses of colonial produce which it pours upon the market of the world, cease to be matter of surprise; nor can we wonder at its yielding the Dutch treasury an annual surplus of more than ten millions of florins, after all local expenses have been paid. The annual production of coffee alone amounts to at least 150,000,000 pounds, which are almost exclusively exported to Holland, and chiefly serve for the consumption of Germany.

Within the last forty years the progress of coffee cultivation in Ceylon has been no less remarkable than its rapid extension in Java or Brazil. Though the plant was found growing in the island by the Portuguese, and is even supposed by some to be indigenous, yet it was only after the subjugation of the ancient kingdom of Kandy by the English in 1815, and the opening of roads in the hill country, that it began to be cultivated on a more extensive scale. The first upland plantation was formed about 1825 by Sir Edward Barnes, the energetic governor to whom the colony is indebted for so many works of public utility; and as a concurrence of favourable circumstances — among others the decline of production in the West Indies, and the increasing consumption in England, in consequence of remissions of duty, — rendered the moment propitious, his example was speedily followed by a host of speculators, so that in an incredibly short time the mountain-ranges in the centre of the island became covered with plantations, and rows of coffee-trees began to bloom upon the solitary hills around the very base of Adam's Peak.

Thus coffee, formerly so unimportant, is now the chief production of Ceylon; its exportation having risen from 1,792,448 pounds in 1827, to 67,453,680 pounds in 1857; and should prices in Europe continue remunerative, there is every reason to believe that in less than fifty years the produce will be trebled, equalling or even surpassing that of Java.

While the production of coffee has thus enormously increased in the above-mentioned countries, it has on the other hand greatly declined in the West Indies. Thus Hayti, which previous to the negro rebellion in 1791 exported seventy-six millions of pounds to France, at a time when coffee was much dearer than at present, now produces less than one half that quantity, and is even distanced by Venezuela, which in 1853 exported thirty-eight millions of pounds.

Since the abolition of slavery in the British West Indies, the export of coffee, which in 1827 amounted to 29,419,598 pounds, dwindled to 4,054,028 in 1857, for such is the indolence of the emancipated negro, or so few his wants, that even the highest wages are insufficient to rouse him to exertion.

In Cuba, finally, which exported more than 40,000,000 of pounds in 1833, the produce has now fallen to one half, though from a different cause, as it has been found that slave-labour can be more advantageously employed in the cultivation of sugar.

With regard to quality, Mocha coffee, though comparatively insignificant in point of quantity, is still prominent in flavour and aroma; the finer sorts of Ceylon and Bourbon rank next, while the cheapest and commonest is the ordinary Brazil.

The general production of coffee throughout the world amounts probably to more than 730,000,000 of pounds, and is constantly increasing as civilisation with its growing wants and luxuries spreads over the face of the globe.

When left to the free growth of nature, the coffee-tree attains a height of from fifteen to twenty feet; in the plantations, however, the tops are generally cut off in order to promote the growth of the lower branches, and to facilitate the gathering of the crop. Its leaves are opposite, evergreen, and not unlike those of the bay-tree; its blossoms are white, sitting on short footstalks, and resembling the flower of the jasmine. The fruit which succeeds is a green berry, ripening into red, of the size and form of a large cherry, and having a pale, insipid, and somewhat glutinous pulp, enclosing two hard and oval seeds or beans, which are too well known to require any further description. The tree is in full bearing from its fourth or fifth year, and continues during a long series of seasons to furnish an annual produce of about a pound and a half of beans.

Coffee.

The seeds are known to be ripe when the berries assume a dark red colour, and if not then gathered, will drop from the trees. The planters in Arabia do not pluck the fruit, but place cloths for its reception beneath the trees, which they shake, and the ripened berries drop readily. These are afterwards spread upon mats, and exposed to the sun until perfectly dry, when the husk is broken with large heavy rollers made either of wood or of stone. The coffee, thus cleared of its husk, is again dried thoroughly in the sun, that it may not be liable to heat when packed for shipment.

This method may, in some measure, account for the superior quality of the Arabian coffee, but in the large plantations of Brazil, Java, Ceylon, and other European colonies, it is necessary to follow a more expeditious plan, to pluck the berries from the trees as soon as they ripen, and immediately to pass them through a pulping mill, consisting of a horizontal fluted roller turned by a crank, and acting against a movable breast-board, so placed as to prevent the passage of whole berries between itself and the roller. The pulp is then separated from the seeds by washing them, and the latter are spread out in the sun to dry; after which the membranous skin or parchment which immediately covers the beans, is removed by means of heavy rollers or stamping.

To be cultivated to advantage, the coffee-tree requires a climate where the mean temperature of the year amounts to at least 68°, and where the thermometer never falls below 55°. It is by nature a forest tree requiring shade and moisture, and thus it is necessary to screen it from the scorching rays of the sun by planting rows of umbrageous trees at certain intervals throughout the field. These also serve to protect it from the sharp winds which would injure the blossoms. It cannot bear either excessive heat or a long-continued drought, and where rain does not fall in sufficient quantity, artificial irrigation must supply it with the necessary moisture. From all these circumstances it is evident that the best situations for the growth o coffee are not the sultry alluvial plains of the tropical and subtropical lands, but the mountain-slopes to an elevation o 4,500 feet.

In Java the zone of the coffee-plantations extends betwee 3,000 and 4,000 feet above the level of the sea; and the primi tive forest is constantly receding before them. Frequently, o felling the woods, a part of the original trees is left standing t shade the tender coffee plants; but oftener the rows are made to alternate with those of the sheltering dadab. Thus a new and luxuriant grove, animated by insects, birds, and a number of small four-footed animals, replaces the old thicket of nature's planting. Straight paths, kept carefully clean, lead through the dense dark-green shrubbery, under whose thick cover the wild cock hastily retreats when surprised by the wanderer. When the trees are in flower, the branches seem to bend under a weight of snow, from the number of dazzling white blossoms, which form a most pleasing contrast to the dark lustrous foliage, while high above the dadabs extend their airy crowns, whose light green leaves are agreeably interspersed with flowers of a brilliant red. A few months later, when the fruits are ripening into carmine, a scene of the most bustling animation ensues, for old and young are busily employed in plucking the swelling berries, and hurrying with filled baskets to the nearest pulping mill. At this time, the whitish excrements of the musang, a small gray long-tailed marten-like rodent, are frequently found upon the ground, consisting wholly of agglutinated, but otherwise undamaged coffee-beans. The musang is extremely fond of the ripe berries, whose pulp dissolves in its stomach, while the undigested beans furnish the best coffee, as the fastidious animal selects none but the choicest fruit. For this reason the musang coffee is carefully collected, and sold at a higher price to wealthy Dutch residents.

In Ceylon the native woodmen are singularly expert in felling forest trees preparatory to the cultivation of coffee. Turning to advantage the luxuriance of tropical vegetation, which lashes together whole forests by a maze of interlacing climbers as firm and massy as the cables of a line-of-battle ship, their practice in steep and mountainous places is to cut half-way through each stem in succession, till an area of some acres in extent is prepared for the final overthrow. They then sever some tall group on the eminence, and allow it in its descent to precipitate itself on those below, when the whole expanse is in one moment brought headlong to the ground, the falling timber forcing down those beneath it by its weight, and dragging those behind to which it is harnessed. The crash occasioned by this startling operation is so loud, that it is audible for two or three miles in the clear and still atmosphere of the hills.

Every subsequent operation must be carried on by coolies from Malabar and the Coromandel coast, as no temptation of wages and no prospect of advantage has hitherto availed to overcome the repugnance of the native Singhalese and Kandyans to engage on any work on estates except the first process of felling the forests. The fallen wood having been burnt or removed, and the ground dug up, the young coffee-trees are transplanted from the nurseries when about eight inches high, and set at intervals of four or six feet. April and October are the best time for this operation, on account of the heavy rains. At least once a month the plantation must be carefully weeded, and the pruning of the young tree takes place at the end of the first year.

The chief harvest is in December and January, but many plants bloom the whole year round.

The produce of the plantations in lower situations is more abundant, but the quality improves and the beans become longer as the vertical elevation increases. The best coffee grows at the height of four thousand feet.

The coffee-plant very soon exhausts the alkaline substances of the soil, and as the depth of vegetable mould is not considerable, and liable to be constantly impoverished by transfiltration or draining on the sloping hill-sides, good crops can only be obtained by plentiful manuring. The best soil for coffee is chalk, or a loose decomposed gneiss; clay is not so favourable, as it prevents the roots from penetrating to a sufficient depth.

The plantations generally yield a profit of from 20 to 30 per cent., but as capital in Ceylon is not to be obtained under 15 or 20, the large produce of many estates is merely imaginary.*

Though the free labour of the coolies does not require the large sum which the Brazilian planter is obliged to invest in the purchase of his slaves, yet the emigrant to Ceylon who intends to devote himself to the cultivation of coffee must not come empty handed or seek to begin his operations on credit alone. It is a perfect illusion, fostered by badly informed travellers or interested agents, that colonial produce can be raised to any advantage with small means; and as experience is necessary, Professor Schmarda † advises the future planter,

---

* Tennent's Ceylon, vol. ii. p. 202.
† Reise um die Erde in den Jahren 1853—1857, vol. i. p. 387.

who should be possessed of at least 3,000*l*, first to become acquainted with the country, and practically to study the cultivation of coffee before settling. The nature of the soil, the vicinity of a road or of an inhabited place for the facility of obtaining manure, are important considerations. It is not advisable to purchase more than five or six hundred acres, nor to cultivate in the first year more than 100 or 150 acres, beginning with the lower part of the estate and then ascending the slopes, so as to profit as long as possible by the drainage of the higher grounds. Large plantations of several thousand acres are attended with the serious disadvantage that, as they are too extensive to be overlooked in detail, the owner is obliged to trust a great part of the management to strangers, who of course make him *pay* well for their trouble. Thus coffee-planting, though it may insure a competence to the prudent cultivator, by no means opens a rapid road to wealth, as most of the speculators vainly imagined who, previous to 1845, embarked their fortunes in the enterprise, and whose visions of opulence ended in the disaster and ruin which are the necessary consequences of extravagance and folly.

Like every other plant cultivated by man, the coffee-tree is exposed to the ravages of many enemies. Wild cats, monkeys, and squirrels prey upon the ripening berries, and hosts of caterpillars feed upon the leaves. Since 1847 the Ceylon plantations have been several times invaded by swarms of the Golunda, a species of rat which inhabits the forests, making its nest among the roots of the trees, and, like the lemmings of Norway and Lapland, migrating in vast numbers when the seeds of the nîlloo-shrub (Strobilanthes), its ordinary food, are exhausted. "In order to reach the buds and blossoms of the coffee, the Golunda eats such slender branches as would not sustain its weight, and feeds as they fall to the ground; and so delicate and sharp are its incisors, that the twigs thus destroyed are detached by as clean a cut as if severed with a knife. The Malabar coolies are so fond of its flesh that they evince a preference for those districts in which the coffee-plantations are subject to its incursions, frying the rats in oil or converting them into curry."[*]

Another great plague is the *Lecanium Coffeæ*, known to

---

[*] Tennent's Ceylon, vol. ii. p. 234.

planters as the coffee bug, but in reality a species of *coccus*, which establishes itself on young shoots and buds, covering them with a noisome incrustation of scales, from the influence of which the fruit shrivels and drops off. A great part of the crop is sometimes lost, and on many trees not a single berry forms from the invasion of this pest, which was first observed in 1843 on an estate at Lapalla Galla, and thence spreading eastward through other plantations, finally reached all the other estates in the island. No cheap and effectual remedy has as yet been found to stay its ravages, and the only hope is, that as other blights have been known to do, it may wear itself out, and vanish as mysteriously as it came.

General Fraser's Coffee-estate at Rangbodde, Ceylon.

CHAPTER XVI.

CACAO AND VANILLA.

The Cacao Tree — Mode of preparing the Beans for the Market — Chocolate — The Vanilla Plant.

THEOBROMA,— food for gods,— the Greek name given by Linnæus to the cacao or chocolate tree, sufficiently proves how highly he valued the flavour of its seeds.

Indigenous in Mexico, it had long been in extensive cultivation before the arrival of the Spaniards, who found the beverage which the Indians prepared from its beans so agreeable, that they reckoned it among the most pleasing fruits of their conquest, and lost no time in making their European friends acquainted with its use. From Mexico they transplanted it into their other dependencies, so that in America its present range of cultivation extends from 20° N. lat. to Guayaquil and Bahia. It has even been introduced into Africa and Asia, in return for the many useful trees that have been imported from the old into the new world. The cacao-tree seldom rises above the height of twenty feet, its leaves are large, oblong, and pointed. The flowers, which are of a pale red colour, spring from the large branches, and even from the trunk and roots. "Never," says the illustrious Humboldt, " shall I forget the deep impression made upon me by the luxuriance of tropical vegetation on first seeing a cacao plantation. After a damp night, large blossoms of the theobroma issue from the root at a considerable distance from the trunk, emerging from the deep black mould. A more striking example of the expansive powers of life could hardly be met with in organic nature." The fruits are large, oval, pointed pods, about five or six inches long, and containing in five compartments from twenty to forty seeds — the well-known cacao of commerce — enveloped in a white pithy substance.

The trees are raised from seed, generally in places screened from the wind. As they are incapable of bearing the scorching rays of the sun, particularly when young, bananas, maize, manioc, and other broad-leaved plants are sown between their rows, under whose shade they enjoy the damp and sultry heat which is indispensable to their growth, for the Theobroma Cacao is essentially tropical, and requires a warmer climate than the coffee-tree or the sugar-cane.

Two years after having been sown, the plant attains a height of three feet, and sends forth many branches, of which, however, but four or five are allowed to remain. The first fruits appear in the third year, but the tree does not come into full bearing before it is six or seven years old, and from that time forward it continues to yield abundant crops of beans during more than twenty years. At first the tender plants must be carefully protected from weeds and insects, but in after years they demand but little attention and labour, so that one negro suffices for keeping a thousand trees in order and collecting their produce.

According to Herndon ("Valley of the Amazons"), the annual produce of a thousand full-grown trees, in the plantations along the lower banks of the monarch of streams, amounts to fifty arrobes, that sell in Peru from two and a half to three milrees the arrobe; and Wagner informs us that in Costa Rica a thousand trees yield about 1,250 pounds, worth twenty dollars the hundredweight.

The beans when first collected from the tree are possessed of an acrimony, which requires a slight fermentation to change into the aromatic principle, to which they are indebted for their agreeable flavour. For this purpose they are thrown into pits, where they remain three or four days covered with a light layer of sand, care being taken to stir them from time to time. They are then taken out, cleaned, and laid out upon mats to dry in the sun. The management of the beans requires some caution, for if the fermentation is allowed to continue too long, they acquire a mouldy taste and smell, which they only lose on being roasted. When thoroughly dried (which is known by their hollow sound when shaken, and by the husk easily separating from the seed when pressed), they are packed in sacks or cases, and sent as soon as possible to the market, a rapid sale being extremely desirable, as it is very difficult to preserve them from insects, more particularly from the cockroaches.

Cacao is chiefly used under the form of chocolate. The beans are roasted, finely ground, so as to convert them into a perfectly smooth paste, and improved in flavour by the addition of spices, such as the sweet-scented vanilla, a short notice of which will not be out of place.

Like our parasitical ivy, the Vanilla aromatica, a native of torrid America, climbs the summits of the highest forest-trees, or creeps along the moist rock crevices on the banks of rivulets.

The stalk, which is about as thick as a finger, bears at each joint a lanceolate and ribbed leaf, twelve inches long and three inches broad. The large flowers, which fill the forest with their delicious odours, are white intermixed with stripes of red and yellow, and are succeeded by long and slender pods containing many seeds imbedded in a thick oily and balsamic pulp. These pods seldom ripen in the wild state, for the dainty monkey knows no greater delicacy, and his agility in climbing almost always enables him to anticipate man.

At present the vanilla is cultivated not only in Mexico, where the villages Papantla and Misantla annually produce about 19,000 pounds or two millions of pods (worth at Vera Cruz a shilling the pod), but in Java, where the industrious Dutch have acclimatised it since 1819.

It is planted under shady trees on a damp ground, and grows luxuriantly; but as a thousand blossoms on an average produce but one pod, it must always remain a rare and costly spice.

Had the ancients known vanilla they would, no doubt, have deemed it more worthy to be the food of the Olympic gods than their fabled ambrosia.

CHAPTER XVII.

COCA.

Its immense Consumption in Peru and Bolivia — Mode of chewing the leaves — Its wonderfully strengthening effects — Fatal consequences of its abuse — Reasons which prompted the Spaniards to interdict its use, and finally to allow and encourage it — Its chemical analysis by Professor Wöhler of Göttingen.

ALTHOUGH but little known beyond the confines of its native country, Coca is beyond all doubt one of the most remarkable productions of the tropical zone, and deserves the more to be noticed as the time is, perhaps, not far distant, when it will assume a conspicuous rank in the markets of the world.

The sultry valleys on the eastern slopes of the Peruvian and Bolivian Andes are the seat of the Erythroxylon Coca, which like the coffee-tree bears a lustrous green foliage, and white blossoms ripening into small scarlet berries. These, however, are not used, but the leaves, which when brittle enough to break on being bent, are stripped from the plant, dried in the sun, and closely packed in sacks. The naked shrub soon gets covered with new foliage, and after three or four months its leaves are ready for a second plucking, though in some of the higher mountain-valleys it can only be stripped once a year. Every eight or ten years the plantations require to be renewed, as the leaves of the old shrubs are less juicy, and consequently of inferior quality. Like the coffee-tree, the coca-shrub thrives only in a damp situation, under shelter from the sun; and for this reason maize, which rapidly shoots up, is generally sown between the rows of the young plants. At a later period, when they no longer need this protection, care must be taken to weed the plantation, and to loosen the soil every two or three months.

The local consumption of coca is immense, as the Peruvian

Indian reckons its habitual use among the prime necessaries of life, and is never seen without his leathern pouch or chuspa, filled with a provision of the leaves, and containing besides a small box of powdered unslaked lime. At least three times a day he rests from his work to chew his indispensable coca. Carefully taking a few leaves out of the bag, and removing their midribs, he first masticates them into the shape of a small ball, which is called an acullico; then repeatedly inserting a thin piece of moistened wood like a toothpick into the box of unslaked lime, he introduces the powder which remains attached to it into the acullico until the latter has acquired the requisite flavour. The saliva, which is abundantly secreted while chewing the pungent mixture, is mostly swallowed along with the green juice of the plant.

When the acullico is exhausted, another is immediately prepared, for one seldom suffices. The corrosive sharpness of the unslaked lime requires some caution, and an unskilled coca-chewer runs the risk of burning his lips, as, for instance, the celebrated traveller Tschudi, who, by the advice of his muleteer, while crossing the high mountain-passes of the Andes, attempted to make an acullico, and instead of strengthening himself as he expected, merely added excruciating pain to the fatigues of the journey.

The taste of coca is slightly bitter and aromatic, like that of bad green tea, but the addition of lime or of the sharp ashes of the quinoa, renders it less disagreeable to the European palate.

It is a remarkable fact that the Indians who regularly use coca require but little food, and when the dose is augmented are able to undergo the greatest fatigues, without tasting almost anything else. Professor Pöppig ascribes this astonishing increase of endurance to a momentary excitement, which must necessarily be succeeded by a corresponding collapse, and therefore considers the use of coca absolutely hurtful. Tschudi, however, is of opinion that its moderate consumption far from being injurious, is, on the contrary, extremely wholesome, and cites the examples of several Indians who, never allowing a day to pass without chewing their coca, attained the truly patriarchal age of one hundred and thirty years. The ordinary food of these people consists almost exclusively of roasted maize or barley, which is eaten dry without any other addition:

and the obstinate obstructions caused by these mealy aliments are obviated by the tonic effects of the coca, which thus removes the cause of many maladies. It may be remarked, that a similar reason is assigned for the custom of areka and betel chewing in Southern Asia. As an instance of the wonderful strengthening properties of the coca, Tschudi mentions the case of an Indian called Hatun Huamang or the "Great Vulture," whom he employed during five consecutive days and nights in making the most laborious excavations, and who never ate anything all the time, or slept more than two hours a night. But every three hours he chewed about half an ounce of the leaves, and constantly kept his acullico in his mouth. When the work was finished, this Indian accompanied Tschudi during a ride of twenty-three leagues, over the high mountain-plains, constantly running alongside of the nimbly-pacing mule, and never resting but for the purpose of preparing an acullico. When they separated, the "Great Vulture" told Tschudi that he would willingly do the same work over again, provided only he had a plentiful allowance of coca. He was sixty-two years old, according to the testimony of the village priest, and had never been ill all his life.

Tschudi often found that coca is the best preservative against the asthmatic symptoms which are produced by the rapid ascension of high mountains. While hunting in the Puna, at an elevation of 14,000 feet above the level of the sea, he always drank a strong infusion of coca before starting, and was then able to climb among the rocks, and to pursue his game, without any greater difficulty in breathing than would have been the case upon the coast. Even after drinking a very strong infusion, he never experienced any symptoms of cerebral excitement, but a feeling of satiety, and though he took nothing else at the time, his appetite returned only after a longer interval than usual.

If the moderate use of coca is thus beneficial in many respects, its abuse is attended with the same deplorable consequences as those which are observed in the oriental opium-eaters and smokers, or in our own incorrigible drunkards.

The confirmed coca-chewer, or coquero, is known at once by his uncertain step, his sallow complexion, his hollow, lack-lustre black-rimmed eyes, deeply sunk in the head, his trembling lips, his incoherent speech, and his stolid apathy. His character is irresolute, suspicious, and false; in the prime of life, he has all the appearances of senility, and in later years sinks into complete idiocy. Avoiding the society of man, he seeks the dark forest, or some solitary ruin, and there, for days together, indulges in his pernicious habit. While under the influence of coca, his excited fancy riots in the strangest visions, now revelling in pictures of ideal beauty, and then haunted by dreadful apparitions. Secure from intrusion, he crouches in an obscure corner, his eyes immovably fixed upon one spot; and the almost automatic motion of the hand raising the coca to the mouth, and its mechanical chewing, are the only signs of consciousness which he exhibits. Sometimes a deep groan escapes from his breast, most likely when the dismal solitude around him inspires his imagination with some terrific vision, which he is as little able to banish as voluntarily to dismiss his dreams of ideal felicity. How the coquero finally awakens from his trance, Tschudi was never able to ascertain, though most likely the complete exhaustion of his supply at length forces him to return to his miserable hut.

Sometimes even Europeans give way to this vice; and Tschudi, during his sojourn in Peru, was acquainted with a Biscayan and an Italian who were both of them incorrigible coqueros.

No historical record informs us when the use of the coca was introduced, or who first discovered the hidden virtues of its leaves. When Pizarro destroyed the empire of Atahualpa he found that it played an important part in the religious rites of the Incas, and that it was used in all public ceremonies, either for fumigation or as an offering to the gods. The priests chewed coca while performing their rites, and the favour of the invisible powers was only to be obtained by a present of these highly valued leaves. No work begun without coca could come to a happy termination, and divine honours were paid to the shrub itself.

After a period of more than three centuries, Christianity has not yet been able to eradicate these deeply-rooted superstitious feelings, and everywhere the traveller still meets with traces of the ancient belief in its mysterious powers. To the present day, the miners of Cerro de Pasco throw chewed coca against the hard veins of the ore, and affirm that they can then be more easily worked,—a custom transmitted to them from their forefathers, who were fully persuaded that the Coyas or subterranean divinities rendered the mountains impenetrable unless previously propitiated by an offering of coca. Even now the Indians put coca into the mouths of their dead, to insure them a welcome on their passage to another world; and whenever they find one of their ancestral mummies, they never fail to offer it some of the leaves.

During the first period after the conquest of Peru, the Spaniards endeavoured to extirpate by all possible means the use of coca, from its being so closely interwoven with the Indian superstitions; but the proprietors of the mines soon became aware how necessary it was for the successful prosecution of their undertakings; the planters also found after a time that the Indians would not work without it; private interest prevailed, as it always does in the long run, over religious zeal and despotic interdictions, and in the last century we even find a Jesuit, Don Antonio Julian, regretting that the use of coca had not been introduced into Europe instead of tea or coffee.

When we consider its remarkable properties, it is indeed astonishing that it has so long remained unnoticed. Were it concealed in the interior of Africa, or extremely difficult to procure, this neglect could be more easily accounted for; but hundreds of our vessels annually frequent the harbours of Peru and Bolivia, where it may be obtained in large quantities, and yet, as far as I know, the first *bale* ever imported into Europe has but just arrived in Hamburg.

For many years Coffeïn, Theïn, Theobromin — the vegetable alkaloids to which coffee, tea, and cacao are indebted for their stimulating properties — have been known to chemists, while the analysis of coca, which might naturally be expected to contain a no less remarkable substance, has only within the last year been attempted by Professor Wöhler, the distinguished chemist of Göttingen, to whom we are at length indebted for the discovery of Cocaïn.

CHAPTER XVIII.

COTTON.

Amazing rise of the Cotton Manufactory, unparalleled in the Annals of Commerce— The Cotton Plants — Their culture in the Confederate States and in India — Prospects of Cotton cultivation in India — Brazilian and Egyptian Cotton — Prospects held forth by Africa — Cotton-seed Oil.

UNDER the Plantagenets and the Tudors, wool formed the chief export of England. The pastoral races that inhabited the British Isles, unskilled in weaving, suffered the more industrious Flemings to convert their fleeces into tissues; and the dominions of the Duke of Burgundy, enriched by manufactures and by the stimulus they gave to agriculture, became the most prosperous part of Europe. At length the islanders began to discover the sources of the wealth which rendered Ghent and Bruges, Ypres and Louvain, the marvel and envy of the mediæval world; and gradually learning to keep their wool at home, invited the Flemings to the shores of England.

Sir Walter Raleigh wrote three centuries ago that the greatest question of the hour was, "Shall we export our great staple to enrich Flanders, or weave it at home?" The bigoted oppression of Spain came in aid to the more enlightened policy of England: wool ceased to be the chief of our exports, and English cloth was now sent abroad in place of the unwrought material. Since this beneficial change, although the devotion of land to pasturage, increased capital, and the cultivation of green crops, have greatly multiplied English flocks and fleeces, the domestic supply has proved insufficient. Drafts have been made on Spain, Saxony, and the Levant; the merino sheep has been naturalised in Australia and at the Cape; and England, in addition to her own vast produce, now annually imports at least 150,000,000 pounds of wool.

But in spite of this wonderful growth, the wool-manufactory

has almost within the memory of living man
great staple of our industry; and vegetable wo
plant totally unknown to our forefathers, now
of all the world-wide importations of England

The rapid growth of the cotton trade to its
importance is indeed unparalleled in the whole
merce. We read, it is true, in Herodotus, of
plant of the Indians, and Pliny tells us of the v
Egyptian priests, for which they were indebted
haired fleeces of their flocks, but to a shrub.
many centuries have passed since the cultivati
plant was introduced into Asia Minor and th
and that the manufacture of the muslins of Hi
remotest antiquity; but not quite a century a
and weaving of cotton was still so limited in
to be noticed in comparison with the nation
the wool-manufactory.

The spinning-jenny, invented about the yea
graves, a carpenter of Blackburn, and which
far improved that by its help a little girl could
into motion, gave the first impulse to the ri
industry, which soon assumed still larger proj
Arkwright's spinning frame, Crompton's mule-
Cartwright's power-loom, and made gigantic st
since 1820, when the power of steam was first
called in aid, to establish those wonderful ma
fit locality independently of the moving power
a few men of genius, most of them be
humbler ranks of life, have contributed perhap
of their contemporaries to the power and gr
country, by enriching it with an industry v
branches, directly and indirectly gives occcup
five millions of our population, which furnishe
third of our colossal exportation, and by provid
nation of the globe with cheap and durable clot
essentially to the comfort and the progress of
A few figures will tend to show the enormous
the cotton trade.

In the year 1781 England imported 28,882
in 1820, 571,731; in 1840, 1,600,370; in 1

and in 1860 at least 3,000,000; a truly prodigious ratio of increase, which is, however, very far from having reached its zenith, and will keep on rising and rising as long as Britain remains the great emporium of the growing commerce of the world.*

In the year 1777 our cotton manufactories occupied no more than 7,900 workmen; in 1839 the number of artisans employed in the mills had increased to 259,385; and in 1856 it amounted to 379,213, in spite of the constant improvement of machinery, another proof, if such were needed, that every progress of mechanical art, by rendering production cheaper and consequently increasing the demand, enlarges instead of restricting the field for manual labour.

In 1857 twenty-eight millions of spindles were set in motion by the 2,210 cotton-mills at that time existing in Great Britain, nearly one for every inhabitant of the realm.

The total value of the production of 1856 was estimated at 64,484,000*l.*, and after deducting the value of the raw material, 37,526,000*l.* remained for expenses, wages, interest of capital and profits. Never since the world stands has one single branch of trade poured forth such streams of gold over a land.

After Great Britain, the cotton industry of France ranks next in importance, requiring about 400,000 bales, and occupying about five millions of spindles; but latterly the German States have made considerable progress, and bid fair soon to overtake their Gallic neighbours, as, in the year 1859, the Zollverein consumed 236,000 bales, and Austria above 200,000. By reason of prohibitions or high protective duties, every country in Europe has endeavoured to naturalise the manufactory of cotton, but while most of them but partially cover their own wants, England's produce inundates the markets of the world.

Liverpool, with a trade eclipsing that of London itself, and Manchester with her train of minor stars, undoubtedly the scene of the most gigantic industry known in the history of man, are chiefly indebted to the cotton-plant for their colossal rise; and on turning our eyes to America, which in the year 1793 grew but a few tons of upland cotton, and in 1859 produced a total of four and three quarters millions of bales, we find a scarcely less remarkable developement of flourishing cities, so that Arkwright and his

* The present American crisis can, of course, produce but a temporary depression.

brother inventors may truly be said to have called forth a new world on both sides of the Atlantic, and to have founded towns whose wealth and activity far eclipse the commerce of Carthage or of Tyre.

There are many different species of the cotton-plant, herbaceous, shrubby, and arboreal. Their original birthplace is the tropical zone, where they are found growing wild in all parts of the world; but the herbaceous species still thrive under a mean temperature of from 60° to 64° F., and are capable of being cultivated with advantage as far as the 40° or even 46° N. lat. The five-lobed leaves have a dark green colour, the flowers are yellow with a purple centre, and produce a pod about the size of a walnut, which, when ripe, bursts and exhibits to view the fleecy cotton in which the seeds are securely embedded.

It is almost superfluous to mention that the United States are the first cotton-producing country in the world. The area suitable for cotton south of the 36° of latitude, comprises more than thirty-nine million acres, of which less than one-sixth part is now devoted to the plant. The yield depends in part upon the length of the season. Seven months are required for an average crop, and the average periods in which the last killing frost of spring and the first killing frost of autumn occur are March 23, and October 26. Cotton is cultivated in large fields, and when the soil is superior, the plant rises to a height of six or eight feet, although in the richest cane-brake soil, exhausted by successive crops, it dwindles down to a height of three or four feet only. The aspect of a cotton field is most pleasing in the autumn, when the dark-coloured foliage and bright yellow flowers, intermingling with the snow-white down of the pods when burst, produce a charming contrast. Unfortunately the poor slaves have but little time to enjoy the scene. They are then overloaded with work, for it is important to pluck as much as possible during the first hours of morning, since the heat of the sun injures the colour of the cotton, and the overripe capsules shed their contents upon the ground, or allow the wind to carry them away. We can easily imagine, even if we had not been taught by " Uncle Tom," how actively the driver plies his whip at harvest time.

The collected produce is immediately carried to the steam-mill to be cleansed of the seeds, and then closely packed in bales weighing on an average 443 lbs., which in the seaports are further reduced by hydraulic presses to half of their previous volume, thus causing a great saving in the freight. Large clippers frequently carry eight or ten thousand of these bales to Liverpool, whence, perhaps, on the day of their arrival they are conveyed by rail to the next manufacturing town, and return in a few days to the port, ready to clothe the Australian gold-digger or the labourer on the banks of the Ganges.

India, which still in the last century provided Europe with the finest cambrics and muslins, now yearly receives from England cotton goods to a large amount. Thus the stream of trade may be said to have rolled backwards to its source, for though the wants of the Hindoo are easily satisfied, and cotton grows at his very door, yet his hand-loom is unable to compete with the machinery and the capital of England. Even in the exportation of the raw material he labours under great disadvantage, when compared with the rival American. Instead of steam-mills for the cleansing of his cotton, he makes use of his hands, which scarcely enable him to pick a pound a day. Wretched beasts of burden then carry it on wretched roads, and often over high mountain-passes, to the coast, where it arrives, sometimes only after fifty or sixty days, in a soiled and deteriorated condition. What a contrast with the States, where the cotton is so much better cleansed at a far inferior cost, where it finds its way so easily to the nearest river, and is then so rapidly conveyed by steam to Savannah, Mobile, or New Orleans!

If to all these disadvantages we add the poor quality, short staple, and consequent low price of the cotton, the tenure of the soil under which half or two-thirds of the crop is required for rent or taxes, the frequent want of irrigation, and the imperfect state of cultivation in general, it is less to be wondered at that the exportation of India amounts only to five or six hundred thousand bales a year, than that the country is able to export it at all.

Much, however, has lately been done, both by government and private exertions, to remedy these defects, and it is to be hoped that the time is not far distant when India, the parent land of cotton industry, will take at least the lead in the export of the raw material, and supplant, by the produce of her free labour, the slave-grown vegetable wool of the Western world.

A vast railway system, extending over more than 2,000 miles, is actively progressing. A central line from Calcutta to Delhi, of 1,100 miles, leads from the great sea-port to the ancient capital of the Moguls. Another of 600 miles from Madras to Belary is soon to be opened. A third railroad of 200 miles, from Bombay into the interior, is in progress; and a fourth of still greater length than the last, through Scinde and the Punjaub, is also commenced. Our fancy delights to dwell on the tide of prosperity ready to flow along these lines of rapid communication, over the plains of India; of the vast improvements in every respect which they are destined to call forth, and of their favourable reaction on our own country, for every progress of the Indian labourer is a boon to the hardworking artisan of England.

In the climate of India, where copious rains are succeeded by long intervals of drought, an artificial supply of water is required for the successful cultivation of the land, and to quote the language of Colonel Grant, a most competent judge, "Nothing appears more susceptible of improvement from culture and a regular supply of water than cotton." In fact the cotton of the common field and that of the irrigated bed cannot be recognised as the same plant; not only do the shrubs attain to a greater size, and bear many more pods, but each pod is much larger and contains a much greater quantity of fibre.

In ancient times large sums were applied by the enlightened sovereigns of India to irrigation, but in the earlier days of British rule, many of their works were suffered to decay. More recently some of their structures have been repaired, and several new canals have been opened, which have not only changed the aspect of the country, but have yielded a direct return of from ten to twenty per cent. Down to 1848 the average annual outlay on public works in India did not exceed 100,000*l*., but in 1857 it had risen to 2,220,000*l*. In Madras alone, ten works of irrigation were begun in 1853 which will irrigate and enrich 450,000 acres of land. In 1857 a joint-stock company was formed in London to construct a canal for irrigation and navigation through Madras, Berar, and Mysore, and another from the Malabar to the Coromandel coast, which will open 400,000 square miles of cotton-growing land—a much larger area than is now devoted to cotton in the United States. The aggregate expenditure of Great Britain on railways, canals, and other works of irrigation in India now exceeds 5,000,000*l.* annually; and we may safely predict, that when all these great and beneficial undertakings shall have been accomplished, the produce of India will astonish the world. Even as it is, the quantity of cotton grown in that country for home consumption is enormous. Competent judges* inform us that each individual requires on an average twenty pounds a year, and this will not seem too much when we reflect that almost all the inhabitants of India exclusively dress in cotton. This gives us an annual internal consumption of about 3,000,000,000 lbs., or a five times larger quantity than that which is spun and woven by all the looms and jennies of Great Britain. Thus India needs only to increase her production by *one-fifth* to supply us with all the cotton we actually want.

The civil war in the United States, with all its gloomy prospects of servile insurrection and agricultural ruin, cannot fail to hasten the work of progress so happily begun on the banks of the Ganges and Nerbuddah, for however it may end, it will have shown the folly and the danger of depending upon one country alone for the supply of an article almost as essential to Great Britain as corn itself.

But India is not the only country which bids fair to destroy the monopoly of the United States, and we may fairly hope that Africa will also ere long weigh more heavily in the balance. In 1857 a Cotton Supply Association was formed in Liverpool for the purpose of stimulating African production. Mr. J. Clegg of Manchester, who acts in concert with this society, reported under date of March 18, 1858, that 407 cotton-gins had been sent out by himself and others to the western coast of Africa, destined principally for the interior, and that he had corresponded with 76 native traders, including 22 chiefs, who were embarking in the culture of cotton. In the Yarriba country between the Niger and the sea, Dr. Barkie found large plantations of cotton. He saw from 1,000 to 2,000 bags of 80 pounds each exposed together in the markets, and realising large profits at a very low

---

* Capper: History of British India.

price. In the great district of the Soudan, which extends from the valley of the Niger to the sources of the Nile, and comprises at least sixteen degrees of latitude, cotton grows spontaneously, and is spun and woven by the native females. England is using every effort to divert the chiefs of this region from the slave-trade to the culture of cotton. With one hand she invites them to produce and sell the raw material, and with the other to receive the fabrics of her varied manufactures. Even now Dr. Livingstone, in his steam-launch, well supplied with cotton seed, is probably ascending the Zambesi towards the equator; a little to the north Messrs. Burton and Speke, with like objects in view, have penetrated into the interior of Ethiopia from the coast of Zanzibar; while the enterprising German, Dr. Barkie, has ascended the Niger, and doubtless distributed cotton-seed and cotton-gins as he advanced. Every friend of humanity will sincerely approve these laudable efforts, as their result must be to strike slavery at the root, and to abolish the most disgraceful trade that ever sullied the annals of our race.

At the present day, Brazil and Egypt rank immediately after the United States and India in the supply of the cotton market. The former exported in the year 1858, 29,341 bales to England, and the latter 94,650. The finest Brazilian and Egyptian qualities are nearly equal to the best Sea-island, which, as we all know, grows on the main coast, and also in the swampy regions bordering on most of the great rivers of the American cotton-producing states, while that which grows further from the sea, and at a higher level, has acquired the name of upland, which, though of inferior quality, is far more important from its vastly greater quantity.

In China, cotton is extensively grown for home consumption, but the exportation of the raw material is very trifling. The yellow colour of nankeen is not an artificial dye, but the intrinsic colour of the cotton of which it is fabricated.

Since the last twenty-five years cotton seeds are used for the production of a good lamp-oil. Formerly, the fibres that remained attached to the seeds after cleansing used to imbibe so much oil as notably to diminish the produce; but a machine invented in America, which strips off the husk from the kernels before submitting them to pressure, has entirely removed this difficulty, and imparted a considerable value to what was nearly a worthless article. Even the solid residue is of use, as a nutritious cattle food. Thus an enterprising people know how to turn the gifts of nature to account, and to rise in prosperity and riches, while so many other nations, though possessing the fairest regions of the globe, remain sunk in poverty and sloth, the victims of want in the midst of abundance.

Caoutchouc Trees — (Indians incising them).

CHAPTER XIX.

CAOUTCHOUC AND GUTTA PERCHA.

The Caoutchouc-Tree — Siphonia Elastica — Manner in which the Resin is collected — Urceola Elastica — Ficus Elastica — Has but recently become important — Mackintosh — Vulcanised Caoutchouc — Multiplicity of uses to which Caoutchouc may be applied — Marine Glue — The Gutta Percha Tree — Properties of the Resin — Its importance for Marine Telegraphy — Will the Tropical Zone be able to satisfy the growing demand?

WHEN we consider the luxuriance of vegetation in the tropical zone, it is not to be wondered at that so many plants of those climes abound with juices of a variety and richness unknown to those of the temperate latitudes. The resins and gums which our indigenous trees produce, either in smaller quantities or fit only for common uses, are there endowed with higher virtues, and ennobled, as it were, by the rays of a more powerful sun. Sometimes they exude spontaneously through the rind and harden in the atmosphere; more frequently a slight incision is required to make the sap gush forth in which they are dissolved, but in every case they require but trifling labour for their collection. Many of them have medicinal qualities, others are esteemed for their aromatic odour, but none ranks higher in a commercial and technical point of view than caoutchouc or India rubber, which was first brought from South America to Europe as a great curiosity at the beginning of the last century, and is now absolutely indispensable for a thousand different uses. Nothing was known even of its origin until the year 1736, when the French naturalist La Condamine, while exploring the banks of the Amazon, discovered that it was produced by a tree which the Indians called *Neve*, and which the learned have since introduced into the family of the Euphorbias under the name of Siphonia elastica. It grows to the height of about sixty feet, acquires a diameter of about twenty-four inches, and is found everywhere scattered through the primitive forests along the borders of the rivers in Guiana and North Brazil.

The resin is collected by the Indians in the following simple manner. With a small hatchet they make deep and long incisions in the rind, from which a milky sap abundantly exudes. A small wooden peg is then fixed into each aperture to prevent its closing, and a cup of moist clay fastened underneath, which in about four or five hours is filled with as many table-spoonfuls of the juice. The produce of a number of incisions having been gathered in a large earthen vessel, is then carried to the hut, where it is spread in thin coatings upon moulds made of clay, and dried, layer after layer, over a fire, until the whole has acquired a certain thickness. When perfectly dry, the clay form within is broken into small fragments, and the pieces are extracted through an aperture, which is always left for the purpose. In the caoutchouc the original pure white colour of the resin naturally changes into black, from the smoke to which it is exposed while drying. It is generally imported in the form of bottles, or in large flat pieces; frequently also in a liquid state, in hermetically closed vessels.

The Indian caoutchouc-gatherers or *seringueros* could easily collect sixteen pounds daily, but as they are extremely indolent, the average produce hardly amounts to one-fourth of that quantity.

Besides the Siphonia elastica, many other American trees, chiefly belonging to the families of the Euphorbiaceæ and Urticeæ, afford excellent kinds of caoutchouc; and since it is become so valuable an article of commerce, the East Indies, and Java likewise, yield considerable quantities, chiefly from the Urceola elastica and the Ficus elastica.

In spite of its many valuable properties, particularly its

wonderful elasticity and utter impermeability by water, it is only within the last forty years that caoutchouc has risen into importance, as before that time no cheap solvent had been found out which, on evaporating, would allow it to assume a variety of useful forms, with the retention of all its useful qualities.

In the year 1820, a Mr. Nadler discovered the art of cutting it into thin threads that could be woven with wool, silk, or cotton, into elastic tissues, but this invention was soon eclipsed by that of Mr. Mackintosh, who, by dissolving caoutchouc in petroleum and spreading it between two pieces of wool or cotton-stuff, united them by pressure into a water-tight cloth, which, though heavy, expensive, and of a disagreeable smell, was justly considered as a most useful fabric for a rainy climate. Since that time the manufacture of caoutchouc tissues has made rapid strides to perfection, particularly since the volatile oil, which is produced by the dry distillation of the resin itself, was found to be its best solvent.

About twenty years ago, overshoes or galoshes of India-rubber first began to be manufactured, but as they had the great fault of becoming soft and adhesive when exposed to warmth, and of hardening and contracting in cold weather, this branch of industry would never have acquired a great extension, if, about the year 1845, Messrs. Hancock and Broding had not found out that, by the addition of a small quantity of sulphur to caoutchouc, the latter, without losing any of its elastic qualities, acquires the property of retaining the same consistency through every change of atmospherical temperature. By this discovery, to which the strange name of *vulcanisation* has been given, a new field was opened for the use of caoutchouc; but the march of progress did not stop here, for a few years since, in America, Mr. Goodyear, by adding about twenty per cent. of sulphur to caoutchouc, converted it into a hard substance, capable of being used, in many instances, instead of wood, horn, or tortoise shell.

If fifty years ago one might ask what caoutchouc was good for besides effacing pencil-marks, its uses are now so manifold that it would almost take a volume to describe them. A small quantity of it dissolved in rape seed oil makes an excellent mixture for greasing machinery, as it remains fluid at every temperature. Melted and mixed with finely powdered slaked lime, it forms an excellent cement or putty, which will keep vessels air-tight for years. The famous marine glue, which is at present of such extensive use in ship-building, from its wonderful powers of soldering blocks of wood together, is merely a mixture of gumlac and India-rubber. In one word, chemistry may point to the various transformations and uses of caoutchouc as to some of its proudest technical triumphs.

The Icosandra Gutta, which furnishes the gutta percha, or gutta tuban, is a native of the Eastern Archipelago and the adjacent lands. A few years since this substance, now so celebrated and of such wide extended use, was totally unknown in Europe, for though from time immemorial the Malays employed it for making the handles of their hatchets and creeses, it was only in the year 1843 that Mr. Montgomery, an English surgeon, having casually become acquainted with its valuable properties, sent an account of it, with samples, to the Royal Society, for which he was most justly rewarded with its gold medal. The fame of the new article spread rapidly throughout the world; science and speculation seized upon it with equal eagerness; a thousand newspapers promulgated its praise; it was immediately analysed, studied, and tried in every possible way, so that it is now as well known and as extensively used as if it had been in our possession for centuries.

The Icosandra Gutta is a large high tree, with a dense crown of rather small dark green leaves, and a round smooth trunk. The white blossoms change into a sweet fruit, containing an oily substance fit for culinary use. The wood is soft, spongy, and contains longitudinal cavities filled with brown stripes of gutta percha. The original method of the Malays for collecting the resin consisted in felling the tree, which was then placed in a slanting position, so as to enable the exuding fluid to be collected in banana leaves. This barbarous proceeding, which from the enormous demand which suddenly arose for the gutta would soon have brought the rapidly rising trade to a suicidal end, fortunately became known before it was too late, and the resin is now gathered in the same manner as caoutchouc, by making incisions in the bark with a chopping knife, collecting the thin, white milky fluid which exudes in large vessels, and allowing it to evaporate in the sun, or over a fire. The solid residuum, which is the gutta percha of commerce, is finally softened in hot water, and pressed into the form of slabs or flat pieces, generally a foot broad, a foot and a half long, and three inches thick.

Gutta percha has many properties in common with caoutchouc, being completely insoluble in water, tenacious, but not elastic, and an extremely bad conductor of caloric and electricity. The name of vegetable leather which has been applied to it, gives a good idea both of its appearance and tenacity.

The uses of gutta percha are manifold. It serves for waterpipes, for vessels fit for the reception of alkaline or acid liquids which would corrode metal or wood, for surgical implements, for boxes, baskets, combs, and a variety of other articles. The wonder of the age, submarine telegraphy, could hardly have been realised without it, as it is only by being cased in so isolating a substance, and one so impermeable by water, that the metallic wire is able to transmit the galvanic stream through the depths of ocean from land to land.

Few articles have risen so rapidly to a great commercial importance as caoutchouc and gutta percha. The importation, which in 1830 amounted to no more than 52,000 pounds, rose in 1860 to no less than 43,039 hundredweight; and the consumption in France has advanced with no less giant strides, as above two millions of pounds were imported in the year 1855. At the present prices the value of the caoutchouc and gutta percha annually consumed in Europe is certainly not far short of a million of pounds sterling, and the quantity made use of in the now Disunited States of America is scarcely less enormous. If to the value of the raw article we add that which is imparted to it by the manufacturer's skill, we can form some idea of the vast importance of these gifts of the tropical zone, and of the extent to which they are made to contribute to the comforts of mankind. But will the supply be able to meet the constantly growing demand, or must the exhausted woods and plantations soon cease to provide the market at a moderate price? Fortunately, of this there is no fear, as the colossal Ficus elastica, which furnishes an abundance of caoutchouc, and is widely diffused over the East Indies, enjoys in Assam alone an area of more than 1,000 square miles. The Urceola elastica, which produces the best quality of the elastic resin, is no less frequent in the islands of the Eastern Archipelago, and its growth is moreover so rapid, that in the space of five years the trunk acquires a thickness of from twenty to thirty inches. Without injury to the tree, from fifty to sixty pounds of caoutchouc may be tapped from it annually, so that, even setting aside America, where the area of the Siphonia elastica is still greater, the East alone would be able to meet any demand. With regard to gutta percha, we may safely reckon upon the same inexhaustible abundance, as the rapidly growing Icosandra has a very wide range, and plantations of it, which might be multiplied to any extent all over the Eastern Islands, have already been formed at Singapore.

CHAPTER XX.

TROPICAL SPICES.

The Cinnamon Gardens of Ceylon — Immense profits of the Dutch — Decline of the Trade — Neglected state of the Gardens — Mode of preparing the Rind — Nutmegs and Cloves — Cruel monopoly of the Dutch — A Spice Fire in Amsterdam — The Clove Tree — Beauty of an Avenue of Clove Trees — The Nutmeg Tree — Mace — The Pepper Vine — The Pimento Tree.

ALTHOUGH the beautiful laurel, whose bark furnishes the most exquisite of all the spices of the East, is indigenous in the forests of Ceylon, yet as no author previous to the fourteenth century mentions its aromatic rind among the productions of the island, there is every reason to believe that the cinnamon, which in the earlier ages was imported into Europe through Arabia, was obtained first from Africa, and afterwards from India. That the Portuguese, who had been mainly attracted to the East by the fame of its spices, were nearly twenty years in India be-

Cinnamon.

fore they took steps to obtain a footing at Colombo, proves that there can have been nothing very remarkable in the quality of the spice at the beginning of the sixteenth century, and that the high reputation of the Ceylon cinnamon is comparatively modern, and attributable to the attention bestowed upon its preparation for market by the Portuguese, and afterwards on its cultivation by the Dutch.

Long after the appearance of Europeans in Ceylon, cinnamon was only found in the forests of the interior, where it was cut and brought away by the *Chalias*, an emigrant tribe which, in consideration of its location in villages, was bound to go into the woods to cut and deliver, at certain prices, a given quantity of cinnamon properly peeled and ready for exportation.

This system remained unchanged so long as Portugal was master of the country, but the forests in which the spice was found being exposed to constant incursions from the Kandyans, who sought every opportunity to obstruct and harass the Chalias and peelers, the Dutch were compelled to form enclosed plantations of their own within range of their fortresses. The native chieftains, fearful of losing the profits derived from the labour of the Chalias, who were attached as serfs to their domains, and whose work they let out to the Dutch, were at first extremely opposed to this innovation, and endeavoured to persuade the Hollanders that the cinnamon would degenerate as soon as it was artificially planted. The withering of many of the young trees seemed to justify the assertion, but on a closer examination it was found that boiling water had been poured upon the roots. A law was now passed declaring the wilful injury of a cinnamon plant a crime punishable with death, and by this severity the project was saved.

The extent of the trade during the time of the Dutch may be inferred from the fact, that the five principal cinnamon-gardens around Nejombo, Colombo, Barberyn, Galle, and Maduro were each from fifteen to twenty miles in circumference. Although they were only first planted in the year 1770, yet before 1796, when Colombo was taken by the English, their annual produce amounted to more than 400,000 lbs. of cinnamon, as much as the demands of the market required.

The profits must have been enormous, for cinnamon was then at least ten times dearer than at present, the trade being exclusively in the hands of the Dutch East Indian Company, which, in order to keep up the price, restricted the production to a certain quantity, and watched over its monopoly with the most jealous tyranny. No one was allowed to plant cinnamon or to peel it, and the selling or importing of a single stick was punished as a capital offence.

When the English took possession of the island, the monopoly was ceded to the East India Company for an annual sum of 60,000*l.* until 1823, when the colonial government undertook the administration of the cinnamon-gardens for its own account.

In 1831 the produce sank to 16,000*l.* sterling, and in the following year the ancient monopoly was abandoned; the government ceased to be the sole exporters of cinnamon, and thenceforward the merchants of Colombo and Galle were permitted to take a share in the trade, on paying to the crown an export duty of three shillings a pound. This was afterwards reduced to one shilling, and ultimately totally abolished; as not alone India and Java, but also Martinique, Guiana, and Mauritius, where the cinnamon tree had been introduced, were found capable of producing the spice; and the cheap substitute of Cassia, a still more formidable competitor, was arriving in Europe in large quantities from South China and the Trans-Gangetic peninsula. In Java alone the export of cinnamon, which in the year 1835 amounted only to 2,200 lbs., increased so rapidly that in 1845 it had already risen to 134,000 lbs., and as it can there be more cheaply produced, and the Dutch government was wise enough to limit the export duty to one halfpenny a pound, an unrestricted free trade was evidently the only means for preventing Ceylon from being entirely supplanted in the markets of the world. Under these circumstances, the Singhalese cinnamon has lost its ancient excellence, less care has been given of late years to the production of the finest qualities for the European market, and the coarser and less valuable shoots have been cut and peeled in larger proportions than formerly. Hence the gross quantity exported from Ceylon in 1857 (887,959 lbs.) was nearly double that of 1841 (452,039 lbs.); but from the joint effects of competition and the deterioration of quality, the prices have proportionally declined, and the total value of the export now hardly amounts to 50,000*l.*

The cinnamon-gardens, whose beauty and luxuriance has been so often vaunted by travellers, have partly been sold, partly leased to private individuals, and though less than a century has elapsed since they were formed by the Dutch, they are already becoming a wilderness.

Those which surround Colombo on the land side exhibit the

effects of a quarter of a century of neglect, and produce a feeling of disappointment and melancholy.

The beautiful shrubs which furnish this spice have been left to the wild growth of nature, and in some places are entirely supplanted by an undergrowth of jungle, while in others a thick cover of climbing plants, bignonias, ipomœas, the quadrangular vine (whose fleshy stem, when freshly cut, yields to the wild elephant a copious draught of pure tasteless fluid), the pitcher plant, and other parasites and epiphytes, conceals them under heaps of verdure and blossom. One most interesting creeper which encumbers the cinnamon tree is the night-blowing Alango, the moon flower of the Europeans, which never opens its petals until darkness comes on, and attracts the eye through the gloom by its pure and snowy whiteness.

It would, however, be erroneous to suppose that the cinnamon gardens have been universally doomed to the same neglect. Thus Professor Schmarda, who visited Mr. Stewart's plantation two miles to the south of Colombo, admired the beautiful order in which it was kept. A reddish sandy clay and fine white quarz sand form the soil of the plantation. White sand is considered as the best ground for the cinnamon tree to grow on, but it requires an abundance of rain, which is never wanting in the south-western part of the island; much sand, much water, much sun, and many termites, are, according to the Singhalese, the chief requirements of the plant. For as these otherwise so destructive creatures do not injure the cinnamon trees, but rather render themselves useful by destroying many other insects, they remain unmolested, and everywhere raise their high conical mounds or nests in the midst of the plantation. The aspect of a well-conditioned cinnamon-garden is rather monotonous, for though the trees when left to their full growth attain a height of forty or fifty feet, and a thickness of from eighteen to twenty inches, yet, as the best spice is furnished by the shoots that spring from the roots after the chief stem has been removed, they are kept as a kind of coppice, and not allowed to rise higher than ten feet.

The shrubs planted in regular rows, four or five feet apart, consist of four or five shoots whose slender stems, very much resembling those of the hazel tree, are leafed from top to bottom.

The leaf when first developed is partly of a bright red, and partly of a pale yellow; it soon, however, assumes a green hue, and when at its full growth is on the upper surface of a dark olive-colour, and on the under side of a lighter green; it somewhat resembles that of the bay, but is longer and narrower. Both sides are frequently deformed by galls produced by the puncture of a small Cynips. The flowers bloom in January, and grow on foot-stalks rising from the axillæ of the leaves and the extremities of the branches, clustering in bunches, which resemble in size and shape those of the lilac, but they are white with a brownish tinge in the centre. Though their smell has been frequently extolled as very fragrant, yet, according to Professor Schmarda, it is weak, and by no means agreeable, resembling that of animal albuminous liquids. The flowers are followed by one-seeded berries, of the shape of an acorn, but not so large as a common pea.

The plants are propagated by seeds or saplings. In two years the shoots are fit for cutting, being then about half an inch thick. Many of the shrubs in Mr. Stewart's plantations were already fifteen years old, but as the shoots are continually cut as soon as they have obtained the proper size, a full-grown trunk never forms, so that the more or less voluminous root-stalk is the only criterion of age.

The peeling of the rind takes place twice a year, from May to June, and in November, as at that time, in consequence of the heavier rains, and the increase of sap, it can be more easily detached from the wood. The epidermis having been scraped off, the bark is placed on mats to dry in the sun, when it curls up, and acquires a darker tint. The smaller pieces are then put inside the larger, and the whole closes together into the tubular form, in which it is sold in the shops. The finer sort is as thin as parchment, light brown, and extremely aromatic.

The Chalias, who are still the only natives employed, and have, from long hereditary practice, acquired a wonderful dexterity, are paid by the weight, receiving fourpence halfpenny per pound of the first quality and thus downwards, the average being about threepence three farthings. Their pay and treatment is good, and nothing is omitted to attach them to the gardens and to prevent them peeling the wild cinnamon trees, which would spoil the market. During the four months of the two harvests, they earn enough to maintain themselves for the whole year. They and their overseers have a great practice in distinguishing the different sorts of cinnamon by the taste; and during the old flourishing times of the monopoly, there were special cinnamon-tasters, who were able to distinguish eight different qualities, and who, during their functions, were obliged to live on rice, bread, or fruit, so as not to irritate their palate, and spoil its sense of discrimination.

The coarser rinds, the offals and fragments broken off in packing, are used for the distillation of cinnamon oil. Thirteen pounds of bark produce an ounce, worth about two shillings.

Another kind of cinnamon oil is distilled from the leaves, and is very different from that which is furnished by the rind. It is dark brown, and resembles in smell the essential oil of cloves, for which it is frequently sold in Europe, as its cost of production is much cheaper.

Nutmegs and cloves, the costly productions of the remotest isles of the Indian Ocean, were known in Europe for centuries before the countries where they grow had ever been heard of. Arabian navigators brought them to Egypt, where they were purchased by the Venetians and sold at an enormous profit to the nations of the west. But, as is well known, the commercial grandeur of the city of the Lagunes was suddenly eclipsed after Vasco de Gama discovered the new maritime road to the East Indies, round the Cape of Good Hope (1498); and when a few years later the countrymen of the great navigator conquered the Moluccas (1511), they for a short time monopolised the whole spice trade much more than their predecessors had ever done before. But here also as in Ceylon the Portuguese were soon obliged to yield to a stronger rival; for the Dutch now appeared upon the scene, and by dint of enterprise and courage soon made themselves masters of the Indian Ocean. In 1605 they drove the Portuguese from Amboina, and before 1621 had elapsed, the whole of the Moluccas were in their possession. Five-and-twenty years later, Ceylon also fell into their hands, and thus they became the sole purveyors of Europe, with cinnamon, cloves, and nutmegs. Unfortunately, the scandalous manner in which they misused their power throws a dark shade over their exploits. For the better to secure the monopoly of the spice trade, they declared war against nature itself, allowed the trees to grow only in particular places, and extirpated them everywhere else. Thus the planting of the nutmeg tree was confined to the small islands of Banda, Lonthoir, and Pulo Aij, and that of the clove to Amboina. Wherever the trees were seen to grow in a wild state, they were unsparingly rooted out, and the remainder of the Moluccas were occupied and subjugated for no other reason.

The natives were treated with unmerciful cruelty, and blood flowed in torrents to keep up the prices of cloves and nutmegs at an usurious height. When these spices accumulated in too large a quantity for the market, they were thrown into the sea or destroyed by fire.

Thus M. Beaumare, a French traveller, relates that on June 10, 1760, he beheld near the Admiralty at Amsterdam a blazing pile of these aromatics, valued at four millions of florins, and an equal quantity was to be burnt the next day. The air was perfumed with their delicious fragance, the essential oils freed from their confinement distilled over, mixing in one spicy stream, which flowed at the feet of the spectators; but no one was suffered to collect any of this, or, on pain of heavy punishment, to rescue the smallest quantity of the spice from the flames.

Fortunately these distressing scenes — for it is painful to see man, under the impulse of an insatiable greed, thus wilfully destroying the gifts of nature — belong to the history of the past. The reign of monopoly has ceased even in the remote Moluccas, and their ports are now, at length, thrown open to the commerce of all nations; for the spice trees having been transplanted into countries beyond the control of the Dutch, the ancient system could not possibly be maintained any longer.

The clove tree belongs to the far-spread family of the myrtles; the small lanceolate evergreen leaves resemble those of the laurel, the flowers growing in bunches at the extremity of the branches. When they first appear, which is at the beginning of the rainy season, they are in the form of elongated greenish buds, from the extremity of which the corolla is expanded, which is of a delicate peach-blossom colour.

When the corolla begins to fade, the calyx turns yellow, and then red; the calyces with their embryo-seed are in this stage of their growth beaten from the tree, and, after being dried in the sun, are known as the cloves of commerce. If the fruit be allowed to remain on the tree after arriving at this period, the calyx gradually swells, the seed enlarges, and the pungent properties of the clove are in great part dissipated.

The whole tree is highly aromatic, and the foot-stalks of the leaves have nearly the same pungent quality as the calyx of the flowers. "Clove trees," says Sir Stamford Raffles, "as an avenue to a residence, are perhaps unrivalled — their noble height, the beauty of their form, the luxuriance of their foliage, and, above all, the spicy fragrance with which they perfume the air, produce, on driving through a long line of them, a degree of exquisite pleasure only to be enjoyed in the clear light atmosphere of these latitudes."

Clove.

Cloves contain a very large proportion of essential oil, which combined with a peculiar resin gives them their pungent aroma. It seems, however, to require a combination of favourable circumstances of climate and soil for the full developement of their virtues, for, though the tree is found in the larger islands of Eastern Asia, and in Cochin China, it has there little or no flavour, and the Moluccas seem to be the only place where the clove comes to perfection without being cultivated. Though it is at present planted in Zanzibar, Cayenne, Bourbon, Trinidad, and other places, yet Amboina still furnishes the best quality and the largest quantity, exporting annually about a million of pounds.

In spite of the endeavours of the Dutch to confine the nutmeg-tree to the narrow precincts of Banda, it has likewise extended its range not only over Sumatra, Mauritius, Bourbon, and Ceylon, but even over the western hemisphere.

It is of a more majestic growth than the clove, as it attains a height of fifty feet, and the leaves, of a fine green on the upper surface, and grey beneath, are more handsome in the outline, and broader in proportion to the length. When the trees are about nine years old, they begin to bear. They are dioecious, having male or barren flowers upon one tree, and female or fertile upon another. The flowers of both are small, white, bell-shaped, without any calyx; the embryo-fruit appearing at the bottom of the female flowers in the form of a little reddish knob. When ripe, it resembles in appearance and size a small peach, and then the outer rind, which is about half an inch thick, bursts at the side, and discloses a shining black nut, which seems the darker from the contrast of the leafy network of a fine red colour with which it is enveloped. The latter forms the Mace of commerce, and having been laid to dry in the shade for a short time, is packed in bags and pressed together very tightly.

Nutmeg.

The shell of the nut is larger and harder than that of the filbert, and could not, in the state in which it is gathered, be broken without injuring the nut. On that account the nuts are successively dried in the sun and then by fire-heat, till the kernel shrinks so much as to rattle in the shell, which is then easily broken. After this the nuts are three times soaked in sea-water and lime; they are then laid in a heap, where they heat and get rid of their superfluous moisture by evaporation. This process is pursued to preserve the substance and flavour of the nut, as well as to destroy its vegetative power.

The kernel contains both a fixed oil, which is obtained by pressure, a pound generally yielding three ounces, and a transparent volatile oil, which may be obtained by distillation in the proportion of one thirty-second part of the weight of nutmeg used.

Banda still furnishes the most and the best nutmegs, though the tree is now cultivated in many other tropical countries. The yearly produce is estimated at about 500,000 pounds of nutmegs and 150,000 pounds of mace.

The outer rinds are likewise not without use to the natives. They are laid in large heaps, and allowed to putrefy, when they get covered with a blackish mushroom, which is esteemed as a great delicacy.

Although not so costly as cloves or cinnamon, pepper is of a much greater commercial value, as its consumption is at least a hundred times greater. It grows on a beautiful vine, which, incapable of supporting itself, twines round poles prepared for it; or, as is more common in the Travancore plantations, the pepper vines are planted near mango and other trees of straight high stems. As these are stripped of the lower branches, the vine embraces the trunk, covering it with elegant festoons and rich bunches of fruit in the style of the Italian vineyards.

The leaf of the pepper plant is large, resembling that of the ivy, and of a bright green; the blossoms appear in June, soon after the commencement of the rains; they are small, of a greenish white, and are followed by the pungent berries, which hang in large bunches, resembling in shape those of grapes, but the fruit grows distinct on little stalks like currants.

Pepper Plant.

In Malabar, pepper is gathered in February, and has the same appearance as in Europe. Although the vine begins to bear in the fourth or sixth year, it is not in perfection before the ninth or tenth, and continues bearing as many years longer, if in a congenial soil.

Assiduity and cleanliness are essentially necessary in a pepper garden; not a weed is permitted to grow; the produce, however, amply compensates this trouble, as a plant in full growth is able to furnish six or seven pounds of berries.

This valuable spice grows chiefly on the Malabar coast, in Sumatra, Borneo, Java, Singapore; its cultivation has also been introduced in Cayenne and the West Indies. The black and white sorts of pepper are both the produce of the same plant.

The best white peppers are supposed to be the finest berries which drop from the tree, and, lying under it, become somewhat bleached by exposure to weather; the greater part of the white pepper used as a condiment is, however, the black merely steeped in water, and decorticated, by which means the pungency and real value of the spice are diminished; but having a fairer and more uniform appearance when thus prepared, it fetches a higher price.*

Jamaica is the chief seat of the magnificent myrtle (Myrtus pimenta), which furnishes the pimento of commerce. This beautiful tree grows to the height of about thirty feet, with a smooth, brown trunk, and shining green leaves resembling those of the bay. In July and August a profusion of white flowers, filling the air with their delicious odours, forms a very pleasing contrast to the dark foliage of its wide-spreading branches. It grows spontaneously in many parts of the island, particularly on the northern side, in high spots near the coast.

When a new plantation is to be formed, no regular planting or sowing takes place, for, as Edwards ("History of Jamaica") observes, "the pimento tree is purely a child of nature, and seems to mock all the labours of man in his endeavours to extend or improve its growth; not one attempt in fifty to propagate the young plants, or to raise them from the seeds in parts of the country where it is not found growing spontaneously, having succeeded. For this reason, a piece of land is chosen, either in the neighbourhood of a plantation already formed, or in a part of the woodland where the pimento-myrtles are scattered in a native state. The land is then cleared of all wood but these trees, which are left standing, and the felled timber is allowed to remain, where it falls to decay, and perishes. In the course of a year, young pimento plants are found springing up on all parts of the land, produced, it is supposed, in consequence of

---

* Importations of Spices in Great Britain during the first eight months of the year 1861 : —

Cinnamon	.	540,203lbs.
Nutmegs	.	301,215 ,,
Cloves	.	41,780 ,,
Pepper	.	5,742,627 ,,
Pimento	.	2,016,300lbs.
Ginger	.	778,000 ,,
Cassia Lignea	.	192,087 ,,

Prices of the best qualities —

	s.	d.			s.	d.	
Cinnamon	2	5	per lb.	Ginger	114	0	per cwt.
Cloves (Amboina)	1	4	,,	Pepper	0	5¼	per lb.
Nutmegs	3	6	,,	White Pepper	1	2	,,
Mace	1	9	,,	Pimento	0	3⅛	,,
Cassia Lignea	92	0	per cwt.				

Economist, October.

the ripe berries having been scattered there by the birds, while the prostrate trees protect and shade the tender seedlings. At the end of two years the land is thoroughly cleared, and none but the most vigorous plants, which come to maturity in about seven years, are left standing."

The berries are carefully picked while yet green, since, when suffered to ripen, they lose their pungency. One person on the tree gathers the small branches, and three others, usually women and children, find full employment in picking the berries from them. The produce is then exposed to the sun for about a week, when the berries lose their green hue and become of a reddish brown. When perfectly dry, they are in a fit state for exportation. In favourable seasons, which, however, seldom occur above once in five years, the pimento crop is enormous, a single tree having been known to yield one hundredweight of the dried spice. From its combining the flavour and properties of many of the oriental aromatics, pimento has derived its popular name of allspice, and, from its being cheaper than black pepper, its consumption is very great.

Though but a lowly root, ginger almost vies in commercial importance with the aromatic rind of the cinnamon-laurel, or the pungent fruit of the nutmeg-myrtle. The plant which produces this valuable condiment belongs to the tropical family of the Scitamineæ, or spice lilies, which also reckons among its members the Cardomum and the Curcuma. Its jointed tubers creep and increase under ground, and from each of them springs up an annual stem about two feet and a half high, with narrow and lanceolate leaves. The flowering stalk rises directly from the root, ending in an oblong, scaly spike; from each of these scales a single white and blue flower is produced. Ginger is imported into this country, under the form of dried roots and as a preserve. We receive it both from the East and West Indies, but that from the latter is much superior in quality to the former.

Cutting the Indigo Plant (from " Rural Life in Bengal ").

CHAPTER XXI.

TROPICAL VEGETABLE DYES.

Indigo — Indigofera tinctoria — Mode of its Cultivation, and preparation of the Dye for the Market — Indigo Factories in Bengal.
Logwood — The British Logwood Cutters in Honduras — Their mode of living and disputes with the Spaniards — Brazil Wood — Red Saunders — Arnatto — Fustic — Turmeric.

OF all the dyeing substances which the tropical zone produces in such endless variety, none is more important in a commercial point of view than indigo.

Cultivated from time immemorial in India, and employed by the Mexicans to give a beautiful hue to their cotton fabrics, long before Cortez with his band of adventurers destroyed the empire of Montezuma, it was first imported into Europe by the Dutch, about the year 1631. For a long time its use met with great opposition both in Germany and France, and it was classed among the pernicious substances that were strictly prohibited as *devil's dyes*, the name of his Satanic majesty being here, as in so many cases, called in to aid the cause of prejudice and ignorance. Finally, however, its intrinsic merits triumphed over every obstacle, and it has now almost entirely superseded the use of woad, which for many centuries had served to dye the blue broad cloths and linens of the European weavers.

Various species of indigo plants or indigoferas are found growing spontaneously in the warmer countries of both hemispheres, but the indigofera *tinctoria* is most generally cultivated, from its yielding a greater quantity of pulp. The knotty shrubby plant rises about two feet from the ground; the leaves are winged like those of the acacia, smooth and soft to the touch, furrowed above, and of a darker colour on the upper than the under side. The small reddish flowers which grow in ears from the axillæ of the leaves have no smell, and are succeeded by long crooked brown pods, which contain small yellow seeds. The plant requires a smooth rich soil, well tilled, and neither too dry nor too moist. A child of the sun, it cannot be advantageously cultivated anywhere except within the tropics, a higher mean temperature than 60° being absolutely necessary for its vegetation. The seed is sown in little furrows about the breadth of the hoe, and two or three inches in depth. These furrows are made a foot apart from each other, and in as straight a line as possible. Soon after sowing, continual attention is required to pluck the weeds, which would quickly choke up the plant and impede its growth. Sufficient moisture causes it to shoot above the surface in three or four days, and it is usually fit for gathering at the end of two months. When it begins to flower it is cut with a sickle a few inches above the roots, and furnishes, after six or eight weeks, a second crop. In the countries most favourable to its growth the planter is even able to cut it four times in one year; but as the produce diminishes fast after the second cutting, the seeds must annually be sown afresh.

The cultivation of indigo would thus seem to be extremely profitable, but the sun, which so rapidly improves and invigorates the plant, calls forth at the same time a multitude of insects and caterpillars, that prey upon the valuable leaves, and frequently disappoint the planter's expectations.

The colouring matter of indigo is not ready-formed in the plant, but requires fermentation and contact with the oxygen of the air to attain the deep cerulean tint which renders it so valuable. The herb is first put into a vat or cistern called the steeping-trough, and there covered with water. The matter begins to ferment, sooner or later, according to the warmth of the weather, and the maturity of the plant; sometimes in six or eight hours, and sometimes in not less than twenty. The liquor grows hot and throws up a plentiful copper-coloured froth which passes into violet towards the end, but the pulp and liquor remain green. At this time, without touching the herb, the liquor impregnated with its tincture is let out by cocks in the bottom into another vat, placed for that purpose so as to be commanded by the first. The process must be conducted with great caution and technical ability, for if the fermenting liquor is abstracted too soon, part of the colouring matter remains in the plant, while by a longer delay one runs the risk of partly losing it through putrefaction.

In the second vat, called the beating vat, the liquor is strongly and incessantly beaten with a kind of buckets fastened to poles, till the colouring matter, which during this process changes from green to deep blue, is united into a body, and acquires a tendency to precipitate. This operation, which generally lasts a couple of hours, must also be conducted with great caution, for by ceasing it too soon, colouring particles not sufficiently saturated with oxygen remain in the liquid, while by continuing it for too long a time, a new fermentation is brought on, which changes and destroys the pigment.

As soon as it is judged that the beating is sufficient, the mass is left at rest for two hours, after which the clear liquor is drawn off by cocks in the side of the vat, and the blue part is discharged by another cock into a boiler, where, having been brought to that degree of consistence, which is safely practicable, it is conveyed into bags of cloth to strain off the remaining moisture, and lastly exposed to the air in the shade, in shallow wooden boxes, till it is thoroughly dry.

Since the last forty years, great changes have taken place in the indigo trade. Formerly that from Guatemala was considered the best, and very little was imported from the East Indies, while now Bengal not only produces the finest indigo, but also brings by far the largest quantity to the market. This change is entirely owing to a few energetic and enlightened men, who, by introducing improvements in the cultivation and manufacture of indigo, have raised it to be the chief export article of the East Indies; and, if we may judge from this example, there is every reason to hope that a similar degree of attention will have the same effect on the growth of East Indian cotton, and soon render us entirely independent of Georgia and Carolina, whether these states escape the dangers of a social revolution, or be engulphed in the ruin provoked by their own folly. The indigo factories in the East Indies are conducted very differently from those in America, on account of the dissimilar circumstances of the population of the two countries. In America the indigo plantations, and the works connected with its preparation, are all the same property, and under the same superintendence, while in Bengal the cultivation is exclusively left to the native farmers, who are provided with seed by the factor, and bound to deliver, at a certain rate of prices, the whole of the plants produced from these seeds. The cultivators, in consequence of failures in crops or other accidents, too frequently require advances from their employer; and thus, though nominally free, they are in reality subjected to him, and compelled to raise the indigo exclusively for the supply of his factory.

These establishments are generally on a very large scale, and nearly always remote from the seat of the English presidencies. The superintendence is seldom intrusted to any but one of the proprietors, who, entirely excluded from the society of his countrymen, consents to many privations, with the hope that in a few years he may reap sufficient wealth to enable him to spend the remainder of his life in comfort, and having accomplished this end, usually resigns his situation to a junior partner, who pursues the same course. But though "man proposes, God disposes," and thus these brilliant expectations are far from being invariably realised; for though in good years the profits of an indigo property are immense, yet these periods of prosperity are frequently followed by such disastrous casualties, that in the West Indies the cultivation of indigo had been almost entirely given up, even before the negro emancipation,—many planters, who had hoped to open a gold mine in its pursuit, having been utterly ruined by the devastations of insects, the great mortality among their slaves, in consequence of the noxious exhalations of the fermentation, and the spoiling of the colouring matter by a faulty method of preparation.

All the intermediate shades of violet and purple may be obtained from the mixture of red and blue, varying according to the different proportions wherein these colours are applied. There are, however, some few vegetable substances which yield a violet or purple dye, without being combined with another colour, and of these logwood is the most important. The stately tree which furnishes this valuable article of commerce is a native of the western world, having been first discovered in the swampy forests of Yucatan, and in the low alluvial grounds that girdle the Bays of Campeachy and Honduras.

Logwood was known as early as the reign of Elizabeth, but its use was forbidden by act of parliament, in consequence of a similar prejudice to that which prevailed against indigo, and the prohibition was rigorously enforced until the year 1661. Logwood now became in great request; and as the indolent Spaniards failed to supply the market, several English adventurers, without first asking permission, settled or squatted on the uninhabited coast of Yucatan, and made the woods near Laguna de Terminos ring with the sound of their industrious axe.

Many years passed without the Spaniards taking any notice of the intruders; but as these, growing bolder by sufferance, began to penetrate farther into the country, to build houses and form plantations, as if they had been masters of the soil, their jealousy was at length aroused, and in 1680 the English settlers, after suffering much annoyance, were forcibly ejected. This triumph on the part of their adversaries was, however, but transitory; and a few months after our sturdy countrymen were again cutting their logwood as busily as ever, in spite of the enmity of man and the innumerable hardships of their laborious occupation.

Their mode of life is thus quaintly described by the famous Dampier in his Voyage to the Bay of Campeachy:—"The logwood-cutters inhabit the creeks of the lagunes in small companies, building their huts by the creeks' sides for the benefit of the sea-breeze, as near the logwood groves as they can, and often removing to be near their business. Though they build their huts but slightly, yet they take care to thatch them very well with palmetto leaves, to prevent the rains, which are there very violent, from soaking in.

"For their bedding, they raise a wooden frame, three feet and a half above ground on one side of the house, and stick up four stakes at each corner, one to fasten their curtains, out of which there is no sleeping for mosquitoes. Another frame they raise, covered with earth, for a hearth, to dress their victuals; and a third to sit at, when they eat it. During the wet season, the land where the logwood grows is so overflowed that they step from their beds into the water, perhaps two feet deep, and continue standing in the wet all day till they go to bed again; but, nevertheless, account it the best season for doing a good day's labour in.

"Some fell the trees, others saw and cut them into convenient logs, and one chips off the bark, and he is commonly the principal man; and when a tree is so thick that after it is logged it remains still too great a burden for one man, it is blown up with gunpowder. The logwood-cutters are generally sturdy strong fellows, and will carry burthens of three or four hundred-weight.

"In some places they go a-hunting wild cattle every Saturday to provide themselves with beef for the week following. When they have killed a beef they cut it into quarters, and taking out the bones, each man makes a hole in the middle of his quarter, just big enough for his head to go through, then puts it on like a frock and trudgeth home; and, if he chanceth to tire, he cuts off some of it and throws it away."

The entire freedom from all restraint which accompanied this wild and adventurous life had such charms for a bold and roving spirit, that Dampier himself sojourned for about a year among the rude wood-cutters of Campeachy, and left them with the intention of again returning for a longer stay.

The intrusion of the British continued to give rise to perpetual disputes and hostilities, until at length, by the treaty of Utrecht in 1713, the privilege of cutting logwood was confirmed to the English, in plain and express terms, so that it was supposed the question was now set at rest for ever. Yet it still continued to be a subject of constant dispute between the parties; and in 1717 the Spanish ambassador delivered a memorial to the British government against the English settlements at Laguna de Terminos, declaring that if, in the space of eight months, they were not abandoned, the inhabitants should be considered and treated as pirates. The British government, however, returned a short and positive answer, that its subjects were fully entitled to remain where they were, and to cut as much logwood as they pleased; and Spain, which seemed to have forgotten that weakness should avoid to raise pretensions which it is unable to enforce, was obliged to abide by this decision, and the settlement continued without being matter of further dispute or treaty for more than forty years. During this long period the British settlers had not been idle. As they fortified themselves against the assaults of the Spanish Americans, their colony assumed a more important and imposing aspect; and though by the treaty of 1763 the two countries came to a compromise on the subject (the English government consenting that the fortifications should be demolished, while the Spanish government engaged that the subjects of Great Britain should not be molested in cutting or shipping logwood, the sovereignty of the country remaining with Spain), yet they had become so used to consider themselves as the rightful owners of the land, that on the French attempting to share their privilege, they drove them away with all the irascibility of offended proprietors.

Most of the red dye-woods of commerce are furnished by the Cæsalpinias, a genus of plants belonging to the widespread family of the Leguminosæ, and indigenous in both hemispheres.

The *Cæsalpinia crista*, which furnishes the best quality, commonly known under the name of Brazil wood, grows profusely in the forests of that vast empire, preferring dry places and a rocky ground. Its trunk is large, crooked, and full of knots; at a short distance from the ground innumerable branches spring forth and extend in every direction in a straggling, irregular, and unpleasing manner. Trees of the largest growth attain to thirty or forty feet high, but they are rarely met with of so great dimensions. The branches are armed with short strong upright thorns, the leaves are small, and never appear in luxuriant foliage. The flowers are of a beautiful red colour, and emit a fragrant smell. Both the thick bark and the white pithy part of the trunk are useless, the hard close-grained heart being the only portion impregnated with colouring matter. The wood is sometimes used in turning, and is susceptible of a good polish, but its chief use is as a red dye. By the addition of acids it produces a permanent orange or yellow colour, while the crimson tints which it imparts are very fleeting.

The red-dye woods known in commerce under the names of Saint Martha, Nicaragua, and Lima, proceed from different varieties of the *Cæsalpinia echinata* and *brasiliensis*. Saint Martha wood is chiefly exported from Rio Hacha, a port of the republic of New Grenada on the Caribbean Sea, being the exclusive product of that country; Lima wood grows along the coast of the Pacific from Panama to Chili, and is exported to Europe from the ports of Mazatlan, Arica, Iquique, and Valparaiso. The eastern hemisphere also furnishes several red-dye woods to commerce. *Cæsalpinia Sappan* and *bimas* grow in the eastern archipelago, and are frequently mixed with the more valuable Brazil wood.

Red Saunders, a solid compact wood, is furnished by the *Pterocarpus santalinus*, a large tree growing chiefly on the coast of Coromandel; and a variety of the same plant produces the Caliatour wood, which is not only used for dyeing, but is highly prized by cabinet makers for its hardness, fine polish, and variously-striped beautiful red colour.

The first Europeans that settled on the banks of the Amazon found that several of the Indian tribes that roamed about in their vicinity painted their bodies with a showy orange-red colour. Their attention was by this means attracted to the Arnatto (*Bixa orellana*), which attains about the size of our hazel tree. The heart-shaped leaves are about four inches long, of a lighter green on the upper surface, and divided by fibres of a reddish-brown colour; the rosy flowers are succeeded by bristled pods somewhat resembling those of a chesnut, which, bursting open when ripe, display a splendid crimson farina or pulp, in which are contained thirty or forty seeds, in shape similar to raisin stones. As soon as they have arrived at maturity the pods are gathered, divested of their husks, bruised, immersed in water, and after a few weeks beaten with wooden paddles to promote the separation of the pulp from the seeds. The turbid liquor is then strained, boiled to a consistent paste, and finally formed into cakes, which are left to dry in the sun. In England arnatto is generally used by the dyer to give a deeper shade to the simple yellow. Being perfectly soluble in spirits of wine, it is much used in this state for lacquering and for giving an orange tint to the yellow varnishes. It is likewise employed in large quantities as a colouring ingredient for cheese, to which it gives the required tinge without imparting any unpleasant flavour or unwholesome quality.

The Yagua Indians prepare arnatto in a peculiar manner, producing the finest scarlet dye, which, according to Castelnau, is not inferior to cochenille itself.

A species of mulberry (*Morus tinctoria*), spontaneously growing in Brazil and in many of the West India islands, where it attains to a considerable size, provides our dyers with fustic. This wood, which is of a sulphur colour, with orange veins, is imported in large blocks chiefly from Cuba and Jamaica, the former being very superior in quality. It is much employed in combination with indigo to dye what is called Saxon green. Other yellow dye-woods, but inferior to fustic, are imported from Tampico, Tuspan, Zapote, Carthagena, Maracaibo, San Domingo, and the East Indies.

Turmeric or Indian saffron is a yellow dye, obtained from the roots of *Curcuma longa*, a plant belonging to the family of the *Scitamineæ*, or spice lilies. It is indigenous in South Asia, Madagascar, and many of the South Sea islands, and is now also cultivated in the West Indies. The roots spread far under the surface of the ground; they are long and succulent, and about half an inch in thickness, having many circular knots, from which arise four or five spear-shaped leaves standing upon long foot-stalks. The yellowish red flowers grow in loose scaly spikes, surmounting the foot-stalks which spring from the larger knots of the roots, and attain to about a foot in height. The roots, which are very hard, externally grey but internally of a deep lively yellow, and not unlike either in figure or size to ginger, must be reduced to powder previously to being employed as a dye. Though turmeric possesses no durability, it is used in considerable quantities for the dyeing of woollen and silk stuffs. With indigo it produces a beautiful green colour, which is frequently employed by confectioners, and it is said that the wily Chinese not seldom make use of this mixture to change their black tea into green.

PART III.

TROPICAL ANIMALS

CHAPTER XXII.

THE INSECT PLAGUES OF THE TROPICAL WORLD.

The Universal Dominion of Insects — Mosquitoes — Stinging Flies — *Œstrus Hominis* — The Chegoe or Jigger — The *Filaria Medinensis* — The Bête-Rouge — Blood-sucking Ticks — Garapatas — The Land-leeches in Ceylon — The Tsetsé Fly — The Tsalt-Salya — The Locust — Its dreadful Devastations — Cockroaches — The Drummer — The Cucarachas and Chilicabras.

THE insect tribes may, without exaggeration, be affirmed to hold a kind of universal empire over the earth and its inhabitants, for nothing that possesses, or has possessed, life is secure from their attacks. They vanquish the cunning of the fox, the bulk of the elephant, the strength of the lion; they plague the reindeer of the northern *tundras,* and the antelope of the African wilds; and all the weapons with which nature has furnished the higher orders of animals against their mightier foes prove ineffectual against these puny persecutors, whose very smallness serves to render them invulnerable. How numerous are the sufferings they entail on man! how manifold the injuries they inflict on his person or his property! To secure himself from their attacks, a perpetual warfare, an ever-wakeful vigilance, is necessary; for, though destroyed by thousands, new legions ever make their appearance, and to repose after a victory is equivalent to a defeat.

In our temperate zone, where a higher cultivation of the ground tends to keep down the number of the lower animals, their persecutions, though frequently annoying, may still be borne with patience; but in the tropical regions, where man is generally either too indolent or not sufficiently numerous to set bounds to their increase, the insects constitute one of the great plagues of life.

Along the low river-banks, and everywhere on hot and swampy grounds, the blood-thirsty Mosquitoes appear periodi-

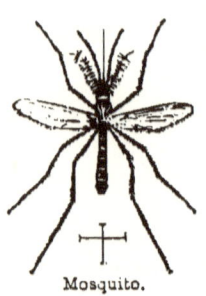

Mosquito.

cally in countless multitudes, the dread of all who live in warm climates. Scarcely has the sun descended below the horizon, when these insects arise from the morass to disturb the rest of man and to render existence a torment.

Not satisfied with piercing the flesh with their sharp proboscis, which at the same time forms a kind of syphon through which the blood flows, these malignant gnats, of which there are many species, inject a poison into the wound, which causes inflammation, and prolongs the pain.

In Angola, Dr. Livingstone found the banks of the river Seuza infested by legions of the most ferocious mosquitoes he ever met with during the course of his long travels. "Not one of our party could get a snatch of sleep. I was taken into the house of a Portuguese, but was soon glad to make my escape, and lie across the path on the lee-side of the fire, where the smoke blew over my body. My host wondered at my want of taste, and I at his want of feeling; for, to our astonishment, he and the other inhabitants had actually become used to what was at least equal to a nail through the heel of one's boot, or the tooth-ache."

Mr. Edwards, on his voyage up the Amazon, was no less tormented by these troublesome pests. "Soon after dark, we crossed the mouth of the Zingu, much to the displeasure of the Indians, who wished to stop upon the lower side; and they were very right, for scarcely had we crossed when we were beset by such swarms of mosquitoes as put all sleep at defiance. Nets were of no avail, even if the oppressive heat would have allowed them; for those which could not creep through the meshes would in some other way find entrance, in spite of every precaution. Thick breeches they laughed at, and the interior of the cabin seemed a bee-hive. This would not do, so we tried the deck, but fresh swarms continually poured over us, and all night long we were foaming with vexation and rage."

On the eastern slopes of the Andes, the mosquitoes or

Sancudos are so frequent that even the Indians consider many districts on the Marañon and Huallaga as absolutely uninhabitable. During his first sojourn in the Peruvian forests, Tschudi lay for several days almost motionless, with a swollen head and limbs, in consequence of the bite of these intolerable flies; and although by degrees the skin became more accustomed to the nuisance, and swelling no longer followed, yet their sting never failed to cause great pain. During three months of the year they infest the province of Maynas to such a degree, that even the stoical Indians utter loud complaints, and the dogs endeavour to escape them by burying themselves in the sand. Breeding cattle in these parts is impossible, for the sancudos would be the inevitable death of the calves.

In no season of the year, at no hour of the day or night, do they desist from their attacks; but they are more peculiarly abundant in the second half of the rainy season. One of the most remarkable circumstances of their history is that they not only totally avoid certain limited districts, but even small places in the middle of a village. It is a fact, well known to every traveller on the Marañon, that certain sand islands or tracts along the banks are deserted by the sancudos, while others, to all appearances quite similar, are infested by them. The so-called *black* water (a still unexplained natural wonder), which resembles a polished surface of black marble, and sometimes even marks the entire course of considerable rivers — as, for instance, the Rio Negro — seem to be particularly unfavourable to them, so that many places in Solimoes are thus converted into a painless paradise.

Most of the independent tribes of Maynas protect themselves against the stings of the sancudos by a kind of tent, made of the leaves of the Mauritia palm, and even the poorest Indians of the missions possess cotton *toldos*, which have an opening only on the ground, and, though insupportably hot, yet keep off the still greater plague of these tormentors.

Not content with a passing attack, a South American gadfly (*Œstrus hominis*) deposits its eggs under the human skin, where the larvæ continue for six months. If disturbed, they penetrate deeper, and produce troublesome ulcers, which sometimes even prove fatal. Thus, in tropical America, we find the same insect tribe which plagues our oxen and horses, and reduces the northern reindeer to desperation, settle on man himself, and render even the lord of the creation subject to its power.

The Chegoe, Pique, or Jigger of the West Indies (*Pulex penetrans*), is another great torment of the hot countries of America. It looks exactly like a small flea, and a stranger would take it for one. However, in about four and twenty hours he would have several broad hints that he had made a mistake in his ideas of the animal. Without any respect for colour, it attacks different parts of the body, but chiefly the feet, betwixt the toe-nails and the flesh. There it buries itself and causes an itching, which at first is not unpleasant, but after a few days gradually increases to a violent pain. At the same time a small white tumour, about the size of a pea, and with a dark spot in the centre, rises under the skin. The tumour is the rapidly growing nest of the chegoe, the spot the little plague itself. And now it is high time to think of its extirpation, an operation in which the negro women are very expert. Gently removing with a pin the skin from the little round white ball or nest, precisely as we should peel an orange, and pressing the flesh all round, they generally succeed in squeezing it out without breaking, and then fill the cavity with snuff or tobacco, to guard against the possibility of a fresh colony being formed by some of the eggs remaining in the wound. New comers are particularly subject to these creatures. Mr. Waterton, who by practise appears to have become very expert in eradicating chegoes' nests, once took four out of his feet in the course of the day, and a negress extracted no less than eighty-three out of Richard Schomburgk's toes in one sitting. "Every evening," says the venerable naturalist of Walton Hall, "before sundown, it was part of my toilet to examine my feet and see that they were clear of chegoes. Now and then a nest would escape the scrutiny, and then I had to smart for it a day or two after. A chegoe once bit upon the back of my hand: wishful to see how he worked, I allowed him to take possession. He immediately set to work head foremost, and in about half an hour he had completely buried himself in the skin. I then let him feel the point of my knife, and exterminated him."

"In the plantations of Guiane there is generally an old negress known by the name of Granny, who loiters about the negro yard, and is supposed to take charge of the little negroes who are too young to work. Towards the close of day you will sometimes hear the most dismal cries of woe coming from that quarter. Old Granny is then at work grubbing the chegoe nests out of the feet of the sable urchins, and filling the holes with lime-juice and Cayenne pepper. This scorching compound has two duties to perform : first, it causes death to any remaining chegoe in the hole, and secondly, it acts as a kind of birch-rod to the unruly brats, by which they are warned to their cost not to conceal their chegoes in future ; for, afraid of encountering old Granny's tomahawk, many of them prefer to let the chegoes riot in their flesh rather than come under her dissecting hand." If the prompt extraction of the chegoe's nests is neglected, the worm-like larvæ creep out, continue the mining operations of their parent, and produce a violent inflammation, which may end in the mortification of a limb. It not unfrequently happens that negroes from sheer idleness or negligence in the first instance have been lamed for life and become loathsome to the sight. In such a state, these miserable objects are incurable, and death only puts an end to their sufferings.

Fortunately, the chegoe is incapable of jumping like the common flea, or else it would have rendered tropical America uninhabitable. It generally frequents sandy places, and attaches itself also to pigs and dogs.

A still more dangerous plague, peculiar to the coast of Guinea and the interior of tropical Africa, to Arabia, and the adjacent countries, is the *Filaria medinensis* of Linnæus. This dreadful worm comes to the herbage in the morning dew, from whence it pierces the skin, and enters the feet of such as walk without shoes, causing the most painful irritations, succeeded by violent inflammation and fever. The natives extract it with the greatest caution by twisting a piece of silk round one extremity of the body and withdrawing it very gently. When we consider that this insidious worm is frequently twelve feet long, although not thicker than a horse-hair, we can readily imagine the difficulty of the operation. If unfortunately the animal should break, the part remaining under the skin grows with redoubled vigour, and frequently occasions a fatal inflammation.

One of these most unwelcome intruders once entered the ankle of the celebrated navigator Dampier. "I was in great torment," says this entertaining traveller, "before it came out. My leg and ankle swelled, and looked very angry, and I kept on a plaster to bring it to head. At last, drawing off my plaster, out came three inches of the worm, and my pain abated. Till that time I was ignorant of my malady, and a gentleman at whose house I was took it for a nerve; but I knew well what it was, and presently rolled it up on a small stick. After this I opened the place every morning and evening, and strained the worm out gently, about two inches at a time — not without some pain — till I had at length got out about two feet."

He afterwards had it entirely discharged by one of the negroes, who applied to it a rough powder, not unlike tobacco leaves, dried and crumbled very small.

Among the plagues of Guiana and the West Indies we must not forget a little insect in the grass and on the shrubs, which the French call Bête-rouge. It is of a beautiful scarlet colour, and so minute that you must bring your eye close to it before you can perceive it. It abounds most in the rainy season. Its bite causes an intolerable itching, which, according to Richard Schomburgk, who writes from personal experience, drives by day the perspiration of anguish from every pore, and at night makes one's hammock resemble the gridiron on which Saint Lawrence was roasted. The best way to get rid of the plague is to rub the part affected with lemon-juice or rum. "You must be careful not to scratch it," says Waterton. "If you do so and break the skin, you expose yourself to a sore. The first year I was in Guiana the bête-rouge and my own want of knowledge, and, I may add, the little attention I paid to it, created an ulcer above the ankle which annoyed me for six months, and if I hobbled out into the grass, a number of bête-rouges would settle on the edges of the sore and increase the inflammation."

The blood-sucking Ticks are also to be classed among the intolerable nuisances of many tropical regions. A large American species called Garapata (*Ixodes sanguisuga*) fixes on the legs of travellers, and gradually buries its whole head in the skin, which the body, disgustingly distended with blood, is unable to follow. On being violently removed, the former remains in the wound, and often produces painful sores. The Indians returning in the evening from the forest or from their field labour generally bring some of these creatures along with them, swollen to the size of hazel-nuts. These ticks seem to have no predilection for any particular animal, but indiscriminately fasten on all, not even sparing the toad or the lizard.

Though countless hosts of ticks infest the Ceylonese jungle, though mosquitoes without number swarm over the lower country, yet the land-leeches which beset the traveller in the rising grounds are a still more detested plague. "They are not frequent in the plains," says Sir E. Tennent, "which are too hot and dry for them; but amongst the rank vegetation in the lower ranges of the hill-country, which is kept damp by frequent showers, they are found in tormenting profusion. They are terrestrial, never visiting ponds or streams. In size they are about an inch in length, and as fine as a common knitting needle, but capable of distention till they equal a quill in thickness and attain a length of nearly two inches. Their structure is so flexible that they can insinuate themselves through the meshes of the finest stocking, not only seizing on the feet and ankles, but ascending to the back and throat, and fastening on the tenderest parts of the body. The coffee planters who live amongst these pests are obliged, in order to exclude them, to envelope their legs in ' leech gaiters,' made of closely woven cloth.

" In moving, the land-leeches have the power of planting one extremity on the earth and raising the other perpendicularly to watch for their victim. Such is their vigilance and instinct that, on the approach of a passer-by to a spot which they infest, they may be seen amongst the grass and fallen leaves, on the edge of a native path, poised erect, and preparing for their attack on man and horse. On descrying their prey they advance rapidly by semicircular strides, fixing one end firmly and arching the other forwards, till by successive advances they can lay hold of the traveller's foot, when they disengage themselves from the ground and ascend his dress in search of an aperture to enter. In these encounters the individuals in the rear of a party of travellers in the jungle invariably fare worst, as the leeches once warned of their approach congregate with singular celerity. Their size is so insignificant, and the wound they make so skilfully punctured, that both are certainly imperceptible, and the first intimation of their onslaught is the trickling of the blood, or a chill feeling of the leech when it begins to hang heavily on the skin from being distended by its repast. Horses are driven wild by them, and stamp the ground in fury to shake them from their fetlocks, to which they hang in bloody tassels. The bare legs of the palankin-bearers and coolies are a favourite resort, and their hands being too much engaged to be spared to pull them off, the leeches hang like bunches of grapes round their ankles; and I have seen the blood literally flowing over the edge of a European's shoe from their innumerable bites. In healthy constitutions the wounds, if not irritated, generally heal, occasioning no other inconvenience than a slight inflammation and itching; but in those with a bad state of body, the punctures, if rubbed, are liable to degenerate into ulcers, which may lead to the loss of limb or of life. Both Marshall and Davy mention that, during the march of the troops in the mountains, when the Kandyans were in rebellion, in 1818, the soldiers, and especially the Madras Sepoys, with the pioneers and coolies, suffered so severely from this cause that numbers of them perished."

Dr. Hooker also mentions the land-leech as infesting the southern slopes of the Himalaya, so that it ranges over a vast extent of territory.

Among the many noxious insects destructive to the property of man, there is, perhaps, none more remarkable than the South African Tsetsé-fly (*Glossina morsitans*), whose peculiar buzz when once heard can never be forgotten by the traveller whose means of locomotion are domestic animals; for it is well known that the bite of this poisonous insect is certain death to the ox, horse, and dog. Fortunately it is limited to particular districts, frequently infesting one bank of a river while the other contains not a single specimen, or else travelling in South Africa would be utterly impossible, and we should now know no more of Lake Ngami or the Zambesi than we did thirty years since. In one journey Dr. Livingstone lost no less than forty-three fine oxen by the bite of the tsetsé. A party of Englishmen once attempted to reach Libebé, but they had only proceeded seven or eight days' journey to the north of the Ngami, when both horses and cattle were bitten by the fly, and the party were in consequence compelled to make a hasty retreat. One of the number was thus deprived of as many as thirty-six horses, excellent hunters, and all sustained heavy losses in cattle.

A most remarkable feature in the bite of the tsetsé is its perfect harmlessness in man and wild animals, and even calves, so long as they continue to suck the cow. The mule, ass, and goat enjoy likewise the same immunity, and many large tribes on the Zambesi can keep no domestic animals except the latter, in consequence of the scourge existing in their country. Dr. Livingstone's children were frequently bitten, yet suffered no harm, and he saw around him numbers of zebras, buffaloes, pigs, pallahs and other antelopes, feeding quietly in the very habitat of the tsetsé, yet as undisturbed by its bite as oxen are when they first receive the fatal poison, which acts in the following manner. After a few days the eyes and nose begin to run, the coat stares as if the animal were cold, a swelling appears under the jaw, and, though the animal continues to graze, emaciation commences, accompanied with a peculiar flaccidity of the muscles; and this proceeds unchecked until, perhaps months afterwards, purging comes on, and the animal, no longer able to graze, perishes in a state of extreme exhaustion. Those which are in good condition often perish, soon after the wound is inflicted, with staggering and blindness, as if the brain were affected by it. Sudden changes of temperature, produced by falls of rain, seem to hasten the progress of the complaint, but in general the emaciation goes on uninterruptedly for months; and do what one may, the poor animals perish miserably, as there is no cure yet known for the disease.

Had any one of our indigenous flies similar poisonous qualities we should never have been able to escape from barbarism; if, by any fatal chance, the tsetsé were to settle among us, our prosperity would soon be at an end, and our civilisation imperilled! Reflections such as these are well calculated to humble our pride and check our presumption!

The Abyssinian Tsalt-salya or Zimb, described by Bruce, seems identical with the tsetsé, or produces at least similar symptoms. At the season when this plague makes its appearance, all the inhabitants along the sea-coast, from Melinde to

Cape Gardafui, and to the south of the Red Sea, are obliged to retire with their cattle to the sandy plains to preserve them from destruction.

The French traveller, D'Escayrac, tells us of a fly in Soudan which leaves the ox uninjured, but destroys the dromedary. On account of this plague the camel is confined to the northern boundary of the Soudan, while the oxen graze in safety throughout the whole country. This fly has caused more migrations among the Arabs of the Soudan than all their wars; and in the dry season it even drives the elephant from Lake Tsad by flying into its ears.

Locust.

Though the locusts not seldom extend their ravages to the steppes of southern Russia, though they have been known to burst like a cloud of desolation over Transylvania and Hungary, and stray stragglers now and then even find their way to England, yet their chief habitat and birthplace is the torrid zone. They wander forth in countless multitudes, and at very irregular periods; but how it comes that they are multiplied to such an excess in particular years and not in others, has never yet been ascertained, and perhaps never will be. They are armed with two pairs of strong mandibles; their stomach is of extraordinary capacity and power; they make prodigious leaps by means of their muscular and long hind legs; and their wings even carry them far across the sea. On viewing a single locust, one can hardly conceive how they can cause such devastations, but we cease to wonder on hearing of their numbers.

Mahomet — so say his followers — once read upon the wing of a locust:

"We are the army of God; we lay ninety-nine eggs; and if we laid a hundred, we should devour the whole earth and all that grows upon its surface."

"O, Allah!" exclaimed the terrified prophet, "Thou who listenest patiently to the prayers of Thy servant, destroy their young, kill their chieftains, and stop their mouths, to save the Moslems' food from their teeth!"

Scarce had he spoken when the angel Gabriel appeared,

saying, "God grants thee a part of thy wishes." And, indeed, as all true believers know, this prayer of their prophet, written on a piece of paper, and enclosed in a reed which is stuck in the ground, is sure to preserve a field or an orchard from locust-devastation.

From 1778 to 1780 the whole empire of Morocco was so laid waste by swarms of these insects, that a dreadful famine ensued. Mr. Barrow, in his travels, states that in the southern parts of Africa the whole surface of the ground might literally be said to be covered with them for an area of nearly 2,000 square miles. When driven into the sea by a north-west wind, they formed upon the shore, for fifty miles, a bank three or four feet high; and when the wind was south-east, the stench was such as to be smelt at the distance of 150 miles.

Major Moore observed at Poonah an army of locusts, which devastated the whole country of the Mahrattas, and most likely came from Arabia. Their columns extended in a width of 500 miles, and were so dense as to darken the light of the sun. It was a red species (not the common *Gryllus migratorius*), whose bloody colour added to the terror of their appearance.

On his way to Rehoboth, a German missionary station in Central Africa, near the tropic of Capricorn, Anderson met with vast numbers of the larvæ of the locust, commonly called by the Boers "Voet-gangers," or pedestrians. In some places they might be seen packed in layers several inches in thickness, and myriads were crushed and maimed by the waggon and cattle. Towards nightfall they crawled on the bushes and shrubs, many of which, owing to their weight and numbers, were either bowed down or broken short off. They were of a reddish colour, with dark markings; and as they hung thus suspended, they looked like clusters of rich fruit. These larvæ are justly dreaded by the colonists, as nothing seems capable of staying their progress. Even rivers form no barrier to their march, as the drowning multitudes afford the survivors a temporary bridge; endeavours to diminish their numbers would appear like attempting to drain the ocean by a pump.

On travelling on, next morning, the locust itself was encountered, and in such masses as literally to darken the air, and fully to justify Southey's description in the epic of Thalaba: —

> "Onward they came—a dark continuous cloud
> Of congregated myriads, numberless;
> The rushing of whose wings was as the sound
> Of a broad river headlong in its course,
> Plunged from a mountain summit; or the roar
> Of a wild ocean in the autumn storm,
> Shattering its billows on a shore of rocks!"

The waggon, or any other equally conspicuous object, could not be distinguished at the distance of one hundred paces. In a particular spot within the circumference of a mile they had not left a particle of any green thing. The noise of their wings was very great—not unlike that caused by a gale of wind whistling through the shrouds of a ship at anchor. It was interesting to witness at a distance the various shapes and forms that these columns assumed, more especially when crossing mountain-ranges. At one time they would rise abruptly in a compact body, as if propelled by a strong gust of wind; then, suddenly sinking, they would disperse into smaller battalions, not unlike vapours floating about a hill-side at early morn, and when slightly agitated by a breeze; or they would resemble huge columns of sand or smoke, changing their shape every minute. During their flight numbers were constantly alighting—an action which has not inaptly been compared to the falling of large snow-flakes. It is, however, not until the approach of night that the locusts encamp. Woe to the spot they select as a resting-place! The sun sets on a landscape green with all the luxuriance of tropical vegetation; it rises in the morning over a region naked as the waste of the Sahara!

The tropical plague of the cockroaches has been introduced into England; but, fortunately, the giant of the family, the *Blatta gigantea*, a native of many of the warmer parts of Asia, Africa, and South America, is a stranger to our land: and the following truthful description of this disgusting insect gives us every reason to be thankful for its absence:—"They plunder and erode all kinds of victuals, dressed and undressed, and damage all sorts of clothes, especially such as are touched with powder, pomatum, and similar substances; everything made of leather; books, paper, and various other articles, which if they do not destroy, at least they soil, as they frequently deposit a drop of their excrement where they settle, and, some way or other, by that means damage what they cannot devour. They fly into the flame of candles, and sometimes into the dishes; are very fond of ink and of oil, into which they are apt to fall and perish, in which case they soon turn most offensively putrid—so that a man might as well sit over the cadaverous body of a large animal as write with the ink in which they have died. They often fly into persons' faces or bosoms, and their legs being armed with sharp spines, the pricking excites a sudden horror not easily described. In old houses they swarm by myriads, making every part filthy beyond description wherever they harbour, which in the daytime is in dark corners, behind clothes—in trunks, boxes, and, in short, every place where they can lie concealed. In old timber and deal houses, when the family is retired at night to sleep, this insect, among other disagreeable properties, has the power of making a noise which very much resembles a pretty smart knocking with the knuckle upon the wainscoting. The *Blatta gigantea* in the West Indies is therefore frequently known by the name of *the drummer*. Three or four of these noisy creatures will sometimes be impelled to answer one another, and cause such a drumming noise that none but those who are very good sleepers can rest for them. What is most disagreeable, those who have not gauze curtains are sometimes attacked by them in their sleep; the sick and dying have their extremities attacked; and the ends of the toes and fingers of the dead are frequently stripped both of the skin and flesh."

According to Tschudi, the cucaracha and chilicabra—two large species of the cockroach—infest Peru in such numbers, as almost to reduce the inhabitants to despair. Greedy, bold, cunning, they force their way into every hut, devour the stores, destroy the clothes, intrude into the beds and dishes, and defy every means that is resorted to for their destruction. Fortunately, they are held in check by many formidable enemies, particularly by a small ant, and a pretty little bird (*Troglodytes audax*) belonging to the wagtail family, which has some difficulty in mastering the larger cockroaches. It first of all bites off their head, and then devours their body, with the exception of their membranaceous wings. After having finished his repast, the bird hops upon the nearest bush, and there begins his song of triumph.

Many other insect plagues might be added to the list, but those I have already enumerated suffice to reconcile us to our misty climate, and to diminish our longing for the palm groves of the torrid zone.

The Tsetse.

CHAPTER XXIII.

TROPICAL INSECTS DIRECTLY USEFUL TO MAN.

The Silk-worm — The Tusseh and Arandi — The Cochineal Insect — The Gum-lack Insect — The Locust used as Food — Other edible Insects — Insects used as Ornaments — The Diamond Beetle.

AFTER having described the miseries which the tropical insects inflict upon man — how they suck his blood, destroy his rest, exterminate his cattle, devour the fruits of his fields and orchards, ransack his chests and wardrobes, feast on his provisions, and plague and worry him wherever they can — I now proceed to the more agreeable task of recounting their services, and relating the benefits for which he is indebted to them.

Among the insects which are of *direct* use to us, the silk-worm (*Bombyx mori*) is by far the most important. Originally a native of tropical or sub-tropical China, where the art of making use of its filaments seems to have been discovered at a very early period, it is now reared in countless numbers far and wide over the western world, so as to form a most important feature in the industrial resources of Europe. Thousands of skilful workmen are employed in spinning and weaving its lustrous threads, and thousands upon thousands, enjoying the fruits of their labours, now clothe themselves, at a moderate price, in silken tissues which but a few centuries back were the exclusive luxury of the richest and noblest of the land.

Besides the silk-worm, we find many other moths in the tropical zone whose cocoons might advantageously be spun, and only require to be better known to become considerable articles of commerce. The tusseh-worm (*Bombyx mylitta*) of Hindostan, which lives upon the leaves of the Rhamnus jujuba, furnishes a dark-coloured, coarse, but durable silk; while the Arandi (*B. cynthia*), which feeds upon the foliage of the castor-oil plant (*Ricinus communis*), spins remarkably soft threads, which serve the Hindoos to weave tissues of uncommon strength.

In America there are also many indigenous moths whose filaments might be rendered serviceable to man, and which seem destined to great future importance, when trade, quitting her usual routine, shall have learnt to pry more closely into the resources of Nature.

While the Cocci, or plant bugs, are in our country deservedly detested as a nuisance, destroying the beauty of many of our garden plants by their blighting presence; while, in 1843, the Coccus of the orange trees proved so destructive in the Azores that the island of Fayal, which annually exported 12,000 chests of fruit, lost its entire produce from this cause alone, two tropical members of the family, as if to make up for the misdeeds of their relations, furnish us—the one with the most splendid of all scarlet dyes, and the other with gum-lack, a substance of hardly inferior value.

The English gardener spares no trouble to protect his hot and greenhouse plants from the invasion of the *Coccus hesperidum*; but the Mexican *haciendero* purposely lays out his Nopal plantations that they may be preyed upon by the *Coccus cacti*, and rejoices when he sees the leaves of his opuntias thickly strewn with this valuable parasite. The female, who from her form and habits might not unaptly be called the tortoise of the insect world, is much larger than the winged male, and of a dark-brown colour, with two light spots on the back, covered with a white powder. She uses her little legs only during her first youth, but soon she sucks herself fast, and henceforward remains immovably attached to the spot she has chosen, while her mate continues to lead a wandering life. While thus fixed like an oyster, she swells or grows to such a size that she looks more like a seed or berry than an insect; and her legs, antennæ, and proboscis, concealed by the expanding body, can hardly be distinguished by the naked eye. Great care is taken to kill the insects before the young escape from the

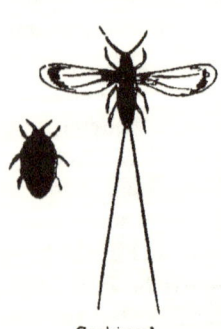

Cochineal.

eggs, as they have then the greatest weight, and are most impregnated with colouring matter. They are detached by a blunt knife dipped in boiling water to kill them, and then dried in the sun, when they have the appearance of small, dry, shrivelled berries, of a deep-brown purple or mulberry colour, with a white matter between the wrinkles. The collecting takes place three times a year in the plantations, where the insect, improved by human care, is nearly twice as large as the wild coccus, which in Mexico is gathered six times in the same period. Although the collecting of the cochineal is exceedingly tedious—about 70,000 insects going to a single pound —yet, considering the high price of the article, its rearing would be very lucrative, if both the insect and the plant it feeds upon were not liable to the ravages of many diseases, and the attacks of numerous enemies.

The conquest of Mexico by Cortez first made the Spaniards acquainted with cochineal. They soon learnt to value it as one of the most important products of their new empire; and in order to secure its monopoly, prohibited, under pain of death, the exportation of the insect, and of the equally indigenous Nopal, or *Cactus cochinellifer.* In the year 1677, however, Thierry de Meronville, a Frenchman, made an effort to deprive them of the exclusive possession of the treasure they guarded with such jealous care. Under a thousand dangers, and by means of lavish bribery, he succeeded in transporting some of the plants, along with their costly parasite, to the French colony of San Domingo; but, unfortunately, his perseverance did not lead to any favourable results, and more than a century elapsed after this first ineffectual attempt before the rearing of cochineal extended beyond its original limits.

In the year 1827, M. Berthelot, director of the botanical garden at Orotava, introduced it into the Canary Islands, where it thrives admirably upon the Opuntia ficus indica; so that in 1838 the exportation amounted to 18,000 lbs., and has since then been continually increasing. Cochineal is now reared near Valencia, Cadiz, and Malaga; in various parts of the West Indies and the United States; in Brazil, East Indies, and Java; and though Mexico still continues to furnish it in the greatest abundance, yet in point of quality it is distanced by its youthful rival, Teneriffe.

In the year 1856 more than 800,000 lbs. of cochineal were imported into France, of which the Canary Islands alone furnished near one-half — a proof, among others, how much the wealth of a country may be increased by the introduction of a new article of commerce.

The Coccus which produces lac, or gum-lac, is a native of India, and thrives and multiplies best on several species of the fig-tree. A cheap method having been discovered within the last years of separating the colouring matter which it contains from the resinous part, it has greatly increased in commercial importance.*

In the tropical zone we find that not only many birds and several four-footed animals live chiefly, or even exclusively, on insects, but that they are even consumed in large quantities, or eaten as delicacies, by man himself. The locust-swarms are welcomed with delight by the Arab of the Sahara and the South African bushman. After being partially roasted, they are either eaten fresh, or dried in hot ashes and stored away. The natives reduce them also to powder or meal, which, eaten with a little salt, is palatable even to Europeans; so that Dr. Livingstone, who, during his residence among the Bakwains, was often obliged to put up with a dish of locusts, says he should much prefer them to shrimps, though he would avoid both if possible.

They evidently contain a great deal of nourishment, as the bushmen thrive wonderfully on them, and hail their appearance as a season of plenty and good living.

Several of the large African caterpillars are edible, and considered as a great delicacy by the natives. On the leaves of the Mopané tree, in the Bushman country, the small larvæ of a winged insect, a species of Psylla, appear covered over with a sweet gummy substance, which is collected by the people in great quantities, and used as food. Another species in New Holland, found on the leaves of the Eucalyptus, emits a similar secretion, which, along with its insect originator, is

---

* During the first eight months of the year 1861, 16,204 cwts. of cochineal and 4,506 cwts. of lac dye were imported into Great Britain. The best quality of Teneriffe cochineal is quoted at 3s. 3d. per lb., while the best Mexican is only 2s. 10d. Lac dye varies in price from 1s. 3d. to 2s. 7d.—*The Economist*.

scraped off the leaves and eaten by the aborigines as a saccharine dainty.

The chirping *Cicadæ*, or frog-hoppers, which Aristotle mentions as delicious food, though maccaroni has long supplanted them in the estimation both of the modern Greeks and of the Italians, are still in high repute among the American Indians. With the exception of one species (*Cicada Anglica*), these insects, equally remarkable for the rapidity of their flight and their faculty of emitting a loud noise, are unknown in our misty islands. Several of the exotic species, when their wings are expanded, measure six inches in extreme length — a size superior to that of many of the humming-birds.

Cicada.

The Chinese, who allow nothing edible to go to waste, after unravelling the cocoon of the silk-worm, make a dish of the pupæ, which the Europeans reject with scorn.

In the chapter on Palms, I have already mentioned that the grubs of several insects which thrive and increase in the Sago-tree, the Areca, and the Cocoa, are considered as great delicacies; and many similar examples might be cited.

The Goliath beetles of the coast of Guinea are roasted and eaten by the natives, who doubtless, like many other savages, not knowing the value of that which they are eating, often make a *bonne bouche* of what an entomologist would most eagerly desire to preserve.

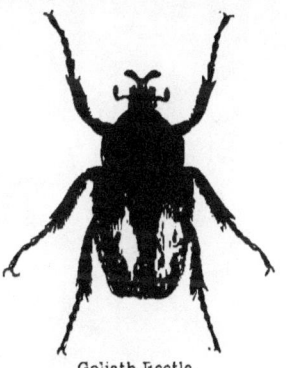

Goliath Beetle.

Several of the more brilliant tropical beetles are made use of as ornaments, not only by the savage tribes, but among nations which are able to command the costliest gems of the East. The golden elytra of the Sternocera chrysis and Sternocera sternicornis serve to enrich the embroidery of the Indian zenana, while the joints of the legs are strung on silken threads, and form bracelets of singular brilliancy.

The ladies in Brazil wear necklaces composed of the azure

green and golden wings of lustrous Chrysomelidæ and Curculionidæ, particularly of the Diamond beetle (*Entimus nobilis*); and in Jamaica, the elytra of the Buprestis gigas are set in earrings, whose gold-green brilliancy rivals the rare and costly Chrysopras in beauty.

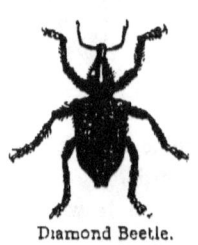

Diamond Beetle.

Buprestis gigas.

CHAPTER XXIV.

THE ENTOMOLOGICAL WONDERS OF THE TROPICS.

Gradual Decrease of Insect-life on advancing towards the poles — Vast number of Beetles in Brazil — The Hercules Beetle — The Goliath — The Inca Beetle — Other colossal Insects — The Walking-leaf and Walking-stick Insects — The Soothsayer — Luminous Beetles — The Cocujas.

ON advancing from the temperate regions to the pole, we find that insect-life gradually diminishes in the same ratio as vegetable life declines.

Thus, on Melville Island (75° N. lat.), where the flora of the ice-bound soil is reduced to the scantiest proportions, but six insects were found during the whole eleven months of Parry's sojourn in the solitude of Winter Harbour; and so great is their scarcity in Nowaja Semlja, that the Russian academician, Von Baer, could not discover a single grub or worm in the carcase of a walrus, which had evidently been lying a long time upon the beach.

The inverse takes place on advancing towards the equator; for as the sun rises more and more to the zenith, we find the insects gradually increasing with the multiplicity of plants, and at length attaining the greatest variety of form, and the highest developement of number, in those tropical lands where moisture combines with heat in covering the ground with a dense and everlasting vegetation. Thus, while not a single species of beetle is found on Melville Island, Greenland (60°-70° N. lat.) boasts of 11; Lapland (64°-71° N. lat.) of 813; and Sweden (56°-60°) of 2,083 Coleoptera. In the milder climate of England (50°-60°) their number increases to 2,263; in sunnier France (41°-51°) it rises to 4,200; and the hothouse temperature of Brazil, from Rio Janeiro to Bahia, fosters no less than 7,500 specific forms of beetle-life. Thus, also, while the whole of Europe and Siberia hardly possess more than 250 diurnal Lepidoptera, or butterflies, the explored parts of Brazil, which are very inferior in extent, have already furnished the naturalist with no less than 600 species, and no doubt contain many more.

The carrion-feeding beetles, however, form a remarkable exception to this general rule, as they are most numerous, both absolutely and relatively, in the temperate regions of the northern hemisphere—most likely from the extreme rapidity of decomposition in the torrid zone not allowing the developement of their larvæ to keep pace with the progress of putrefaction. Thus the families of the Carabi and Hydrocanthari have only 591 and 65 species in South America and Mexico, while they respectively number 1,090 and 156 in Europe and Southern Russia.

Another peculiarity of beetle-life in the equatorial zone is, that in tropical America most Carabi live on the trees, while in our country they all, with rare exceptions, dwell upon the ground—a circumstance probably due to the legions of ants which there take possession of the soil, and so effectually perform the duties of scavengers as to leave but little for others.

In the countries which, from the never-failing abundance of food, and constant warmth, are most favourable to the multiplication of insects, these creatures may naturally be expected to attain the greatest size.

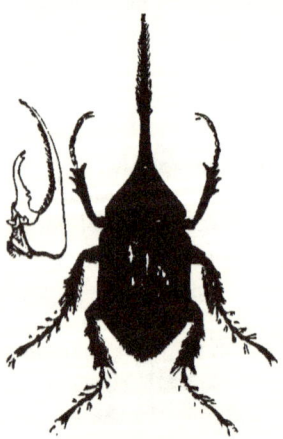
Hercules Beetle.

Thus the European rhinoceros beetle (*Oryctes nasicornis*), though an inch and a quarter long, is far surpassed by the Megasominac of torrid America.

The colossal Hercules beetle (*M. hercules*), which is sometimes seen in great numbers on the Mammee tree (*Mammæa Americana*), attains a length of five or even six inches, and is distinguished, like the other species of the genus, by the singular horn-shaped processes rising from the head and thorax, which give it a very grotesque and even formidable appearance. Though but little is yet known of its economy, it most likely subsists upon putrescent wood, and evidently leads a tree life, like the other members of the family—the Elephant, the Neptune, the Typhon, the Hector, the Mars—whose very names indicate that they are "first-rate liners" in the insect world. These beetles excavate burrows in the earth, where they conceal themselves during the day, or live in the decomposed trunks of trees, and are generally of a dark rich brown or chestnut colour. On the approach of night they run about the footpaths in woods, or fly around the trees to a great height with a loud humming noise. Resembling the large herbivorous quadrupeds by their comparative size and horn-like processes, they are still further like them in their harmless nature, and thus deserve in more than one respect to be called the elephants among the insect tribes.

The Goliaths of the coast of Guinea are nearly as large as the American Megasominae, and surpass them in brilliancy of colouring. Some years ago these huge beetles, which live exclusively on the juice of trees, were very rare, and fetched extravagant prices. Thus Mr. Swainson mentions £30 having been offered and refused for a single specimen, the proprietor demanding £50. It was, however, subsequently sold for £10; and the market having latterly been better supplied with Goliaths, a further reduction has taken place, though some of the species are still worth £5 or £6.

The South American Inca beetles greatly resemble the African Goliaths, equalling them in size and beauty.

Many of the tropical dragon-flies, grasshoppers, butterflies, and moths, are of no less colossal dimensions in their several orders than the giants among

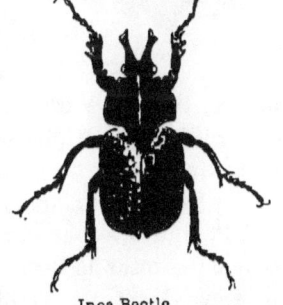

Inca Beetle.

the beetles. The Libellula lucretia, a South American dragon-fly, measures five inches and a half in length; the giant Phasma is a span long; and the cinnamon-eating Atlas-moth of Ceylon often reaches the dimensions of nearly a foot in the stretch of its superior wings. The names of many other species conspicuous by their size might be added; but these examples suffice to show the enormous proportions attained by insects in the warmer regions of the globe.

In the tropical zone, where the prodigality of life multiplies the enemies which every creature has to encounter, we may naturally expect to find the insects extremely well provided with both passive and active means of defence.

Many so closely resemble in colour the soil or object on which they are generally found, as to escape even the eye of a hungry enemy. The wings of several Brazilian moths appear like withered leaves that have been gnawed round their margins by insects; and when these moths are disturbed, instead of flying away, they fall upon the ground like the leaf which they resemble, so that it is difficult, if not impossible, on such occasions to know what they really are.

Phyllium.

The illusion is still more complete when the likeness of form is joined to that of colour, as in the walking-leaf and walking-stick insects. Some, of an enormous length, look so exactly like slender dead twigs covered with bark, that their insect nature can only be discovered by mere accident— upon being handled they feign death, and their legs are often knobbed, like the withered buds of trees; some resemble living twigs, and are green; others such as are decayed, and are therefore coloured brown; the wings of many put on the resemblance of dry and crumpled leaves, while those of others are vivid green—in exact accordance with the plants they respectively inhabit. This highly remarkable family consists of the herbivorous Phasmas and Phylliums — the former of which have a thin twig-like shape, while the latter have an enlarged body — and of the carniverous Mantes, or soothsayers. As the Mantis is slow and without much muscular energy, and its organisation requires a large supply of food, Nature has disguised it under the form of a plant, the better to deceive its victims. Like a cat approaching a mouse, it moves almost imperceptibly along, and steals towards its prey, fearful of putting it to flight. When sufficiently near, the fore legs are suddenly darted out to their full length, and seize the doomed insect, which vainly endeavours to extricate itself; the formation of the fore leg enabling the tibia to be so closed on the sharp edge of the thigh as to amputate any slender substance brought within its grasp, and to make even an entomologist repent a too hasty seizure of his prize.

The Mantis, by the attitude it assumes when lurking for its prey or advancing upon it—which is done by the support of the four posterior legs only, whilst the head and prothorax are raised perpendicularly from the body, and the exterior legs are folded in front—greatly resembles a person praying. Hence, in France it is called Le Prêcheur, or Le Pric Dieu; the Turk says it points to Mecca; and several African tribes pay it religious observances. In reality, however, its ferocity is great, and the stronger preying on the weaker of their own species, unmercifully cut them to pieces. Thus, two Mantes which Sir E. Tennent enclosed in a box were both found dead a few hours after, severed limb from limb in their deadly fight.

Mantis.

Within the space of a week, Professor Burmeister saw a Mantis devour daily some dozens of flies, and occasionally large grasshoppers and young frogs, consuming, now and then, lizards three times its own length, as well as many large fat caterpillars. Hence it may be judged what ravages these strange-formed creatures must cause among all weaker beings which incautiously approach them, and that, far from being the saints, they are, in reality, the tigers of the insect world. Among the organic marvels of the innocent herbivorous Phylliums, their seed-like eggs must be mentioned; for the wonderful provision of Nature in giving the parents a plant-like form extends even to their progeny, in order to secure them from similar dangers. Though generally tropical, yet Van Diemen's Land possesses a gigantic walking-stick, or Phasma, the body of which is eight inches long; and the *Mantis religiosa* is found all over Southern Europe.

The leaf-like form which renders the Phylliums one of the wonders of entomology, appears likewise in other insects. Thus, in the *Diactor bilineatus*, a native of Brazil, the hind legs have singular leaf-like appendages to their tibial

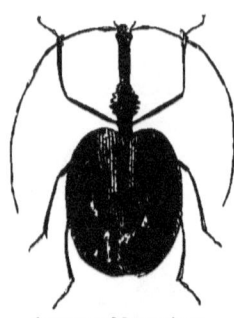

Javanese Mormolyce.

joints; and in the *Javanese Mormolyce*, a carabidous insect remarkable for its extreme flatness and the elongation of its head, we find the elytra spreading out in the form of broad leaves.

The long hairs, stiff bristles, sharp spines, and hard tubercular prominences with which many caterpillars are bristled, and studded, are a most effectual means of defence, and often prove a grievous annoyance to the entomologist, from their poisonous or stinging properties. Mr. Swainson once finding in Brazil a caterpillar of a beautiful black colour, with yellow radiated spines, and being anxious to secure the prize, incautiously took hold of it with the naked hand; but so instantaneous and so violent was the pain which followed, that he was obliged to return home. Warm fomentations — placing the hand in tepid water — every device that could be thought of to allay the itching produced by the venomous hairs of this creature were in turn resorted to, with little or no effect, for several hours, nor had it entirely ceased on the following morning.

Though the great majority of luminous animals are marine, frequently lighting up the breaking wave with millions of moving atoms, or spreading over the beach like a sheet of fire,* yet several insects are also endowed with the same wonderful property. The European glow-worms and fire-flies, sparkling on the hedge-rows, or flying in the summer air, afford a charming spectacle. But this brilliancy is far surpassed by that of the phosphorescent beetles of the torrid zone. Thus the Cocujas of South America, which emits its light from two little transparent tubercles on the sides of the thorax, while our Lampyrides shine from the hinder part of the abdomen, is said to glow with such intensity that a person may with great ease read the smallest print by the phosphorescence of one of these insects, if held between the fingers and gradually moved along the lines with the luminous spots above the letters; but if eight or ten of them are put into a phial the light will be sufficiently good to admit of writing by it.

Cocujas.

* "The Sea and its Living Wonders," ch. xx.

In the woods of Sarawak Mr. Adams observed a splendid glow-worm (*Lampyris*), each segment of the body illuminated with three lines of tiny lamps, the luminous spots on the back being situated at the posterior part of the segmentary rings on the median line, while those along the sides of the animal were placed immediately below the stomates or spiracula, each spiraculum having one bright spot. This very beautiful insect was found shining as the darkness was coming on, crawling on the narrow pathway, and glowing among the dead damp wood and rotten leaves. When placed around the finger, it resembled in beauty and brilliancy a superb diamond ring.

The sparkling effulgence of the tropical Elaters is frequently made use of by the fair sex, as an equally singular and striking ornament. The ladies of the Havana attach them to their clothes on occasions of festivity, and the Indian Bayaderes often wear them in their hair.

In Prescott's "Conquest of Mexico," we are told that in 1520, when the Spaniards visited that country, the wandering sparks of the Elater, "seen in the darkness of the night, were converted by the excited imaginations of the besieged into an army with matchlocks;" and on another occasion their phosphorescence caused British troops to retreat: for when Sir John Cavendish and Sir Robert Dudley first landed in the West Indies, and saw at night an innumerable quantity of lights moving about, they fancied that the Spaniards were approaching with an overwhelming force, and hastily re-embarked before their imaginary foe.

Above two hundred species of tropical insects are known to possess the luminous property in a greater or smaller degree; but the phosphorescence of the famous Lantern Fly (*Fulgora laternaria*), as described by Madam Merian, in her "Insects of Surinam," and repeated upon her authority in numerous works on natural history, seems to be entirely fabulous, no modern traveller having ever seen it emit the least light. The enormous transparent prolongation of the forehead with which this beautiful and large insect is endowed, may, from its fanciful resemblance to a lantern, have given rise to the fable.

Lantern Fly.

Ants and Termites.

CHAPTER XXV.

ANTS AND TERMITES.

Vast numbers of Ants in the Tropical Zone — Excruciating pain caused by the Sting of the Ponera Clavata — The Black Fire-Ant of Guiana. — The Dimiya of Ceylon — The Kaddiya — The Red Ant of Angola — Devastations of the Viviagua in the West Indian Coffee Plantations — The Atta Cephalotes, or the Umbrella Ant — Household Plagues — Difficulty of preserving Sugar from their attacks — The Ranger Ants — Wonderful construction of Tropical Ants — Slave-making Ants — Cow-keeping Ants — The Mexican Honey Ant — Devastation of the Termites — Their Services and Uses — Their marvellous Buildings — Formation of a Termite Colony — Amazing Fecundity of the Termite Queen — Consequence of an Attack upon a Termite Hill — Wars between Termites and Black Ants — American Termites — Termites esteemed as a Delicacy — Marching White Ants — Mysteries of Termite Life.

THE family of Ants is undoubtedly the most numerous of any in the whole circle of winged insects, as its colonies are not confined to one particular region, but are thickly planted over the greatest part of the habitable world. There is scarcely a field in Britain that does not contain millions; we cannot rest upon a bank without reclining upon the walls of their cities; their chief quarters, however, are established in the torrid zone, where they may truly be said to hold a despotic sway over the forest and the savanna, over the thicket and the field. It is hardly possible to penetrate into a tropical wood without being reminded, by their stings and bites, that they consider the visit as an intrusion, while they themselves unceremoniously invade the dwellings of man, and lay ruinous contributions on his stores. The inconceivable number of their species defies the memory of the naturalist, to whom many are even still entirely unknown. From almost microscopical size to an inch in length, of all colours and shades between yellow, red, brown, and black, of the most various habits and stations, the ants of a single tropical land would furnish study for years to a zealous entomologist. Every family of plants has its peculiar species, and many trees are even the exclusive dwelling-place of some ant nowhere else to be found. In the scathes of leaves, in the corollas of flowers, in buds and blossoms, over and under the earth, in and out of doors, one meets these ubiquitous little creatures, which are undoubtedly one of the great plagues of the torrid zone.

While our indigenous ants cause a disagreeable burning on the skin, by the secretion of a corrosive acid peculiar to the race, the sting or bite of many tropical species causes the most excruciating tortures. "I have no words," says Schomburgk, "to describe the pain inflicted upon me by the mandibles of the *Ponera clavata*, a large, and, fortunately, not very common ant, whose long black body is beset with single hairs. Like an electric shock the pain instantly shot through my whole body, and soon after acquired the greatest intensity in the breast, and over and under the armpits. After a few minutes I felt almost completely paralysed, so that I could only with the greatest difficulty, and under the most excruciating tortures, totter towards the plantation, which, however, it was impossible for me to reach. I was found senseless on the ground, and the following day a violent wound fever ensued."

The *Triplaris Americana*, a South American tree, about sixty or eighty feet high, the branches of which are completely hollow and transversely partitioned at regular intervals, like the stems of the bamboo, is the retreat of one of the most ferocious ants. Woe to the naturalist who, ignorant of the fact, endeavours to break off a shoot of the Triplaris, or only knocks against the tree, for thousands will instantly issue from small round lateral openings in the plant, and fall upon him with inconceivable fury. The touch of a hot iron is not more painful than their bite, and the inflammation and pain last for several days after.

The black fire-ant of Guiana, though very small, is capable of inflicting excessive pain. "These insects," says Stedman, "live in such amazing multitudes together, that their hillocks have sometimes obstructed our passage by their size, over which, if one chances to pass, the feet and legs are instantly covered with innumerable hosts of these creatures, which seize the skin with such violence in their pincers, that they will sooner suffer the head to be parted from the body than let go their hold. The burning pain which they occasion cannot, in my opinion, proceed from the sharpness of the pincers only, but must be owing to some venomous fluid, which they infuse, or which the wound imbibes from them. I can aver that I have seen them make a whole company hop about as if they had been scalded with boiling water."

Of the more than seventy species of ants which occur in Ceylon alone, Sir E. Tennent describes the Dimiya, or great red ant, as the most formidable. "Like all their race, these ants are in perpetual motion, forming lines on the ground, along which they pass in continual procession to and from the trees on which they reside. They are the most irritable of the whole order in Ceylon, biting with such intense ferocity as to render it difficult for the unclad native to collect the fruit from the mango-trees, which the red ants especially frequent. They drop from the branches upon travellers in the jungle, attacking them with venom and fury, and inflicting intolerable pain both upon animals and man. On examining the structure of head through a microscope, I found that the mandibles, instead of meeting in contact, are so hooked as to cross each other at the points, whilst the inner line is sharply serrated throughout its entire length, thus occasioning the intense pain of their bite, as compared with that of the ordinary ant."

Another species, called the Kaddiya, is so much dreaded by the Singhalese, that, according to one of their legends, the Cobra invested it with her own venom in admiration of its singular courage.

"Having, while in Angola, accidentally stepped upon a nest of red ants," says Livingstone, "not an instant seemed to elapse before a simultaneous attack was made on various unprotected parts, up the trousers from below, and on my neck and breast above. The bites of these furies were like sparks of fire, and there was no retreat. I jumped about for a second or two, then in desperation tore off all my clothing, and rubbed and picked them off seriatim as quickly as possible. Ugh! they would make the most lethargic mortal look alive! Fortunately, no one observed this rencontre, or word might have been taken back to the village that I had become mad. It is really astonishing how such small bodies can contain so large an amount of ill nature. They not only bite, but twist themselves round after the mandibles are inserted, to produce laceration and pain more than would be effected by the single wound. Frequently, while sitting on the ox, as he happened to tread near a band, they would rush up his legs to the rider, and soon let him know that he had disturbed their march. They possess no fear, attacking with equal ferocity the largest as well as the smallest animals. When any person has leaped over the band, numbers of them leave the ranks and rush along the path, seemingly anxious for a fight."

But however formidable the weapons of the ants may be, yet the injuries they inflict upon the property of man, pouring over his plantations like a flood, and sweeping away the fruits of his labours, are of a much more lasting and serious nature than their painful bite or venomous sting.

In the West Indies, the brown-black Viviagua, about one-third of an inch long, and with a prickly thorax, is the greatest enemy of the coffee plantations. In one day it will rob a full-grown tree of all its leaves. It digs deep subterranean passages of considerable dimensions and irregular forms, with a great number of hand-high galleries branching out from the sides, and does even more harm to the coffee-plants by its mining operations, than by robbing them of their foliage.

Attacked in their roots, they fall into what may be called a consumptive state, bear no fruit, and die after a few months' lingering. The complete extirpation of the nest, and keeping up for some time a strong fire in the excavation, is the only means to subdue the evil, which leads to incalculable losses, when, through negligence, the Viviagua has once been allowed to multiply its numbers.

Other species are no less destructive to the sugar plantations, either by settling in the interior of the stalks (like the *Formica analis*), or by undermining the roots (like the *Formica saccharivora*), so that the plant becomes sickly and dies. About eighty years ago the island of Grenada was overrun by hosts of these devastating insects. Many household animals died from their attacks, and they effectually cleared the land of rats, mice, and reptiles. Streams of running water failed to interrupt their progress, and fire was vainly used to stop them, for millions rushed into the flames, and served as a bridge for the myriads that followed. All the means employed to save the sugar plantations from their fury proved ineffectual, until in the year 1780 the plague was swept away at once by a dreadful tornado, accompanied by a deluge of rain.

The Atta cephalotes, a species of ant distinguished by its large head, is the most formidable enemy of the banana and cassava fields. It lies in the ground and multiplies amazingly; in a very short time it will strip off the leaves of an entire field, and carry them to its subterranean abodes. Even where their nest is a mile distant from a plantation, these arch depredators know how to find it, and soon form a highway, about half a foot broad, on which they keep up the most active communications with the object of their attack. In masterly order, side by side, one army is seen to move onwards towards the field, while another is returning to the nest. In the last column each individual carries a round piece of leaf, about the size of a sixpence, horizontally over its head—a circumstance from which the insect has also been named the Umbrella ant. If the distance is too great, a party meets the weary carriers half way, and relieves them of their load. Although innumerable ants may thus be moving along, yet none of them will ever be seen to be in the other's way; and all goes on with the regularity of clock-work.

A third party is no less actively employed on the scene of destruction, cutting out circular pieces of the leaves, which, as soon as they drop upon the ground, are immediately seized by the attentive and indefatigable carriers. Neither fire nor water can prevent them from proceeding with their work. Though thousands may be killed, yet in less than an hour all the bodies will have been removed. Should the highway be closed by an insurmountable obstacle, another is soon laid out, and after a few hours the operations, momentarily disturbed, resume their former activity. The ants themselves, particularly the winged females, are considered a great delicacy by the Indians, who eat the abdomen, either raw or roasted. The taste is said to be agreeably saccharine.

Not satisfied with devouring his harvests, the tropical ants, as I have already mentioned, leave man no rest even within doors, and trespass upon his household comforts in a thousand various ways.

In Mainas, a province on the Upper Amazon, Professor Pöppig counted no less than seven different species of ants among the tormenting inmates of his hut. The diminutive red Amache was particularly fond of sweets. Favoured by its smallness, it penetrates through the imperceptible openings of a cork, and the traveller was often obliged to throw away the syrup which in that humid and sultry country replaces the use of crystallised sugar, from its having been changed into an ant-comfit. This troublesome lover of sweets lives under the corner-posts of the hut, so that it is quite impossible to dislodge him. The number of the Puca ticse, a red ant, of the ordinary size, was still greater; the trunks and papers were swarming with it, in spite of every precaution, so that it was quite incomprehensible how it found means to overcome all the obstacles that had been devised against it.

"The only possible way," says Stedman, "of keeping the ants from the refined sugar is by hanging the loaf to the ceiling by a nail, and making a ring of dry chalk around it, very thick, which crumbles down the moment the ants attempt to pass it. I imagined that placing my sugar-boxes in the middle of a tub, and on stone surrounded by deep water, would have kept back this formidable enemy; but to no purpose: whole armies of the lighter sort, to my astonishment, marched over the surface, and but very few of them were drowned. The main body constantly scaled the rock, and, in spite of all my efforts, made their entry through the keyholes, after which, the only way to clear the garrison is, to expose it to a hot sun, which the invaders cannot bear, and all march off in a few minutes."

The devastations of the house-ants are peculiarly hateful to the naturalist, whose collections, often gathered with so much danger and trouble, they pitilessly destroy. Richard Schomburgk suspended boxes with insects from the ceiling by threads strongly rubbed over with arsenic soap; but when, on the following morning, he wished to examine his treasures, instead of his rare and beautiful specimens he found nothing but a set of infamous red ants, who, crawling down the threads, had found means to invade the boxes and utterly to destroy their valuable contents.

In countless multitudes the Ranger ants break forth from the primeval forest, marching through the country in compact order, like a well-drilled army. Every creature they meet in their way falls a victim to their dreadful onslaught — rats, mice, lizards, and even the huge python, when in a state of surfeit from recent feeding. If a house obstructs their route, they do not turn out of the way, but go quite through it. Though they sting cruelly when molested, the West Indian planter is not sorry to see them in his house, for it is but a passing visit, and their appearance is the death-warrant for every spider, scorpion, cockroach, or reptile that pollutes his dwelling. Unfortunately, this thorough cleansing is but of short duration, as in less than a week tropical life calls forth a new generation of vermin.

A very formidable species of Ranger ant is found on the West Coast of Africa, attacking any animal whatever that impedes its progress, so that there is no escape but by immediate flight, or instant retreat to the water. The inhabitants of the negro villages are frequently obliged to abandon their dwellings, taking with them their children, and to wait till the scourge has passed.

The wonderful societies of the ants, their strength and perseverance, their unwearied industry, their astonishing intelligence, are so well known, and have been so often and so admirably described,* that it would be trespassing on the patience of my readers were I to enter into any lengthened details on the subject. And yet, the observations of naturalists have chiefly been confined to the European species, while the economy of the infinitely more numerous tropical ants, confined to countries or places hardly ever visited, or even unknown to civilised man, remains an inexhaustible field for future inquiry.

---

* Kirby and Spence's "Introduction to Entomology;" Swainson's "Habits and Instincts of Animals."

The study of their various buildings alone, from the little we know of them, would occupy a zealous entomologist for years. Here we have an American species that forms its globular nest of the size of a large Dutch cheese, of small twigs artistically interlaced; here another, which constructs its dwelling of dried excrements, attaching it to a thick branch; while a third (*Formica bispinosa*) uses the cotton of the Bombaceæ for its building material, and through the chemical agency of its pungent secretion converts it into a spongy substance.

On the west coast of Borneo, Mr. Adams noticed two kinds of ants' nests—one species of the size of a man's hand, adhering to the trunk of trees, resembling, when cut through, a section of the lungs; the other was composed of small withered bits of sticks and leaves, heaped up in the axils of branches, somewhat in the form of flattened cylinders and compressed cones. A third species, still more ingenious, constructs its domicile out of a large leaf, bending the two halves by the weight of united millions till the opposite margins meet at the under surface of the mid-rib, where they are secured by a gummy matter. The stores and larvæ are conveyed into the nest so made by regular beaten tracks along the trunk and branches of the tree.

On the large plains near Lake Dilolo, where water stands so long annually as to allow the lotus and other aqueous plants to come to maturity, Dr. Livingstone had occasion to admire the wonderful sagacity of the ants, whom he declares to be wiser than some men, as they learn by experience. When all the ant horizon is submerged a foot deep, they manage to exist by ascending to little houses, built of black tenacious loam, on stalks of grass, and placed higher than the line of inundation. This must have been the result of experience, for if they had waited till the water actually invaded their habitations on the ground, they would not have been able to procure materials for their higher quarters, unless they dived down to the bottom for every mouthful of clay. Some of these upper chambers are about the size of a bean, and others as large as a man's thumb. They must have been built in anticipation, " and if so," says the celebrated traveller, " let us humbly hope that the sufferers by the late inundations in France may be possessed of as much common sense as the little black ants of the Dilolo plains."

Two species of continental Europe, the *Formica rubescens* and *sanguinea*, are remarkable or infamous for their slave-making expeditions. Unable or unwilling to work themselves, they make war upon others for the sole purpose of procuring bondsmen, who literally and truly labour for them, and perform all the daily domestic duties of the community. Mr. Lund ("Annales des Sciences Naturelles," tome xxiii. p. 113) has found among the Brazilian ants similar mixed societies, and mentions the *Myrmica paleata* as a species whose nest contains labourers of the nearly related *Myrmica erythrothorax*. These are evidently workers by compulsion, though it must be confessed that they are far better treated than most "niggers" by the model republicans of Carolina or Louisiana.

The Aphides, or plant lice, eject a sweet, honey-like fluid, which may be correctly termed their milk, and which is so grateful to the ants, that they attend on the honey-flies for the sole purpose of gathering it, and literally milk them as we do our cows, forcing them to yield the fluid, by alternately patting them with their antennæ. But the most extraordinary part of these proceedings is, that the ants not only consider the Aphides as their property, but actually appropriate to themselves a certain number, which they enclose in a tube of earth or other materials near their nest, so that they may be always at hand to supply the nourishment which they may desire. The yellow ant, the most remarkable "cow-keeper" among our indigenous species, pays great attention to its herds, plentifully supplying them with proper food, and tending their young with the same tenderness which it exhibits towards its own. With the same provident care a large black ant of India constructs its nest at the root of the plant upon which its favourite species of aphis resides.

The ants of tropical America, where no Aphides are found, derive their honey from another family of insects, the numerous and grotesquely-formed Membracidæ, which are most abundant in the regions of Brazil.

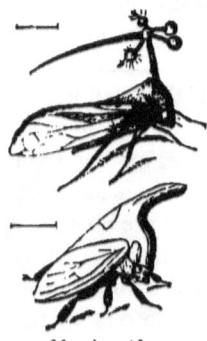

Membracidæ.

According to Mr. Swainson, who first stated the fact, many of these little Membracidæ live in families of twenty or thirty, all clustered together on the panicles of grasses,

and on the tops of other plants, like the European plant-lice. These are regularly visited by parties of a little black ant, which may be seen going and coming to their heads, and attending them with the same care which the European ants bestow on the Aphides. To render the similarity with cattle still more complete, the Membracidæ possess horns growing out of their heads, or are otherwise armed, while their large abrupt heads remind the entomologist of the bull or cow. Mr. Swainson's remarks did not extend to the particular mode by which these insects eject their secretion; but the surrounding leaves of the stalk which they inhabit are very clammy, like those of the plants infested with the Aphides of Europe, and the circumstance of their always being attended by ants places the fact beyond all doubt.

The Mexican honey ants (*Myrmecocystus Mexicanus*) are, if possible, still more remarkable, for here we see an animal rearing others of the same species for the purpose of food. Some of these ants namely are distinguished by an enormous swelling of the abdomen, which is converted into a mass like honey, and being unable, in their unwieldy condition, to seek food themselves, are fed by the labourers, until they are doomed to die for the benefit of the community. Whether this vast distension is the result of an intestinal rupture, caused by an excessive indulgence of the appetite, or whether they are purposely selected, confined, and over-fed, or wounded for the purpose, has not yet been ascertained.

These honey ants are brought to the Mexican markets, and considered a great delicacy.

The termites, or white ants, as they are commonly called, though they in reality belong to a totally different order of insects, are spread in countless numbers over all the warmer regions of the earth, emulating on the dry land the bore-worm in the sea; for when they have once penetrated into a building, no timber except ebony and iron-wood, which are too hard, or such as is strongly impregnated with camphor and aromatic oils, which they dislike, is capable of resisting their attacks. Their favourite food is wood, and so great are their multitudes, so admirable their tools, that in a few days they devour the timber work of a spacious apartment. Outwardly, the beams and rafters may seem untouched, while their core is completely consumed, for these destructive miners work in the dark, and seldom attack the outside until they have previously concealed themselves and their operations by a coat of clay. Scarcely any organic substance remains free from their attacks; and forcing their resistless way into trunks, chests, and wardrobes, they will often devour in one night all the shoes, boots, clothes, and papers they may contain. It is principally owing to their destructions, says Humboldt, that it is so rare to find papers in tropical America of an older date than fifty or sixty years. Smeathman relates, that a party of them once took a fancy to a pipe of fine old Madeira, not for the sake of the wine, almost the whole of which they let out, but of the staves, which, however, may not have proved less tasteful from having imbibed some of the costly liquor. On surveying a room which had been locked up during an absence of a few weeks, Forbes, the author of the "Oriental Memoirs," observed a number of advanced works in various directions towards some prints and drawings in English frames; the glasses appeared to be uncommonly dull, and the frames covered with dust. On attempting to wipe it off, he was astonished to find the glasses fixed to the wall, not suspended in frames as he left them, but completely surrounded by an incrustation cemented by the white ants, who had actually eaten up the deal frames and backboards and the greater part of the paper, and left the glasses upheld by the incrustation or covered way which they had formed during their depredations.

On the small island of Goree, near Cape Verde, the famous naturalist, Adanson, lived in a straw hut, which, though quite new at the time he took up his residence in it, became transparent in many places before the month was out. This might have been endured, but the villainous termites ravaged his trunk, destroyed his books, penetrated into his bed, and at last attacked the naturalist himself. Neither sweet nor salt water, neither vinegar nor corrosive liquids, were able to drive them away, and so Adanson thought it best to abandon the premises, and to look out for another lodging.

One night, Professor von Martius ("Travels in Brazil") was awakened by a disagreeable feeling of cold across his body. Groping in the dark, he found a cool greasy mass crawling right over the bed, and on a light being brought, saw to his astonishment that his rest had been disturbed by an innumerable host of white ants. The room having been uninhabited for some time, they had formed a clay nest in one of the corners, communicating with similar constructions under the roof, and the whole colony was now busy migrating. They formed a column about a foot and a half broad, and their multitudes poured along in one continuous stream, regardless of the fate of thousands of their companions, whom the naturalist scalded to death with boiling water. Their march ceased only with the dawn of day, and several baskets were filled with the bodies of the slain.

Even Europe is not free from the devastations of the white ants. Thus *Termes flavicollis* has been introduced from Northern Africa into the neighbourhood of Marseilles; the North American *Termes flavipes* has found its way to Portugal, and even o the imperial gardens of Schönbrunn, near Vienna; and in western France, particularly in the Département de la Charente Inferieure, we find a small but extremely voracious species (*Termes lucifuga*), which destroys houses to their very foundations, and once even devoured part of the archives in the Prefect's palace.

But if the greedy termite destroys like the bore-worm many a useful work of man, its ravages are perhaps more than compensated by its services in removing decayed vegetable substances from the face of the earth, and thus contributing to the purity of the air and the beauty of the landscape. If the forests of the tropical world, where thousands of gigantic trees succumb to the slow ravages of time, or are suddenly prostrated by lightning or the hurricane, still appear in all the verdure of perpetual youth, it is chiefly to the unremitting labours of the Termites that they are indebted for their freshness.

Though belonging to a different order of the insect world, the economy of the termites is very similar to that of the real ants. They also form communities, divided into distinct orders: labourers (larvæ), soldiers (neuters), perfect insects — and they also erect buildings, but of a far more astonishing structure. Several of their species (*T. atrox, bellicosus*, Smeathman) erect high dome-like edifices, rising from the plain, so that at first sight they might be mistaken for the hamlets of the negroes;

others (*T. destructor arborum*) build on trees, often at a considerable height above the ground. These sylvan abodes are frequently the size of a hogshead, and are more generally found in the new world.

The clay-built citadels or domes of the *Termes bellicosus*, a common species on the West Coast of Africa, attain a height of twelve feet, and are constructed with such strength that the traveller often ascends them to have an uninterrupted view of the grassy plain around. Only the under part of the mound is inhabited by the white ants, the upper portion serving principally as a defence from the weather, and to keep up in the lower part the warmth and moisture necessary to the hatching of the eggs and cherishing of the young ones. In the centre, and almost on a level with the ground, is placed the sanctuary of the whole community—the large cell, where the queen resides with her consort, and which she is doomed never to quit again, after having been once enclosed in it, since the portals soon prove too narrow for her rapidly-increasing bulk. Encircling the regal apartment, extends a labyrinth of countless chambers, in which a numerous army of attendants and soldiers is constantly in waiting. The space between these chambers and the external wall of the citadel is filled with other cells, partly destined for the eggs and young larvæ, partly for store-rooms. The subterranean passages which lead from the mound are hardly less remarkable than the building itself. Perfectly cylindrical, and lined with a cement of clay, similar to that of which the hill is formed, they sometimes measure a foot in diameter. They run in a sloping direction, under the bottom of the hill, to a depth of three or four feet, and then ramifying horizontally into numerous branches, ultimately rise near to the surface at a considerable distance. At their entrance into the interior of the hill, they are connected with a great number of smaller galleries, which ascend the inside of the outer shell in a spiral manner, and winding round the whole building to the top, intersect each other at different heights, opening either immediately into the dome in various places, and into the lower half of the building, or communicating with every part of it by other smaller circular passages. The necessity for the vast size of the main galleries underground, evidently arises from the circumstance of their being the great thoroughfare for the inhabitants, by which they fetch their clay, wood, water, or provisions, and their gradual ascent is requisite, as the Termites can only with great difficulty climb perpendicularly.

It may be imagined that such works require an enormous population for their construction; and, indeed, the manner in which an infant colony of termites is formed and grows, until becoming, in its turn, the parent of new migrations, is not the least wonderful part of this wonderful insect's history.

At the end of the dry season, as soon as the first rains have fallen, the male and female perfect termites, each about the size of two soldiers, or thirty labourers, and furnished with four long narrow wings folded on each other, emerge from their retreats in myriads. After a few hours their fragile wings fall off, and on the following morning they are discovered covering the surface of the earth and waters, where their enemies—birds, reptiles, ants—cause so sweeping a havoc that scarce one pair out of many thousands escapes destruction. If by chance the labourers, who are always busy prolonging their galleries, happen to meet with one of these fortunate couples, they immediately, impelled by their instinct, elect them sovereigns of a new community, and, conveying them to a place of safety, begin to build them a small chamber of clay, their palace and their prison — for beyond its walls they never again emerge.

Termite.

Soon after the male dies, but, far from pining and wasting over the loss of her consort, the female increases so wonderfully in bulk that she ultimately weighs as much as 30,000 labourers, and attains a length of three inches, with a proportional width. This increase of size naturally requires a corresponding enlargement of the cell, which is constantly widened by the indefatigable workers. Having reached her full size, the queen now begins to lay her eggs, and as their extrusion goes on uninterruptedly, night and day, at the rate of fifty or sixty in a minute, for about two years, their total number may probably amount to more than fifty millions! A wonderful fecundity, which explains how a termite colony, originally few in number, increases in a few years to a population equalling or surpassing that of the British empire.

This incessant extrusion of eggs necessarily calls for the attention of a large number of the workers in the royal chamber, to take them as they come forth, and carry them to the nurseries, in which, when hatched, they are provided with food, and carefully attended till they are able to shift for themselves, and become in their turn useful to the community.

In widening their buildings according to the necessities of their growing population, from the size of small sugar-loaves to that of domes which might be mistaken for the hovels of Indians or negroes, as well as in repairing their damages, the termite workers display an unceasing and wonderful activity; while the soldiers, or neuters, which are in the proportion of about one to every hundred labourers, and are at once distinguished by the enormous size of their heads, armed with long and sharp jaws, are no less remarkable for their courage and energy.

Soldier.

When any one is bold enough to attack their nest and make a breach in its walls, the labourers, who are incapable of fighting, immediately retire, upon which a soldier makes his appearance, obviously for the purpose of reconnoitring, and then also withdraws to give the alarm. Two or three others next appear, scrambling as fast as they can one after the other; to these succeed a large body, who rush forth with as much speed as the breach will permit, their numbers continually increasing during the attack. These little heroes present an astonishing, and at the same time a most amusing spectacle. In their haste they frequently miss their hold, and tumble down the sides of their hill; they soon, however, recover themselves, and being blind, bite everything they run against. If the attack proceeds, the bustle increases to a tenfold degree, and their fury is raised to its highest pitch. Woe to him whose hands or legs come within their reach, for they will make their fanged jaws meet at the very first stroke, drawing their own weight in blood, and never quitting their hold, even though they are pulled limb from limb. The courage of the bulldog is as nothing compared to the fierce obstinacy of the termite-soldier.

So soon as the injury has ceased, and no further interruption is given, the soldiers retire, and then you will see the labourers hastening in various directions towards the breach, each carrying in his mouth a load of tempered mortar half as big as himself, which he lays on the edge of the orifice, and immediately hastens back for more. Not the space of the tenth part of an inch is left without labourers working upon it at the same moment; crowds are constantly hurrying to and fro; yet, amid all this activity, the greatest order reigns — no one impedes the other, but each seems to thread the mazes of the multitude without trouble or inconvenience. By the united labours of such an infinite host the ruined wall soon rises again; and Mr. Smeathman has ascertained that in a single night they will restore a gallery of three or four yards in length.

In numbers and architectural industry the American Termites are not inferior to those of the old world. In the savannas of Guiana their sugar-loaf or mushroom-shaped, pyramidal or columnar hills are everywhere to be seen, impenetrable to the rain, and strong enough to resist even a tropical tornado. On the summits of these artificial mounds a neat little falcon (*Falco sparverius*) often takes his station, darting down, from time to time, like lightning upon some unfortunate lizard, and then again speedily returning to his look-out. The large caracara eagle (*Polyborus caracara*) likewise chooses these eminences as an observatory from whence he rushes robber-like on his prey; there also an ugly black lizard (*Ecchymotes torquatus*) loves to sun itself, but disappears immediately in the grass as soon as a traveller approaches.

In many parts of the Brazilian campos or savannas the termite-hills, which are there generally of a more flattened form, are so numerous that one is almost sure to meet with one of them at the distance of every ten or twenty paces. The great ant-bear digs deep holes into their sides, where afterwards small owls build their nests. Similar termite structures, of a dark-brown colour, and a round form, are attached to the thick branches of the trees, and you will scarcely meet with a single specimen of the tall candelabra-formed cactuses (*Cerei*), so common on those high grass-plains, that is not loaded with their weight.

In spite of their working in the dark, in spite of their subterranean tunnels, their strongholds, and the fecundity of their queens, the termites, even when their swarms do not expose themselves to the dangers already mentioned, are subject to the attacks of innumerable foes — myrmecophagi, orycteropi, birds, and a whole host of insects — that do man no little service by keeping them within bounds.

One of their most ferocious enemies is a species of black ant, which, on the principle of setting one thief to catch another, is used by the negroes of Mauritius for their destruction. When they perceive that the covered ways of the termites are approaching a building, they drop a train of syrup as far as the nearest encampment of the hostile army. Some of the black ants, attracted by the smell and taste of their favourite food, follow its traces and soon find out the termite habitations. Immediately part of them return to announce the welcome intelligence, and after a few hours a black army, in endless columns, is seen to advance against the white-ant stronghold. With irresistible fury (for the poor termites are no match for their poisonous sting and mighty mandibles), they rush into the galleries, and only retreat after the extirpation of the colony.

Mr. Baxter ("Eight Years' Wanderings in Ceylon") once saw an army of black ants returning from one of these expeditions. Each little warrior bore a slaughtered termite in his mandibles, rejoicing no doubt in the prospect of a comfortable meal, or a quiet dinner-party at home. Even man is a great consumer of termites, and they are esteemed a delicacy by negroes and Indians, both in the old and in the new world.

"The Bayeiye chief Palani, visiting us while eating," says Dr. Livingstone, "I gave him a piece of bread and preserved apricots, and as he seemed to relish it much, I asked him if he had any food equal to that in his country. 'Ah!' said he, 'did you ever taste white ants?' As I never had, he replied: 'Well, if you had, you never could have desired to eat anything better.'"

In some parts of the East Indies the natives have an ingenious way of emptying a termite-hill, by making two holes in it, one to the windward and the other to the leeward, placing at the latter opening a pot rubbed with an aromatic herb to receive the insects, when driven out of their nest by a fire of stinking materials made at the former breach. Thus they catch great quantities, of which they make, with flour, a variety of pastry. In South Africa the general way of catching them is to dig into the ant-hill, and when the builders come forth to repair the damage, to brush them off quickly into the vessel, as the anteater does into his mouth. They are then parched in iron pots over a gentle fire, stirring them about as is done in roasting coffee, and eaten by handfuls, without sauce or any other addition, as we do comfits. According to Smeathman, they resemble in taste sugared cream, or sweet almond paste, and are, at the same time, so nutritious, that the Hindoos use them as a restorative for debilitated patients.

While most termites live and work entirely under covered galleries, the marching white ant (*T. viarum*) exposes itself to the day. Mr. Smeathman on one occasion, while passing through a dense forest, suddenly heard a loud hiss like that of a serpent; another followed, and struck him with alarm; but a moment's reflection led him to conclude that these sounds proceeded from white ants, although he could not see any of their huts around. On following this noise, however, he was struck with surprise and pleasure at perceiving an army of these creatures emerging from a hole in the ground, and marching with the utmost swiftness. Having proceeded about a yard, this immense host divided into two columns, chiefly composed of labourers, about fifteen abreast, following each other in close order, and going straight forward. Here and there was seen a soldier, carrying his vast head with apparent difficulty, at a distance of a foot or two from the columns; many other soldiers were to be seen, standing still or passing about, as if upon the look-out lest some enemy should suddenly surprise their unwarlike comrades. But the most extraordinary and amusing part of the scene was exhibited by some other soldiers, who having mounted some plants, ten or fifteen inches from the ground, hung over the army marching below, and by striking their jaws upon the leaves at certain intervals, produced the noise above mentioned; to this signal the whole army immediately returned a hiss, and increased their pace. The soldiers at these signal-stations sat quite still during these intervals of silence, except now and then making a slight turn of the head, and seemed as solicitous to keep their posts as regular sentinels. After marching separately for twelve or fifteen paces, the two columns of this army again united, and then descended into the earth by two or three holes. Mr. Smeathman watched them for more than an hour, without perceiving their numbers to increase or diminish. Both the labourers and soldiers of this species are furnished with eyes.

One of the many unsolved mysteries of termite life is whence they derive the large supplies of moisture with which they not only temper the clay for the construction of their long covered ways above ground, but for keeping their passages uniformly damp and cool below the surface. Yet their habits in this particular are unvarying, in the seasons of drought as well as after rain; in the most arid positions; in situations inaccessible to drainage from above, and cut off by rocks and impervious strata from springs from below. Struck with this wonderful phenomenon, Dr. Livingstone raises the question whether the termites may not possess the power of combining the oxygen or hydrogen of their vegetable food by vital force, so as to form water; and indeed it is highly probable that they are endowed with some such faculty, which, however wonderful, would still be far less astonishing than the miracles of their architectural instinct.

CHAPTER XXVI.

TROPICAL SPIDERS AND SCORPIONS.

Immense Webs of several Tropical Spiders—Their Means of Defence—Beautiful Colouring of the Epeiras—The Trap-door Spider—Wonderful Maternal Instincts of Spiders—Enemies of the Spiders—Their Usefulness—Mortal Combat between a Spider and a Cockroach—Scorpions—Dreadful Effects of the Sting of Tropical Scorpions—A Scorpion Battle—The Galeodes—Combat of a Galeode and a Lizard—Formidable character of the Tropical Centipedes.

AN insect, half of whose body is generally fixed to the other by a mere thread, whose soft skin is unable to resist the least pressure, and whose limbs are so loosely attached to the body as to be torn off by the slightest degree of force, would seem utterly incapable of protecting its own life and securing that of its progeny. Such, however, is the physical condition of the spiders, who would long since have been extirpated, if nature had not provided them with the power of secreting two liquids, the one a venom ejected by their mandibles, the other of a glutinous nature, transuded by papillæ at the end of their abdomen. These two liquids amply supply the want of all other weapons of attack or defence, and enable them to hold their own against a host of enemies. With the former they instantly paralyse insects much stronger and much more formidable in appearance than themselves; while with the latter they spin those threads which serve them in so many ways, to weave their wonderful webs, to traverse the air, to mount vertically, to drop uninjured, to construct the hard cocoons intended to protect their eggs against their numberless enemies, or to produce the soft down which is to preserve them from the cold.

Preying on other insect tribes, which they attack with the ferocity of the tiger, or await in their snares with the patient artifice of the lynx, the spiders may naturally be expected to be most numerous in the torrid zone, where nature has provided them with the greatest abundance of food. There also, where so many beetles, flies, and moths attain a size unknown in temperate regions, we find the spiders growing to similar gigantic dimensions, and forming webs proportioned to the bulk of the victims which they are intended to ensnare.

In some parts of Makololo, Dr. Livingstone saw great numbers of a large beautiful yellow-spotted spider, the webs of which were about a yard in diameter. The lines on which these webs were spun, extended from one tree to another, and were as thick as coarse thread. The fibres radiated from a central point, where the insect waited for its prey. The webs were placed perpendicularly, and a common occurrence in walking was to get the face enveloped in them, as a lady is in a veil.

By means of their monstrous webs many giant-spiders of the tropical zone are enabled to entangle not only the largest butterflies and moths, but even small birds. Tremeyer tells us that there are spiders in Mexico which extend such strong nets across the pathways, that they strike off the hat of the passer by; and at Goree and in Senegal several spiders spin threads so strong as to be able to bear a weight of several ounces, and which no doubt would be made use of for twine, if the negroes did not already possess vegetable fibres in abundance fit for the purpose. In the forests of Java, Sir George Staunton saw spider-webs of so strong a texture that it required a sharp knife to cut one's way through them; and many other similar examples might be mentioned.

These large spiders so temptingly suspended in mid-air in the forest glades, seem very much exposed to the attacks of birds, but in many cases it has pleased nature to invest them with large angular spines sticking out of their bodies in every kind of fashion. Some are so protected by these long prickles that their bodies resemble a miniature "chevaux de frise," and could not by any possibility be swallowed by a bird without producing a very unpleasant sensation in his throat. One very remarkable species (*Gasteracantha arcuata*, Koch) has two enormous recurved conical spines, proceeding upwards from the posterior part of the body, and several times longer than the entire spider.

Other araneæ, to whom these means of defence have been denied, are enabled by their colour to escape the attacks of many enemies, or to deceive the vigilance of many of their victims. Thus, those that spend their lives among the flowers and foliage of the trees are, in general, delicately and beautifully marked with green, orange, black, and yellow, while those which frequent gloomy places are clothed with a dark-coloured and dingy garb, in accordance with their habits. In the forests about Calderas, in the Philippine Archipelago, Mr. Adams saw handsomely coloured species of theridia crouching among the foliage of the trees: while numbers of the same genus of a black colour were running actively about among the dry dead leaves that strewed the ground, looking, at a little distance, like odd-shaped ants, and no doubt deceiving many an antagonist by this appearance. One species, which knew it was being watched, placed itself upon a diseased leaf, where it remained quite stationary until after the departure of the naturalist, who, had he not seen the sidelong movement of the cunning little creature in the first instance, would not have been able to distinguish its body from the surface of the leaf. While, in this case, dulness of colour served as a defence, the vividly-coloured spiders that live among the foliage and flowers no doubt attract many flies and insects by reason of their gaudily-tinted bodies.

One of the most remarkable instances of the harmony of colour between the araneæ and their usual haunts was noticed by Mr. Adams among dense thickets formed by the Abrus precatoria, where he found a spider with a black abdomen marked on each side with scarlet, thus resembling the colours of the seeds of the Abrus, so well known to children under the name of "black-a-moor beauties."

An exception to the general rule is, however, found in those very large and powerful species, which, if not rendered somewhat conspicuous to the sight of other insects, might do too much damage to the tribes which they keep in check. Most of these, therefore, have the thorax and abdomen margined with a light colour that contrasts strongly with that of their bodies, and, in many cases, gives timely warning of their approach.

The European spiders have generally a very repulsive appearance, while many of the tropical species are most splendidly ornamented, or rather illuminated, many of them by the vividness of their colours resembling the gaudy missals painted by monks in the Middle Ages. Thus, among the epeiras of the Philippian Isles, are found white figures on a red ground; red, yellow, and black, in alternate streaks; orange marbled with brown, light green with white ocelli, yellow with light brown festoons, or ash-coloured and chesnut bodies with crescents, horse-shoes, Chinese characters, and grotesque hieroglyphics of every description. Unfortunately, these colours, lustrous and metallic as the feathers of the humming-bird, are, unlike the bright colours of the beetle, totally dependent on the life of the insect which they beautify, so that it is impossible to preserve them.

While most spiders obtain their food either by patiently waiting in ambush or by catching it with a bound, the enormous mygales, or trap-door spiders, run about with great speed in and out, behind and around every object, searching for what they may devour, and from their size and rapid motions exciting the horror of every stranger. Their body, which sometimes attains a length of three inches, while their legs embrace a circle of half a foot in diameter, is covered all over with brown, reddish brown, or black hair, which gives them a funereal appearance, while their long feelers armed with sharp hooks proclaim at once what formidable antagonists they must be to every insect that comes within their reach. Though some species are found in Southern Europe, in Chili, or at the Cape, yet they are chiefly inhabitants of the torrid zone, both in the old and the new world. Some of them weave cells between the leaves, in the hollows of trees or rocks, while others dig deep tubular holes in the earth, which they cover over with a lid, or rather with a door formed of particles of earth cemented by silken fibres, and closely resembling the surrounding ground. This door or valve is united by a silken hinge to the entrance at its upper side, and is so balanced, that when pushed up, it shuts again by its own weight; nay, what is still more admirable, on the interior side opposite to the hinge a series of little holes may be perceived, into which the mygale introduces its claws to keep it shut, should any enemy endeavour to open it by force. The interior of the nest, which is sometimes nine inches deep, is lined with a double coat of tapestry, the one nearest the wall, which is of a coarser tissue, being covered with a pure white silken substance like paper.

"In the forests of Brazil," says Mr. Swainson, " we once met with a most interesting little spider, which sheltered itself in the same manner. Its case was suspended in the middle of its web. Upon being disturbed, the little creature ran to it with swiftness. No sooner had it gained its retreat than the door closed as if by a spring, and left us in silent admiration — too great to lead us to capture the ingenious little creature for our collection."

At Caldera, Mr. Adams observed a dingy little species of spider of the genus Clubiona, concealing itself in very snug retreats formed out of a dead leaf, rolled round in the shape of a cylinder, lined with a soft silken tissue, and closed at one end by means of a strong woven bolt-door. When hunted, it was amusing to see the frightened little creatures run for protection into their tiny castles, where they would doubtless be safe from the attacks of birds, owing to the leaves not being distinguishable from others that strew the ground.

All species of spiders are gifted with the same maternal instinct, and resort to various methods for the purpose of securing their cocoons. The Theridion, when a seizure of the precious burden is threatened, tumbles together with it to the ground, and remains motionless, while the Thorinsa covers it with its body, and when robbed of it, wanders about disconsolate. In a forest of the Sooloo Islands, Mr. Adams found the ground literally overrun with a small black agile species of Lycosa, many of which had a white flattened globose cocoon affixed to the ends of their abdomen. It was most amusing to watch the care with which these jealous mothers protected the cradles of their little ones, allowing themselves to fall in to the hands of the enemy rather than be robbed of the silken nests that contained them.

If the spiders are at war with all other insects, and contribute to keep them within bounds by the destruction they cause among their ranks, they in their turn have to suffer from the attacks of many enemies. Several species of monkeys, squirrels, lizards, tortoises, frogs, and toads catch and devour them wherever they can. In Java and Sumatra, we even find several birds belonging to the order of sparrows that have been named Arachnotheræ, from their living almost exclusively on spiders. Armed with a prodigiously long recurved and slender beak, they know how to pursue them and drag them forth from the most obscure recesses.

It is amongst the insects themselves, however, that the spiders have to fear their most numerous and formidable enemies. Independently of those which they find in their own class, the centipedes seize them beyond the possibility of escape; while several species of philanthus, pompilius, and sphex, more savage and poisonous than themselves, will rush upon spiders eight times their size and weight, and benumbing them with a sting, bear them off to their nests, to serve as food for their larvæ.

But the insects which in appearance are the tiniest and most delicate, are perhaps those which most cruelly wound the spiders, by attacking them in their eggs, which they watch over with such affection, as to be ever ready for them to make the sacrifice of their own lives. The *Pimpla Arachnitor* pierces with its invisible gimlet the tender skin of the spider's egg, and, without tearing it, introduces its own eggs into the liquid. The pimpla's egg soon comes to maturity, and the larva devours the substance of that of the spider, from whence a winged insect bursts forth — a phenomenon which made some naturalists, too hasty to judge from appearances, believe that spiders were able to procreate four-winged flies.

Notwithstanding the disgust or horror which they generally inspire, the spiders are, with very rare exceptions, by no means injurious to man. However promptly their venom may act upon insects, even that of the largest species of Northern Europe produces, on coming into contact with our skin, no pain or inflammation equalling in virulence that of the wasp, the bee, the gnat, or other insects of a still smaller size. The giant spiders of a sunnier sky, armed with more formidable weapons, naturally produce a more painful sting; but even here the effects have been much exaggerated, and the wonderful stories about the Sicilian tarantula's bite, which we read of in Brydone and other authors, are nothing but fables raised upon a very slender foundation of truth. Azara mentions that, several of his negroes having been bitten by the large Avicular mygale (*M. avicalaria*) of South America, a slight ephemeral fever was the only result.

In the country of the Makalolo, Dr. Livingstone feeling something running across his forehead as he was falling asleep, put up his hand to wipe it off, and was sharply stung, both on the hand and head; the pain was very acute. On obtaining a light, he found that it had been inflicted by a light-coloured spider about half an inch in length; but one of the negroes having crushed it with his fingers, he had no opportunity of examining whether the pain had been produced by poison from a sting, or from its mandibles. No remedy was applied, and the pain ceased in about two hours.

The Bechuanas believe that there is a small black spider in the country whose bite is fatal, but Dr. Livingstone never met with an instance in which death could be traced to this insect.

If thus, among the many species of spiders, hardly a single one may be said to be formidable to man, the indirect services which they render him — by diminishing the number of noxious insects, or keeping in check the legions of gnats which irritate and annoy him by their attacks — are far from inconsiderable.

Nor are they entirely without direct use. Several savage nations eat spiders, and the inhabitants of New Caledonia reckon a large species of epeira amongst the choicest delicacies of the land. Even in Europe some people enjoy a spider, and the famous astronomer Lalande was far from being singular in this respect. They are said to taste like filberts, and the proper way to eat them is to take off the legs, and to swallow the abdomen, after having washed and rubbed it with butter.

The property of spiders'-webs to stop an hæmorrhage or the bleeding of a wound is a well-known fact, and they have also been recommended as an anti-febrifuge.

In several countries where the insects cause great ravages, the services of the spiders are duly appreciated. Thus in the West Indies, a large and formidable trap-door spider, which would make a European start back with horror, is looked upon with pleasure by the islanders of the torrid zone, who respect it as a sacred animal, by no means to be disturbed or harmed, as it delivers them from the cockroaches, which otherwise would overrun their dwellings. Those who do not possess these spiders take good care to purchase and transport them into their houses, expecting from them similar services to those we derive from a good domestic cat. I wonder whether it has ever been attempted to acclimatise this spider or any of the larger mygales in London, where they no doubt would be equally useful, and find a superabundance of vermin to feast upon.

A trap-door spider bounding on a cockroach, with all the ferocity of a tiger springing on a deer or an antelope, would have all the interest of a bull-fight or a tournament, if the diminutive size of the combatants were swelled to more ample proportions.

Mr. E. Layard has described ("Ann. Mag. Nat. H.," May 1853) one of these encounters which he witnessed near a ruined temple in Ceylon. When about a yard apart, each of the enemies discerned the other and stood still, the spider with his legs slightly bent and his body raised, the cockroach confronting him, and directing his antennæ with a restless undulation towards his enemy. The spider, by stealthy movements, approached to within a few inches, and paused, both parties eyeing each other intently; then suddenly a rush, a scuffle, and both fell to the ground, when the blatta's wings closed; the spider seized it under the throat with his claws, and when he had dragged it into a corner, the action of his jaws was distinctly audible. Next morning, Mr. Layard found the soft parts of the body had been eaten, nothing but the head, thorax, and elytra remaining.

When we consider the large size of many of the tropical spiders, and the strength of their threads, it seems probable that their cocoons might be put to some use. We are told by Azara, in his "Travels to Paraguay," that a spider exists in that country the silk of whose spherical cocoons, measuring an inch in diameter, is spun on account of its permanent orange-colour. The eyes and noses of the women employed in unravelling the cocoons are said to water considerably, though without their perceiving any pungent smell, or feeling any other inconvenience. This spider is perhaps the same as that which, according to M. de Bomare, is known in the interior of South America under the name of the silk spider. Its cocoon is of the size of a pigeon's egg, the silk is soft, and can be easily carded.

Attempts have also been made in Europe to utilise the threads of the large indigenous spiders. About the beginning of the last century, M. Bon, a Frenchman, who seems to have been the first that ever put the idea into practice, collected a sufficient quantity to make some stockings and gloves, which he presented to the king, Louis Quatorze, and to the Academy of Sciences in Paris. His discovery caused a sensation at the time, and his dissertation on the subject was translated into all European languages, and at a later period even into the Chinese, by order of the Emperor Kien Long. The celebrated Reaumur, however, who was commissioned by the Academy to report on M. Bon's discovery, pointed out how difficult it would be to put it to any extensive use, as it would require no less than 55,296 of the epeira diadema to produce a single pound of silk; and how were all these to be provided with flies!

Diadem Spider.

If the extreme fineness of the spider's threads is an obstacle to their being spun and woven, this property, united with their metallic brilliancy, renders them an excellent material for the construction of the micrometers used for astronomical purposes: the finest silver-thread which it is possible to spin having a diameter of $\frac{1}{974}$ of an inch, while spiders' threads measure only $\frac{1}{4000}$ or even $\frac{1}{6000}$. Troughton, an eminent English instrument maker, first thought of substituting them for the silver-threads then in use, and they were found to answer so well that since that time they have been constantly employed.

The scorpions, which even in Europe are reckoned among the most malignant insects, are truly terrific in the torrid zone, where they frequently attain a length of six or seven inches. Closely allied to the spiders, their aspect is still more repulsive. Were one of the largest scorpions menacingly to creep up against you, with extended claws and its long articulated sharply-pointed tail projecting over its head, I think, despite the strength of your nerves, you would start back, justly concluding that a creature of such an aspect must necessarily come with the worst intentions. The poison of the scorpion is discharged like that of the snake. Near the tip of the crooked sting namely, which terminates the tail, we find two or three very small foramina, through which, on pressure, the venom of the gland with which they are connected immediately issues forth. By means of this weapon, even the small European scorpions are able to kill a dog, while the tropical giants of the race inflict wounds that become fatal to man himself. The sting of several South American scorpions produces fever, numbness of the limbs, tumours on the tongue, weakness of the sight, and other nervous symptoms, lasting twenty-four or forty-eight hours; but the African scorpions seem to be still more formidable. Mr. Swainson informs us that the only means of saving the lives of our soldiers who were stung by those of Egypt, was the amputation of the wounded limb; and Professor Ehrenberg, who, while making his researches on the Natural History of the Red Sea, was stung five times by the Androctonus quinquestriatus, and funestus, says he can well believe, from the dreadful pains he suffered, that the poison of these scorpions may become fatal to women and children.

Scorpion.

A servant of Mr. Russegger's ("Travels in Nubia"), while emptying a trunk, was stung in the breast by a large scorpion, which had concealed itself among the linen. For hours the pain was dreadful, shooting from time to time through the whole nervous system, and almost depriving the patient of consciousness. A cold perspiration covered his brow, and it was only after the internal and external application of ammonia, one of the chief remedies for sustaining the sinking flame of life, that he at length felt some relief, though he had still to suffer for several days from a strong fever.

"The black, or rock scorpion," says Lieutenant Patterson, "is nearly as venomous as any of the serpent tribe. A farmer who resided at a place called the Paarle, near the Cape, was stung by one in the foot, during my stay in the country, and died in a few hours."

The scorpions live mostly on the ground, in gloomy recesses, and even in the nooks and corners of dwelling-houses, so that, in countries where they are known to abound, it is necessary to be very cautious in removing stones, pieces of wood, &c. Of a ferocious cruel disposition, they are not only the foes of all other animals, but carry on a war of extermination among themselves, and are even said to kill and devour their own progeny, without pity, as soon as they are born; thus rendering good service to the community at large. Maupertuis once inclosed a hundred scorpions — a select and delightful party — in a box. Immediately a furious battle ensued — one against all, all against one — and in an hour's time scarcely one of the combatants survived the conflict.

The bite of the galeodes — a tropical insect closely related to the scorpion (and resembling it in form), but without a sting — is said to be fatal to man and beast. Captain Hutton, who has described the *Galeodes vorax* of India,[*] once threw a lizard, six inches long, before this insect. The galeodes immediately sprang upon the reptile, fastening upon its shoulders; and though at first the poor lizard defended itself valiantly, rolling about in all directions, and endeavouring to shake off its tormentor, the galeodes held fast, gnawing deeper and deeper until it reached the vitals of its victim, which it finally devoured, leaving nothing but skin and bone behind. A young sparrow confined under a glass bell with a galeodes was also killed; the insect, however, did not devour the bird, or any part of it, but seemed satisfied with having killed it.

The poison of the scorpion is lodged in its tail, but that of the centipede is in its jaws. These are likewise among the pests of tropical climates; for, although several are found in Europe, or even in Britain, yet, from their small size, they are harmless to man. Those of India and South America, on the other hand, are enormous, frequently six or seven inches long, and their sting is no less painful and virulent than that of the dreaded scorpion itself.

Several of these creatures are phosphorescent; and the elegant colouring of some of the large tropical species would no doubt be admired if their noxious qualities, their strange form, and their resemblance to caterpillars and worms, did not inspire a disgust which it is difficult to surmount.

Centipede.

[*] "Journal of the Asiatic Society of Bengal," No. 45.

Coral Island.

CHAPTER XXVII.

THE TROPICAL OCEAN.

Wanderings of an Iceberg — The Tropical Ocean — The Cachalot — The Frigate Bird — The Tropic Bird — The Esculent Swallow — The Flying-fish — The Bonito — The White Shark — Tropical Fishes — Crustaceans — Land Crabs — Mollusks — Jelly Fish — Coral Islands.

DAY after day the glacier of the north protrudes its mass, farther and farther into the sea, until finally, rent by the tides, and with a crash louder than that of the avalanche, the iceberg rolls into the abyss. The frost-bound waters, that have languished so many years in their Greenland prison, are now drifting to the south, on their way to the tropical ocean; but the sun must still rise and set for many a day before they bid adieu to the fogs of the north.

See there yon dismal ice-blocked shore, with the jagged mountains in the background, their snowy peaks rising high into the sky. Screeching sea-birds—fulmars, gulls, guillemots, auks—mix their hoarse voices with the melancholy tones of the breakers and the winds, and between them all resounds, from time to time, the bellowing of the walrus or the roar of the polar bear.

The weak rays of the sun, just dipping over the horizon, have called forth these symptoms of life; but as soon as the great luminary disappears, animal creation becomes mute, and the voices of the air and ocean are again the only sounds which break the silence of the arctic night.

The crystal mass floats along, buried in deep darkness; but soon a new and wondrous sight is seen, for the flaming swords of the northern light flit through the heavens, casting a magic gleam, here on the desert shore, there on the dark bosom of the sea.

Advancing farther and farther to the south, the iceberg loses one after another the witnesses of its first migrations, and wasting more and more, at length entirely merges in the tepid Gulf-stream. The enthralled waters are now all liberated, but many on their western passage are again diverted to the north, and the others reach, only after a long circuit, the mighty equatorial stream, which carries them along, through the torrid ocean, from one hemisphere to the other.

The animal life they meet with in these sunny regions is very

Sperm Whale.

different from that which witnessed their passage through the higher latitudes.

The large whalebone whale, the rorqual and narwhal of the north, have disappeared, but *pods* of the mighty sperm whale rapidly traverse the equatorial seas.

The birds also exhibit new types of being. The royal albatross avoids the torrid zone, but the high-soaring frigate-bird hovers over the waters, where it is seen darting upon the flying-fish, and like the skua gull of the north, attacking the weaker sea-birds in order to make them disgorge their prey.

"He is almost always a constant attendant upon our fishermen," says Dr. Chamberlain,* "when pursuing their vocation on the sand-banks in Kingston Harbour, or near the Palisados. Over their heads it takes its aërial stand, and watches their motions with a patience and a perseverance the most exemplary. It is upon these occasions that the pelicans, the gulls, and other sea-birds become its associates and companions. These are also found watching with equal eagerness and anxiety the issue of the fishermen's progress, attracted to the spot by the sea of living objects immediately beneath them. And then it is, when these men are making their last haul, and the finny tribes are fluttering and panting for life, that this voracious bird exhibits his fierce propensities. His hungry companions have scarcely secured their prey by the side of the fishermen's canoes, when, with the lightning's dart, they are pounced upon with such violence, that to escape its rapacious assaults they readily, in turn, yield their hard-earned booty to this formidable opponent. The lightness of its trunk, the short torso and vast spread of wing, together with its long slender and forked tail, all conspire to give it a superiority over its tribe, not only in length and rapidity of flight, but also in the power of maintaining itself, on outspread pinions, in the regions of its aërial habitations amidst the clouds; where, at times, so lofty are its soarings, that its figure becomes almost invisible to the spectator in this nether world."

Frigate Bird.

The beautiful tropic birds, whose name implies the limit of their abode—for they are seldom seen but a few degrees south or north of either tropic—hover at such a distance from the nearest land, that it is still an enigma where they pass the night—whether they sleep upon the waters, or whether their extraordinary length of wing bears them to some isolated rock. Nothing can be more graceful than their flight. They glide

---

* "Jamaica Almanac," 1843.

along, most frequently without any motion of their outstretched pinions, but at times this smooth progression is interrupted by sudden jerks. When they see a ship, they never fail to sail round it, and the mariner bound to the equatorial regions, hails them as the harbingers of the tropics. The two long straight narrow feathers of which their tail consists, are employed by the natives of the greater part of the South Sea Islands as ornaments of dress, and serve to distinguish the chieftains from the multitude.

The esculent swallow (*Colocalia esculenta*)—whose edible nest, formed by a secretion which hardens in the air, is one of the greatest dainties of the Chinese epicure—may almost be considered as a sea-bird, as it chiefly inhabits marine caves in various islands of the Indian Archipelago, and exclusively seeks its food in the teeming waters.

The steep sea-walls along the south coast of Java are clothed to the very brink with luxuriant woods, and screw-pines strike everywhere their roots into their sides or look down from the margin of the rock upon the sea below. The surf of ages has worn deep caves into the chalk cliffs, and here the swallow builds her nest. When the sea is most agitated, whole swarms are seen flying about, and purposely seeking the thickest wave-foam, where no doubt they find their food. From a projecting cape, or looking down upon the play of waters, may be seen the mouth of the cave of Gua Rongkop, sometimes completely hidden under the waves, and then again opening its black recesses, into which the swallows vanish, or from which they dart forth with the rapidity of lightning. While at some distance from the coast the blue ocean sleeps in peace, it never ceases to fret and foam against the foot of these mural rocks, where the most beautiful rainbows glisten in the rising vapour.

Who can explain the instinct which prompts the birds to glue their nests to the high dark vaults of those apparently inaccessible caverns? Did they expect to find them a safe retreat from the persecutions of man? Then surely their hopes were vain, for where is the refuge to which his greediness cannot find the way? At the cavern of Gua Gede the brink of the coast lies eighty feet above the level of the sea at ebb-tide. The wall first bends inwards, and then at a height of twenty-five feet from the sea throws out a projecting ledge, which is of great use to the nest-gatherers, serving as a support for a rattan ladder let down from the cliff. The roof of the cavern's mouth lies only ten feet above the sea, which even at ebb-tide completely covers the floor of the cave, while at flood-tide the opening of the vast grotto is entirely closed by every wave that rolls against it. To penetrate into the interior is thus only possible at low water, and during very tranquil weather, and even then it could not be done if the roof were not perforated and jagged in every direction.

The boldest and strongest of the nest-gatherers wedges himself firmly in the hollows, or clings to the projecting stones while he fastens rattan ropes to them, which then hang four or five feet from the roof. To the lower end of these ropes long rattan cables are attached, so that the whole forms a kind of suspension bridge, throughout the entire length of the cavern, alternately rising and falling with its inequalities. The cave is 100 feet broad and 150 feet long, as far as its deepest recesses. If we justly admire the intrepidity of the St. Kildans, who, let down by a rope from the high level of their rocky birthplace, remain suspended over a boisterous sea, we needs must also pay a tribute of praise to the boldness of the Javanese nest-gatherers, who, before preparing their ladders for the plucking of the birds' nests, first offer solemn prayers to the goddess of the south coast, and deposit gifts on the tomb where the first discoverer of the caves and their treasures is said to repose.

While traversing the tropical ocean, the mariner often sees whole shoals of flying-fishes (*Exocoetus volitans, Pterois volitans*) dart out of the waters to escape the jaws of the bonito, and the coryphaena. But while avoiding the perils of the deep, new dangers await them in the air, for before they can drop into the sea, the frigate-bird frequently pounces upon them, and draws them, head-foremost, into his maw.

Coryphaena.

The bonito and coryphaena in their turn are often transpierced by the lance of the sword-fish, who, like the saw-snouted pristis, is said to engage even the sperm whale, and to put this huge leviathan to flight.

But of all the monsters of the tropical seas, there is none more dreaded by man than the white shark.

Woe to the sailor that falls overboard while one of these tyrants of the ocean is prowling about the ship; but woe also to the shark who, caught by a baited hook, is drawn on board, for a slow and cruel death is sure to be his lot. Mutilated and hacked to pieces, his torments are protracted by his uncommon tenacity of life.

Such, besides herds of playful dolphins, are the members of the finny creation most commonly met with on the high seas, but in general the waters at a greater distance from the land are poor in fishes. The tropical fishes chiefly abound near the coasts, in the sheltered lagoons, and in the channels which wind through numberless reefs or islands.

As the colibris dart from flower to flower in the Brazilian woods, thus the gorgeous balistinae and glyphodons sport about the submerged coral-gardens, and enhance the brilliancy of their fairy bowers.

While these lustrous fishes, belted with azure, red, and gold, defy the imagination of the poet to describe their beauty, others remind one by their deformity of the chimeras engendered by the diseased brain of a delirious patient. Here we see the hideous frog-fish creeping along like a toad upon his hand-like fin, there the sun-fish swimming about like a vast head severed from its trunk. Cased like the armadillo in an inflexible coat of mail, into which every movable part can be withdrawn, the trunk-fish derides the attack of many an enemy; and inflating its spiny body, the diodon, like the hedgehog of the land, bids defiance to his foes. .

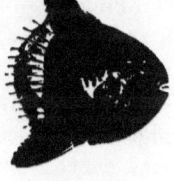

Sun Fish.

On examining the crustacean world, we find that it has established its head-quarters in the tropical zone. There a multitude of wondrous types unknown to the colder regions of the globe attract the attention of the naturalist: the transparent phyllosomas, not thicker than the thinnest wafer, and the strange sword-tails, whose body is covered by a double shield, and terminates in a long horny process, used by the

Limulus.

Malays as a point for their arrows. The crabs and lobsters of the tropical waters are not only more numerous than in our colder seas, but they attain a far greater size than those of the temperate regions of the globe.

The decapod crustaceans (cray-fish) which inhabit our rivers and brooks, are long-tailed like the lobster, but in the torrid zone the river species all belong to the order of the short-tailed crabs, the most perfect and highly developed of the class. Some species even entirely forsake the water and spend their days on shore, not only on the beach, but far inland on the hills. When the season for spawning arrives, vast armies of these land-crabs set out from their mountainous abodes, marching in a direct line to the sea-shore, for the purpose of depositing their eggs in the sand. On this expedition nothing is allowed to deter them from their course. With unyielding perseverance they surmount every obstacle, whether it be a house, rock, or other body, never thinking of going round, but scaling and passing over it in a straight line. Having reached the limit of their journey, they deposit their eggs in the sand, and recommence their toilsome march towards their upland retreats. They set out after nightfall and steadily advance, until the dawn warns them to seek concealment in the inequalities of the ground or among any kind of rubbish, where they lie, until the stars again invite them to pursue their course. On their seaward journey they are in full vigour and fine condition, and this is the time when they are caught in great numbers for the table, their flesh being held in high estimation; but on returning from the coast, they are exhausted, poor, and unfit for use.

The mollusks are no less profusely scattered over the tropical seas and coasts than the higher organised crustaceans. There we find those mighty cephalopods whose long fleshy processes, as thick as a man's thigh, are able, it is said, to seize the fisherman in his boat and drag him into the sea; and there is the abode of the tridacna, whose colossal valves, measuring five feet across, attain a weight of five hundred pounds, and serve both as receptacles for holy water in Catholic churches and to collect the rain in the South Sea Islands.

The rarest and most beautiful of shells, the royal spondylus, the carinaria vitrea, the scalaria pretiosa, the cypraca aurora, and a host of volutes, harps, marginelles, cones, &c., of the most exquisite colouring, are all inhabitants of the warmer waters; and the most costly gift of the sea, the oriental pearl, is the produce of a mollusk which is found scattered over many parts of the Indian and Pacific Oceans.

On descending still lower in the scale of marine life, we find the jelly-fish disporting in the tropical waves in hosts as brilliant as the skies. Some are formed like a mushroom, others assume the shape of a belt or girdle; others are globular, while some are circular, flat, or bell-shaped; and others again resemble a bunch of berries. In colour, perhaps the most delicate is the lovely velella, with its pellucid crest, its green transparent body and fringe of purple tentacles; but it is surpassed in size and gorgeousness by the physalia, or "Portuguese man of war," whose large air-sack, with its splendid vertical comb, shines in every shade of purple and azure. The greatest marvels of the tropical ocean are, however, beyond comparison, the wondrous buildings of the lithophytes, or stone polyps, the reefs and coral islands. Here we see them forming vast barriers which fringe the shores for hundreds and hundreds of miles; there they rise in circular atolls over the blue waves, like bridal rings dropped from the heavens upon the surface of the seas. All is wonderful in these amazing constructions — their puny architects, the lagoons they encircle, the power with which they resist the most furious breakers, the little world of plants drifted over the waters, which ultimately covers them with a verdant crown, and invites man to settle on these gardens of the ocean. There the tall cocoa-palm rocks its feathered crest in the beeeze, affording both shade and fruit to the islander, and there the sea-bird finds a resting-place after its wide flight over the deserts of the equatorial sea.

The Uropeltis Philippinus.

CHAPTER XXVIII.

SNAKES.

First Impression of a Tropical Forest — Exaggerated Fears — Comparative rareness of Venomous Snakes — Their Habits and External Characters — Anecdote of the Prince of Neu Wied — The Bite of the Trigonocephalus — Antidotes — Fangs of the Venomous Snakes described — The Bush-Master — The Echidna Ocellata — The Rattlesnakes — Extirpated by Hogs — The Cobra de Capello — Indian Snake-Charmers — Maritime Excursions of the Cobra — The Egyptian Haje — The Cerastes — Boas and Pythons — The Jiboya — The Anaconda — Enemies of the Serpents — The Secretary — The Adjutant — The Mungoos — A Serpent swallowed by another — The Locomotion of Serpents — Anatomy of their Jaws — A Python-Meal — Serpents feeding in the Zoological Gardens — Domestication of the Rat-Snake — Water-Snakes.

ON penetrating for the first time into a tropical forest, the traveller is moved by many conflicting emotions. This luxuriance of vegetation revelling in ever-changing forms, these giants of the wood clasped by the python-folds of enormous creepers, and bearing whole hosts of parasites on their knotty arms; this abundance of blossoms, equally captivating to the eye and delicious to the smell, all unite in raising the soul to the fullest enjoyment of the moment; and yet the heart is, at the same time, chilled with vague fears, that mix like a discordant sound with the harmonies of this sylvan world. For in the hollows of the tangled roots and in the dense underwood of the forest a brood of noxious reptiles loves to conceal itself, and who knows whether a snake, armed with poisonous fangs, may not dart forth from the rustling foliage.

Gradually, however, these reflections wear away, and time and experience convince one that the snakes in the tropical woods are hardly more to be feared than in the forests of Germany or France, where also the viper will sometimes inflict a deadly wound. These reptiles are, indeed, far from being of so frequent occurrence as is generally believed; and on meeting with a snake, there is every probability of its belonging to the harmless species, which show themselves much more frequently by day, and are by far more numerous. Even in India and Ceylon, where serpents are said to abound, they make their appearance so cautiously that the surprise of long residents is invariably expressed at the rarity with which they are to be seen. Dr. Russell, who particularly studied the serpents of India, found that, out of forty-three species which he himself examined, not more than seven were found to possess poisonous fangs; and Davy, whose attention was carefully directed to the snakes of Ceylon, came to the conclusion that but four out of the twenty species he could collect, were venomous, and that, of these, only two were capable of inflicting a wound likely to be fatal to man.

Sir E. Tennent, who frequently performed journeys of two to five hundred miles through the jungle without seeing a single snake, never heard, during his long residence in Ceylon, of the death of a European which was caused by the bite of one of these reptiles; and in almost every instance accidents to the natives happened at night, when the animal, having been surprised or trodden on, had inflicted the wound in self-defence. Thus, to avoid danger, the Singhalese, when obliged to leave their houses in the dark, carry a stick with a loose ring, the noise of which, as they strike it on the ground, is sufficient to warn the snakes to leave their path.

During his five years' travels through the whole breadth of tropical America, from the Atlantic to the Pacific, M. de Castelnau, although ever on the search, collected no more than ninety-one serpents, of which only twenty-one were poisonous;

a proof that they are not more frequently met with in the primitive forests of Brazil, than in the jungles of India or Ceylon.

The habits of the venomous snakes, and the external characters by which they are distinguished from the harmless species, likewise tend to diminish the danger to be apprehended from them. Thus, their head is generally flat, broad, lanceolate; they have an aperture or slit on each cheek, behind the nostrils, and an elongated vertical pupil like many other nocturnal animals.

They are also generally slower and more indolent in their motions, and thus are more easily avoided. No venomous snake will ever be found on a tree, and on quietly approaching one in the forest or in the savanna, it will most likely creep away without disputing the path, as it is not very anxious uselessly to squander the venom which nature gave it as the only means for procuring itself food.

"There is not much danger in roving amongst snakes," says Waterton, who, from spending many a month in tropical wilds, may justly be called an excellent authority, "provided only that you have self-command. You must never approach them abruptly; if so, you are sure to pay for your rashness; because the idea of self-defence is predominant in every animal, and thus the snake, to defend himself from what he considers an attack upon him, makes the intruder feel the deadly effect of his envenomed fangs. The labarri snake is very poisonous, yet I have often approached within two yards of him without fear. I took care to advance very softly and gently, without moving my arms, and he always allowed me to have a fine view of him, without showing the least inclination to make a spring at me. He would appear to keep his eye fixed on me, as though suspicious, but that was all. Sometimes I have taken a stick ten feet long and placed it on the labarri's back; he would then glide away, without offering resistance. However, when I put the end of the stick abruptly to his head, he immediately opened his mouth, flew at it, and bit it." But although accidents from venomous snakes are comparatively rare, yet the consequences are dreadful when they do take place, and the sight of a cobra or a trigonocephalus preparing for its fatal spring may well appal the stoutest heart.

Prince Maximilian of Neu Wied, having wounded a tapir,

was following the traces of his game along with his Indian hunter, when suddenly his companion uttered a loud scream. He had come too near a labarri snake, and the dense thicket prevented his escape. Fortunately the first glance of the distinguished naturalist fell upon the reptile, which with extended jaws and projecting fangs was ready to dart upon the Indian, but at the same moment, struck by a ball from the prince's rifle, lay writhing on the ground. The Indian, though otherwise a strong-nerved man, was so paralysed by fear, that it was some time before he could recover his self-possession — a proof, among others, that it is superfluous to attribute a fascinating power to the venomous snakes, as the effects of terror are quite sufficient to explain why smaller animals, unable to flee the impending danger, become their unresisting victims, and even seem, as it were, wantonly to rush upon destruction. Thus Pöppig saw on the banks of the Huallaga an unfortunate frog, which, after being for some time unable to move, at length made a desperate leap towards a large snake that was all the time fixing its eye upon it, and thus paid the confusion of its senses with the loss of its life.

A poor Indian girl that accompanied Schomburgk on his travels through the forests of Guiana was less fortunate than the Prince of Neu Wied's companion. She was bitten by a trigonocephalus, and it was dreadful to see how soon the powers of life began to ebb under the fatal effects of the poison. The wound was immediately sucked, and spirits of ammonia, the usual remedy, profusely applied both externally and inwardly, but all in vain. In less than three minutes, a convulsive trembling shook the whole body, the face assumed a cadaverous aspect, dreadful pains raged in the heart, in the back, less in the wound itself; the dissolved blood flowed from the ears and nose, or was spasmodically ejected by the stomach; the pulse rose to 120-130 in the minute; the paralysis which first benumbed the bitten foot spread farther and farther, and in less than eight minutes the unfortunate girl was no longer to be recognised. The same day the foot swelled to shapeless dimensions, and she lay senseless until, after an agony of sixty-three hours, death relieved her from her sufferings.

A great many antidotes have been recommended against serpentine poison, but their very number proves their inefficacy. One of the most famous is the juice of a Peruvian climbing plant, the vejuco de huaco (*Mikania Huaco*, Kunth), the remarkable properties of which were first discovered by a negro, who observed that when the huaco, a kind of hawk which chiefly feeds on snakes, has been bitten by one of them, it immediately flies to the vejuco and eats some of its leaves.

It is a well-known fact that serpentine poison may be swallowed with impunity; it shows its effects only on mixing directly with the blood. A tight ligature above the wound, along with sucking, burning, or cutting it out, are thus very rational remedies for preventing the rapid propagation of the venom. Suction is, however, not always unattended with danger to the person who undertakes the friendly office. Thus Schomburgk relates the misfortunes of a poor Indian, whose son had been bitten in the cheek. The father instantly sucked the wound, but a hollow tooth conveyed the poison into his own body. His cheek swelled under excruciating pains, and without being able to save his son, his own health and vigour were for ever lost. For such are the dreadful consequences of this poison, that they incurably trouble the fountains of life. The wound generally breaks open every year, emitting a very offensive odour, and causes dreadful pains at every change of the weather.

Although all the venomous snakes produce morbid symptoms nearly similar, yet the strength of the poison varies according to the species of the serpent, and to the circumstances under which it is emitted. It is said to be most virulent during very hot weather, when the moon changes, or when the animal is about to cast its skin. The effects are naturally more powerful and rapid when a larger quantity of poison flows into the wound, and a snake with exhausted supplies from repeated bitings will evidently strike less fatally than another whose glands are inflated with poison after a long repose.

Before describing some of the most conspicuous of the venomous serpents, a few words on the simple but admirable mechanism of their delicate but needle-like fangs will not be out of place. Towards the point of the fang, which is invariably situated in the upper jaw, there is a little oblong aperture on the convex side of it, and through this there is a communication down the fang to the root, at which lies a little bag containing the poison. Thus, when the point of the fang is pressed, the root of the fang also presses against the bag and sends up a portion of the poison it contained. The fangs being extremely movable, can be voluntarily depressed or elevated; and as from their brittleness they are very liable to break, nature, to provide for a loss that would be fatal, has added behind each of them smaller or subsidiary fangs ready to take their place in case of accident.

Unrivalled in the display of every lovely colour of the rainbow, and unmatched in the effects of his deadly poison, the bushmaster or counacutchi (*Lachesis rhombeata*) glides on, sole monarch of the forests of Guiana or Brazil, as both man and beast fly before him. In size he surpasses most other venomous species, as he sometimes grows to the length of fourteen feet. Generally concealed among the fallen leaves of the forest, he lives on small birds, reptiles, and mammalians, whom he is able to pursue with surprising activity. Thus, Schomburgk once saw an opossum rushing through the forest, and closely followed by an enormous bush-master. Frightened to death and utterly exhausted, the panting animal ascended the stump of an old tree, and thence, as if rooted to the spot, looked with staring eyes on its enemy, who, rolled in a spiral coil, from which his head rose higher and higher, slowly and leisurely, as if conscious that his prey could not possibly escape him, prepared for his deadly spring. This time, however, the bush-master was mistaken, for a shot from Schomburgk's rifle laid him writhing in the dust, while the opossum, saved by a miracle, ran off as fast as he could. Fortunately for the planter and negroes, the bushmaster is a rare serpent, frequenting only the deepest shades of the thicket, where in the day-time he generally lies coiled upon the ground.

Still rarer, though if possible yet more formidable, is a small brown viper (*Echidna ocellata*), which infests the Peruvian forests. Its bite is said to be able to kill a strong man within two or three minutes. The Indian, when bitten by it, does not even attempt an antidote against the poison, but stoically bids adieu to his comrades, and lays himself down to die.

The ill-famed wide-extended race of the rattlesnakes, which ranges from South Brazil to Canada, belongs exclusively to the new world. They prefer the more elevated, dry, and stony regions, where they lie coiled up in the thorny bushes, and only attack such animals as come too near their lair. Their bite is said to be able to kill a horse or an ox in ten or twelve minutes; but, fortunately, they are afraid of man, and will not venture to attack him unless provoked. When roused to anger they are, however, very formidable, as their fangs penetrate through the strongest boot. One of the most remarkable features of their organisation is a kind of rattle terminating the tail, and consisting of a number of pieces inserted into each other, all alike in shape and size, hollow, and of a thin, elastic, brittle substance, like that of which the scales are externally formed. When provoked, the strong and rapid vibratory motions imparted to the rattle produce a sound which has been compared to that of knife-grinding, but is never loud enough to be heard at any distance, and becomes almost inaudible in rainy weather.

Rattlesnake.

Naturalists distinguish at least a dozen different species of rattlesnakes, the commonest of which are the Boaquira (*Crotalus horridus*), which frequents the warmest regions of South America, and the Durissus (*C. durissus*), which has chosen the United States for its principal home. The chief enemy of this serpent is the hog, whom it dreads so much that on seeing one it immediately loses all its courage, and instantly takes to flight. But the hog, who smells it from afar, draws nearer and nearer, his bristles erected with excitement, seizes it by the neck, and devours it with great complacency, though without touching the head. As the hog is the invariable companion of the settler in the backwoods, the rattlesnake everywhere disappears before the advance of man, and it is to be hoped that a century or two hence it will be ranked among the extinct animals. The American Indians often regale on the rattlesnake. When they find it asleep, they put a small forked stick over its neck, which they keep immovably fixed to the ground, giving the snake a piece of leather to bite, and this they pull back several times with great force, until they perceive that the poison-fangs are torn out. They then cut off the head, skin the body, and cook it as we do eels. The flesh is said to be white and excellent.

None of the American snakes inhabit the old world, but in the East Indies and Ceylon other no less dangerous species appear upon the scene, among which the celebrated Cobra di Capello is one of the most deadly. A few years since, as many of my readers will remember, a cobra in the Zoological Gardens destroyed its keeper. In a fit of drunkenness, the man, against express orders, took the reptile out of its cage, and placing its head inside his waistcoat, allowed it to glide round his body. When it had emerged from under his clothes on the other side, apparently in good humour, the keeper squeezed its tail, when it struck him between his eyes; in twenty minutes his consciousness was gone, and in less than three hours he was dead.

As long as the cobra is in a quiet mood, its neck is nowhere thicker than its head or other parts; but as soon as it is excited, it raises vertically the anterior part of its trunk, dilates the hood on each side of the neck, which is curiously marked in the centre in black and white, like a pair of spectacles, and then swells out to double its former proportions, and advances against the aggressor by the undulating motion of the tail. It is not only met with in the cultivated grounds and plantations, but will creep into the houses and insinuate itself among the furniture. Bishop Heber heard at Patna of a lady who once lay a whole night with a cobra under her pillow. She repeatedly thought during the night that she felt something move, and in the morning when she snatched her pillow away, she saw the thick black throat, the square head, and the green diamond-like eyes of the reptile advanced within two inches of her neck. Fortunately the snake was without malice; but alas for her if she had during the night pressed him a little too roughly.

This is the snake so frequently exhibited by the Indian jugglers, who contrive by some unknown method to tame them so far as to perform certain movements in cadence, and to dance to the sound of music, with which the cobra seems much delighted, keeping time by a graceful motion of the head, erecting about half its length from the ground, and following the few simple notes of the conjuror's flute with gentle curves like the undulating lines of a swan's neck. It has been naturally supposed, before this could be done, that the poisonous fangs had been extracted; but Forbes, the author of "Oriental Memoirs," had nearly been taught at his cost that this is not always practised. Not doubting but that a cobra, which danced for an hour on the table while he painted it, had been disarmed of its fatal weapons, he frequently handled it to observe the beauty of the spots, and especially the spectacles on the hood. But the next morning his upper servant, who was a zealous Mussulman, came to him in great haste and desired he would instantly retire and praise the Almighty for his good fortune. Not understanding his meaning, Forbes told him that he had already performed his devotions, and had not so many stated prayers as the followers of his prophet. Mahomet then informed him that while purchasing some fruit in the bazaar, he observed the man who had been with him on the preceding evening entertaining the country people with the dancing snakes; they, according to their usual custom, sat on the ground around him, when, either from the music stopping too suddenly or from some other cause irritating the snake, which he had so often handled, it darted at the throat of a young woman, and inflicted a wound of which she died in about half an hour. Mahomet once more advised him to thank Allah, and recorded his master in his calendar as a lucky man. That the snake-charmers control the cobra not by extracting its fangs, but by courageously availing themselves of its timidity and reluctance to use them, was also proved during Sir E. Tennent's residence in Ceylon by the death of one of these performers, whom his audience had provoked to attempt some unaccustomed familiarity with the cobra; it bit him on the wrist, and he expired the same evening.

There are many varieties and several species of the Cobra di Capello, among which the minelle, though the smallest of all the Indian serpents, is the most dangerous, its bite occasioning a speedy and painful death. Well may it be called

> "The small, close-lurking minister of fate,
> Whose high concocted venom through the veins
> A rapid lightning darts, arresting swift
> The vital current."

The deserted nests of the termites are the favourite retreat of the sluggish and spiritless cobra, which watches from their apertures the toads and lizards on which it preys. On coming upon it, its only impulse is concealment; and when it is unable to escape, a few blows from a whip are sufficient to deprive it of life.

It is a curious fact that, though not a water-snake, the cobra sometimes takes considerable excursions by sea. When the *Wellington,* a Government vessel employed in the inspection of the Ceylonese pearl-banks, was anchored about a quarter of a mile from land, a cobra was seen, about an hour before sunset, swimming vigorously towards the ship. It came within twelve yards, when the sailors assailed it with billets of wood and other missiles, and forced it to return to land.

The Egyptian Haje (*Naja Haje*), a near relation of the Indian cobra, is most likely the asp of ancient authors, which the celebrated Cleopatra chose as the instrument of her death, to avoid figuring in the triumph of Augustus. Like the cobra, it inflates its neck when in a state of excitement, and as it raises its head on being approached, as if watchful for its safety, it was venerated by the ancient Egyptians as a symbol of divinity, and as the faithful guardian of their fields. Divine honours have, however, much more frequently been paid to the venomous snakes from the terror they inspire, than from far-fetched notions of beneficence. Several Indian tribes in North America adore the rattlesnake; and in the kingdom of Widah, on the coast of Guinea, a viper (*Viper idolum*) has its temple and ministers, and is no less carefully provided for than if it were an inmate of the Zoological Gardens.

The Cerastes, or horned-viper, one of the most deadly serpents of the African deserts, is frequently exhibited by Egyptian jugglers, who handle and irritate it with impunity: they are supposed to render themselves invulnerable by the chewing of a certain root, but most likely, as in the case of the cobra-charmers, their secret consists in their courage and perfect knowledge of the animal's nature.

Although the Boas and Pythons are unprovided with venomous fangs, yet, from their enormous size, they may well be ranked among the deadly snakes; for as Waterton justly remarks, "it comes nearly to the same thing in the end whether the victim dies by

Boa.

poison from the fangs, which corrupts his blood and makes it stink horribly, or whether his body be crushed to mummy and swallowed by a Python."

The kingly Jiboya (*Boa constrictor*) inhabits the dry and sultry localities of the Brazilian forests, where he generally conceals himself in crevices and hollows in parts but little frequented by man, and sometimes attains a length of thirty feet. To catch his prey he ascends the trees, and lurks, hidden in the foliage, for the unfortunate agutis, pacas, and capybaras, whom their unlucky star may lead within his reach. When full-grown he seizes the passing deer; but in spite of his large size he is but little feared by the natives, as a single blow of a cudgel suffices to kill him. Prince Maximilian of Neu Wied tells us that the experienced hunter laughs when asked whether the Jiboya attacks and devours man.

The Sucuriaba, Anaconda, or Water Boa (*Eunectes murinus*), as it is variously named, attains still larger dimensions than the constrictor, as some are said to have been found of a length of forty feet. It inhabits the large rivers, lakes, and marshy grounds of tropical America, and passes most of its time in the water, now reposing on a sand-bank, with only its head above the surface of the stream, now rapidly swimming like an eel, or abandoning itself to the current of the river.

Often, also, it suns itself on the sandy margin of the stream, or patiently awaits its prey, stretched out upon some rock or fallen tree. With sharp eye it observes all that swims in the waters as well as all that flies over them, or all that comes to the banks to quench its thirst; neither fish nor aquatic bird is secure from the swiftness of its attack, and woe to the capybara that comes within the grasp of its folds. Such is its voracity, that Firmin ("Histoire Naturelle de Surinam") found in the stomach of an Anaconda a large sloth, an Iguana nearly four feet long, and a tolerably-sized Ant-bear, all three nearly in the same state as when they were swallowed — a proof that their capture had taken place within a short time. As is commonly the case with reptiles, the water-boa is very tenacious of life, and though the head may be nearly severed from the trunk, the entrails taken out of the body, and the skin detached, it will still move about for a considerable time.

The boas principally inhabit America, although some species

are likewise met with in Asia; but the still more formidable pythons are confined to the hot regions of the Old World. They are said to enlace even the tiger or the lion in their fatal embrace, and to judge by their size and strength, this assertion seems by no means improbable.

The various serpent tribes are exposed to the attacks of many enemies, who fortunately keep their numbers within salutary bounds, and avenge the death of the countless insects, worms, toads, frogs, and lizards, that fall a prey to their strength or their venom. Several species of rapacious and aquatic birds live upon snakes, the American ostrich thins their ranks wherever he can, and the African Secretary is renowned for his prowess in serpentine warfare.

Secretary Bird.

"The battle was obstinate," says Le Vaillant, describing one of these conflicts, "and conducted with equal address on both sides. The serpent, feeling the inferiority of his strength, in his attempt to flee, and regain his hole, employed that cunning which is ascribed to him, while the bird, guessing his design, suddenly stopped him, and cut off his retreat by placing herself before him at a single leap. On whatever side the reptile endeavoured to make its escape, his enemy was still found before him. Then, uniting at once bravery and cunning, he erected himself boldly to intimidate the bird, and hissing dreadfully, displayed his menacing throat, inflamed eyes, and a head swelled with rage and venom. Sometimes this threatening appearance produced a momentary suspension of hostilities, but the bird soon returned to the charge, and covering her body with one of her wings as a buckler, struck her enemy with the horny protuberances upon the other, which, like little clubs, served the more effectually to knock him down as he raised himself to the blow; at last he staggered and fell, the conqueror then despatched him, and with one stroke of her bill laid open his skull."

The secretary-eagle has now been successfully acclimatised in the West Indies, where he renders himself useful by the destruction of the venomous snakes with which the plantations are infested.

Gravely, "with measured step and slow," like a German philosopher cogitating over the nature of the absolute, but, as

Adjutant.

we shall presently see, much more profitably engaged, the adjutant wanders among the reeds on the banks of the muddy Ganges. The aspect of this colossal bird, measuring six feet in height and nearly fifteen from tip to tip of the wings, is far from being comely, as his enormous bill, his naked head and neck, except a few straggling curled hairs, his large craw hanging down the forepart of the neck like a pouch, and his long naked legs, are certainly no features of beauty. Suddenly he stops, dips his bill among the aquatic plants, and immediately raises it again triumphantly into the air, for a long snake, despairingly twisting and wriggling, strives vainly to escape from the formidable pincers which hold it *in carcere duro*. The bird throws back his head, and the reptile appears notably diminished in size; a few more gulps, and it has entirely disappeared. And now the sedate bird continues his stately promenade with the self-satisfied mien of a merchant who has just made a successful speculation, and is engaged in the agreeable calculation of his gains. But, lo! again the monstrous bill descends, and the same scene is again repeated. The good services of the giant heron in clearing the land of noxious reptiles, and the havoc he is able to make among their ranks, may be judged of by the simple fact, that on opening the body of one of them, a land-tortoise ten inches long and a large black cat were found entire within it, the former in the pouch, as a kind of stock in trade, the latter in the stomach, all ready for immediate consumption.

Trusting to his agility and the certainty of his eye, the Indian ichneumon or mongoos attacks without hesitation the most venomous serpents. The cobra, which drives even the leopard to flight, rises before the little creature with swelling hood and fury in its eye; but swift as thought, the ichneumon, avoiding the death-stroke of the projecting fangs, leaps upon its back, and fastening his sharp teeth in the head, soon despatches the helpless reptile.

Mongoos.

The serpents sometimes even feed upon their own brethren. Thus a rat-snake in the Zoological Gardens was once seen to devour a common Coluber Natrix, but not having taken the measure of his victim, he could not dispose of the last four inches of his tail, which stuck out rather jauntily from the side of his mouth, with very much the look of a cigar. After a quarter of an hour the tail began to exhibit a retrograde motion, and the swallowed snake was disgorged, nothing the worse for his living sepulchre with the exception of the wound made by his partner when first he seized him.

A python in the same collection, who had lived for years on friendly terms with a brother nearly as large as himself, was found one morning sole tenant of his den. As the cage was secure, the keeper was puzzled to know how the serpent had escaped. At last it was observed that the remaining inmate had swollen remarkably during the night, when the truth came out. It was, however, the last meal of the fratricide, for in some months he died.

When we consider that the snakes have neither legs, wings, nor fins, and are indeed deprived of all the usual means of locomotion, the rapidity of their progress is not a little surprising. On examining the anatomical structure of their body, however, it will be remarked that while we have only twelve pair of ribs united in front by the breast-bone and cartilage, the snake has often more than three hundred, unconnected in front, and consequently much more free in their motions, a faculty which is still increased by the great mobility of the spondyli of the backbone. Between the ribs and the broad transverse scales or plates which exist on the belly of all such serpents as move rapidly, we find numerous muscles connecting them one with another, and thus, amply provided with a whole system of strong pulleys and points of attachment, the reptile, bringing up the tail towards the head, by bending the body into one or more curves, and then again resting upon the tail and extending the body, glides swiftly along, not only upon even ground, but even sometimes from branch to branch, as the smallest hold suffices for its stretching out its body at a foot's length into the air, and thus reaching another sallying point for further progress.

The anatomy of the serpent's jaws is no less remarkable than the mechanism of its movements. In spite of their proverbial wisdom, snakes would not be able to exist unless they were able to swallow large animal masses at a time. For, however rapid their motions may be, those of their prey are in general still more active, and thus they are obliged to wait in ambush till a fortunate chance provides them with a copious meal. The victim is often much more bulky than the serpent itself, but still, without tearing it to pieces, it is able to engulph it in its swelling maw. For the two halves of its lower jaw do not coalesce like ours into one solid mass, but are merely connected in front by a loose ligament, so that each part can be moved separately. The bones of the upper jaw and palate are also, loosely attached or articulated one with the other, and thus the whole mouth is capable of great distension. By this mechanism, aided by the numerous sharp teeth which are so many little hooks with the point curved backwards, each side of the jaws and mouth being able to act as it were independently of the other, alternately hooks itself fast to the morsel, or advances to fasten itself farther on in a similar manner, and thus the reptile draws itself over its prey, somewhat in the same way as we draw a stocking over our leg, after having first, by breaking the bones, fashioned it into a convenient mass, and rendered its passage more easy by lubricating it with its saliva. Slowly the huge lump disappears behind the jaws, descends lower and lower beneath the scales which seem ready to burst asunder with distension, and then the satisfied monster coils himself up once more to digest his meal in quiet. The time required for this purpose varies of course according to the size of the morsel; but often weeks or even months will pass before a boa awakens from the lethargic repose in which — the image of disgusting gluttony — he lies plunged after a superabundant meal.

A huge python in the Zoölogical Gardens fasted the almost incredible time of twenty-two months, having probably prepared himself for his abstinence by a splendid gorge; and Duméril mentions a rattlesnake in the Jardin des Plantes which likewise took no nourishment during twenty-one months, but then, as if to make up for lost time, swallowed three hares within five days.

The reptiles in the Zoological Gardens are offered food once a week, but even then their appetites are frequently not yet awakened, though great care is taken never to spoil their stomachs by excess.

This is the time for visiting the Reptile House, which otherwise offers but little amusement, as the great snakes have either retired from public life under their blankets, or lie coiled upon the branches of the trees in their dens. Three o'clock is the feeding-time, and the reptiles which are on the look-out, seem to know full well the errand of the man who enters with the basket, against the side of which they hear the fluttering wings of the feathered victims, and the short stamp of the doomed rabbits. The keeper opens the door at the back of the den of the huge pythons, for these he need not fear, takes off their blanket, and drops a rabbit, who hops from side to side, curious to inspect his new habitation, and probably finding it to his taste, sits on his haunches and leisurely begins to wash his face. Silently the python glides over the stones, uncurling his huge folds, looks for an instant upon his unconscious victim, and the next has seized him with his jaws. His contracting folds are twisted as swiftly as a whiplash round his shrieking prey, and for ten minutes the serpent lies still, maintaining his mortal knot until his prey is dead, when, seizing it by the ears, he draws it through his vice-like grip, crushing every bone, and elongating the body preparatory to devouring it.

The arrangement for feeding the venomous kinds is, of course, more cautious. The door opens at the top instead of at the side of the dens, and with good reason; for no sooner does the keeper remove with a crooked iron rod, the blanket from the cobra, than the reptile springs with inflated hood into an S like attitude and darts laterally at his prey, whose sides have scarcely been pierced, when it is seized with tetanic spasms, and lies convulsed in a few seconds.

These instantaneous effects, almost as rapid as those of a mortal shot or of lightning itself, might at first sight seem to warrant the conclusion that the genius of evil had formed the venomous serpents to be his chosen agents of destruction; but at a nearer view, they afford but another proof of the beneficence of the Creator in providing weak, sober, and by no means cruel creatures with a weapon which makes up to them for the want of speed, and at the same time abridges the torments of their victims.

Though generally the objects of abhorrence and fear, yet serpents sometimes render themselves useful or agreeable to man. Thus the rat-snake of Ceylon (*Coryphodon Blumenbachii*) in consideration of its services in destroying vermin, is often kept as a household pet, and so domesticated by the natives as to feed at their table.

" I once saw an example of this in the house of a native," says Wolf in his " Life and Adventures in Ceylon ; " " It being meal-time, he called his snake, which immediately came forth from the roof under which he and I were sitting. He gave it victuals from his own dish, which the snake took of itself from off a fig leaf that was laid for it, and ate along with its host. When it had eaten its full, he gave it a kiss, and bade it go to its hole."

The beautiful coral-snake (*Elaps corallinus*) is fondled by the Brazilian ladies, but the domestication of the dreaded cobras as protectors in the place of dogs, mentioned by Major Skinner, on undoubtedly good authority,* is still more re-

---

* Sir E. Tennent's Ceylon, vol. i. p. 193.

markable. They glide about the house, going in and out at pleasure, a terror to thieves, but never attempting to harm the inmates.

The agility of the Rat-snake in seizing its nimble-footed prey is truly wonderful. One day Sir Emerson Tennent had an opportunity of surprising a coryphodon which had just seized on a rat, and of covering it suddenly with a glass shade, before it had time to swallow its prey. The serpent, which appeared stunned with its own capture, allowed the rat to escape from its jaws, which cowered at one side of the glass in an agony of terror. On removing the shade, the rat, recovering its spirits, instantly bounded towards the nearest fence, but quick as lightning, it was followed by its pursuer, which seized it before it could gain the hedge, through which the snake glided with its victim in its jaws.

The Tree-snakes offer many beautiful examples of the adaptation of colour to the animal's pursuits, which we have already had occasion to admire in our brief review of the tropical insect world. They are frequently of an agreeable green or bluish hue, so as hardly to be distinguishable from the foliage among which they seek their prey, or where they themselves are liable to be seized upon by their enemies. They are often able vertically to ascend the smoothest trunks and branches, in search of squirrels and lizards, or to rifle the nests of birds.

Water-snake.

The Water-snakes which infest some parts of the tropical seas, though far from equalling in size the vast proportions of the fabulous sea-serpent, are very formidable from their venomous bite. They have the back part of the body and tail very much compressed and raised vertically, so as to serve them as a paddle with which they rapidly cleave the waters.

Toad and Anolis.

CHAPTER XXIX.

LIZARDS, FROGS AND TOADS.

Their Multitude within the Tropics — The Geckoes — Anatomy of their Feet — The Anolis — Their Love of Fight — The Chameleon — Its wonderful Changes of Colour — Its Habits — Peculiarities of its Organisation — The Iguana — The Teju — The Water-Lizards — Lizard Worship on the Coast of Africa — The Flying Dragon — The Basilisk — Frogs and Toads — The Pipa — The Bahia Toad — The Giant Toad — The Musical Toad — Brazilian and Surinam Tree-Frogs.

THE equatorial regions may well be called the head-quarters of the lizard race, as these reptiles nowhere else appear in such a multitude of genera, species, and individuals. The stranger is struck with their numbers as soon as he sets his foot on a tropical shore, for on all sides, on the sands and in the forests, on banks and rocks, on the trees and on the ground, innumerable varieties of lizards are seen basking, rustling, crawling, climbing, or rapidly darting along.

The *Geckoes* might even claimed to be ranked among the domestic animals, as they take up their abode in the dwellings of man, where they make themselves useful by the destruction of flies, spiders, and other noxious or disagreeable insects, which they almost always swallow entire, their throat being as broad as the opening of their jaws. During the day time they generally remain concealed in some dark crevice or chink, but towards evening they may be seen running along the steepest walls with marvellous rapidity, in keen pursuit of their prey, frequently standing still, nodding with their head, and uttering shrill tones, most likely by smacking their tongue against the palate. Their flattened flexible body seems to mould itself into the hollows, in which they often remain motionless for hours, and their generally dull colour harmonizes so well with their resting-places, as to render them hardly distinguishable, a circumstance which answers the double purpose of masking their presence from the prey for which they lie in wait, and from the enemies that might be inclined to feast upon them. Among these, some of the smaller birds of prey — hawks and owls — are the most conspicuous, not to mention man, the arch-persecutor of almost every animal large enough to attract his notice.

Gecko.

How comes it that these nocturnal lizards, seemingly in defiance of the laws of gravitation, are thus able to adhere to ceilings or any other overhanging surfaces? An inspection of the soles of their broad feet will soon solve the enigma, for all their toes are considerably dilated on their margins, and divided beneath into a number of transverse lamellæ, parallel to each other, and generally without any longitudinal furrow. From these a fluid exudes which serves to attach the animal to the surface. They are also generally provided with sharp and crooked claws, retractile and movable, like those of a cat, and which render them good service in climbing the trees.

In spite of their harmless nature, the Geckoes, their real utility being forgotten over imaginary grievances, nowhere enjoy a good reputation, probably in consequence of their ugliness and the wild expression of their large eyes. They are accused of tainting with a virulent secretion every object they touch, and of provoking an eruption on the skin merely by running over it — a popular prejudice which naturally causes many a poor inoffensive Gecko's death. They abound all over the torrid zone, even in the remote islands of the Pacific, such as

Otaheite and Vanikoro. Duméril enumerates fifty-five different species, only two of which are indigenous in Southern Europe, while India monopolises no less than thirteen for her share.

Mr. Adams once witnessed in Borneo a desperate struggle between a Gecko and a large Tarantula spider. After a long and doubtful contest, the Gecko proved at length victorious, and succeeded in swallowing the insect, whose enormous legs, protruding from the lizard's mouth, gave the animal the look of some monstrous cuttle-fish.

The graceful *Anolis* are peculiar to America. By the structure of their feet, provided with long unequal toes, they are related to the Geckoes, but are distinguished from them by a more slender form of body, by their extremely long thin tail, and a large neck-pouch, which dilates under the influence of excitement. These small and nimble creatures, the largest species seldom exceeding eight inches in length, are as touchy as fighting-cocks. On approaching them, they instantly blow up their pouch, open widely their diminutive jaws, and spring upon the aggressor, striving to bite him with their teeth, which, however, are too tiny to do much harm. Among each other they live in a perpetual state of warfare. As soon as one Anolis sees another, he makes a rapid advance, while his adversary awaits him with all the courage of a gallant knight. Before beginning the conflict, they make all sorts of menacing gestures, convulsively nodding their heads, puffing up their pouches, until finally they close in desperate struggle.

Anolis.

> "The meeting of these champions proud
> Seems like the bursting thunder-cloud."

If they are of equal strength, the battle remains for some time undecided. At length the vanquished Anolis turns and runs away, but he may think himself fortunate if he escapes with the loss of his tail. Many of them are thus deprived of this ornamental appendage, which they voluntarily leave behind to avoid a still greater disaster, and then they become timid, melancholy, and fond of retirement, as if ashamed of being seen, only regaining their spirits when, by a wonderful power of reproduction, the amputated tail has been replaced by another.

Like many other lizards, the Anolis possesses the faculty of changing colour when under the influence of excitement, but of all animals, whether terrestrial or marine, none is more famous or remarkable in this respect than the *Chameleon*. It frequently happens that man, not satisfied with the wonders which nature everywhere exposes to his view, adds to their marvels others of his own invention, and thus many a fable has been told about the Chameleon. It has been said, for instance, that it could emulate all the colours of the rainbow, but the more accurate observations of Hasselquist and other naturalists have shown that the whole change, which takes place most frequently when the Chameleon is exposed to full sunshine or under the influence of emotion, consists in its ordinary bluish-ash colour, turning to a green or yellowish hue with irregular spots of a dull red. Like many other reptiles, the Chameleon has the power of inflating its lungs and retaining the air for a long time so as one moment to appear as fat and well-fed as an alderman, and the next as lean and bony as a hungry disciple of the muses. These alternating expansions and collapses seem to have a great influence on the change of colour, which, however, according to Milne Edwards, is principally owing to the skin of the animal consisting of two differently coloured layers, placed one above the other, and changing their relative positions under the influence of excitement.

Chameleon.

In our cold and northern regions the captive Chameleon cuts but a sorry figure; but in his own sunny regions, which extends from southern Spain and Sicily to the Cape, and eastwards from Arabia and Hindostan to Australia, it is said to be by no means deficient in beauty, in spite of its strangely-formed carinated head, its enormously projecting eyes, and its granulated skin. Its manner of hunting for the little winged insects, that form its principal food, is very peculiar. Although the movements of its head are very limited, on account of the shortness of its neck, this deficiency is amply supplied by the wide range of its vision, each eye being able to move about in all directions independently of the other. Thus while one of them attentively gazes upon the heavens, the other minutely examines the ground, or while one of them rolls in its orbit, the other remains fixed; nay, their mobility is so great, that without even moving its stiff head, this wonderful lizard, like Janus, the double-faced god of ancient Rome, can see at the same time all that goes on before and behind it. When an insect comes flying along, the chameleon, perched on a branch, and half concealed between the foliage, follows it in all its movements by means of his powerful telescopes, until the proper moment for action appears. Then, quick as thought, he darts forth, even to a distance of five or six inches, his long fleshy glutinous tongue, which is moreover furnished with a dilated and somewhat tubular tip, and driving it back with the same lightning-like velocity, engulphs his prey. This independence of the eyes is owing to the imperfect sympathy which subsists between the two lobes of the brain and the two sets of nerves which ramify throughout the opposite sides of its frame. Hence also one side of the body may be asleep while the other is vigilant, one may be green while the other is ash-blue, and it is even said that the Chameleon is utterly unable to swim, because the muscles of both sides are incapable of acting in concert.

Destined for an arboreal life, he is provided with organs beautifully adapted for supporting himself on the flexible branches; for besides the cylindrical tail nearly as long as his body which he coils round the boughs, his five toes are united two and three by a common skin, so as to form, as it were, a pair of pincers or a kind of hand, admirably suited for a holdfast.

Among the *Iguanas*, a huge lizard tribe, characterised by a carinated back and tail, and a large denticulated gular pouch,

Iguana.

the common or great American Guana (*Iguana Tuberculata*) deserves particular notice, as its white flesh is considered a great delicacy in Brazil and the West Indies. Notwithstanding its large size, for it not seldom attains a length of four or five feet, and the formidable appearance of its serrated back, it is in reality by no means of a warlike disposition, and so stupid that, instead of endeavouring to save itself by a timely flight, it merely stares with its large eyes, and inflates its pouch, while the noose is passing round its neck to drag it forth from its hole.

The Bahama islands abound with Guanas, which form a great part of the subsistence of the inhabitants. They are caught by dogs, trained for the purpose, in the hollow rocks and trees where they nestle, and either carried alive for sale to Carolina, or kept for home consumption. They feed wholly on vegetables and fruit, particularly on a kind of fungus, growing at the roots of trees, and on the fruits of the different kinds of ananas, whence their flesh most likely acquires its delicate flavour.

The famous South American monitory lizard or Teju (*Tejus monitor*) is one of the largest and most beautiful of the whole race, as he measures no less than five feet from the snout to the tip of the tail, which is nearly twice as long as the body, while his black colour, variegated with bright yellow bands and spots, produces an agreeable and pleasing effect. The head is small, the snout gradually tapers, the limbs are slender, and the tail, which is laterally compressed, gradually decreases towards the extremity. The Teju lives in cavities and hollows, frequently under the roots of trees. When pursued, he runs rapidly straight forward to his burrow, but when his retreat is intercepted, he defends himself valiantly, and proves a by no means contemptible antagonist, as he is able to bite through a thick boot, and a stroke with his strong and muscular tail will completely disable a dog. Schomburgk frequently found the leathery eggs of the Teju in the large deserted nests of the Termites, not only in the forests, but in the plantations. Though the Monitor generally lives on land, he is an excellent swimmer, and catches many a fish in its native element. His chief food, however, consists in various fruits, rats, mice, birds, and he also devours a large number of the eggs and young of the alligator. The attachment to man which is universally attributed to him in Brazil, and the warning which, like his relation the Monitor of the Nile, he is said to give to him of the approach of the cayman or the crocodile, by emitting a peculiar and shrill sound, are idle fables

Monitor.

which hardly required the contradiction of Prince Maximilian of Neu Wied, who in all his travels never once heard the Teju's monitory cry, although occasions were not wanting when it might have been of service. As this lizard's flesh is nearly equal in flavour to that of the Guana, it is frequently hunted down by dogs, so that its supposed friendliness to man, " requiting good for evil," would be meritorious indeed, and might be held up as a model to all Christians.

The large Water-lizards (*Hydrosauri*) frequent the low river banks or the margins of springs, and although they may be seen basking on rocks or on the dead trunk of some prostrate tree in the heat of the sun, yet they appear more partial to the damp weeds and undergrowth in the neighbourhood of water. Their gait has somewhat more of the awkward lateral motion of the crocodile, than of the lively action of the smaller saurians. When attacked, they lash violently with their tail, swaying it sideways with great force like the cayman. These modern types of the Mososaurus and Iguanodon have a graceful habit of extending the neck, and raising the head to look about them, and as you follow them leisurely over the rocks, or through the jungle, they frequently stop, turn their heads round, and take a deliberate survey of the intruder. They are by no means vicious, though they bite severely when provoked, acting, however, always on the defensive. On examining their stomachs, crabs, locusts, beetles, the remains of jumping fish, the scales of snakes, and bones of frogs and other small animals are discovered. Like that of the Iguanas, their flesh is delicate eating, resembling that of a very young sucking-pig. Mr. Adams gives us an amusing description of his contests with a gigantic Water-lizard (*Hydrosaura giganteus*): " Throwing myself on him, I wounded him with a clasp knife in the tail, but he managed to elude my grasp and made for the woods. I succeeded, however, in tracking his retreating form, on hands and knees, through a low covered labyrinth in the dense undergrowth, until I saw him extended on a log, when, leaving the jungle, I called my servant, a marine, who was shooting specimens for me, and pointing out the couchant animal, desired him to shoot him in the neck, as I did not wish the head to be injured, which he accordingly did. Entering the jungle, I then closed with the wounded saurian, and seizing him by the throat, bore him in triumph to our quarters. Here he soon recovered, and hoping to preserve him alive to study his habits, I placed him in a Malay wicker hen-coop. As we were sitting, however, at dinner, the black cook, with great alarm depicted in his features, reported that 'Alligator got out his cage!' Seizing the carving knife, I rushed down, and was just in time to cut off his retreat into the adjoining swamp. Turning sharply round, he made a snap at my leg, and received in return a 'Rowland for his Oliver' in the shape of an inch or so of cold steel. After wrestling on the ground, and struggling through the deserted fire of our sable cook, I at length secured the runaway, tied him up to a post, and to prevent further mischief, ended his career by dividing the jugular. The length of this lizard from actual measurement was five feet ten inches and a half."

These semi-aquatic, dingy-hued saurians are admirably adapted to the hot moist swamps and shallow lagoons that fringe the rivers of the tropical alluvial plains. As we watch their dark forms, plunging and wallowing in the water, or sluggishly moving over the soft and slimy mud, the imagination is carried back to the age of reptiles, when the muddy shores of the primeval ocean swarmed with their uncouth forms. The huge lizard, six or seven feet long, to which divine honours are paid at Bonny on the coast of Guinea, belongs most likely to this amphibious class. Undisturbed, the lazy monsters crawl heavily through the streets, and as they pass, the negroes reverentially make way. A white man is hardly allowed to look at them, and hurried as fast as possible out of their presence. An attempt was once made to kidnap one of these dull lizard-gods for the benefit of a profane museum, but the consequences were such as to prevent a repetition of the offence, for all trade and intercourse with the ships in harbour was immediately stopped, and affairs assumed so hostile an aspect, that the foreigners were but too glad to purchase peace with a considerable sacrifice of money and goods. When one of these lizards crawls into a house, it is considered a great piece of good fortune; and when it chooses to take a bath, the Bonnians hurry after it in their canoes. After having allowed it to swim a stretch, and to plunge several times, they seize it for fear of danger, and carry it back again to the land, well pleased at once more having the sacred reptile in their safe possession.

The formidable name of Flying Dragons has been given to a

genus of small lizards, remarkable for the expansible cutaneous processes with which the sides are furnished, and by whose means they are enabled to spring with more facility from branch to branch, and even to support themselves for some time in the air, like the bat or flying-squirrel. The tiny painted Dragon of the East, the Flying Lizard of the woods, is fond of clinging with its wings to the smooth trunks of trees, and there remaining immovable, basking in the sun. When disturbed, it leaps and shuffles away in an awkward manner. One Mr. Adams had in his possession, reminded him of a bat when placed on the ground. Sometimes the strange creature would feign death, and remain perfectly motionless, drooping its head, and doubling its limbs, until it fancied the danger over, then cautiously raising its crouching form, it would look stealthily around, and be off in a moment. The dragon consumes flies in a slow and deliberate manner, swallowing them gradually; its various species belong exclusively to India and the islands of the Eastern Archipelago.

Flying Dragon.

Who has not heard of the fatal glance of the *basilisk*, which, according to poetical fancy, obliged all other poisonous animals to keep at a respectful distance

"from monster more abhorr'd than they?"

The truth is, that the lizards that bear this dreaded name, which has been given them from the fanciful resemblance of their pointed occipital crest to a regal crown, are quite as harmless and inoffensive as the flying dragon. They are chiefly inhabitants of South America, where they generally lead a sylvan life, feeding on insects.

Basilisk.

Among the lizards of every size, colour, and variety of form with which the warmer regions of the globe abound, from the size of a span to that of a cubit, dull brown or clothed in a livery of vivid tints, of elegant shape or of frightful ugliness, crested or smooth-backed, terrestrial or aquatic, many more might be mentioned; but as their way of life is very similar, and their numerous modifications of structure are subjects of interest only to the professed naturalist, I may well pass them over in silence, and terminate the chapter with a few words on the toads and frogs of the torrid zone.

Of the former there is none more famous than the hideous *Pipa Surinamensis*, which considerably exceeds in size the common toad, and whose deformity is often aggravated by a phenomenon unexampled in the rest of the animal world, namely, the young in various stages of exclusion, proceeding from cells dispersed over the back of the parent. It was for a long time supposed that the ova of this extraordinary reptile were produced in the dorsal cells without having been first excluded in the form of spawn; but it is now thoroughly ascertained that the female Pipa deposits her eggs or spawn at the brink of some stagnant water, and that the male collects or amasses the heap of ova, and deposits them with great care on the back of the female, where, after impregnation, they are pressed into the cellules, which are at that period open for their reception, and afterwards close over them; thus retaining them till the period of their second birth, which happens in somewhat less than three months, when they emerge from the back of the parent in their complete state. This species inhabits the obscure nooks of houses in Cayenne and Surinam, avoiding the light of day as if conscious of its unrivalled hideousness.

Surinam Toad.

Mr. Darwin thus describes a remarkable species of toad he noticed at *Bahia*. "Amongst the Batrachian reptiles, I found only one little toad, which was most singular from its colour. If we imagine, first, that it had been steeped in the blackest ink, and then, when dry, allowed to crawl over a board freshly painted with the brightest vermilion, so as to colour the sides of its feet and parts of its stomach, a good idea of its appearance will be gained.

Bahia Toad.

If it is an unnamed species, surely it ought to be called *diabolicus*, for it is a fit toad to preach in the ear of Eve. Instead of being nocturnal in its habits as other toads are, and living in damp and obscure recesses, it crawls during the heat of the day about the dry sand hillocks and arid plains, where not a single drop of water can be found. It must necessarily depend on the dew for its moisture, and this probably is absorbed by the skin, for it is known that these reptiles possess great powers of cutaneous absorption. At Maldonado I found one in a situation nearly as dry as at Bahia Blanca, and, thinking to give it a great treat, carried it to a pool of water; not only was the little animal unable to swim, but I think without help would soon have been drowned."

The giant-toad (*Bufo gigas, agua*), frequents the Brazilian campos in such numbers that in the evening or after a shower of rain, when they come forth from their hiding-places to regale on the damp and murky atmosphere, the earth seems literally to swarm with them. They are double the size of our common toad, and are even said to attain, with their outstretched hind legs, a foot's length, with a proportionate girth. Covered with unsightly warts, and of a dull grey colour, their aspect is repulsive, and when excited, they eject a liquid which is very much feared by the natives. Their voice is loud and disagreeable, while Guinea possesses, in the *Breviceps gibbosus*, a small toad which is said to sing delightfully, "charming the swamps with its melodious note."

A Brazilian tree-frog, (*Hyla crepitans*) which adheres to the large leaves, not merely with its widened toes, but with its constantly viscid body, has a voice which sounds like the cracking of a large piece of wood, and generally proceeds from many throats at a time. On wandering through the forests of Brazil, Prince Maximilian of Neu Wied was often surprised by this singular concert issuing from the dark shades of the forest.

A Surinam tree-frog (*Hyla micans*) has the singular property of secreting a luminous slime, so as to look in the dark like a yellowish will-o'-the-wisp. Its voice is most disagreeable, and is said at times completely to overpower the orchestra of the theatre in Paramaribo, thus emulating the stentorian achievements of the Virginian bull-frog.

CHAPTER XXX.

TORTOISES AND TURTLES.

The Galapagos—The Elephantine Tortoise—The Marsh-Tortoises—Mantega—River-Tortoises—Marine Turtles—On the Brazilian Coast—Their numerous Enemies—The Island of Ascension—Turtle-Catching at the Bahama and Keeling Islands—Turtle caught by means of the Sucking-Fish—The Green Turtle—The Hawksbill Turtle—Turtle-Scaling in the Feejee Islands—Barbarous mode of selling Turtle-Flesh in Ceylon—The Coriaceous Turtle—Its awful Shrieks.

IN the South Sea, exposed to the vertical beams of the equatorial sun, lies a large group of uninhabited islands, on whose sterile shores you would look in vain for the palms, bananas, or bread-fruit trees of more favoured lands, as rain falls only upon the heights, and never descends to call forth plenty on the arid coasts.

And yet, this desolate group offers many points of interest to the naturalist, for the Galapagos or Tortoise Islands represent, as it were, a little world in themselves, a peculiar creation of animals and plants, reminding us, more strongly than the productions of any other land, of an earlier epoch of planetary life. Here are no less than twenty-six different species of landbirds, which, with one single exception, are found nowhere else. Their plumage is homely, like the flora of their native country; their tameness so great that they may be killed with a stick. A sea-mew, likewise peculiar to this group, mixes its shriek with the hoarse-resounding surge; lizards, existing in no other country, swarm about the shore, and the gigantic land-tortoise (*Testudo indica, elephantina*), although now spread over many other countries, is supposed by Mr. Darwin to have had its original

Amblyrhyne.

seat in the Galapagos, where it was formerly found in such vast numbers as to have given the group its Spanish name. If the seafarer visits these treeless shores, which as yet produce nothing else worth gathering, it is chiefly for the purpose of catching a few of these huge animals, which, in spite of frequent persecutions, still amply reward a short sojourn with a rich supply of fresh meat. Their capture costs nothing but the trouble, for man has not yet drawn the boundary marks of property over the tenantless land.

The elephantine tortoise inhabits as well the low and sterile country, where it feeds on the fleshy leaves of the cactus, as the mountainous regions where the moist trade-wind calls forth a richer vegetation of ferns, grasses, and various trees. On this meagre food, which seems hardly sufficient for a goat, it thrives so well that three men are often scarcely able to lift it, and it not seldom furnishes more than 200 pounds of excellent meat.

"The tortoise," says Mr. Darwin, "is very fond of water, drinking large quantities, and wallowing in the mud. The larger islands alone possess springs, and these are always situated towards the central parts, and at a considerable elevation. The tortoises, therefore, which frequent the lower districts, when thirsty, are obliged to travel from a long distance. Their broad and well-beaten paths radiate off in every direction, from the wells even down to the sea coast, and the Spaniards, by following them up, first discovered the watering places. When I landed at Chatham Island, I could not imagine what animal travelled so methodically along the well-chosen tracks. Near the springs it was a curious spectacle to behold many of these great monsters, one set eagerly travelling onward with outstretched necks, and another set returning after having drank their fill. When the tortoise arrives at the spring, quite regardless of any spectator, it buries its head in the water above the eyes, and greedily swallows great mouthfuls, at the rate of about ten in a minute. The inhabitants* say each animal stays three or four days in the neighbourhood of the water, and then returns to the lower

---

* At the time of Mr. Darwin's visit an attempt, since given up, had been made to colonise the islands, which are once more only tenanted by casual adventurers, and may well be called *uninhabited*.

country; but they differed in their accounts respecting the frequency of these visits. The animal probably regulates them according to the nature of the food which it has consumed. It is, however, certain that tortoises can subsist even on those islands where there is no other water than what falls during a few rainy days in the year. I believe it is well ascertained that the bladder of the frog acts as a reservoir for the moisture necessary to its existence — such seems to be the case with the tortoise. For some time after a visit to the springs, the urinary bladder of these animals is distended with fluid, which is said gradually to decrease in volume, and to become less pure. The inhabitants, when walking in the lower districts and overcome with thirst, often take advantage of this circumstance by killing a tortoise, and if the bladder is full, drinking its contents. In one I saw killed, the fluid was quite limpid, and had only a very slightly bitter taste.

"The tortoises, when moving towards any definite point, travel by night and day, and arrive at their journey's end much sooner than would be expected. The inhabitants, from observations on marked individuals, consider that they can move a distance of about eight miles in two or three days. One large tortoise, which I watched, I found walked at the rate of sixty yards in ten minutes, that is, three hundred and sixty in the hour, or four miles a day, allowing also a little time for it to eat on the road. The flesh of this animal is largely employed, both fresh and salted, and a beautifully clear oil is prepared from the fat.

"When a tortoise is caught, the man makes a slit in the skin near its tail, so as to see inside its body, whether the fat under the dorsal plate is thick. If it is not, the animal is liberated, and it is said to recover soon from this strange operation. In order to secure the tortoises, it is not sufficient to turn them like turtle (their upper buckler being highly arched, while it is more flattened in the aquatic families, for the better adaptation of their forms to motion in a liquid), for they are often able to regain their upright position."

They are said to be completely deaf; so much is certain, that they do not perceive a person even when walking close behind them. Mr. Darwin often amused himself by overtaking the slow and monstrous creatures, who, as soon as he had passed them, instantly withdrew their head and legs, and fell flat down with a loud hiss and heavy noise as if touched by lightning. He then mounted upon their back, and on giving them a smart slap or two on the hind part of their carapace, they rose and leisurely proceeded with their learned freight, the author of "Origin of Species" finding it very difficult to maintain his equilibrium on this strange beast of burthen.

It is a remarkable fact, that though the land-tortoises are scattered in many places over the warmer regions of the globe, and even extend as far as Patagonia and the south of Europe, yet not a single one has hitherto been found in Australia, where, equally strange to say, no indigenous monkey exists.

The marsh tortoises, or *Emydæ*, have their chief seat in tropical America and the Indian Archipelago, where an abundance of swamps, lagoons, lakes, pools, and gently-flowing rivers favours the increase of their numbers. They play an important part in the domestic economy of the Indians along the great streams of the New World, the deep rolling Orinoco or the thousand-armed Amazons.

Marsh Tortoise, (Emys picta.)

During the dry season, all the neighbouring tribes are busy collecting the countless eggs which the cold-blooded creatures confide to the life-awakening powers of the heated sands: partly for their own consumption, and partly for the manufacture of oil. According to Herndon ("Exploration of the Valley of the Amazons, 1850-1853") from five to six thousand jars of mantega, or tortoise-oil are annually gathered on the banks of the Marañon. Each animal furnishes on an average eighty eggs, and forty tortoises are reckoned for each jar, which contains forty-five pounds, and is worth about six shillings on the spot. The manufacturing process, which is carried on in a most primitive manner, exhales an insupportable stench. The eggs, namely, are thrown into a boat, and trodden to pieces with the feet. The shells having been removed, the rest is left for several days to putrify in the sun. The oil which collects on the surface of the decomposing mass is then skimmed off, and boiled in large kettles. The neighbouring strand swarms with carrion vultures, and the smell of the offal attracts a number of alligators, all hoping to come in for their share of the feast.

Similar disgusting scenes occur on the banks of the Orinoco, where annually about 5,000 jars are collected. The number of eggs required for the production of the oil is estimated at about 33,000,000; the produce of at least 300,000 of the reptiles.

The marsh-tortoises may be said to form the connecting link between the eminently aquatic marine, and river chelonians and the land-tortoises, as the formation of their feet, armed with sharp claws or crooked nails, and furnished with a kind of flexible web, connecting their distinct and movable toes, allows them both to advance much quicker on the dry land than the latter, and to swim rapidly either on the surface or in the depth of the waters.

According to the more or less terrestrial habits of the various species, the feet are more or less webbed, for in those that habitually remain on the banks of the lagoons, the connecting membrane is confined to the basis of the toes, while in others, that but rarely come on shore, it sometimes reaches to the extremity of the claws: another beautiful example of the foresight of the Almighty in adapting organic structure to the wants of His creatures.

The marsh-tortoises, being endowed with more rapid power of locomotion, are not vegetarians like the land-tortoises, but chiefly live on mollusks, fishes, frogs, toads, and annelides. Although the eggs are palatable, the flesh is generally too coarse even for the craving appetite of an Indian.

The river-tortoises differ in many respects from the sea-turtles, although formed like them for a purely aquatic life. In both families the extremities are complete fins, serving as oars, but the fore feet of the river-tortoises are not double the length of the hind feet, as we find in the marine chelonians; and while the latter have a short apoplectic neck, that of the river-tortoise is generally very long and surmounted by a small and narrow head. The river-tortoises are exclusively confined to the warmer countries of the globe, and sometimes weigh as much as seventy pounds. It seems that during the night, and when they fancy themselves secure from danger, they repose upon the small river islands, or on rocks and trunks of trees that have fallen on the banks, or are drifted along by the current, and instantly plunge again into the water at the sight of man or at the least alarming noise. They are extremely voracious, and being very active swimmers, kill numbers of fish and reptiles. As their own flesh is held in high estimation, they are caught by hand-lines, a living bait being made use of to attract them, as they will not seize a dead or motionless prey. When they wish to seize their food or to defend themselves, they dart forwards their head and long neck with the velocity of lightning, and are said in this manner to surprise and seize even small birds that incautiously fly too near the surface of the water. They bite lustily with their sharp beak, never quitting their hold till they have fairly scooped out the morsel, so that the fishermen stand in great awe of their powerful mandibles, and generally cut off their head as soon as they are caught, rightly judging this to be the most radical means to prevent any further mischief.

The turtles, which are likewise inhabitants of the warmer latitudes, though sometimes a strange erratic propensity or mischance will carry them as far from their usual haunts as the North Sea, have, as we all know, a far greater commercial and gastronomic value than all the rest of the tortoise tribes.*

During the Brazilian summer (December, January, February), colossal turtles are seen everywhere swimming about along the coast, raising their thick round heads above the water, and waiting for the approach of night to land. The neighbouring Indians are their bitterest enemies, killing them whenever they can. Thus these dreary sand coasts, bounded on one side by the ocean and on the other by gloomy primeval forests, offer on all sides pictures of destruction, for the bones and shells of slaughtered turtles everywhere bestrew the ground. Two parallel grooves indicate the path of the turtle after landing; they are the marks of the four large and long fin-shaped feet or paddles, and between them may be seen a broad furrow where the heavy body trailed along the ground. On following these traces about thirty or forty yards shore-upwards, the huge animal may be found sitting in a flat excavation formed by its circular movements, and in which one half of its body is imbedded. It allows itself to be handled on all sides without making the least attempt to move away, being probably taught by instinct how useless all endeavours to escape would be. A

---

* For more ample details on the Marine Chelonians, see Chap. IX. of "The Sea and its living Wonders."

blowing or snorting like that of a goose when any one approaches its nest, at the same time inflating its neck a little, are the sole signs of defence which it exhibits. Having thus prepared a comfortable seat, the turtle begins to dig with her hind feet a cylindrical hole about a foot and a half deep, in which she deposits about a hundred eggs, covered with a leathery, flexible, and whitish skin. This being accomplished, the hole is again loosely filled up with sand, and slowly as she came, the turtle, leaving her eggs to be hatched by the sun, returns to her accustomed element.

Similar scenes take place during the dry season, throughout the whole of the tropical zone, on every sandy, unfrequented coast: for the same instinct which prompts the salmon to swim stream-upwards, the cod to seek elevated submarine banks, or the penguin to leave the high seas and settle for the summer on some dreary rock, attracts also the turtles from distances of fifty or sixty leagues to the shores of desert islands or solitary bays.

The enemies of the marine chelonians are no less numerous than those of the terrestrial or fluviatile species. While the full-grown turtles, as soon as they leave the water, are exposed to the attacks of many ravenous beasts, from the wild dog to the tiger or jaguar, storks, herons, and other strand- or sea-birds devour thousands upon thousands of the young before they reach the ocean, where sharks and other greedy fishes still further thin their ranks, so that but very few escape from the general massacre, and the whole race can only maintain itself by its great fecundity.

Of all the foes of the turtle-tribe there is, however, none more formidable than man, as even on the most lonely islands the seafarer lies in wait, eager to relieve the monotony of his coarse fare by an abundant supply of their luscious flesh.

On the isle of Ascension, the head-quarters of the finest turtle in the world, all the movements of the poor creatures are carefully watched, and when, after having deposited their eggs in the sand, they waddle again towards the sea, their retreat is often intercepted, for two stout hands running up to the unfortunate turtle after the completion of her task, one seizes a fore-flipper and dexterously shoves it under her belly, to serve as a purchase; whilst the other, avoiding a stroke which might lame him, cants her over on her back, where she lies helpless. From fifteen to thirty are thus turned in a night. In the bays, when the surf or heavy rollers prevent the boats being beached to take on board the turtles when caught, they are hauled out to them by ropes.

In former times, as long as the island had neither master nor inhabitants, every ship's crew that landed helped itself to as many turtles as it could catch; but since England has taken possession of the island, turtle-turning has been converted into a Government monopoly, and £2 10s. is the fixed price for each. They are kept in two large enclosures near the sea, which flows in and out, through a breakwater of large stones. A gallows is erected between the two ponds, where the turtles are slaughtered for shipping, by suspending them by the hind-flippers and then cutting their throats. Often above 300 turtles, of 400 lbs. and 500 lbs. each, are lying on the sand or swimming about in the ponds — a sight to set an alderman mad with delight. The best way to send home turtles from Ascension is to head them up in a cask, and have the water changed daily by the bunghole and a cock. Though the extremes of heat and cold are equally injurious to them, they should always arrive in England during hot weather. Thus an unfortunate captain, on one occasion, took from Ascension 200 turtles, and timing his arrival badly, brought only four of them alive to Bristol.

The way by which the turtles are most commonly taken at the Bahama Islands is by striking them with a small iron peg of two inches long, put in a socket at the end of a staff of twelve feet long. Two men usually set out for this work in a canoe, one to row and gently steer the boat, while the other stands at the end of it with his weapon. The turtles are sometimes discovered by their swimming with their head and back out of the water, but they are more often seen lying at the bottom, a fathom or more deep. If a turtle perceives he is discovered, he starts up to make his escape; the men in the boat, pursuing him, endeavour to keep sight of him, which they often lose and recover again by the turtle putting his nose out of the water to breathe.

On Keeling Island, Mr. Darwin witnessed another highly interesting method of catching turtle, which he describes in the vivid manner which has rendered his works so deservedly popular.

"I accompanied Captain Fitzroy to an island at the head of the lagoon," says this eminent naturalist; "the channel was exceedingly intricate, winding through fields of delicately-branched corals. We saw several turtles, and two boats were then employed in catching them. The method is rather curious: the water is so clear and shallow that, although at first a turtle quickly dives out of sight, yet in a canoe, or boat under sail, the pursuers, after no very long chase, come up to it. A man, standing ready in the bows, at this moment dashes through the water upon the turtle's back; then clinging with both hands by the shell of the neck, he is carried away till the animal becomes exhausted and is secured. It was quite an interesting chase to see the two boats thus doubling about, and the men dashing into the water trying to seize their prey."

On the coast of Mozambique, a large species of Remora, or sucking-fish, is made use of for turtle-catching. A strong cord of palm fibres is attached to the tail of this singular creature, and serves to drag it

Sucking Fish.

out of the water along with its prey, to which it so firmly adheres, by means of the remarkable striated apparatus on its head, that turtles weighing several hundred pounds can in this manner be raised from the bottom.

The Green turtle (*Chelonia midas*), which has been known to attain a length of seven feet, and a weight of 900 lbs., is most prized for its flesh; but the Hawksbill (*Chelonia imbricata*), which hardly reaches one-third of the size, is of far greater commercial value, the plates of its shell being stronger, thicker, and clearer than those of any other species. It is caught all over the tropical seas, but principally near the Moluccas, the West Indian and the Feejee Islands, where it is preserved in pens

Green Turtle.

by the chiefs, who have a barbarous way of removing the valuable part of the shell from the living animal. A burning brand is held close to the outer shell, until it curls up and separates a little from that beneath. Into the gap thus formed a small wooden wedge is then inserted, by which the whole is easily removed from the back. When stripped, the animal is again put into the pen, where it has full time for the growth of a new shell — for though the operation appears to give great pain, it is not fatal.

A similar cruel method of removing the tortoise's shell by heat is resorted to in Ceylon; but the mode in which the flesh of the edible turtle is sold piecemeal, while it is still alive, by the fishermen of that island, is still more repulsive, and a disgrace to the Colonial Government, which allows it to be openly practised. " The creatures," says Sir Emerson Tennent, "are to be seen in the market-place undergoing this frightful mutilation, the plastron and its integuments having been previously removed, and the animal thrown on its back, so as to display all the motions of the heart, viscera, and lungs. A broad knife, from twelve to eighteen inches in length, is first inserted at the left side, and the women, who are generally the operators, introduce one hand to scoop out the blood, which oozes slowly. The blade is next passed round till the lower shell is detached and placed to one side, and the internal organs exposed in full action. Each customer, as he applies, is served with any part selected, which is cut off as ordered, and sold by weight. Each of the fins is thus successively removed, with portions of the fat and flesh, the turtle showing by its contortions that each act of severance is productive of agony. In this state it lies for hours writhing in the sun, the heart and head being usually the last pieces selected; and till the latter is cut off, the snapping of the mouth, and the opening and closing of the eyes, show that life is still inherent, even when the shell has been nearly divested of its contents."

The Coriaceous turtle (*Sphargis coriacea*), of a more elongated form than the other species, and whose outer covering, marked along its whole length by seven distinct, prominent, and tuberculated ridges, is not of a horny substance, but resembles strong leather, grows to the greatest size of all the marine chelonians, some having been taken above eight feet in length, and weighing no less than 1,600 lbs., so that even the crocodile can hardly be compared to it in bulk.

While the land-tortoises can scarcely be said to have a voice, merely hissing or blowing when irritated or seized, the coriaceous turtle, when taken in a net or seriously wounded, utters loud shrieks, or cries, that may be heard at a considerable distance — a power which, in an inferior degree, seems to belong to most of the fluviatile and marine chelonians.

The turtles generally live on marine plants, but the Caouana, or Loggerhead (*Chelonia caouana*), and the Hawksbill (*C. imbricata*), feed on crustaceans and cuttle-fish, which they can easily crush in their strong, horny beak. The Caouana and the coriaceous turtles are frequently found in the Mediterranean, and on the coasts of South America and Africa. Both are of no commercial importance; their shell is almost useless, and their flesh, which, like that of the alligator, exhales a strong smell of musk, is extremely coarse and ill-flavoured.

Loggerhead.

Crocodiles and Alligators.

CHAPTER XXXI.

CROCODILES AND ALLIGATORS.

Their Habits—The Gavial and the Tiger—Mode of Seizing their Prey—Their Voice—Their Preference of Human Flesh—Alligator against Alligator—Wonderful Tenacity of Life—Tenderness of the Female Cayman for her Young—Enemies of the Crocodile—Torpidity of Crocodiles during the Dry Season—Their Awakening from their Lethargy with the First Rains—"Tickling a Crocodile."

THERE was a time, long before man appeared upon the scene, when huge crocodiles swarmed in the rivers of England, and, for aught we know, basked on the very spot where now their grim representatives can hardly be said to adorn the grounds of Sydenham Palace.

But the day when the ferocious, bone-harnessed Saurians lorded it in the European streams has passed, never to return; the diminished warmth of what are now the temperate regions of the globe having long since confined them to the large rivers and lagunes of the torrid zone. The scourge and terror of all that lives in the waters which they frequent, they may with full justice be called the very images of depravity, as perhaps no animals in existence bear in their countenance more decided marks of cruelty and malice. The depressed head, so significant of a low cerebral developement; the vast maw, garnished with formidable rows of conical teeth, entirely made for snatch and swallow; the elongated mud-coloured body, with its long lizard-like tail, resting on short legs, stamp them with a peculiar frightfulness, and proclaim the baseness of their instincts.

The short-snouted, broad-headed Alligators, or Caymen, belong to the New World; the Gavials, distinguished by their straight, long, and narrow jaw, are exclusively Indian; while the oblong-headed Crocodiles are not only found in Africa and Asia, but likewise infest the swamps and rivers of America.

Alligator.

All these animals, however, though different in form and name, have everywhere similar habits and manners; so that, in general, what is remarked of the one may be applied to the others.

Awkward and slow in their movements on the land, they are very active in the water, darting along with great rapidity by means of their strong muscular tail and their webbed hind feet. They sometimes bask in the sunbeams on the banks of the rivers, but oftener float on the surface, where, concealing their head and feet, they appear like the rough trunk of a tree, both in shape and colour, and thus are enabled the more easily to deceive and catch their prey.

In America, many a slow-paced Capybara, or Water-pig, coming in the dusk of evening to slake its thirst in the lagune, has been suddenly seized by this insidious foe; and the Gangetic Gavial is said to make even the tiger his prey. When the latter quits the thick cover of the jungle to drink at the stream, the Gavial, concealed under water, steals along the bank, and, suddenly emerg-

Capybara.

ing, furiously attacks the tiger, who never declines the combat; and though in the struggle the Gavial frequently loses his eyes and receives dreadful wounds on the head, he at length drags his adversary into the water, and there devours him.*

In order to observe the manner in which the Alligator seizes its prey, Richard Schomburgk frequently tied a bird or some large fish to a piece of wood, and then turned it adrift upon the stream. Scarcely had the Cayman perceived his victim, than

* Forbes' "Oriental Memories," vol. i. p. 357.

he slowly and cautiously approached, without even rippling the surface of the water, and then curving his back, hurled his prey, by a stroke of his tail, into his wide-extended jaws.

On the American streams, the stillness of the night is often interrupted by the clacking of the Cayman's teeth, and the lashing of his tail upon the waters. The singular and awful sound of his voice can also readily be distinguished from that of all the other beasts of the wilderness. It is like a suppressed sigh, bursting forth all of a sudden, and so loud as to be heard above a mile off. First, one emits this horrible noise; then another answers him; and far and wide the repetition of the sound proclaims that the Caymen are awake. When these hideous creatures have once tasted the flesh of man, they are said, like the cannibals of the Feejee Islands, to prefer it to that of any animal.

During Humboldt's stay at Angostura, a monstrous Cayman seized an Indian by the leg while he was busy pushing his boat ashore in a shallow lagune, and immediately dragged him down into the deeper water. The cries of the unfortunate victim soon attracted a large number of spectators, who witnessed the astonishing courage with which he searched in his pocket for a knife. Not finding the weapon, he then seized the reptile by the head, and pressed his fingers into its eyes — a method which saved Mungo Park's negro from a similar fate. In this case, however, the monster did not let go his hold, but disappearing under the surface with the Indian, came up again with him as soon as he was drowned, and dragged the body to a neighbouring island.

"One Sunday evening," says Waterton, "some years ago, as I was walking with Don Felipe de Yriarte, Governor of Angostura, on the bank of the Orinoco— 'Stop here a minute or two, Don Carlos,' said he to me, 'while I recount a sad accident. One fine evening, last year, as the people of Angostura were sauntering up and down in the Alameda, I was within twenty yards of this place, when I saw a large Cayman rush out of the river, seize a man, and carry him down, before anybody had it in his power to assist him. The screams of the poor fellow were terrible, as the Cayman was running off with him. He plunged into the river with his prey; we instantly lost sight of him, and never saw or heard him more.'"

Humboldt also relates that, during the inundations of the Orinoco, alligators will sometimes make their appearance in the very streets of Angostura, where they have been known to attack and drag away a human prey.

Even among each other, these ferocious animals frequently engage in deadly conflict. Thus, Richard Schomburgk once saw a prodigiously large Cayman seize one of a smaller species (*Champsa vallifrons*) by the middle of the body, so that the head and tail projected on both sides of its muzzle. Now both of them disappeared under the surface, so that only the agitated waters of the otherwise calm river announced the death-struggle going on beneath; and then again the monsters reappeared, wildly beating the surface; so that it was hardly possible to distinguish here a tail, or there a monstrous head, in the seething whirlpool. At length, however, the tumult subsided, and the large Cayman was seen leisurely swimming to a sand-bank, where he immediately began to feed upon his prey.

The same traveller relates an interesting example of the Cayman's tenacity of life. One of them having been wounded with a strong harpoon, was dragged upon a sand-bank. Here the rays of the sun seemed to infuse new life into the monster, for, awaking from his death-like torpidity, he suddenly snapped about him with such rage that Schomburgk and his assistants thought it prudent to retreat to a safer distance. Seizing a long and mighty pole, the bravest of the Indians now went towards the Cayman, who awaited the attack with wide-extended jaws, and plunged the stake deep into his maw — a morsel which the brute did not seem to relish. Meanwhile two other Indians approached him from behind, and kept striking him with thick clubs upon the extremity of the tail. At every blow upon this sensitive part, the monster bounded in the air and extended his frightful jaws, which were each time immediately regaled with a fresh thrust of the pole. After a long and furious battle, the Cayman, who measured twelve feet in length, was at last slain. Another remarkable instance of the vitality of the common crocodile is mentioned by Sir E. Tennent. A gentleman at Galle having caught on a baited hook an unusually large one, it was disembowelled by his coolies, the aperture in the stomach being left expanded by a stick placed across it. On returning, in the afternoon, with a view to secure the head, they found that the creature had crawled for some distance, and made its escape into the water. We all know the intense hatred which sailors bear to the shark, and with what savage delight they drag one on board, and hack him to pieces with their knives before life is extinct; but the American Indian is a no less inveterate enemy of the Cayman, and, when occasion offers, lets him feel the full extent of his inventive cruelty. Among the Javanese, on the contrary, we find the crocodile considered as a sacred animal, on account of his clearing the rivers and lagunes of putrefying substances; and the friendship even seems to be reciprocal, as Bennett saw Javanese convicts busy working up to their middle in water, quite near the monsters.

Like the sea-turtles, the crocodiles generally deposit their eggs, which are about the size of those of a goose, and covered with a calcareous shell, in holes made in the sand, leaving them to be hatched by the warm rays of the tropical sun. In some parts of America, however, they have been observed to resort to a more ingenious method, denoting a degree of provident instinct which could hardly have been expected in a cold-blooded reptile. Raising a small hillock on the banks of the river, and hollowing it out in the middle, they collect a quantity of leaves and other vegetable matters, in which they deposit their eggs. These are covered with the leaves, and are hatched by the heat extricated during their putrefaction, along with that of the atmosphere.

Callous to every other generous sentiment, the female Cayman continues for some time after their birth to watch over her young with great care. One day, as Richard Schomburgk, accompanied by an Indian, was busy fishing on the banks of the Essequibo, he suddenly heard in the water a strange noise, resembling the mewing of young cats. With eager curiosity he climbed along the trunk of a tree overhanging the river, about three feet above the water, and saw beneath him a brood of young alligators, about a foot and a half long. On his seizing and lifting one of them out of the water, the mother, a creature of prodigious size, suddenly emerged with an appalling roar, making desperate efforts to reach her wriggling and screeching offspring, and increasing in rage every time Schomburgk tantalised her by holding it out to her. Having been wounded with an arrow, she retired for a few moments, and then again returned with redoubled fury, lashing the waters into foam by the repeated strokes of her tail. Schomburgk now cautiously retreated, as in case of a fall into the water below, he would have had but little reason to expect a friendly reception, the monster pertinaciously following him to the bank, but not deeming it advisable to land, as here it seemed to feel its helplessness. The scales of the captured young one were quite soft and pliable, as it was only a few days old, but it already had the peculiar musk-like smell which characterises the full-grown reptile. Like so many other tropical animals, the Cayman awakens to a more active life in the cool of the evening, and loves to bask in the noontide sun on sandbanks and rocks. Sometimes he is found with wide-extended jaws, covered with brilliantly-feathered birds, which are not afraid of resting there — and indeed a strange kind of sympathy, or mutual understanding seems to exist between creatures so dissimilar. On the banks of the Nile, herons are often seen to wait for hours in the neighbourhood of the crocodiles, until they spring into the water and drive the fishes within their reach; and a small plover is even said to extract annoying insects from the monster's mouth, who graciously requites the office by not devouring its benefactor.

The young of the crocodiles have no less numerous enemies than those of the snakes. Many an egg is destroyed in the hot sand by small carnivora, or birds, before it can be hatched; and as soon as the young creep out of the broken shell, and instinctively move to the waters, the Ichneumon — a kind of weasel, to whom, on this account, the ancient

Ichneumon.

Egyptians paid divine homage — or the long-legged Heron gobble up many of them, so that their span of life is short indeed. In the water they are not only the prey of various sharp-toothed fishes, but even of the males of their own species, while the females do all they can to protect them. Even the full-grown crocodile, in spite of its bony harness, is not exempt from attack. Thus, in the river of Tabasco, a tortoise of the genus *Cinyxis*, after having been swallowed by the alligator, and, thanks to its shelly case, arriving unharmed in its stomach, is said to have eaten its way out again with its sharp beak,* thus putting the monster to a most excruciating death.

* Carl Heller's "Travels in Mexico in the Years 1845–1848."

Even man not only kills the hideous reptiles in self-defence, or for the sake of sport, but for the purpose of regaling upon their flesh. In the Siamese markets and bazaars, crocodiles, large and small, may be seen hanging in the butchers' stalls, instead of mutton or lamb; and Captain Stokes,* who more than once supped off alligators' steaks, informs us that the meat is by no means bad, and has a white appearance like veal.

I have already mentioned, in the chapter on the Llanos, that in many tropical countries the aridity of the dry season produces a similar torpidity in reptile life to that which is caused by the cold of winter in the higher latitudes. In Ceylon, when the watercourses begin to fail and the tanks become exhausted, the marsh-crocodiles are sometimes encountered wandering in search of water in the jungle; but generally, during the extreme drought, they bury themselves in the sand, where they remain in a state of torpor, till released by the recurrence of the rains. Sir Emerson Tennent, whilst riding across the parched bed of a tank, was shown the recess, still bearing the form and impress of the crocodile, out of which the animal had been seen to emerge the day before. A story was also related to him of an officer who, having pitched his tent in a similar position, had been disturbed during the night by feeling a movement of the earth below his bed, from which, on the following day, a crocodile emerged, making its appearance from beneath the matting.

Like the rattlesnake, crocodiles seem to possess the power of fascinating their prey, or rather of completely depriving their victims of all presence of mind, by the terror which they inspire. In Sumatra, Marsden once saw a large crocodile in a river, looking up to an overhanging tree, on which a number of small monkeys were sitting. The poor creatures were so beside themselves for fright, that instead of escaping to the land, which they might easily have done, or of quietly remaining where they were, they hurried towards the extremities of the branches, and at length fell into the water, where the dreadful monster was awaiting them.

Crocodiles sometimes indulge in strange wanderings. Chamisso mentions one having been drifted from the Pelew Islands to Eap, one of the Carolines, where it was killed after

* "Discoveries in Australia."

having devoured a woman; and about thirty years ago, the inhabitants of one of the Feejee Islands were equally astonished and alarmed at seeing a large crocodile emerge from the lagune, and lazily creep on shore. At first they took it for some marine deity; but it soon proved that its visit was not of a beneficent nature, as it seized and devoured nine of them at various intervals. After many unavailing attempts to destroy the monster, it was at length caught with a sling passed over the bough of a large tree, the other end of the rope being held at a distance by fourteen men who lay concealed, while one of the party offered himself as a bait to entice the reptile to run into the snare. Captain Fitzroy ("Voyage of the Beagle"), who relates the fact, supposes that the animal must have been drifted by the currents all the way from the East Indies — a voyage which, in fact, is not more surprising than to see a turtle land upon the shores of the North Sea, or a sperm whale flounder about in the Thames.

Like many other of the lower animals, the crocodile, when surprised, endeavours to save himself by feigning death. Sir Emerson Tennent relates an amusing anecdote of one that was found sleeping several hundred yards from the water. "The terror of the poor wretch was extreme when he awoke and found himself discovered and completely surrounded. He was a hideous creature, upwards of ten feet long and evidently of prodigious strength, had he been in a condition to exert it; but consternation completely paralysed him. He started to his feet, and turned round in a circle, hissing and clacking his bony jaws, with his ugly green eye intently fixed upon us. On being struck, he lay perfectly quiet and apparently dead. Presently he looked round cunningly, and made a rush towards the water; but on a second blow he lay again motionless, and feigning death. We tried to rouse him, but without effect; pulled his tail, slapped his back, struck his hard scales, and teased him in every way, but all in vain: nothing would induce him to move, till, accidentally, my son, a boy of twelve years old, tickled him gently under the arm, and in an instant he drew it close to his side, and turned to avoid a repetition of the experiment. Again he was touched under the other arm, and the same emotion was exhibited, the great monster twisting about like an infant to avoid being tickled."

CHAPTER XXXII.

TROPICAL BIRD LIFE IN BOTH HEMISPHERES.

The Toucan—Its Quarrelsome Character—The Humming-birds—Their wide Range over the New World—Their Habits—Their Enemies—Their Courage—The Cotingas—The Campanero—The Tanagras—The Manakins—The Cock of the Rock—The Troupials—The Baltimore—The Pendulous Nests of the Cassiques—The Mocking-bird—Strange Voices of Tropical Birds—The Goat-sucker's Wail—The Organista—The Cilgero—The Flamingos—The Scarlet Ibis—The Jabiru—The Roseate Spoon-bill—The Jacana—The Calao—The Sun-birds—The Melithreptes—The Argus—The Peacock—The Lyre Bird—The Birds of Paradise—The Australian Bower-bird—The Talegalla—The Devil-bird—The Baya—The Tailor-bird—The Republican Gros-beak—The Korwê.

USEFUL in many respects to man, no class of animals is more interesting or agreeable to him than that of the birds, whether we consider the beauty of their plumage, the grace of their movements, the melody of their voice, or the instinct that regulates their migrations and prompts them to construct their nests; so that their study forms, without doubt, one of the most attractive departments in the whole range of natural history.

But it is at the same time one of the most difficult, particularly in countries where man has not yet mastered the powers of vegetation, where numberless creepers and bush-ropes render the forest impenetrable, and the pathless wilderness obstructs the observer at every step. Thus it is by no means surprising that so many secrets still veil the life of the tropical birds — that comparatively so little is known as yet of their economy and mode of existence.

Many families of birds have a wide range over the whole earth: falcons hover over the Siberian fir-woods, as over the forests of the Amazons; in every zone are found woodpeckers, owls, and long-beaked martin-fishers, while thrushes enliven

with their song both the shades of the beech-woods and the twilight of the cocoa-nut groves. In the north and in the south, fly-catchers carry destruction among the numerous insect-tribes; in every latitude, crows cleanse the fields of vermin; and swallows, pigeons, ducks, gulls, petrels, divers, and plovers frequent the fields and lakes, the banks and shores in all parts of the world.

Thus the class of birds shows us a great similarity in the distribution of its various forms all over the earth; and we find the same resemblance extending also to their mode of life, their manners, and their voice. The woodpeckers make everywhere the forest resound with the same clear note, and the birds of prey possess in every clime the same rough screech so consonant to their habits, while a soft cooing everywhere characterises the pigeon-tribes. But, notwithstanding this general uniformity and this wide range of many families of birds, each zone has at the same time its peculiar ornithological features, that blend harmoniously with the surrounding world of plants and animals, and, taking a prominent part in the aspect of nature, at once attract the attention of the stranger.

In this respect, as in so many others, the warmer regions of the globe have a great advantage over those of the temperate and glacial zones; and here, where warmth and moisture call forth an exuberant vegetation, they produce an equal multiplicity of animal forms, among which many birds rival the most gorgeous flowers by the splendour of their plumage.

On turning to each continent in particular, we again find each endowed with its peculiar genera of birds, and thus, though tropical America has many of its feathered tribes in common with the torrid zone of the Old World, it enjoys the exclusive possession of the Toucans, Colibris, Crotophagi, Jacamars, Anis, Dendrocolaptes, Manakins, and Tangaras; while the Calaos, the Souimangas, the Birds of Paradise, and many others, are confined to the eastern hemisphere. A complete review of all these various forms of the feathered creation would fill volumes: my narrow limits necessarily confine me to a brief account of those tribes which are either the most remarkable, or the most widely different from the birds which we are accustomed to see in Europe.

By their enormous bill, which might seem rather adapted to a bird of ostrich-like dimensions than to one not much larger than a crow, the Toucans (*Ramphastidæ*) are distinguished from all the other feathered races of America. Were it of a strong and solid texture, this huge beak would infallibly weigh them to the ground; but being of a light and cellular structure, they carry it easily, and leap with such agility from bough to bough, that it does not then appear preposterously large.

When flying, it gives them, indeed, a very awkward appearance, but the beauty of its colouring soon reconciles the eye to its disproportionate size: for the brightest red, variegated with black and yellow stripes on the upper mandible, and a stripe of the liveliest sky-blue on the lower, contribute to adorn the bill of the Bouradi, as one of the three Toucan species of Guiana is called by the Indians. Unfortunately, these brilliant tints fade after death, and it requires all the art of a Waterton to fix and preserve their evanescent hues.

Toucan.

The plumage of this strange bird rivals the beak in beauty of colouring, and the feathers are frequently used as ornaments by the Brazilian ladies, as well as by the Indian tribes that roam through the vast forests of South America. The Toucans are generally seen in small flocks or troops, and from this it might be supposed they were gregarious; "but upon a closer examination," says the author of "Wanderings in South America," "you will find it has only been a dinner-party which breaks up and disperses towards roosting-time." While thus assembled, discord never ceases to reign, for there is hardly a more quarrelsome and imperious bird than the toucan.

Schomburgk relates an anecdote of a tamed Ramphastos who, by dint of arrogance, assisted by his enormous beak, had made himself despot not only over the domestic fowls, but even over the larger four-footed animals of an estate in Guiana. Large and small willingly submitted to him, so that when a dispute arose among the trumpeters and hoccos of the yard, the combatants all dispersed as soon as he made his appearance, and if by chance he had been overlooked in the heat of the fray, his powerful beak soon reminded them that their lord and master was by no means inclined to tolerate disputes among his subjects.

On bread being thrown among them, none of his two or four-legged subjects would have ventured to seize the smallest morsel before the toucan had liberally helped himself. This domineering spirit even went so far that he inhospitably reminded every strange dog that came near the premises, that none durst enter his domains without his permission. There is no knowing to what lengths he might not have carried his despotism, if a powerful mastiff, one day entering the yard and taking several bones without leave, had not put an end to his tyranny. For scarcely had the toucan perceived the intruder, when angrily rushing upon him, he attacked him with his beak. The dog at first only growled, without suffering himself to be disturbed in his meal, but as the bird continued to bite, he finally lost his patience and, snapping at the toucan, wounded him so severely on the head that he soon after expired.

A bird with so strange a beak must naturally be expected to feed and drink in a strange manner. When the toucan has seized a morsel, he throws it into the air and lets it fall into his throat; when drinking, he dips the point of his mandibles into the water, fills them by a powerful inspiration, and then throws back the head by starts. The tongue is also of a very singular form, being narrow and elongated, and laterally barbed like a feather. The toucans are very noisy birds. In rainy weather their clamour is heard at all hours of the day, and in fair weather at morning and evening. The sound which the Bourodi makes is like the clear yelping of a puppy dog, and you fancy he says " pia-po-o-co," and thus the South American Spaniards call him Piapoco.

To paint the Humming-bird with colours worthy of its beauty, would be a task as difficult as to fix on canvas the glowing tints of the rainbow, or the glories of the setting sun. Unrivalled in the metallic brilliancy of its plumage, it may truly be called the bird of paradise; and had it existed in the old world it would no doubt have claimed the title instead of the splendid bird which has now the honour to bear it. See with what lightning speed it darts from flower to flower; now hovering for an instant before you, as if to give you an opportunity of admiring its surpassing beauty, and now again vanishing with the rapidity of thought. But do not fancy that these winged jewels of the air, buzzing like bees round the blossoms less gorgeous than themselves, live entirely on the honey-dew collected within their petals; for on opening the stomach of the Trochilus dead insects are almost always found there, which its long and slender beak, and cloven extensile tongue, like that of the woodpecker, enable it to catch at the very bottom of the tubular corollas.

The torrid zone is the chief seat of the humming-birds, but in summer they wander far beyond its bounds, and follow the sun in his annual declensions to the poles. Thus, in the north, they appear as flying visitors on the borders of the Canadian lakes, and on the southern coast of the peninsula of Aljaschka; while in the southern hemisphere they roam as far as Patagonia, and even as Tierra del Fuego; visiting in the northern hemisphere the confines of the walrus, and reaching in the south the regions of the penguins and the lion-seal; advancing towards the higher latitudes with the advance of summer, and again retreating at the approach of autumn.

Humming Bird.

The nest of the humming-bird is as elegant and neat as its tiny constructor, a little capsule formed externally of grey lichens so as to avoid notice, and lined internally with the soft down of the cotton tree. In this fragile cradle, suspended from a branch, or leaf, or even a blade of the straw which covers the hut of the Indian, the female lays two white eggs, the size of peas, which are hatched in about twelve days by the alternate incubation of the male and female, producing young no bigger than a common fly, naked, blind, and so weak, as hardly to be able to raise their little bills for the food provided for them by their parents.

Nothing can exceed the tenderness which the male evinces during breeding time for his lovely companion, nor the courage which he displays for her protection. On the approach of an intrusive bird, though ten times bigger than himself, he will not hesitate a moment to attack the disturber of his nest, his bravery adds a tenfold increase to his powers, the rapidity of his movements confounds his enemy, and finally drives him to flight.

Proud of this success, the little champion returns to his partner, and flaps triumphantly his tiny wings. But with all his activity and courage, he is not always able to avert disaster from his nest, for an enormous bush spider, covered all over with black hair (*Mygale*), too often lurks in the vicinity, watching for the moment when the little birds shall creep out of the shell. With sudden attack it then invades the nest, and sucks their life-blood. Against this enemy neither courage nor despair are of any avail, and if the poor humming-bird endeavours to avenge the slaughter of his young, he only shares their fate. When the dark long-legged monster entwines his brilliant prey, one might almost fancy an angel of light bleeding under the talons of a demon.

From the chivalrous character of the humming-birds it is not surprising that the most violent passions agitate their little breasts; so that in their desperate contests, they will tilt against each other with such fury, as if each meant to transfix his antagonist with his long bill. It may indeed be truly said that these little creatures are sadly prone to quarrel over their cups, not of wine, but nectareous flowers. Frequently four or five of them may be seen engaged in a flying fight when disputing the possession of a blossoming tree in the forests of Brazil, and then they dart so swiftly through the air that the eye can scarcely follow them in their meteorlike evolutions.

What a splendid addition, even to the magnificence of a regal drawing-room, the humming-birds would be, if they allowed themselves to be confined in a cage; but perpetual movement in the air is to them a necessity, and to deprive them of liberty is to rob them of life. All attempts to transport them alive to Europe have hitherto been fruitless. The celebrated ornithologist Latham relates that a young man cut off the branch on which a humming-bird was breeding, and took it on board the ship which conveyed him to England. The mother soon grew tame, and took the biscuit and honey, that was offered her; she also continued to breed during the passage, but died as soon as the young crept out of the shell. These came alive to England, and withstood during two months our uncongenial climate. They grew so tame as to feed from the lips of Lady Howard, to whom they had been presented: a lovely picture worthy of Anacreon Moore, —

> "To whom the lyre was given,
> With all the trophies of triumphant song."

As the smallest shot would blow the tiny humming-birds to pieces, and inevitably destroy the beauty of their plumage, they are taken by aspersing them with water from a syphon, or by means of a butterfly net.

There are many species of humming-birds, various in size and habits, with straight or curved bills, with a naked or a crested head, with a short or a long tail; some constantly concealing themselves in the solitudes of the forest; while others hover round the habitations of man, and frequently during their disputes pursue each other into the apartments whose windows are left open, taking a turn round the room, as flies do with us, and then suddenly regaining the open air.

The Karabamite, the largest of the humming-birds of Guiana, is all red and changing gold green, except the head, which is black. He has two long feathers in the tail which cross each other, and these have gained him the name of Ara humming-bird from the Indians. He alone of all the tribe never shows his beauty to the sun, and were it not for his shining colours one might almost be inclined to class him with the goat-suckers on account of his habits. He keeps close by the side of woody fresh-water rivers, and dark and lonely creeks. He leaves his retreat before sunrise to feed on the insects over the water, but retires to it again as soon as the sun's rays cause a glare of light—is sedentary all day long, and comes out again for a short time after sunset. He builds his nest on a twig over the water in the unfrequented creeks; it looks like tanned cow-leather, and has no particle of lining. The rim of the nest is doubled inwards, for instinct has taught the bird to give it this shape, in order that the eggs may be prevented from rolling out.

The smallest of the humming-birds (*Trochilus minimus*), golden green, brown and bluish black, weighs but from twenty to forty-five grains, and is surpassed in weight and dimensions by more than one species of bee. Well may we exclaim with Pliny, that Nature is most admirable in her most diminutive works—"Maxima miranda in minimis."

Next to the humming-birds the Cotingas display the gayest plumage. They are, however, not often seen, for they lead a solitary life in the moist and shadowy forests, where they feed on the various seeds and fruits of the woods. One species is attired in burning scarlet, others in purple and blue, but they are all so splendidly adorned that it would be difficult to say which of them deserved the prize for beauty. Most of the Cotingas have no song; the nearly related snow-white Campanero or bell-bird, however, amply makes up for the deficient voice of his cousins, by the singularity and sweetness of his note. He is about the size of a jay. On his forehead rises a singular spiral tube nearly three inches long. It is jet black, dotted all over with small white feathers. It has a communication with the palate, and when filled with air looks like a spire, when empty it becomes pendulous. His note is loud and clear, like the sound of a bell, and may be heard at the distance of three miles. "In the midst of these extensive wilds," says Waterton, "generally on the dried top of an aged mora, almost out of gun reach, you will see the Campanero. No sound or song from any of the winged inhabitants of the forest causes such astonishment as his toll. With many of the feathered race he pays the common tribute of a song to early morn, and even when the meridian sun has shut in silence the mouths of almost the whole of animated nature, the Campanero still cheers the forest; you hear his toll, and then a pause for a minute; then another toll, and then a pause again, and then a toll and again a pause. Then he is silent for six or eight minutes, and then another toll, and so on. Acteon would stop in mid-chase, Maria would defer her evening song, and Orpheus himself would drop his lute to listen to him, so sweet, so novel and romantic, is the toll of the pretty snow-white Campanero. He is never seen to feed with the other Cotingas, nor is it known in what part of Guiana he makes his nest."

Campanero.

The Tangaras resemble our finches, though they are far more splendidly attired. Their plumage is very rich and diversified, some of them boast six separate colours; others have the blue, purple, green, and black so finely blended into each other that it would be impossible to mark their boundaries; while others again exhibit them strong, distinct, and abrupt. The flight of the Tangaras is rapid, their manners lively. They live upon insects, seeds, berries, and many of them have a fine song. Among their numerous species, spread over all the warmer regions of America, the scarlet Piranga is pre-eminent for beauty, and when in the blooming thickets, along the woody river's banks, the meridian sun shows off his plumage in all its splendour, the huntsman pauses to admire the magnificent bird, and delays his murderous aim.

In the deep forests, which they never quit for the open plains, reside the Manakins (*Pipra*), pretty little birds, whose largest species scarcely attain the dimensions of the sparrow, while the smallest are hardly equal to the wren. The plumage of the full-grown male is always black, enlivened by brilliant colours, that of the female and of the young birds greenish. Their flight is rapid but short, and they generally roost on the middle branches of the trees. In the morning they unite in little troops, and seek their food, which consists of insects, and small fruit, uttering at the same time their weak but melodious notes. As the day advances they separate and seek the deepest forest-shades, where they live in solitude and silence.

The famous orange-coloured Cock of the Rock of Guiana (*Rupicola aurantia*), which owes its name to its comb-like crest, is nearly related to the manakins. It is a great rarity, even in its own country, and as it dwells in the most secluded forests, is but seldom seen by travellers. Richard Schomburgk relates the following wonderful story of the bird, which, if not proceeding from so trustworthy a source, might almost be considered fabulous. "A troop of these beautiful birds was celebrating its dances on the smooth surface of a rock; about a score of them were seated on the branches as spectators, while one of the male birds, with proud self-confidence, and spreading tail and wings, was dancing on the rock. He scratched the ground or leaped vertically into the air, continuing these saltatory movements until he was tired, when another male took his place. The females, meanwhile, looked on attentively, and applauded the performance of the dancers with laudatory cries. As the feathers are highly prized, the Indians lay in wait with their blow-pipes near the places where the Rupicolas are known to dance. When once the ball has begun, the birds are so absorbed by their amusement, that the hunter has full time to shoot down several of the spectators with his poisoned arrows, before the rest take the alarm."

On penetrating into the wilds of Guiana, the pretty songsters called Troopials, (*Icterus, Xanthornus*) pour forth a variety of sweet and plaintive notes. Resembling the starling by their habits, they unite in troops, and live on insects, berries, and seeds.

The variegated Troopial (*Oriolus varius*) displays a wonderful instinct in the construction of his nest, which he generally builds on fruit-trees; but when circumstances force him to select a tree whose branches have far less solidity, as, for instance, the weeping-willow, his instinct almost rises to a higher intelligence. First, he binds together, by means of bits of straw, the small and flexible branches of the willow, and thus forms a kind of conical basket in which he places his nest, and instead of the usual hemispherical form, he gives it a more elongated shape, and makes it of a looser tissue, so as to render it more elastic and better able to conform to the movements of the branches when agitated by the wind.

The neat little black and orange Baltimore (*Icterus Baltimore*) constructs a still more marvellous nest on the tulip trees, on whose leaves and flowers he seeks the caterpillars and beetles which constitute his principal food. When the time comes for preparing it, the male picks up a filament of the *Tillandsia usneoides* and attaches it by its two extremities to two neighbouring branches. Soon after, the female comes, inspects his work, and places another fibre across that of her companion. Thus by their alternate labours a net is formed, which soon assumes the

Baltimore Bird.

shape of a nest, and as it advances towards its completion, the affection of the tender couple seems to increase. The tissue is so loose as to allow the air to pass through its meshes, and as the parents know that the excessive heat of summer would incommode their young, they suspend their nest so as to catch the cooler breeze of the north-east when breeding in Louisiana; while in more temperate regions, such as Pennsylvania and New York, they always give it a southern exposition, and take care to line it with wool or cotton. Their movements are uncommonly graceful; their song is sweet; they migrate in winter towards more southerly regions, Mexico or Brazil, and return after the equinox to the United States.

The Cassiques, which are nearly related to the troopials or orioli, are no less remarkable for their architectural skill. They suspend their large pendulous nests, which are often above four feet long, at the extremities of branches of palm trees, as far as possible from all enemies that might by climbing reach the brood, often choosing, for still further protection, trees on which the wasps or maribondas have already built their nests, as these are adversaries whose sharp stings no tiger-cat or reptile would desire to face. The nest of the *Cassicus cristatus* is artificially woven of lichens, bark-fibres and the filaments of the tillandsias, while that of the tupuba (*Cassicus ruber*), which is always suspended over the water, consists of dry grasses, and has a slanting opening in the side, so that no rain can penetrate it. On passing under a tree, which often contains hundreds of cassique nests, one cannot help stopping to admire them, as they wave to and fro, the sport of every storm and breeze, and yet so well constructed as rarely to be injured by the wind. Often numbers of one species may be seen weaving their nests on one side of a tree, while numbers of another species are busy forming theirs on the opposite side of the same plant, and what is, perhaps, even still more wonderful than their architectural skill, though such near neighbours, the females are never observed to quarrel!

The *Cassicus persicus*, a small black and yellow bird, somewhat larger than the starling, has been named the mocking-bird, from his wonderful imitative powers. He courts the society of man, and generally takes his station on a tree close to his house, where for hours together he pours forth a succession of ever-varying notes. If a toucan be yelping in the neighbourhood, he immediately drops his own sweet song, and answers him in equal strain. Then he will amuse his audience with the cries of the different species of the woodpecker, and when the sheep bleat he will distinctly answer them. Then comes his own song again, and if a puppy dog or a guinea fowl interrupt him, he takes them off admirably, and by his different gestures during the time, you would conclude that he enjoys the sport.

Wild and strange are the voices of many of the American forest-birds. In the Peruvian woods the black Toropishu (*Cephalopterus ornatus*) makes the thicket resound with his hoarse cry, resembling the distant lowing of a bull; and in the same regions the fiery-red and black-winged Tunqui (*Rupicola Peruviana*) sends forth a note, which might readily be mistaken for the grunting of a hog, and strangely contrasts with the brilliancy of his plumage.

But of all the startling cries that issue from the depths of the forest, none is more remarkable than the Goatsucker's lamentable wail. "Suppose yourself in hopeless sorrow," says Waterton, "begin with a high, loud note, and pronounce ha, ha, ha, ha, ha! each note lower and lower till the last is scarcely heard, pausing a moment or two between every note, and you will have some idea of the mourning of the largest goatsucker in Demerara.

"Four other species of goatsucker articulate some words so distinctly, that they have received their names from the sentences they utter, and absolutely bewilder the stranger on his arrival in these parts. The most common one sits down close by your door, and flies and alights three or four yards before you, as you walk along the road, crying, 'Who are you, who-who-who-who are you?' Another bids you, 'Work away, work-work-work away.' A third cries mournfully, 'Willy come go, Willy-Willy-Willy come go.' And high up in the country, a fourth tells you to, 'Whip-poor-Will, whip-whip-whip-poor-Will.'"

You will never persuade the negro to destroy the birds, or get the Indian to let fly his arrow at them, for they are held to be the receptacles for departed souls, who came back again to earth, unable to rest for crimes done in their days of nature, or expressly sent to haunt cruel and hardhearted masters, and retaliate injuries received from them. If the largest goatsucker chance to cry near the white man's door, sorrow and grief will soon be inside, and they expect to see the master waste away with a slow consuming sickness. If it be heard close to the negro's or Indian's hut, from that night misfortune sits brooding over it, and they await the event in terrible suspense.

During the daytime, the goatsucker, whose eyes, like those of the owl, are too delicately formed to bear the light, retires to the deepest recesses of the forest, but when the sun has sunk behind the western woods, he may, on moonlight nights, be seen silently hovering in the forest glades, or hopping about among the herds. This poor bird has the character of a nocturnal thief, but never has a more unjust accusation been made, as, far from robbing the flocks of their milk, he does all he can to free them from insects. "See how the nocturnal flies are tormenting the herd," says Waterton, "and with what dexterity he springs up, and catches them, as fast as they alight on the belly, legs, and udder of the animals. Observe how quiet they stand, and how sensible they seem of his good offices, for they neither strike at him, nor hit him with their tail, nor tread on him; nor try to drive him away as an uncivil intruder. Were you to dissect him, and inspect his stomach, you would find no milk there: it is full of the flies which have been annoying the herd."

The large tropical nocturnal butterflies, or moths, form the chief food of the wide-beaked goatsucker, and the number of their wings that may be seen lying about, give proof of the ravages he commits among their ranks. For as the bat with his hooked thumb cuts off the wings of the moths and cockchafers which he catches on his twilight excursions, thus, also, the goatsucker refrains from swallowing these parts, and his hooked and incurvated upper mandible seems purposely intended for clipping them.

While the goatsucker makes the forest resound with his funereal tones, other birds of the forest pour forth the sweetest notes. Dressed in a sober cinnamon brown robe, with blackish olive-coloured head and neck, the Organist (*Troglodytes leucophrys*) enlivens the solitude of the Peruvian forests. The astonished wanderer stops to listen to the strain, and forgets the impending storm.

The Cilgero, a no less delightful songster, frequents the mountain regions of Cuba, and the beauty of his notes may be inferred from the extravagant price of several hundred dollars, which the rich Havanese are ready to pay for a captive bird.

Wagner ("Travels in Costa Rica," 1853, 1854) tells us that our nightingale is far inferior to the Cilgero, who entertains his mate with the softest tones of the harmonica, and in Guiana the flute-bird (*Cyphorinus cantans*) delights the ear with his melodious song. All these lonely musicians of the grove belong to the extensive finch tribe, and, like their European cousins, appear in a simple unostentatious garb.

The same beauty of plumage which characterises so many of the American forest-birds, adorns, likewise, the feathered tribes of the swamp and the morass, of the river and the lake. Nothing can exceed in beauty a troop of deep red Flamingoes (*Phœnicopterus ruber*) on the green margin of a stream. Raised on enormous stilts, and with an equally disproportionate length of neck, the flamingoes would be reckoned among the most uncouth birds, if their splendid robe did not entitle them to rank among the most beautiful.

They always live in troops, and range themselves, whether fishing or resting, like soldiers, in long lines. One of the number acts as sentinel, and on the approach of danger gives a warning scream, like the sound of a trumpet, when, instantly, the whole troop, expanding their flaming wings, rise loudly clamouring into the air.

These strange-formed birds build in the swamps high conical nests of mud, in the shape of a hillock with a cavity at top, in which the female generally lays two white eggs of the size of those of a goose, but more elongated. The rude construction is sufficiently high to admit of her sitting on it conveniently, or rather riding, as the legs are placed on each side at full length. Their mode of feeding is no less remarkable. Twisting their neck in such a manner that the upper part of their bill is applied to the ground, they at the same time disturb the mud with one of their webbed feet, thus raising up from the water insects and spawn, on which they chiefly subsist.

The Rose-coloured Flamingo, with red wings and black quills (*P. antiquorum*), adorn the creeks and rivers of tropical Africa and Asia, and in warm summers extends his migrations as far northwards as Strasburg on the Rhine. The sight of a troop of flamingoes approaching on the wing, and describing a great fiery triangle in the air, is singularly majestic. When about to descend, their flight becomes slower, they hover for a moment, then their evolutions trace a conical spire, and, finally alighting, they immediately arrange themselves in long array, place their sentinels, and begin their fishing operations.

Egyptian Ibis.

The scarlet American Ibis, with black tipped wings, though inferior in size to his celebrated cousin, the sacred bird of the Egyptians, far surpasses him in beauty.

Six feet high, stately as a grenadier of the guards, the American Jabiru stalks along the banks of the morasses. His plumage is white, but his neck and head are black, like his long legs; his conical, sharp, and powerful black bill, is a little recurved, while that of the stork, to whom he is closely related, is straight. He destroys an incredible number of reptiles and fishes, and, being very sly, is difficult to kill. Two similar species, respectively inhabit Western Africa and Australasia.

The roseate American Spoon-bill (*Platalea Ajaja*) is particularly remarkable for his curious large beak, dilating at the top into a broad spoon or spatula, which, though not possessed of great power, renders him excellent service in disturbing the mud and seizing the little reptiles and worms he delights to feed on.

The Jacana possesses enormously long and slender toes, armed with equally long spine-like claws. While pacing the ground they seem as inconvenient as the snow shoes of a Laplander, and yet nothing can be more suitable for a bird destined to stalk over the floating leaves of the Nelumbos and Nymphæas, and to seek for water insects on this unstable foundation. The jacana is found all over tropical America, and is also called the surgeon, from the nail of his hinder toe being sharp and acuated like a lancet.

All these strange and wondrous birds, and numberless others, whose mere enumeration would be fatiguing to the reader, justify the ornithological reputation of the woods and swamps of tropical America. And indeed the feathered races nowhere find a richer field for their developement than here, where the vegetable world revels in luxuriant growth; and myriads of insects, peopling the forest, the field, and the water, furnish each kind according to its wants with an inexhaustible supply of food. The circumstance that man but thinly inhabits these wilds, is another reason which favours the multiplication of the feathered tribes; for, in Europe also, birds would no doubt be far more numerous, if the farmer, the sportsman, and so many other enemies were not continually thinning their ranks. To these elements of destruction they are far less exposed in tropical America, and being comparatively but little disturbed, they reign, as it were, over the forest and the open field, over the mountain and the plain, over the river and the lake.

It is evident that, in countries where a uniform temperature renders winter and summer almost undistinguishable, the birds are not compelled to wide migrations. Thus, in Brazil, the swallow and the cuckoo are sedentary, the storks never leave the land where they breed, and the singing birds tune their notes the whole year round. The search for food or other local causes alone compel the Brazilian birds to wander from one district to another. Thus the rainy season produces in the vast forests a damp coolness uncongenial to their feathered tenants; who then eagerly seek the open country, where the bananas, oranges, and guavas, laden with ripe fruit, invite them to an abundant banquet. This is the time for the Brazilian settler to seize his rifle, and to shoot without trouble a number of the finest birds:—parrots, toucans, cotingas—that are almost inaccessible at other seasons. The war he wages with the winged visitors of his plantation and garden, is partly for the sake of their delicate flesh, but chiefly for the protection of his property, as the parrots are particularly fond of maize, while the cassiques and the toucans have a predilection for guavas and bananas, and the finches for the rice field. At this time also the Indians provide themselves with the beautifully coloured feathers of the aras, which serve them to plume their arrows, and to decorate their persons, for these birds, otherwise so shy, are then attracted far beyond the skirts of the forest by the favourite fruits of the Sapucaya (*Lecythis ollaria*) or of the prickly spinea.

Thus man knows how to profit by the annual wanderings of

the birds; but in this respect the settler of European extraction is far inferior to the wild Indian. Accustomed to rely on himself alone, the savage trusts to his own unfailing powers, and his hardened constitution, his falcon eye, his sharp ear and swift foot, are sure to provide him with food, where the European would fall a victim to despair.

The breeding season of the tropical birds is not restricted to a few months. Many Brazilian birds build their nests almost at all times of the year, but the greater number at the end of the rainy season, in September, October, November, and December, which, in the southern parts of that vast country, are the warmest months of the year, and consequently the most congenial to the young brood.

It is a remarkable circumstance that rarely more than two eggs are found in the nests of the forest birds, and even in those of many of the aquatic tribes, and this is the more to be wondered at, when we consider the vast number of enemies which in the equatorial zone menace the existence of the feathered races. But against these dangers we have seen that many of them guard themselves and their young by wonderful nests.

Although in the torrid zone we hardly ever meet with a single aboriginal species of plant or animal common to both hemispheres, yet the analogy of climate everywhere produces analogous organic forms, and when on surveying the feathered tribes of America, we are struck by any bird remarkable for its singularity of shape or mode of life, we may expect to find its representative in Asia, Africa, or Australia.

Rhinoceros Hornbill.

Thus the enormous beak of the toucan is emulated or surpassed by that of the Indian Calao, or Rhinoceros Hornbill (*Buceros rhinoceros*), whose twelve-inch long, curved, and sharp-pointed bill is, moreover, surmounted with an immense appendage, in the form of a reverted horn, the use of which belongs as yet to the secrets of nature. While the toucans are distinguished by a gaudy plumage, the calaos are almost entirely decked with a robe black as that of the raven, and enhancing the beautiful red and orange colours of their colossal beak. Generally congregating in small troops like the toucans, they inhabit the dense forests, where they chiefly live on fruits, seeds, and insects, which they also swallow whole, throwing them up into the air and catching them as they fall. The clapping together of their mandibles causes a loud and peculiar noise, which towards evening interrupts the silence of the forest. The flight of a bird burdened with such a load must naturally be short: they hop upon their thick clumsy feet, and generally roost upon the highest trees.

The brilliant Sun-birds or Suimangas (*Cinnyris*), belonging to the order of the Certhias or creepers, are the Colibris of the old world, equally ethereal, gay, and sprightly in their motions, flitting briskly from flower to flower, and assuming a thousand lively and agreeable attitudes. As the sunbeams glitter on their bodies, they sparkle like so many gems. As they hover about the honey-laden blossoms, they vibrate rapidly their tiny pinions, producing in the air a slight whirring sound, but not so loud as the humming noise produced by the wings of the colibris. Thrusting their slender beaks into the deep-cupped flowers, they probe them with their brushlike tongues for insects and nectar. Some are emerald green, some vivid violet, others yellow with a crimson wing, and rivalling the colibris by the metallic lustre of their plumage, they surpass them by their musical powers, for while the latter can only hum, the sun-birds accompany their movements with an agreeable chirp.

The nearly-related *Melithreptes*, or Honey-eaters of the South Sea Islands, distinguished by a very long curved beak, and a tongue split into two slender filaments, furnish the chief ornaments of the Polynesian kings and chieftains. Thus the famous royal mantle of Taméhaméha the Great is completely covered with the golden plumage of the *Melithreptes pacificus*, and as this not very common brown-coloured bird has only three or four yellow feathers in each wing, it may easily be conceived that the most costly brocades of Lyons are far from equalling in value this splendid robe of state, which is no less than ten feet long and seven feet broad. Even the small diadems made of the feathers of this bird, which are worn by the ladies of rank in the Sandwich Islands, are worth several hundred dollars. Idols or mantles of the Polynesians, decorated with the scarlet feathers of the *Melithreptes vestiarius*, are frequently met with in ethnographical museums.

Argus Pheasant.

While the superb ocellated turkey of Honduras (*Meleagris ocellata*) displays, with all the pride of a peacock, the eye-like marks of his tail and upper-coverts, the no less beautifully spotted Argus, a bird nearly related to the gold and silver pheasants which have been introduced from China into the European aviaries, conceals his splendour in the dense forests of Java and Sumatra. The wings of this magnificent creature, whose plumage is equally remarkable for variety and elegance, consist of very large feathers, nearly three feet long, the outer webs being adorned with a row of large eyes, arranged parallel to the shaft; the tail is composed of twelve feathers, the two middle ones being about four feet in length, the next scarcely two, and gradually shortening to the outer ones. Its voice is plaintive and not harsh, as in the Indian peacock, which Alexander the Great is said to have first introduced into Europe, though its feathers had many centuries before been imported by the Phœnicians.

Javanese Peacock.

The Peacock is still found wild in many parts of Asia and Africa, but more particularly in the fertile plains of India. Another species, nearly similar in size and proportions, but distinguished by a much longer crest, inhabits the Javanese forests.

Though of less dazzling splendour than this peacock's tail, that of the Menura, or Lyre-bird, is unrivalled for its elegance. Fancy two large, broad, black and brown striped feathers, curved in the form of a Grecian Lyre, and between both, other feathers, whose widely-distanced silken barbs envelope and surmount them with a light and airy gauze. No painter could possibly have imagined anything to equal this masterpiece of nature, which its shy possessor conceals in the wild bushes of Australia.

"Of all the birds I have ever met with," says Mr. Gould, "the menura is by far the most difficult to procure. While among the brushes, on the coast or on the sides of the mountains in the interior, I have been surrounded by those birds pouring forth their loud and liquid calls for days together, without being able to get a sight of them, and it was only by the most determined perseverance and extreme caution that I was enabled to effect the desired object."

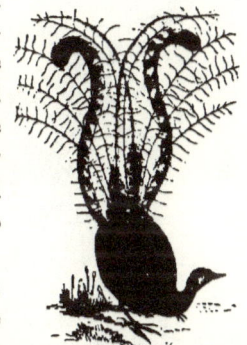

Lyre-bird.

The lyre-bird is constantly engaged in traversing the brush from mountain-top to the bottom of the gullies, whose steep and rugged sides present no obstacle to its long legs and powerful muscular thighs. When running quickly through the brush, it carries the tail horizontally, that being the only position in which it could be borne at such times. Besides its loud, full cry, which may be heard at a great distance, it has an inward and varied song, the lower notes of which can only be heard when you have stealthily approached to within a few yards of the bird when it is singing. Its habits appear to be solitary, seldom more than a pair being seen together. It constructs a large nest, formed on the outside of sticks and twigs, like that of a magpie, and lined with the inner bark of trees and fibrous roots.

In the neighbouring regions of Papua or New Guinea, and the small isles in their immediate vicinity, extending only a few degrees on each side of the Equator, we find the seat of the wondrous Paradiseidæ, distinguished in most species by that peculiar union of splendour and elegance which seems to render them more worthy of the gardens of Eden than of a terrestrial home.

The great Bird of Paradise (*P. apoda*) may justly be said to surpass in beauty the whole of the feathered creation. The throat is of the brightest emerald, and the canary-coloured neck blends gradually into the fine chocolate of the other parts of the body. From under the short chestnut-coloured wings project the long delicate and gold-coloured feathers, whose beautiful and graceful tufts are equally valued by the princes of the East and the ladies of England. The chocolate-coloured tail is short, but two very long shafts of the same hue considerably exceed in length even the long, loose plumes of the sides.

Bird of Paradise.

Unable to fly with the wind, which would destroy their loose plumage, the birds of Paradise take their flight constantly against it, being careful not to venture out in hard blowing weather. The Papuas climb, during the night, upon the high forest trees, where they have observed the birds to roost, and patiently await the dawn to catch them in nooses, or to shoot them with blunted arrows. The Portuguese first found these birds on the island of Gilolo, and as the Papuas tear off their legs before bringing them to market, it was for a long time supposed that they were destitute of these organs. The most absurd fables were founded on this imaginary deficiency: it was said that they passed their whole life sailing in the air, dew being their only food; that they never took rest except by suspending themselves from the branches of trees by the shafts of their two elongated tail feathers; that they never touched the earth till the moment of their death; and the Malays still believe that they retire for breeding to the groves of Paradise. It is almost superfluous to add that the researches of modern travellers have fully proved the utter fallacy of these ridiculous tales.

There are no less than six different genera of Paradise birds, each comprehending several species, and were it possible to penetrate into the forests of New Guinea, no doubt many more would be found.

The ornithological wonders of Australia are inferior to those of no other part of the world. Can anything, for instance, be more extraordinary than the constructions of the Bower-birds, which are built not for the useful purpose of containing the young, but purely as a playing place or an assembly room? "The structures of the spotted bower-bird," says Mr. Gould, "are in many instances three feet in length. They are outwardly built of twigs, and beautifully lined with tall grasses, so disposed that their heads nearly meet; the decorations are very profuse, and consist of bivalve shells, crania of small mammalia, and other birds. Evident and beautiful indications of design are manifest throughout the whole of the bower and decorations formed by this species, particularly in the manner in which the stones are placed within the bower, apparently to keep the grasses with which it is lined fixed firmly in their places. These stones diverge from the mouth of the run on each side, so as to form a little path, while the immense collection of decorative materials, bones, shells, &c., are placed in a heap before the entrance of the avenue, this arrangement being the same at both ends. I frequently found these st uctures at a considerable distance from the rivers, from the borders of which they alone could have procured the shells and small round pebbly stones; their collection and transportation must, therefore, be a task of great labour and difficulty. As these birds feed almost entirely upon seeds and fruits, the shells and bones cannot have been collected for any other purpose than ornament; besides, it is only those that have been bleached perfectly white in the sun, or such as have been roasted by the natives, and by this means whitened, that attract their attention."

For what purpose these curious bowers are made is not yet, perhaps, fully understood; they are certainly not used as a nest, but as a place of resort, where the assembled birds run through and about the bower in a playful manner, and that so frequently that it is seldom entirely deserted. The proceedings of these birds have not been sufficiently watched to render it certain whether the runs are frequented throughout the whole year or not, but it is highly probable that they are merely resorted to as a rendezvous or playing ground at the pairing time, and during the period of incubation.

Three satin bower-birds, thus called from their deep shining blue-black plumage, were brought to the Zoological Gardens in 1849. Immediately upon their arriving, they commenced the construction of one of their bowers or " runs," for which purpose they made use of the twigs of an old besom, bending them into a shape like the ribs of a man of war, the top being open and the length varying from six to twelve inches. Against the sides they placed bright feathers, or whatever else containing colour they could get hold of. Unfortunately, two of them dying, the surviving male, dispirited and forlorn, neglected his assembly-room, which was now rendered useless. The satin bower-bird, like the magpie, is well known by the natives to be a terrible thief, and they always search his abode for anything they may have lost.

The Talegalla or Brush-turkey is no less interesting. In appearance it is very like the common black turkey, but is not quite so large: the extraordinary manner in which its eggs are hatched constitutes its singularity. It collects together a great heap of decaying vegetables as the place of deposit of its eggs, thus making a hot-bed, arising from the decomposition of the collected matter, by the heat of which the young are hatched. This mound varies in quantity from two to four cartloads, and is of a perfectly pyramidical form: it is not, however, the work of a single pair of birds, but is the result of the united labour of many, and the same site appears to be resorted to for several years in succession. "The mode," says Mr. Gould, " in which the materials composing these mounds are accumulated is equally singular, the bird never using its bill, but always grasping a quantity in its foot, throwing it backwards to one common centre, and thus clearing the surface of the ground to a considerable distance so completely that scarcely a leaf or blade of grass is left." The heap being accumulated and time allowed for a sufficient heat to be engendered, the eggs, each measuring not less than four inches in length — an enormous size, considering the bulk of the bird — are deposited, not side by side, as is ordinarily the case, but planted at the distance of nine or twelve inches from each other, and buried at nearly an arm's depth perfectly upright, with the large end upwards; they are covered up as they are laid, and allowed to remain until hatched. After six weeks of burial, the eggs, in succession and without any warning, give up their chicks — not feeble, but full-fledged and strong, so that at night they scrape holes for themselves, and lying down therein are covered over by the old birds and thus remain until morning. The extraordinary strength of the newly-hatched

Latham Talegalla.

birds is accounted for by the size of the shell, since in so large a space it is reasonable to suppose that the young ones would be much more developed than is usually found in eggs of smaller dimensions. Other Australian birds, such as the Jungle-fowl (*Megapodius tumulus*), Duperrey's Megapodius (*M. Duperreyii*), which inhabits the forests of New Guinea, and the Leipoas or native pheasants, construct similar mound-like nests. Those of the jungle-fowl, observed at Port Essington, are described as fifteen feet high, and sixty in circumference at the base, and so enveloped in thickly foliaged trees as to preclude the possibility of the sun's rays reaching any part of it.

It is not to be wondered at that in the tropical world, where lizards, snakes, and frogs attain such extraordinary dimensions, the cranes or stork tribes, which chiefly live upon these reptiles, should also grow to a more colossal size than their European representatives. Thus, while torrid America boasts of the jabiru, Africa and India possess the still larger Argala, or adjutant, whose feeding exploits and ugliness have already been mentioned in the chapter on Snakes.

His beak, measuring sixteen inches in circumference at the base, corresponds with his appetite. He is soon rendered familiar with man, and when fish or other food is thrown to him, he catches it very nimbly and immediately swallows it entire. A young bird of this kind, about five feet in height, was brought up tame and presented to a chief on the coast of Guinea, where Mr. Smeathman lived. It regularly attended the hall at dinner-time, placing itself behind its master's chair, frequently before any of the guests entered. The servants were obliged to watch it carefully, and to defend the provisions by beating it off with sticks; still it would frequently snatch off something from the table, and one day purloined a whole boiled fowl, which it swallowed in an instant. It used to fly about the island, and roost very high among the silk-cotton trees; from this station, at the distance of two or three miles, it could see when the dinner was carried across the court, when, darting down, it would arrive early enough to enter with some of those who carried in the dishes. Sometimes it would stand in the room for half an hour after dinner, turning its head alternately as if taking a deep interest in the conversation. These birds are found in companies, and when seen at a distance near the mouths of rivers advancing towards an observer, it is said that they may be easily mistaken for canoes on the surface of a smooth sea; and when on the sand-banks, for men and women picking up shell-fish on the beach.

The tropical forests of the eastern hemisphere resound with bird-cries no less appalling, wild, or strange than those of the western world. In the close jungle of Ceylon one occasionally hears the call of the Copper-smith (*Megalasara Indica*), whose din resembles the blows of a smith hammering a cauldron, or the strokes of the great orange-coloured Woodpecker (*Brachypterus aurantius*), as it beats the decaying trees in search of insects; but of all the yells that fancy can imagine there is none to equal that of the Singhalese Devil-bird, or Gualama. "Its ordinary cry," says Mr. Mitford, "is a magnificent clear shout like that of a human being, and which can be heard at a great distance, and has a fine effect in the silence of the closing night. It has another cry like that of a hen just caught, but the sounds which have earned for it its bad name, and which I have heard but once to perfection, are indescribable; the most appalling that can be imagined, and scarcely to be heard without shuddering. I can only compare it to a boy in torture, whose screams are being stopped by being strangled. On hearing this dreadful note the terrified Singhalese hurries from the spot, for should he chance to see the bird of ill omen he knows that his death is nigh. A servant of Mr. Baker's,[*] who had the misfortune of seeing the dreaded gualama, from that moment took no food, and thus fell a victim to his superstitious despair. This horror of the natives explains the circumstance that it is not yet perfectly ascertained whether the devil-bird is an owl (*Syrnium*) or a night hawk.

As if to make amends for this screech, the robin of Neueraellia, the long-tailed thrush, the oriole, the dayal-bird, and some others equally charming, make the forests and savannas of the Kandyan country resound with the rich tones of their musical calls.

---

[*] Baker's "Eight Years' Wanderings in Ceylon," vol. i. p. 167.

In general the tropical woods and fields are by no means so deficient in agreeable songsters as is commonly imagined; and while the Americans rejoice in the voices of the Banana Bird (*Icterus xanthornus*), or of the *Tryothorus plateus*, which is as fond of the company of man as the latter of its lovely note, the African and the Indian listen with delight to the tiny Senegalis and Bengalis, which, like the canary-bird, belong to the extensive and sprightly tribe of the finches.

The wonderful pendulous nests of the American cassiques are equalled, if not surpassed, by those of the Indian Baya. These birds are found in most parts of Hindostan; in shape they resemble the sparrow, as also in the brown feathers of the back and wings; the head and breast are of a bright yellow, and in the rays of a tropical sun have a splendid appearance when flying by thousands in the same grove. They make a chirping noise, but have no song; they associate in large communities, and cover clumps of palmyras, acacias, and date-trees with their nests. These are formed in a very ingenious manner by long grass woven together in the shape of a bottle, and suspended by so slender a thread to the end of a flexible branch that even the squirrel dare not venture his body on so fragile a support, however his mouth may water at the eggs and prey within. These nests contain several apartments, appropriated to different purposes: in one the hen performs the office of incubation; another, consisting of a little thatched roof and covering a perch without a bottom, is occupied by the male, who cheers the female with his chirping note. The Hindoos are very fond of these birds for their docility and sagacity; when young, they teach them to fetch and carry, and at the time the young women resort to the public fountains their lovers instruct the baya to pluck the tica or golden ornament from the forehead of their favourite and bring it to their master.

The Tailor-bird of Hindostan (*Sylvia sutoria*) is equally curious in the structure of its nest, and far superior in the elegance and variety of its plumage, which in the male glows with the varied tints of the colibri. The little artist first selects a plant with large leaves and then gathers cotton from the shrub, spins it to a thread by means of its long bill and slender feet, and then, as with a needle, sows the leaves neatly together to conceal its nest. Who, on witnessing these miracles of instinct, would not exclaim with the poet:

> " Behold a bird's nest!
> Mark it well, within, without!
> No tool had he that wrought, no knife to cut;
> No nail to fix, no bodkin to insert,
> No glue to join: his little beak was all!
> And yet how neatly finish'd! What nice hand,
> With every implement and means of art,
> Could compass such another!"

On turning to the wilds of Africa the Gros-beak affords us a no less wonderful example of nest-building, for here we find, not one single pair, but hundreds living under the same roof, perfectly resembling that of a thatched house, and with a projecting ridge, so that it is impossible for any reptile to approach the entrances concealed below. "Their industry," says Paterson, their first discoverer ("Travels in Africa"), "seems almost equal to that of the bee; throughout the day they appear to be busily employed in carrying a fine species of grass, which is the principal material they employ for the purpose of erecting this extraordinary work, as well as for additions and repairs. Though my short stay in the country was not sufficient to satisfy me, by ocular proof, that they added to their nest as they annually increased in numbers, still from the many trees which I have seen borne down by the weight, and others which I have observed with their boughs completely covered over, it would appear that this really was the case. When the tree which is the support of this aërial city is obliged to give way to the increase of weight, it is obvious they are no longer protected, and are under the necessity of rebuilding in other trees. One of these deserted nests I had the curiosity to break down, so as to inform myself of its internal structure, and I found it equally ingenious with that of the external. There are many entrances, each of which forms a separate street with nests on both sides, at about two inches distant from each other. The grass with which they are built is called the Boshman's grass, and I believe the seed of it to be their principal food; though, on examining their nests, I found the wings and legs of different insects. From every appearance, the nest which I dissected had been inhabited for many years; and some parts of it were much more complete than others. This, therefore, I conceive nearly to amount to a proof that they added to it at different times as they found necessary, from the increase of the family, or rather, I should say, of the nation or community."

The tree usually selected for these nests is the kameel-doorn or giraffe-thorn, which derives its name from constituting the chief food of the beautiful cameleopard, and, on account of its size and peculiar growth, having the foliage disposed from the top downwards in umbrella-shaped manner, is a great ornament to the arid wastes of South Africa.

The instinct of the birds seems to have pointed out to them that it is peculiarly adapted for the purpose, as its smooth and polished bark effectually secures them from the attacks and injuries of all the snakes, lizards, and other reptiles which swarm around their habitations, and if they could ascend the stem, would be but too happy to suck the eggs and destroy the young. Captain Harris having lost his way in the desert, and observing large thatched houses resembling haystacks in many of the trees, imagined that they had been erected by the natives, as a defence against the lions, whose recent tracks he distinguished in every direction, and ascended more than one in the hope of at least finding some vessel containing water.

Though far less ingenious, yet the nest of the Korwê (*Tockus erythrorynchus*) is too curious to be passed over in silence. The female having entered her breeding-place, in one of the natural cavities of the mopane tree, a species of bauhinia, the male plasters up the entrance, leaving only a narrow slit by which to feed his mate, and which exactly suits the form of his beak. The female makes a nest of her own feathers, lays her eggs, hatches them, and remains with the young till they are fully fledged. During all this time, which is stated to be two or three months, the male continues to feed her and the young family. The prisoner generally becomes quite fat, and is esteemed a very dainty morsel by the natives, while the poor slave of a husband gets so lean and weak, that on the sudden lowering of the temperature, which sometimes happens after a fall of rain, he is benumbed, falls down, and dies.

The first time Dr. Livingstone saw this bird was at Kolobeng, where he had gone to the forest for some timber. Standing by a tree, a native looked behind him and exclaimed, "There is the nest of a korwê." Seeing a slit only about half an inch wide and three or four inches long in a slight hollow of the tree, and thinking the word korwê denoted some small animal, he waited with interest to see what the Bechuana would extract. The latter, breaking the clay which surrounded the slit, put his arm into the hole and brought out a tockus, or red-beaked hornbill, which he killed.

The same habit had been previously observed in a species of Buceros, in Java; and the natives of Ceylon assert that the large Malabar Hornbill (*Buceros malabaricus*) likewise builds in holes in the trees, where the female successfully guards her treasures from the monkey tribes, her bill filling up nearly the whole entrance.

The Condor.

CHAPTER XXXIII.

TROPICAL BIRDS OF PREY.

The Condor—His Marvellous Flight—His Cowardice—Various Modes of Capturing Condors—Ancient Fables circulated about them—Comparison of the Condor with the Albatross—The Carrion Vultures—The King of the Vultures —Domestication of the Urubu—Its Extraordinary Memory—The Harpy Eagle—Examples of his Ferocity—The Oricou—The Bacha—His Cruelty to the Klipdachs—The Fishing Eagle of Africa—The Musical Sparrow-hawk— The Secretary Eagle.

THE flight of the Condor is truly wonderful. From the mountain-plains of the Andes, the royal bird, soaring aloft, appears only like a small black speck on the sky, and a few hours afterwards he descends to the coast and mixes his loud screech with the roar of the surf. No living creature rises *voluntarily* so high, none traverses in so short a time all the climates of the globe. He rests at night in the crevices of the rocks, or on some jutting ledge; but as soon as the first rays of the sun light the summits of the mountains, while the darkness of night still rests upon the deeper valleys, he stretches forth his neck, shakes his head as if fully to rouse himself, stoops over the brink of the abyss, and flapping his wings, dives into the aërial ocean. At first his flight is by no means strong, he sinks as if borne down by his weight, but soon he ascends, and sweeps through the rarefied atmosphere without any perceptible vibratory motion of his wings. "Near Lima," says Mr. Darwin, "I watched several condors for nearly half an hour without once taking off my eyes. They moved in large curves, sweeping in circles, descending and ascending without once flapping. As they glided close over my head, I intently watched from an oblique position the outlines of the separate and terminal feathers of the wing; if there had been the least vibratory movement these would have blended together, but they were seen distinct against the blue sky. The head and neck were moved frequently and apparently with force, and it appeared that the extended wings formed the fulcrum on which the movements of the neck, body, and tail acted. If the bird wished to descend the wings were for a moment collapsed, and then, when again expanded with an altered inclination, the momentum gained by the rapid descent seemed to urge the bird upwards with the even and steady movement of a paper kite."

According to Humboldt and D'Orbigny, the condor is a contemptible coward, whom the stick of a child is able to put to flight. Far from venturing to attack any full-grown, larger animal—the lama, the ox, or even man, as former travellers asserted—he feeds, like other vultures, only upon dead carcases, or on new-born lambs and calves, whom he tears from the side of their mothers. He thus does so much damage to the herds, that the shepherds pursue and kill him whenever they can. As even a bullet frequently glances off from his thick feathery coat, the natives never use fire-arms for his destruction, but make use of various traps, of the sling, or of the *bolas*, which they are able to throw with such marvellous dexterity.

In the Peruvian province of Abacay, an Indian provided with cords conceals himself under a fresh cow's skin, to which some pieces of flesh are left attached. The condors soon pounce upon the prey, but while they are feasting, he fastens their legs to the skin. This being accomplished, he suddenly comes forth, and the alarmed birds vainly flap their wings, for other Indians hurry towards them, throw their mantles or their lassos over them, and carry the condors to their village, where they are reserved for the next bull-fight.

For a full week before this spectacle is to take place, the bird gets nothing to eat, and is then bound upon the back of a bull which has previously been scarified with lances. The bellowing of the poor animal, lacerated by the famished vulture, and vainly endeavouring to cast off its tormentor, amuses what may well be called the " swinish multitude."

In the province of Huarochirin there is a large natural funnel-shaped excavation, about sixty feet deep, with a diameter of about eighty feet at the top. A dead mule is placed on the brink of the precipice. The tugging of the condors at the dead carcase causes it to fall into the hole; they follow it with greedy haste, and having gorged themselves with food, are unable again to rise from the narrow bottom of the funnel. Tschudi saw the Indians kill at once, with sticks, twenty-eight of the birds which had been thus entrapped. In a somewhat similar manner condors are caught in Peru, Bolivia, and Chili, as far as their range extends, and are frequently brought to Valparaiso and Callao, where they are sold for a few dollars to the foreign ships, and thence conveyed to Europe.

The condor, though a very large bird, about four feet long, and measuring at least three yards from tip to tip of his extended wings, is far from attaining the dimensions assigned to him by the earlier writers and naturalists, who, emulating Sindbad the Sailor, in his account of the roc, described him as a giant whose bulk darkened the air. Fortunately the works of nature do not require the exaggerations of fiction to be rendered interesting, and the marvels of organic nature which scientific inquiries reveal are far more wonderful than any which romancers may invent.

The condor reminds us of the Albatross. As the former sweeps in majestic circles high above the Andes, the latter soars gracefully over the ocean, "and without ever touching the water with his wings, rises with the rising billow and falls with the falling wave."* If the wonderful power of wing which bears the condor, often within the space of a few hours, from the sea-shore into the highest regions of the air, and the strength of breast which is able to support such changes of atmospheric pressure, may well raise our wonder, the indomitable pinions of the albatross are no less admirable. Both are unable to take wing from a narrow space, and both finally, so lordly in their movements,

* "The Sea and its Living Wonders," p. 139.

feed in the same ignoble manner, the condor pouncing from incredible distances upon the carcase of the mule or lama, while the albatross gorges upon the fat of the stranded whale.

While the condor is considered an enemy to man, the Gallinazos, Turkey-buzzards, or common American Carrion Vultures

Turkey Buzzard.

(*Vultur aura, V. urubu*), are very serviceable to him, by consuming the animal offals which, if left to putrefaction, would produce a pestilence. Thus they generally, in tropical America, enjoy the protection of the law, a heavy fine, amounting in some towns to 300 dollars, being imposed upon the offender who wantonly kills one of these scavengers. It is consequently not to be wondered at that, like domestic birds, they congregate in flocks in the streets of Lima, and sleep upon the roofs of the houses.

In 1808, Waterton saw the vultures in Angostura as tame as barn-fowls; a person who had never seen one would have taken them for turkeys. They were very useful to the citizens; had it not been for them, the refuse of the slaughter-houses would have caused an intolerable nuisance.

The Aura is dark-brown black, with a red and naked head and neck, covered with wrinkles and warts; the Urubu is very similar, only the head and neck are grey-black, but equally wrinkled and ugly. The latter ranges over South America in countless numbers, as D'Orbigny witnessed on a visit to a hacienda on the river Plata, where 12,000 oxen had been killed for salting. During this wholesale massacre, which lasted several months, the bones and entrails were cast along the banks of the stream, where at least 10,000 urubus had congregated to enjoy the banquet.

King Vulture.

It is a remarkable fact that, though hundreds of gallinazos may be feeding upon a carcase, they immediately retire when the King of the Vultures (*Sarcoramphus papa*) makes his appearance, who yet is not larger than themselves. Perching on the neighbouring trees, they wait till his majesty

— a beautiful bird, with head and neck gaudily coloured with scarlet, orange, blue, brown and white — has sufficiently gorged himself, and then pounce down with increased voracity upon their disgusting meal.

According to Humboldt, they are intimidated by the greater boldness of the sarcoramphus. The true reason of their homage, however, seems to be the fear they entertain for the more powerful beak of the "king," who, from a similar motive, gives way to the still mightier condor.

Among the sea-birds we find the same phenomenon. On the ice-bound coast of Nowaja Semlja, suffering no other bird in his vicinity, dwells the fierce Burgomaster (*Larus glauca*). None of its class dares dispute the authority of this lordly bird when it descends on its prey, though in the possession of another. It is the general attendant on the whale-fisher, whenever spoils are to be obtained. Then it hovers over the scene of action, and having marked out its morsel, descends upon it and carries it off on the wing.*

The Indians of Guiana sometimes amuse themselves with catching one of the urubus by means of a piece of meat attached to a hook, and decking him with a variety of strange feathers, which they attach to him with soft wax. Thus travestied, they turn him out again among his comrades, who, to their great delight, fly in terror from the nondescript; and it is only after wind and weather have stripped him of his finery that the outlaw is once more admitted into urubu-society.

When full of food this vulture, like the other members of his tribe, certainly appears an indolent bird. He will stand for hours together on a branch of a tree, or on the top of a house, with his wings drooping, or after rain, spreading them to catch the rays of the sun. But when in quest of prey, he may be seen soaring aloft on pinions which never flutter, and which at the same time carry him with a rapidity equal to that of the golden eagle. Scarcely has he espied a piece of carrion below, when, folding his broad wings, he descends with such speed as to produce a whistling sound, resembling that of an arrow cleaving the air.

The gallinazos when taken young can be so easily tamed

---

* "The Sea and its Living Wonders," 2nd edit. p. 136.

that they will follow the person who feeds them for many miles. D'Orbigny even mentions one of these birds that was so attached to its master that it accompanied him, like a dog, wherever he went. During a serious illness of its patron, the door of the bedroom having been left open, the bird eagerly flew in, and expressed a lively joy at seeing him again.

Relying on their inviolability, the gallinazos, like chartered libertines, are uncommonly bold, and during the distributions of meat to the Indians, which regularly take place every fortnight in the South American Missions, they not seldom come in for their share by dint of impudence. In Concepcion de Mojos, an Indian told M. D'Orbigny, who was present on one of those occasions, that he would soon have the opportunity of seeing a most notorious thief, well known by his lame leg; and the bird, making his appearance soon after, completely justified his reputation. The traveller was also informed that this urubu knew perfectly well the days of distribution in the different missions; and eight days later, while witnessing a similar scene at Magdalena, twenty leagues distant, he heard the Indians exclaim, and looking up saw his lame acquaintance of Concepcion hurrying to the spot, with the anxious mien of one that is afraid of missing a meal. The padres in both missions assured him that the vulture never failed to make his appearance at the stated time; a remarkable instance of memory, or highly developed instinct in a bird.

"If you dissect a vulture," says Waterton, "that has just been feeding on carrion, you must expect that your olfactory nerves will be somewhat offended with the rank effluvia from his craw, just as they would be were you to dissect a citizen after the lord-mayor's dinner. If, on the contrary, the vulture be empty at the time you commence the operation, there will be no offensive smell, but a strong scent of musk."

The Harpy Eagle (*Thrasaëtus harpyia*) is one of the finest of all the rapacious birds. The enormous developement of his beak and legs, and his consequent strength and power in mastering his prey, correspond with his bold and noble bearing, and the fierce lustre of his eye. His whole aspect is that of formidably organised power, and even the crest adds much to his terrific appearance.

"Among many singular birds and curiosities," says Mr.

Edwards, in his "Voyage up the Amazon," "that were brought to us, was a young harpy eagle, a most ferocious looking character, with a harpy's crest and a beak and talons in correspondence. He was turned loose into the garden, and before long gave us a sample of his powers. With erected crest and flashing eyes, uttering a frightful shriek, he pounced upon a young ibis, and quicker than thought had torn his reeking liver from his body. The whole animal world there was wild with fear."

Harpy Eagle.

The harpy attains a greater size than the common eagle. He chiefly resides in the damp lowlands of tropical America, where Prince Maximilian of Neu Wied met with him only in the dense forests, perched on the high branches. The monkey, vaulting by means of his tail from tree to tree, mocks the pursuit of the tiger-cat and boa, but woe to him if the harpy spies him out, for, seizing him with lightning-like rapidity, he cleaves his skull with one single stroke of his beak.

Fear seems to be totally unknown to this noble bird, and he defends himself to the last moment. D'Orbigny, relates that one day, while descending a Bolivian river in a boat with some Indians, they severely wounded a harpy with their arrows, so that it fell from the branch on which it had been struck. Stepping out of the canoe, the savages now rushed to the spot where the bird lay, knocked it on the head, and tearing out the feathers of its wings, brought it for dead to the boat. Yet the harpy awakened from his trance, and furiously attacked his persecutors. Throwing himself upon D'Orbigny he pierced his hand through and through with the only talon that had been left unhurt, while the mangled remains of the other tore his arm, which at the same time he lacerated with his beak. Two men were hardly able to release the naturalist from the attacks of the ferocious bird.

On turning from the New to the Old World, we find other but not less interesting raptorial birds sweep through the higher regions of the air in quest of prey. The gigantic oricou, or Sociable Vulture (*Vultur auricularis*), inhabits the greater part of Africa, and builds his nest in the fissures of rocks on the peaks of inaccessible mountains. In size he equals the condor, measuring upwards of ten feet across the wings expanded, and his flight is not less bold; leaving his lofty cavern at dawn,

Sociable Vulture.

he rises higher and higher, till he is lost to sight; but, though beyond the sphere of human vision, the telescopic eye of the bird is at work. The moment any animal sinks to the earth in death, the unseen vulture detects it. Does the hunter bring down some large quadruped, beyond his powers to remove, and leave it to obtain assistance? — on his return, however speedy, he finds it surrounded by a band of vultures, where not one was to be seen a quarter of an hour before.

Le Vaillant having once killed three zebras, hastened to his camp, at about a league's distance, to fetch a wagon; but on returning he found nothing but the bones, at which hundreds of oricous were busy picking. Another time, having killed a gazelle, he left the carcase on the sand, and retired into the bushes to observe what would happen. First came crows, who with loud croakings wheeled round the dead animal: then after a few minutes; kites and buzzards appeared, and finally he saw the oricous descending in spiral lines from an enormous height. They alighted upon the gazelle, and soon hundreds of raptorial birds were assembled. Thus the small robbers had first pointed out the way to those of middle size, who in their turn roused the attention of the bandits of a higher order; and none of them came too short, for after the powerful oricous had dismembered the carcase, some very good morsels remained for the buzzards, and the bones furnished excellent pickings for the crows.

The Bacha (*Falco bacha*, Daudin) inhabits India and Africa, where he sits for days on the peak of precipitous cliffs, on the look out for rock-rabbits (*Hyrax Capensis*). These poor animals, who have good reason to be on their guard, venture only with the greatest caution to peep out of their caves and crevices in which they take up their abode, and to which they owe their Dutch name of "klipdachs." Meanwhile the bacha remains immovable, as if he were part of the rock on which he perches, his head muffled up in his shoulders, but watching with a sharp eye every movement of his prey, until, finally, some unfortunate klipdachs venturing forth, he darts upon him like a thunderbolt. If this rapid attack proves unsuccessful, the bacha slinks away, ashamed, like a lion that has missed his spring, and seeks some new observatory, for he is well aware that no rock-rabbit in the neighbourhood will venture to stroll out during the remainder of the day.

But if he succeeds in seizing the klipdachs before it has time to leap away, he carries it to a rocky ledge, and slowly tears it to pieces. The terrible cries of the animal appear to sound like music in his ears, as if he were not only satisfying his hunger but rejoicing in the torments of an enemy. This scene of cruelty spreads terror far and wide, and for a long time no klipdachs will be seen where the bacha has held his bloody repast.

The Fishing Eagle of Africa (*Haliætus vocifer*), first noticed by Le Vaillant, may be seen hovering about the coasts and river-mouths of that vast continent. He is never found in the interior of the country, as the African streams are but thinly stocked with fish, which form his principal food. " Elastic and buoyant, this agile dweller in the air mounts to soaring heights, scanning with sharp and piercing eye the motions of his prey below. Energetic in his movements, impetuous in his appetites, he pounces with the velocity of a meteor on the object of his wishes, and with a wild and savage joy tears it to pieces. His whole sense of existence is the procuring of food, and for this he is ever on the alert, ever ready to combat, to ravage, and destroy."[*] He generally devours his prey on the nearest rock, and loves to return to the same spot where the bones of gazelles and lizards may be seen lying about, a proof that his appetite is not solely confined to the finny tribes. When these birds are sitting, they call and answer each other with a variously-toned shriek which they utter under curious movements of the head and neck.

While all other raptorial birds croak or shriek, the musical Sparrow-Hawk of Africa (*Melierax musicus*, Gray) pours forth his morning and evening notes to entertain his mate while she is performing the business of incubation. Every song lasts a minute, and then the hunter may approach, but during the

---

[*] A. Adams. "Notes of the Natural History of the Islands of the Eastern Archipelago. Narrative of the Voyage of H.M.S. Samarang."

pause he is obliged to remain perfectly quiet, as then the bird hears the least noise and immediately flies away.

The prowess of the Secretary-eagle *(Serpentarius cristatus)* attacking the most venomous serpents has already been mentioned in the chapter on these noxious reptiles. The long legs of this useful bird, which owes its name of secretary to the crest on the back of its head, reminding one of the pen stuck behind the ear, according to the custom of writing-clerks, might give one reason to reckon it, at first sight, among the cranes or storks, but its curved beak and internal organisation prove it to belong to the falcon tribe. Its feet being incapable of grasping, it keeps constantly on the ground in sandy and open places, and runs with such speed as to be able to overtake the most agile reptiles. The destruction it causes in their ranks must be great indeed, for Le Vaillant mentions that having killed one of these birds he found in its crop eleven rather large lizards, three serpents of an arm's length, and eleven small tortoises, besides a number of locusts, beetles, and other insects, swallowed most likely by way of dessert. What enviable powers of digestion!

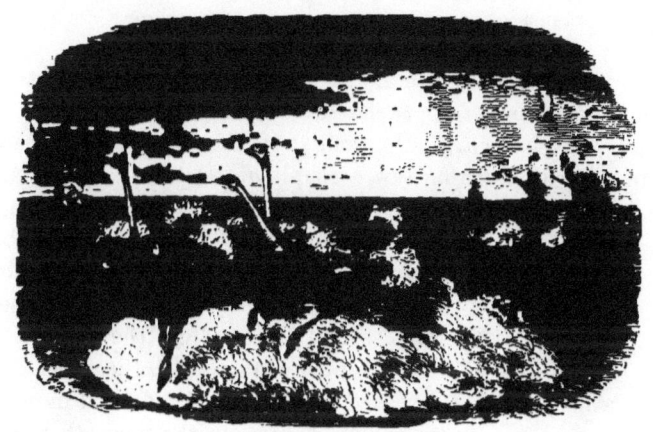

Ostrich catching.

CHAPTER XXXIV.

THE OSTRICH AND THE CASSOWARY.

Size of the Ostrich—Its astonishing Swiftness—Ostrich Hunting—Stratagem of the Ostrich for protecting its Young—The poisoned Arrow of the Bushmen—Enemies of the Ostrich—The White Vulture—Points of Resemblance with the Camel—Voice of the Ostrich said to resemble that of the Lion—Its Voracity—Ostrich Feathers—Bechuana Parasols—Domestication of the Ostrich in Algeria—The American Rheas—The Cassowary—The Australian Emu.

IN the African plains and wildernesses, where the lion seeks his prey, where the pachyderms make the earth tremble under their weighty strides, where the giraffe plucks the high branches of the acacia, and the herds of the antelope bound along: there also dwells the Ostrich, the king of birds, if size alone give right to so proud a title; for neither the condor nor the albatross can be compared in this respect to the ostrich, who raises his head seven or eight feet above the ground, and attains a weight of from two to three hundred pounds. His small and weak wings are incapable of carrying him through the air, but their flapping materially assists the action of his legs, and serves to increase his swiftness when, flying over the plain, he "scorns the horse and its rider." His feet appear hardly to touch the ground, and the length between each stride is not unfrequently from twelve to fourteen feet, so that for a time he might even outstrip a locomotive rushing along at full speed.

In Senegal, Adanson saw a couple of ostriches so tame that two negro boys could sit upon the largest of them. "Scarce had

Ostrich.

he felt the weight," says the venerable naturalist, "when he began to run with all his might, and thus they rode upon him several times round the village. I was so much amused with the sight, that I wished to see it repeated; and in order to ascertain how far the strength of the birds would reach, I ordered two full-grown negroes to mount upon the smallest of them and two others upon the strongest. At first they ran in a short gallop with very small strides, but after a short time they extended their wings like sails, and scampered away with such an amazing velocity that they scarcely seemed to touch the ground. Whoever has seen a partridge run knows that no man is able to keep up with him, and were he able to make greater strides his rapidity would undoubtedly be still greater. The ostrich, who runs like a partridge, possesses this advantage, and I am convinced that these two birds would have distanced the best English horses. To be sure they would not have been able to run for so long a time, but in running a race to a moderate distance they would certainly have gained the prize."

Not only by his speed is the ostrich able to baffle many an enemy, the strength of his legs also serves him as an excellent means of defence; and many a panther or wild dog coming within reach of his foot has had reason to repent of its temerity. But in spite of the rapidity of his flight, during which he frequently flings large stones backwards with his foot, and in spite of his strength, he is frequently obliged to succumb to man, who knows how to hunt him in various ways.

Unsuspicious of evil, and enjoying the full liberty of the desert, a troop of ostriches wanders through the plain, the monotony of which is only relieved here and there by a clump of palms, a patch of candelabra shaped tree-euphorbias, or a vast and solitary baobab. Some leisurely feed on the sprouts of the acacias, the hard dry leaves of the mimosas, or the prickly *Naras* whose deep orange-coloured pulp forms one of their favourite repasts; others agitate their wings and ventilate the delicate plumage, the possession of which is soon to prove so fatal to them. No other bird is seen in their company — for no other bird leads a life like theirs; but the zebra and the antelope are fond of associating with the ostrich, desirous perhaps of benefiting by the sharpness of his eye, which is capable of discerning danger at the utmost verge of the horizon. But in spite of its vigilance, misfortunes are already gathering round the troop, for the Bedouin has spied them out, and encircles them with a ring of his fleetest coursers. In vain the ostrich seeks to escape. One rider drives him along to the next, the circle gradually grows narrower and narrower, and, finally, the exhausted bird sinks upon the ground, and receives the death-blow with stoical resignation.

But the exertion of a long protracted chase is not always necessary to catch the ostrich, for before the rainy season, when the heat is at its height, he is frequently found upon the sand with outstretched wings and open beak, and allows himself to be caught after a short pursuit by a single horseman, or even by a swift-footed Betchuan.

To surprise the cautious seal the northern Eskimo puts on a skin of the animal, and imitating its motions mixes among the unsuspicious herd; and, in South Africa, we find the Bushman resort to a similar stratagem to outwit the ostrich. He forms a kind of saddle-shaped cushion, and covers it over with feathers, so as to resemble the bird. The head and neck of an ostrich are stuffed, and a small rod introduced. Preparing for the chase, he whitens his black legs with any substance he can procure, places the saddle on his shoulders, takes the bottom part of the neck in his right hand, and his bow and poisoned arrows in his left. Under this mask he mimics the ostrich to perfection, picks away at the verdure, turns his head as if keeping a sharp look out, shakes his feathers, now walks, and then trots, till he gets within bow-shot, and when the flock runs, from one receiving an arrow, he runs too. Sometimes, however, it happens that some wary old bird suspects the cheat, and endeavours to get near the intruder, who then tries to get out of the way, and to prevent the bird from catching his scent, which would at once break the spell. Should one of the birds happen to get too near in pursuit, he has only to run to windward, or throw off his saddle, to avoid a stroke from a wing which would lay him prostrate.

The Bushman frequently has recourse to a much simpler plan. Having discovered the nest of an ostrich, he removes the eggs as the first fruits of conquest, and then, concealing himself in the empty cavity, patiently waits for the return of the bird, which he generally despatches with one of those poisoned arrows which make incredible havoc among the wild herds of the bush or the savannah. According to Dr. Livingstone, the venom most generally employed is the milky juice of the tree euphorbia, which is particularly hurtful to the equine race. When it is mixed with the water of a pond, a whole herd of zebras will fall dead from the effects of the poison before they have moved away two miles; while on oxen and men it acts as a drastic purgative only. This substance is used all over the country, though in some places the venom of serpents and a certain bulb, *Amaryllis toxicaria*, are added, in order to increase the virulence. A slender reed only slightly barbed with bone or iron, but imbued with this poison, is sufficient to destroy the most powerful animal. Thus we find the African savage subdue the beasts of the field by similar means to those which are used by the wild nations on the banks of the Orinoco or the Amazon.

The ostrich generally passes for a very stupid animal, yet to protect its young it has recourse to the same stratagems which we admire in the plover, the oyster-catcher,* and several other strand-birds. Thus MM. Andersson and Galton, while traversing a barren plain, once hit upon a male and female ostrich, with a brood of young ones about the size of ordinary barn-door fowls. This was a sight they had long been looking for, having been requested by Professor Owen to procure a few craniums of the young ostrich, in order to settle certain anatomical questions; so forthwith dismounting from their oxen, they gave chase, which proved of no ordinary interest.

"The moment the parent birds became aware of our intention, they set off at full speed, the female leading the way, the young following in her wake, and the male, though at some little distance, bringing up the rear of the family-party. It was very

---

* "The Sea and its Living Wonders," p. 119.

touching to observe the anxiety the old birds evinced for the safety of their progeny. Finding that we were quickly gaining upon them, the male at once slackened his pace, and diverged somewhat from his course; but, seeing that we were not to be diverted from our purpose, he again increased his speed, and with wings drooping, so as almost to touch the ground, he hovered round us, now in wide circles, and then decreasing the circumference till he came almost within pistol-shot, when he abruptly threw himself on the ground and struggled desperately to regain his legs, as it appeared, like a bird that had been badly wounded. Having previously fired at him, I really thought he was disabled, and made quickly towards him. But this was only a ruse on his part; for on my nearer approach he slowly rose, and began to run in an opposite direction to that of the female, who by this time was considerably ahead with her charge. After about an hour's severe chase, we secured nine of the brood; and though it consisted of about double that number, we found it necessary to be contented with what we had bagged."*

While breeding, the ostrich likewise resorts to various artifices to remove intruders from its rude nest, which is a mere cavity scooped out a few inches deep in the sand and about a yard in diameter. Thus Professor Thunberg relates that riding past a place where a hen-ostrich sat on her nest, the bird sprang up and pursued him, in order to draw off his attention from her young ones or her eggs. Every time the traveller turned his horse toward her, she retreated ten or twelve paces, but as soon as he rode on, pursued him again. Is it not truly wonderful how parental affection at the approach of danger seems to rouse the intelligence of an animal to higher exertions, and to raise it above its usual sphere!

The instinct of the ostrich in providing food for its young is no less remarkable, for it is now proved that this bird, far from leaving its eggs, like a cold-blooded reptile, to be vivified by the sun, as was formerly supposed, not only hatches them with the greatest care, but even reserves a certain portion of eggs to provide the young with nourishment when they first burst into life: a wonderful provision, when we consider how

* Andersson, Lake Ngami.

difficult it would be for the brood to find any other adequate food in its sterile haunts. In Senegal, where the heat is extreme, the ostrich, it is said, sits at night only upon those eggs which are to be rendered fertile, but in extratropical Africa, where the sun has less power, the mother remains constant in her attentions to the eggs both day and night.

The number of eggs which the ostrich usually sits upon is ten; but the Hottentots, who are very fond of them, upon discovering a nest, seize fitting opportunities to remove one or two at a time; this induces the bird to deposit more, and in this manner she has been known, like the domestic hen, to lay between forty and fifty in a season. Thus also the Icelanders and other Norsemen force the eider-duck, by repeated plunderings of her nest, to divest herself of all her down; and when she has no more to spare, the gander willingly deprives himself of part of his snow-white and rose-red garment.

But the ostrich has other enemies besides the savage or the hungry traveller to fear for its young brood. Thus the natives about the Orange river assert that, when the birds have left their nest in the middle of the day in search of food, a white vulture may be seen soaring in mid air, with a stone between his talons. Having carefully surveyed the ground below him, he suddenly lets fall the stone, and then follows it in rapid descent. On running to the spot you will find a nest of probably a score of eggs, some of them broken by the vulture who used this ingenious device for procuring himself a dainty meal. This reminds one of a similar artifice through which the sea-eagle of the north overpowers even the ox; an artifice which Leopold von Buch * would willingly have doubted, had it not been frequently related to him, too circumstantially and too positively throughout the whole extent of the Norwegian isles. The eagle casts himself into the waves, arises with drooping wings, and rolls himself upon the beach until they are completely covered with sand and gravel. Then he mounts into the air, and soars above his victim. Close above its head he flings sand and pebbles into its eyes, and increases the terror of the brute by repeated strokes with his wings. The bewildered ox runs along with mad speed, and at length falls

---

* Travels in Norway.

down exhausted or precipitates himself from the cliffs, when the eagle quietly tears him to pieces.

Almost as soon as the chicks of the ostrich (which are about the size of pullets) have escaped from the shell, they are able to walk about and to follow the mother, on whom they are dependent for a long time. And here again we find a wonderful provision of nature in providing the young of the ostrich with a colour and a covering admirably suited to the localities they frequent. The colour is a kind of pepper and salt, agreeing well with the sand and gravel of the plains, which they are in the habit of traversing, so that you have the greatest difficulty in discerning the chicks even when crouching under your very eyes. The covering is neither down nor feathers, but a kind of prickly stubble, which no doubt is an excellent protection against injury from the gravel and the stunted vegetation amongst which they dwell.

The ostrich resembles in many respects the quadrupeds, and particularly the camel, so that it may almost be said to fill up the chasm which separates the mammalia from the birds, and to form a connecting link between them. Both the ostrich and the dromedary have warty excrescences on the breast upon which they lean whilst reposing, an almost similarly formed foot, the same muscular neck; and when we consider that they both feed upon the most stunted herbage, and are capable of supporting thirst for an incredibly long time, being, in fact, both equally well formed for living on the arid plains, it is certainly not to be wondered at that the ancients gave the ostrich a name betokening this similitude (*Struthio camelus*), and that the fancy of the Arabs ascribes its original parentage to a bird and to a dromedary.

"The usual cry of the ostrich," says Harris, "is a short roar, but when brought to bay it hisses like the gander;" and both Dr. Livingstone and Anderson affirm that its cry so greatly resembles that of the lion, as occasionally to deceive even the natives. Our celebrated countryman, who is evidently no friend of the lion, tells us that to talk of his majestic roar is mere twaddle, as the silly ostrich makes a noise as loud. He admits, however, that there is a great difference between the singing noise of a lion when full, and his deep, gruff growl when hungry, and that in general the lion's roar seems to come *deeper* from the chest than that of the ostrich. Has this similitude of voice never been noticed in the Zoological Gardens?

It is difficult to ascertain what the tastes of the ostrich may be while roaming the desert, but when in captivity no other bird or animal shows less nicety in the choice of its food, gobbling down with avidity stones, pieces of wood and iron, spoons, knives, and other articles of equally *light* digestion that may be presented to it. Thus it has always been far-famed for the wonderful powers of its stomach, and many amusing anecdotes are told of its voracity.

A batch of these birds having once been brought to a small town, for the inspection of the curious, a respectable matron, anxious to obtain a sight of the strange creatures, hastily shut up her house, and, key in hand, hurried to the spot where they were kept. Scarcely had she arrived, when one of them gravely stalked up, as if to thank her for her visit, and suddenly bending its long neck, to her horror, snatched the key out of her hand, and swallowed it in a trice; so that the indignant old lady — thus shut out of her own house — vowed that if all the beasts of Africa were to pass her door, she would not so much as open it to look at them.

"Nothing," says Methuen, speaking of a domesticated ostrich, "disturbed its digestion — dyspepsia (happy thing) was undreamt of in its philosophy. One day a Muscovy-duck brought a promising race of ducklings into the world, and with maternal pride conducted them forth into the yard. Up with solemn and measured stride walked the ostrich, and, wearing the most mild and benignant cast of face, swallowed them all, one after the other, like so many oysters, regarding the indignant hissings and bristling plumage of the hapless mother with stoical indifference."

Baron Aucapitaine, to whom we owe an interesting account of the ostrich, relates that he every evening used to regale a tame ostrich with a newspaper, which the bird completely swallowed, thus literally stuffing itself with all the knowledge of the day.

The costly white plumes of the ostrich, which are chiefly obtained from the wings, form a considerable article of commerce, having been prized in all ages for the elegance of their long, waving, loose, and flexible barbs. At the Cape the price varies from one or two guineas to twelve for the pound, the latter sum, however, being only paid for very prime feathers. The thinner the quill and the longer and more wavy the plume, the more it is prized. From seventy to ninety feathers go to the pound; but a single bird seldom furnishes more than a dozen, as many of them are spoilt by trailing or some other accident. The vagrant tribes of the Sahara sell their ostrich plumes to the caravans which annually cross the desert, and convey them to the ports of the Mediterranean. Here they were purchased as far back as the twelfth or thirteenth century, by the Pisanese or Genoese merchants, through whose agency they ultimately crossed the Alps to decorate the stately *burggräfinnen* of the Rhine, or the wives of the opulent traders of Augsburg or Nuremberg. At a still more remote period the Phœnicians brought ostrich-feathers from Ophir to Tyre, whence they were distributed among the princes of the Eastern world.

The Damaras and the Bechuanas manufacture handsome parasols from the black body feathers of the ostrich, which, besides affording protection from the sun's rays, not unfrequently prove serviceable in the chase, and being stuck into the ground at the proper moment, divert the attention of a lion from the object of his vengeance, and thus enable the rest of the party to rush in and despatch him with their assegais. Contrary to our European notions, which assign ornament and a tender care for the complexion more particularly to the fair sex, we here find the men, "whose skin," says Harris, "somewhat coarser than the hide of a rhinoceros, might vie in point of colour with a boot," exclusively guard their complexion with these elegant umbrellas.

The thick skin of the ostrich, decked with its coat of feathers, is likewise a prized article in African trade, and, according to Baron Aucapitaine, fetches from seventy-five to ninety francs in Tripoli and Tunis.

In the *Tell*, or the cultivated coast districts of Algeria, the ostrich is often domesticated, particularly on account of its eggs, which weigh three pounds, and are equivalent to twenty-four of the common fowl's eggs. It might be supposed that one of these giant eggs would be too much for the most vigorous appetite, yet Anderson saw two natives despatch five of them in the course of an afternoon, besides a copious allowance of flour and fat. According to the taste of this Swedish Nimrod, they afford an excellent repast; while Dr. Livingstone tells us they have a strong disagreeable flavour, which only the keen appetite of the desert can reconcile one to. "The Hottentots use their trousers to carry home the twenty or twenty-five eggs usually found in a nest; and it has happened that an Englishman, intending to imitate this knowing dodge, comes to the wagons with blistered legs, and, after great toil, finds all the eggs uneatable from having been some time sat upon."

Even the egg-shell has its value, and is an excellent vessel for holding liquids of any kind. The Bushmen have hardly any other household utensil. By covering it with a light network it may be carried slung across the saddle. Grass and wood serve as substitutes for corks.

The flesh of the ostrich is decidedly coarse, but as there is no accounting for tastes, the Romans seem to have prized it; and Firmus, one of their pseudo-emperors, most likely desirous of emulating the gormandising powers of the bird on which he fed, is said to have devoured a whole ostrich at *one sitting.*

Though not possessing the true camel-bird, America has the large *Rheas*, which from their size and similar habits have been styled the ostriches of the New World, though differing in many essential characters. One species, the *Rhea Darwinii*, inhabits Patagonia, while the Emu or Nandu (*Rhea Americana*) is found throughout the whole eastern part of South America, from Buenos Ayres to the Orinoco, wherever open plains, pampas, campos, or savannas, invite it to take up its residence. The nandu is not near so tall as the true ostrich, scarcely rising above four feet, and is of a uniform grey colour except on the back, which has a brown tint. The back and rump are furnished with long feathers, but not of the same rich and costly kind as those which adorn the African ostrich. Its feeble wings merely serve to accelerate its flight, serving it as oars or sails, particularly when running with the wind. "It is not easily caught," says the Prince of Neu Wied, "as it not only runs very fast, but in zigzag lines, so that the horse, rendered giddy by so many evolutions, at length drops down with its rider."

The Indian Archipelago and New Holland have likewise their peculiar struthionidous birds.

THE CASSOWARY AND EMU

The galeated Cassowary (*Casuarius galeatus*), thus called from its head being surmounted by a kind of horny helmet, is a native of Java and the adjacent isles. The skin of the head and upper part of the neck is naked, of a deep blue and fiery red tint, with pendant caruncles similar to those of the turkey-cock. It is much inferior in size to the ostrich, and its wings are reduced to so rudimentary a state, consisting merely of five long bristles, without any plumes, that they are even unable to assist it in running.

Cassowary.

All its feathers are of the same kind, being entirely designed for covering, and resemble at a little distance a coat of coarse or hanging hair. It feeds on fruits, eggs of birds, and tender herbage, and is said to be as voracious as the ostrich.

The cassowary is a very swift runner; striking out alternately with one of its robust and powerful legs, it projects its body violently forward with a bounding motion far surpassing the speed of the horse.

The Australian Emu (*Dromaius Novæ Hollandiæ*) is allied to the cassowary, though differing in many external characters. Both the helmet, and the long pens or quills observable in the wings of the latter, are here wanting;

Emu.

its neck and legs are longer, its feathers, for the most part grey and brown mixed, are not so filiform, and its beak also is differently shaped. In size it more nearly approaches the ostrich, rising to a height of seven feet, and from its great muscular power is able to run so quickly as to distance the swiftest greyhound. Incessant persecutions have driven it far away from the colonised parts of the country; but it has still a vast range in the wilds of the interior. It lives on fruits, eggs, and even small animals, which it swallows entire.

Parrots.

CHAPTER XXXV.

PARROTS.

Their Peculiar Manner of Climbing—Points of Resemblance with Monkeys—Their Social Habits—Their Connubial Felicity—Inseparables—Talent for Mimicry—Wonderful Powers of Speech and Memory—Their Wide Range within the Temperate Zones—Colour of Parrots Artificially Changed by the South American Indians—The Cockatoos—Cockatoo killing in Australia—The Macaw—The Parakeets.

THE parrots have so many points of resemblance with the monkeys in their tastes and habits, that, notwithstanding their different appearance, one might almost be tempted to call them near relations.

As the monkey seldom or never sets his foot on even ground, but climbs or springs from branch to branch, thus also the parrot will rarely be seen walking; his flight is rapid, but generally only of short duration, so that evidently neither the ground nor the air were destined for his habitual abode. In climbing, however, he shows an uncommon expertness and agility, unlike that of any other quadruped or bird, as the organ he chiefly uses for the purpose is his beak.

He first seizes with his powerful mandibles the branch he intends to ascend, and then raises his body one foot after the other; or, if he happens to have a sweet nut in his bill which he is anxious to preserve, he presses his lower mandible firmly upon the branch, and raises himself by the contraction of the muscles of his neck. On descending, he first bends his head, lays the back of his beak upon the branch, and while the extended neck supports the weight of the body, brings down one foot after the other.

While accidentally walking on even ground, he also frequently uses his upper mandible as a kind of crutch, by fixing its point or its back upon the ground; for the formation of his toes is such, that he can walk but very slowly, and consequently requires the aid of that singular support. Thus monkeys and parrots are, in the fullest sense of the word, dendritic animals — the free children of the primitive forest. But if the toes of the parrot are but ill adapted for walking, they render him valuable services, in seizing or grasping his food. They even form a kind of hand, with which, like the monkey, he conveys the morsel to his beak. This easily cracks the hardest nutshells, after which the broad and fleshy tongue adroitly extracts the kernel.

In his free state the parrot lives only upon nuts and seeds; when captive, however, he becomes omnivorous, like man his master, eats bread and meat, sugar and pastry, and is very fond of wine, which has a most exhilarating effect on his spirits.

Like most monkeys, the parrots are extremely social. At break of day they generally rise in large bands, and with loud screams fly away to seek their breakfast. After having feasted together, they retire to the shady parts of the forest as soon as the heat begins to be oppressive, and a few hours before the setting of the sun, reappear in large troops.

"Every day," says Le Vaillant, "the African damask parrots (*Psittacus infuscatus*) fly to the water at the same hour to bathe themselves—in which operation they take great delight: all the flocks of the whole canton assemble towards evening with much noise and animation—and this is the signal for their visit to the water, which is often at a great distance, since no other than the purest water will please them. They are then seen huddling or rolling over each other, pell mell, on the banks of the water, frolicking together, dipping their heads and wings into the water in such a manner as to scatter it over all their plumage, and exhibiting a most entertaining spectacle to the observer. This ceremony being over, they revisit the trees on which they previously assembled, where they sit in order to adjust and clean their feathers; and this being finished, they fly off in pairs, each pair seeking its particular retreat in the wood, where they wait till morning."

If the monkeys are distinguished by a strong affection for their young, the parrots may well be cited as models of connubial love, for when once a pair has been united, its attachment remains unaltered unto death.

Far more than the turtle-dove, the little passerine parrot (*Psittacus passerinus*) of Brazil, or the *Psittacus pullarius*, or love-parrot of Guinea, deserved to be celebrated by poets as the emblem of conjugal affection. Never seen but in each other's company, each delights to imitate the actions of the other, feeding, sleeping, bathing together; and when one dies, the other soon follows its partner. A gentleman who had lost one of a pair of these inseparables, attempted to preserve the other by hanging up a looking-glass in its cage. At first the joy of the poor bird was boundless, as he fancied his mate restored to his caresses; but soon perceiving the deception, he pined away and died.

Another point of resemblance between the parrots and monkeys is their talent for mimicry; but while the latter, favoured by the similarity of their organisation to that of man, strive to copy his gestures and actions, the former endeavour to imitate his voice and to repeat his words, an attempt facilitated by the extreme mobility of their tongue and upper mandible, no less than by the peculiar construction of their larynx or windpipe.

These imitative instincts appear the more remarkable when we consider that both the monkeys and the parrots have no pursuits that necessarily bring them into closer connection with man. They are comparatively useless to him, live almost constantly at a distance from his haunts in the depths of the forests, and are so far from seeking his company, that they retreat as fast as they can on seeing him approach. How comes it, then, that they have been gifted with their wonderful ability to imitate his language and his actions, and of what use is it to them or to us?

The talent of speech has not been given to all the parrots alike. The beautiful American aras, for instance, are in this respect remarkably stupid, while the purple lory of the East Indies, and the grey African parrot (*Psittacus erithacus*), are remarkable for their linguistic attainments. They are often able to retain whole songs and sentences, and to repeat them with astonishing exactness. Thus Le Vaillant mentions a grey parrot he saw at the Cape, who was able to repeat the whole of the Lord's Prayer in Dutch, throwing himself at the same time on his back, and folding the toes of both his feet.

In a town of Normandy the wife of a butcher had beaten her little child so unmercifully that it died in consequence of the ill-treatment it had undergone. No enquiry was made on the subject; but a grey parrot that inhabited the opposite house of a shoemaker constantly repeated, "Why do you beat me? why do you beat me?" in so plaintive and imploring a tone, that the passers-by rushed into the shoemaker's shop and loudly exclaimed against his supposed barbarity. To exculpate himself, the poor shoemaker pointed to his parrot, and told the story of the child; so that after a short time the woman, against whom the indignation of the public was now fully roused, was obliged to leave the town.

Willoughby mentions a grey parrot who, on being told to laugh, immediately burst into a loud fit of laughter, and soon after exclaimed, "How impertinent, to order me to laugh!" Another parrot that was kept in an earthenware shop never failed, when an article was broken or ill-used, to exclaim, "How clumsy!" Buffon mentions a parrot who, having been taught to speak during the passage by an old sailor, had so completely adopted his gruff voice and husky cough, as to be mistaken for the weather-beaten tarpaulin himself. Although the bird was afterwards presented to a young lady, and no more heard the voice of its first instructor, it did not forget his lessons, and nothing could be more ludicrous than to hear it suddenly pass from the sweet tones of its fair mistress to the rough accents of its first teacher.

The grey parrot not only imitates the voice of man, but has also a strong desire to do so, which he manifests by his attention in listening, and by the continuous efforts he makes to repeat the phrases he has heard. He seems to impose upon himself a daily task, which even occupies him during sleep, as he speaks in his dreams. His memory is astonishing, so that a cardinal once gave a hundred gold crowns for one of these birds that correctly repeated a long prayer; and M. de la Borde told Buffon he had seen one that was fully able to perform the duties of a ship's chaplain.

All parrots are more or less susceptible of education, and, particularly when caught young, grow very much attached to the master that feeds them. Those that are sent to Europe are generally taken from the nest, and have thus never experienced the sweets of freedom; but they are also frequently caught full grown. The American Indians know how to strike them with small arrows, whose points are blunted with cotton, so as to stun without killing them; or else, under the trees on which they perch, they light a fire of strong-smelling weeds, whose vapours cause them to drop to the ground. These captives are frequently extremely stubborn; but blowing the fumes of tobacco into their face until they fall asleep, is an infallible remedy to cure them of their obstinacy, this operation being so little to their taste that it need hardly ever be repeated twice.

Parrots are known to attain a very great age. One that was brought to Florence in 1633, and belonged to the Grand Duchess of Tuscany, died in 1743, having thus lived more than a century in exile.

Although preeminently tropical, like the colibris, several parrots range far within the temperate zone, as they are found in the Southern hemisphere at the Straits of Magellan and on the Macquarie Islands, and in the Northern, in the neighbourhood of Cairo and in Kentucky, where the Carolina parrot is often seen in great numbers during the summer.

The parrots are subdivided into numerous groups and species, chiefly according to the various forms of their bills and tails. The short-tailed parrots of the Old World mostly display bright or gaudy colours, such as the Lories, which owe their name to the frequency with which they repeat this word, while the American species are generally green. The Indians have, however, found out an ingenious method to adorn or *illustrate*, as it were, the plumage of the Amazonian parrot (*Psittacus Amazonicus*), which is in great request, from its being easily tamed, and learning to speak with facility. They take a young bird from its nest, pluck the feathers from its back and shoulders, and then rub the naked parts with the blood of a small species of frog. The feathers which grow again after this operation are no longer green, but yellow, or of a bright red colour. Many birds die in consequence of being plucked, and thus these metamorphosed parrots are extremely rare, notwithstanding the high prices which the savages obtain for them. Azara, in his work on South America, relates that the Indians of the warm regions of Paraguay sometimes sell to the Europeans completely yellow parrots with the exception of the few parts that were originally blue and red, and in which that colour is said to be substituted for green by means of arnatto. The parrots whose plumage has thus been artificially changed are ever after silent, melancholy, and of very weak health. By what chance may the Indians have hit upon this strange industry, worthy of the refined arts of our own dealers in natural curiosities?

Lory.

The Cockatoos are distinguished from the other parrots by a crest or tuft of elegant feathers on the head, which they can raise and depress at pleasure. They inhabit the East Indies and Australia, and have generally a white or roseate plumage. Their chief resorts are dense and humid forests, and they frequently cause great devastations in the rice plantations, often pouncing to the number of six or eight hundred upon a single field, and destroying even more than they devour, as they seem to be possessed of the mania to break and tear everything their beak can lay hold of. They walk less awkwardly than most other parrots.

The great white cockatoo (*Cacatua Cristata*), who is able to erect his beautiful yellow crest to the height of five inches, as a cock does his comb, is the species most frequently seen in Europe. This bird is half-domesticated in several parts of India, as it builds its nest under the roofs of houses, and this tameness results from its intelligence, which seems superior to that of other parrots. It listens attentively, but vainly strives to repeat what is said.

As Australia, the land of anomalies in natural history, possesses a black swan, it also gives birth to a splendid black cockatoo (*Cacatua Banksii*), the finest and the rarest of the whole genus.

Captain Grey gives us an animated description of cockatoo-killing in Australia. "Perhaps the finest sight that can be seen, in the whole circle of native sports, is the killing cockatoos with the kiley or boomerang. A native perceives a large flight of cockatoos in a forest which encircles a lagoon: the expanse of water affords an open clear space above it, unencumbered with trees, but which raise their gigantic forms all around, more vigorous in their growth from the damp soil in which they flourish; and in their leafy summits sit a boundless number of cockatoos, screaming and flying from tree to tree, as they make their arrangements for a night's sound sleep. The native throws aside his cloak, so that he may not even have this slight covering to impede his motions, draws his kiley from his belt, and with a noiseless, elastic step, approaches the lagoon, creeping from tree to tree, from bush to bush, and disturbing the birds as little as possible. Their sentinels, however, take the alarm; the cockatoos farthest from the water fly to the trees near its edge, and thus they keep concentrating their forces as the native advances; they are aware that danger is at hand, but are ignorant of its nature. At length the pursuer almost reaches the edge of the water, and the scared cockatoos with wild cries spring into the air; at the same instant the native raises his right hand high over his shoulder, and bounding forward with his utmost speed for a few paces to give impetus to his blow, the kiley quits his hand as if it would strike the water, but when it has almost touched the unruffled surface of the lake, it spins upwards with inconceivable velocity, and with the strangest contortions. In vain the terrified cockatoos strive to avoid it; it sweeps wildly and uncertainly through the air (and so eccentric are its motions, that it requires but a slight stretch of the imagination to fancy it endowed with life), and with fell swoops is in rapid pursuit of the devoted birds, some of whom are almost certain to be brought screaming to the earth. But the wily savage has not yet done with them; he avails himself of the extraordinary attachment which these birds have for one another, and fastening a wounded one to a tree, so that its cries may induce its companions to return, he watches his opportunity, by throwing his kiley or spear, to add another bird or two to the booty he has already obtained."

The magnificent Macaws, or Aras, of South America are distinguished by having their cheeks destitute of feathers, and their tail feathers long. Their size and splendid plumage render them fit ornaments of princely gardens, but their loud and piercing screams would prove a great annoyance to the inmates of humbler dwellings.

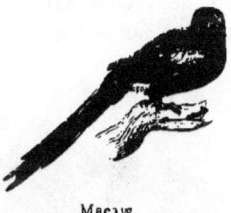

Macaw.

"Superior in size and beauty to every parrot of South America," says Waterton, " the ara (*Macrocercus Macao*) will force you to take your eyes from the rest of animated nature, and gaze at him: his commanding strength; the flaming scarlet of his body; the lovely variety of red, yellow, blue, and green, in his wings; the extraordinary length of his scarlet and blue tail, seem all to form and demand for him the title of emperor of all the parrots. He is scarce in Demerara, till you reach the confines of the Macoushi country; there he is in vast abundance: he mostly feeds on trees of the palm species. When the concourites have ripe fruit on them, they are covered with this magnificent parrot: he is not shy or wary; you may take your blowpipe and quiver of poisoned arrows, and kill more than you are able to carry back to your hut. They are very vociferous, and, like the common parrots, rise up in bodies towards sunset, and fly two and two to their place of rest. It is a grand sight in ornithology to see thousands of aras flying over your head, low enough to let you have a full view of their flaming mantle. The Indians find their flesh very good, and the feathers serve for ornaments in their head-dresses. They breed in the holes of trees, and are easily reared and tamed."

The Paroquets, or Parakeets, are smaller than the common

parrots, and have longer tails. There are numerous species, some distinguished by a very long pointed tail, and collar-like mark round the neck, which inhabit the Asiatic continent and islands; and others, natives of Australia, which are distinguished by their colour being gorgeously variegated and peculiarly mottled on the back, by their tail feathers not being pointed, and by their being furnished with elongated tarsi adapted for running on the ground.

To the former belongs the beautiful ring paroquet, which is supposed to have been the first bird of the parrot kind known to the ancient Greeks, having been brought from the island of Ceylon, after the Indian expeditions of Alexander the Great;

Green Parakeet.

to the latter, the elegant green parakeet, which in the hot seasons congregates about the pools in almost incredible numbers. Though capable of a rapid and even flight, and frequently at great altitudes, it is generally found running over the ground, and treading its way among the grasses to feed on the seeds. It can easily be domesticated, and a more elegant or beautiful pet can scarcely be conceived.

It is a strange fact that the parrots, that will eat nux vomica without danger, expire in convulsions after having tasted parsley, another proof of the truth of the saying that what is poison for one creature is food for another.

Caravan.

CHAPTER XXXVI.

THE CAMEL.

The Ship of the Desert—Paramount Importance of the Camel in the Great Tropical Sand-wastes—Its Organisation admirably adapted to its Mode of Life—Horrors and Beauties of the Desert—The Camel an Instrument of Freedom—The Robber Bedouin—Immemorial Thraldom of the Camel—Its Unamiable Character—Excuses that may be urged in its Behalf.

THERE is a sea without water and refreshing breezes, without ebb and flood, without fishes and algæ! And there is a ship which safely conveys goods and passengers from one shore to the other of that sea, a ship without sails or masts, without keel or rudder, without screw or paddle, without cabin or deck!

This ship so swift and sure is the Dromedary, and that sea is the desert; which none but he, or what he carries, can pass.

In many respects, the vast sandy deserts of Africa and Asia remind one of the ocean. There is the same boundless horizon, the same unstable surface, now rising, now falling with the play of the winds; the same majestic monotony, the same optical illusions, for as the thirsty mariner sees phantom palm-groves rise from the ocean, thus also the sand-waste transforms itself, before the panting caravan, into the semblance of a refreshing lake. Here we see islands, verdant oases of the sea — there, oases, green islands of the desert; here, sand billows — there, water waves, separating widely different worlds of plants and animals; here, the ship, the camel of the ocean — there, the dromedary, the ship of the desert!

But for this invaluable animal, the desert itself would ever have remained impassable and unknown to man. On it alone depends the existence of the nomadic tribes of the Orient, the whole commercial intercourse of North Africa and Southwest Asia, and no wonder that the Bedouin prizes it, along with the fruit-teeming date-palm, as the most precious gift of Allah. Other animals have been formed for the forest, the water, the savannah; to be the guide, the carrier, the companion, the purveyor of all man's wants in the desert, is the camel's destiny.

Wonderfully has he been shaped for this peculiar life, formed to endure privations and fatigues under which all but he would sink. On examining the camel's foot, it will at once be seen how well it is adapted for walking on a loose soil, as the full length of its two toes is provided with a broad, expanded, and elastic sole. Thus the camel treads securely and lightly over the unstable sands, while he would either slip or sink on a muddy ground. He can support hunger longer than any other mammiferous animal, and is satisfied with the meanest food. Frugal, like his lord the wiry Bedouin, the grinding power of his teeth and his cartilaginous palate enable him to derive nutriment from the coarsest shrubs, from thorny mimosas and acacias, or even from the stony date-kernels, which his master throws to him after having eaten the sweet flesh in which they are imbedded.

For many days he can subsist without drinking, as the pouch-like cavities of his stomach — a peculiarity which distinguishes him from all other quadrupeds, perhaps, with the sole exception of the elephant — form a natural cistern or reservoir, whose contents can be forced upwards by muscular contraction, to meet the exigencies of the journey. It is frequently believed that this liquid remains constantly limpid and palatable, and that in cases of extreme necessity camels are slaughtered to preserve the lives of the thirsty caravan, but according to Russegger (Travels in Nubia) these accounts are fabulous, as, particularly after a long abstinence from drinking, the dromedary's supply is nothing but a most nauseous mixture of putrid water and half-digested food, from which even Tantalus would turn away disgusted. But the "ship of the desert" is not only provided with water for the voyage, but also with liberal stores of fat, which are chiefly accumulated in the hump; so that this prominence, which gives it so deformed an appearance, is in reality of the highest utility — for should food be scarce, and this is almost always the case while journeying through the desert, internal absorption makes up in some measure for the deficiency, and enables the famished camel to brave for some time longer the fatigues of the naked waste. Yet all mortal endurance has its limits, and even the camel, though so well provided against hunger and thirst, must frequently succumb to the excess of his privations, and the bleached skeletons of the much-enduring animal strewed along the road mark at once the path of the caravan and the dreadful sufferings of a desert-journey.

But even these horrid wastes, where the glowing Khamsin whirls the sands in suffocating eddies, have beauties of their own. Particularly when the full moon shines in the dark blue sky, bespangled with constellations of a brilliancy unknown to the northern firmament, when the mountains throw their dark shades far away over the yellow sands, and the picturesque effect of the scene is enhanced by the aspect of the tents, the watch-fires, and the reposing animals; then we may well conceive how the wandering Bedouin loves the desert no less than the mariner loves the ocean, or the Swiss peasant his snow-clad mountains, and how it inflames the imagination of the oriental poets to many a song, solacing the tediousness of the encampment, and handed on from one generation to another.

To the camel the vagrant Arab owes his immemorial liberty and independence; when attacked, he places at once the desert between the enemy and himself. Thus he has ever been indomitable, and when in other parts of the world we find that the fatal possession of an animal — the sable, the sea-otter — has entailed the curse of slavery upon whole nations, the dromedary in Arabia appears as the instrument of lasting freedom. Many a conquering horde has been stopped in its career by the desert, and while the false glory of the scourges of mankind that have so often thrown the East into bondage passed like a shadow, one century after another looks down from the heights of Sinai upon the free and unfettered sons of Ismael.

But the Arab too often tarnishes his liberty by crime, and degrades the "ship of the desert" to be the accomplice of a robber. The Bedouin anxious to pursue this base profession inures himself, from an early age, to every fatigue, banishes sleep, patiently endures thirst, hunger, and heat; and in the same manner accustoms his dromedary to every privation. A few days after the animal's birth he folds its legs under its body, forces it to kneel; and loads it with a weight which is gradually increased as it increases in strength. Instead of allowing it to seek its food whenever it pleases, or completely to satisfy its thirst, he accustoms it to perform longer and longer journeys without eating or drinking, trains it to equal the horse in swiftness, as it surpasses him in strength — and when perfectly assured of its fleetness and endurance, loads it with the necessary provisions, rides away upon its back, waylays the traveller, plunders the secluded dwelling, and when pursued and forced to save his booty by a speedy flight, then shows what he and his dromedary can perform. Hurrying on day and night, almost without repose, or eating or drinking, he travels two hundred leagues in a week, and during this whole time his dromedary is allowed but one hour's rest a day, and a handful of meal for food. On this meagre diet the unwearied animal often speeds on seven or eight days without finding any water, and when by chance a pool or a source lies on his way, he smells it at the distance of half a league, his burning thirst imparts new vigour to his speed, and he then drinks at once both for the past and the future, as his journeys often last several weeks, and his privations endure as long as his journeys!

Bactrian Camel.

While the Bactrian Camel with a double hump ranges from Turkestan to China, the single-hump camel or dromedary, originally Arabian, has spread in opposite directions towards the East Indies, the Mediterranean, and the Niger, and is used in Syria, Egypt, Persia, and Barbary, as the commonest beast of burden. It serves the robber, but it serves also the peaceful merchant, or the pilgrim, as he wanders to Mecca to perform his devotions at the prophet's tomb. In long array, winding like a snake, the caravan traverses the desert. Each dromedary is loaded, according to its strength, with from six hundred to a thousand pounds, and knows so well the limits of its endurance, that it suffers no overweight, and will not stir before it be removed. Thus, with slow and measured pace, the caravan proceeds at the rate of ten or twelve leagues a day, often requiring many a week before attaining the end of its journey.

Dromedary.

When we consider the deformity of the camel, we cannot doubt that its nature has suffered considerable changes from the thraldom and unceasing labours of more than one millenium. Its servitude is of older date, more complete, and more irksome, than that of any other domestic animal — of older date, as it inhabits the countries which history points out to us as the cradle of mankind; more complete, as all other domestic animals still have their wild types roaming about in unrestrained liberty, while the whole camel race is doomed to slavery; more irksome, finally, as it is never kept for luxury or state like so many horses, or for the table like the ox, the pig, or the sheep, but is merely used as a beast of transport, which its master does not even give himself the trouble to attach to a cart, but whose body is loaded like a living wagon, and frequently even remains burdened during sleep.

Thus, the camel bears all the marks of serfdom. Large naked callosities of horny hardness cover the lower part of the breast and the joints of the legs, and although they are never wanting, yet they themselves give proof that they are not natural, but that they have been produced by an excess of misery and ill-treatment, as they are frequently found filled with a purulent matter.

The back of the camel is still more deformed by its single or double hump than its breast or legs by their callosities; and as the latter are evidently owing to the position in which the heavily burthened beast is forced to rest, it may justly be inferred that the hump also, which merely consists of an accumulation of fat, did not belong to the primitive animal, but has been produced by the pressure of its load. Even its evident use as a store-house for a desert journey may have contributed to its development, as nature is ever ready to protect its creatures, and to modify their forms according to circumstances; and thus, what at first was a mere casual occurrence, became at length, through successive generations, the badge and heir-loom of the whole race.

Even the stomach may, in the course of many centuries, have gradually provided itself with its water-cistern, since the animal, after a long and tormenting privation, whenever an opportunity of satisfying its thirst occurred, distended the coats of that organ by immoderate draughts, and thus, by degrees, gave rise to its pouch-like cavities.

The hardships of long servitude, which have thus gradually deformed the originally, perhaps, not ungraceful camel, have no doubt also soured its temper, and rendered its character as unamiable as its appearance is repulsive.

"It is an abominably ugly necessary animal," says Mr. Russell, in a letter dated from the camp of Lucknow; "ungainly, morose, quarrelsome, with tee-totalling propensities; unaccountably capricious in its friendships and enmities; delighting to produce with its throat, its jaws, its tongue, and its stomach, the most abominable grunts and growls. Stupidly bowing to the yoke, it willingly submits to the most atrocious cruelties, and bites innocent, well-meaning persons, ready to take its part. When its leader tears its nostril, it will do no more than grunt; but ten against one it will spit at you if you offer it a piece of bread. For days it will march along, its nose close to the tail of the beast that precedes it, without ever making the least attempt to break from the chain; and yet it will snort furiously at the poor European who amicably pats its ragged hide."

The camel seems to have been rather harshly dealt with in this description; at any rate, it may plead for its excuse that it would be too much to expect a mild and amiable temper in a toil-worn slave.

Giraffes and Zebras.

CHAPTER XXXVII.

THE GIRAFFE AND THE ZEBRA.

Beauty of the Giraffe—Its Wide Range of Vision—Use of its Horns—Giraffe Hunting—The Giraffes of the Zoological Gardens—The Quagga—The Douw—The Zebra—Its Lamentable Wailings—Its Inaccessible Retreats.

WHICH of all four-footed animals raises its head to the most towering height? Is it the colossal elephant or the "ship of the desert?" No doubt the former reaches many a lofty branch with its flexible proboscis, and the eye of the long-necked camel sweeps over a vast extent of desert; but the Giraffe embraces a still wider horizon, and plucks the leaves of the *mokaala* at a still greater height. A strange and most surprising animal, almost all neck and leg, seventeen feet high against a length of only seven from the breast to the beginning of the tail, its comparatively small and slanting body resting on long stilts, its diminutive head fixed at the summit of a column; and yet in spite of these apparent disproportions, which seem rather to belong to the world of chimeras than to the realities of nature, of so elegant and pleasing an appearance, that it owes its Arabic name, *Xirapha*, to the graceful ease of its movements.

The beauty of the giraffe is enhanced by its magnificently spotted skin, and by its soft and gentle eyes, which eclipse those of the far-famed gazelle of the East, and, by their lateral projection, take in a wider range of the horizon than is subject to the vision of any other quadruped, so as even to be able to anticipate a threatened attack in the rear from the stealthy lion or any other foe of the desert.

The long black tail, invariably curled above the back, no doubt renders it good service against many a stinging insect; and the straight horns, or rather excrescences of the frontal bone, small as they are, and muffled with skin and hair, are by no means the insignificant weapons they have been supposed to be. "We have seen them wielded by the males against each other with fearful and reckless force," says Maunder, in his excellent "Dictionary of Animated Nature," "and we know that they are the natural arms of the giraffe most dreaded by the keeper of the present living giraffes in the Zoological Gardens, because they are most commonly and suddenly put in use. The giraffe does not butt by depressing and suddenly elevating the head, like the deer, ox, or sheep, but strikes the callous obtuse extremity of the horns against the object of his attack with a sidelong sweep of the neck. One blow thus directed at full swing against the head of an unlucky attendant would be fatal. The female once drove her horns in sport through an inch board."

Skull of Giraffe.

The projecting upper lip of the giraffe is remarkably flexible, and its elongated prehensile tail, performing in miniature the part of the elephant's proboscis, is of material assistance in browsing upon the foliage and young shoots of the prickly acacia, which constitute the animal's chief food.

The bird on its nest is, no doubt, often surprised by the head of the giraffe suddenly peering into its lofty abode; but be not afraid, sweet creature—for the soft-eyed intruder, content with vegetable fare, intends no harm to thee or to thy tender brood.

With feet terminating in a divided hoof, and a ruminant like our ox, the giraffe has four stomachs, and an enormous intestinal length of 288 feet, a formation which bears testimony to the vast and prolonged powers of digestion necessary to extract nutrition from its hard and meagre diet.

Ranging throughout the wide plains of Central Africa, from Cafraria to Nubia, the giraffe, though a gregarious animal, generally roams about only in small herds. It is, indeed, by no means common even at its head-quarters, and Captain Harris, who traversed the desert as far as the Tropic of Capricorn, seldom found giraffes without having followed their trail, and never saw more than five-and-thirty in a day. Notwithstanding the rapidity with which the cameleopard strides along, the fore and hind leg on the same side moving together, instead of diagonally as in most other quadrupeds, yet a full gallop quite dissipates its power; and the hunters, being aware of this, always try to press the giraffes at once to it, knowing that they have but a short space to run before the animals are in their power. In doing this the old sportsmen are careful not to go too close to the giraffe's tail; for this animal, says Dr. Livingstone, "can swing his hind foot round in a way which would leave little to choose between a kick with it and a clap from the arm of a windmill."

The author of the "Wild Sports of Africa," draws a most animated picture of a giraffe hunt, breathing the full life and excitement of the chase, and capable of raising the envy of all his brother Nimrods:—

"Many days had now elapsed since we had even seen the cameleopard, and then only in small numbers, and under the most unfavourable circumstances. The blood coursed through my veins like quicksilver, therefore, as on the morning of the 19th, from the back of ·Breslar, my most trusty steed, with a firm-wooded plain before me, I counted thirty-two of these animals industriously stretching their peacock-necks to crop the tiny leaves which fluttered above their heads in a mimosa grove that beautified the scenery. They were within a hundred yards of me; but having previously determined to try the *boarding* system, I reserved my fire.

"Although I had taken the field expressly to look for giraffes, and had put four of the Hottentots on horseback, all excepting

Piet, had as usual slipped off unperceived in pursuit of a troop of *koodoos*. Our stealthy approach was soon opposed by an ill-tempered rhinoceros, which, with her ugly calf, stood directly in the path, and the twinkling of her bright little eyes, accompanied by a restless rolling of the body, giving earnest of her intention to charge. I directed Piet to salute her with a broadside, at the same moment putting spurs to my horse. At the report of the gun, and the sudden clattering of hoofs, away bounded the giraffes in grotesque confusion, clearing the ground by a succession of frog-like hops, and soon leaving me far in the rear. Twice were their towering forms concealed from view by a park of trees, which we entered almost at the same instant, and twice, in emerging from the labyrinth, did I perceive them tilting over an eminence immeasurably in advance. A white turban that I wore round my hunting cap, being dragged off by a projecting bough, was instantly charged by three rhinoceroses, and, looking over my shoulder, I could see them long afterwards, fagging themselves to overtake me. In the course of five minutes the fugitives arrived at a small river, the treacherous sands of which receiving their long legs, their flight was greatly retarded; and after floundering to the opposite side, and scrambling to the top of the bank, I perceived that their race was run. Patting the steaming neck of my good steed, I urged him again to his utmost, and instantly found myself by the side of the herd. The stately bull being readily distinguishable from the rest by his dark chesnut robe and superior stature, I applied the muzzle of my rifle behind his dappled shoulder with the right hand, and drew both triggers; but he still continued to shuffle along, and being afraid of losing him, should I dismount, among the extensive mimosa groves with which the landscape was now obscured, I sat in my saddle, loading and firing behind the elbow; and then, placing myself across his path, until the tears trickling from his full brilliant eye, his lofty frame began to totter, and at the seventeenth discharge from the deadly-grooved bore, like a falling minaret bowing his graceful head from the skies, his proud form was prostrate in the dust. Never shall I forget the tingling excitement of that moment. At last, then, the summit of my hunting ambition was actually attained, and the towering giraffe laid low. Tossing my turbanless cap into the air, alone in the wild wood, I hurraed with bursting exultation, and, unsaddling my steed, sank exhausted beside the noble prize I had won.

"When I leisurely contemplated the massive frame before me, seeming as though it had been cast in a mould of brass, and protected by a hide of an inch and a half in thickness, it was no longer matter of astonishment that a bullet, discharged from a distance of eighty or ninety yards, should have been attended with little effect upon such amazing strength. The spell was now broken, and the secret of cameleopard-hunting discovered. The next day Richardson and myself killed three; one, a female, slipping upon muddy ground, and falling with great violence before she had been wounded, a shot in the head despatched her as she lay. From this time we could reckon confidently upon two out of each troop that we were fortunate enough to find, always approaching as near as possible in order to ensure a good start, galloping into the middle of them, boarding the largest, and riding with him until he fell."

After man, the giraffe's chief enemy is the lion, who often waits for it in the thick brakes on the margin of the rivers or pools, and darts upon it with a murderous spring while it is slaking its thirst. Andersson once saw five lions, two of whom were in the act of pulling down a splendid giraffe, while the other three were watching close at hand the issue of the deadly strife; and Captain Harris relates that, while he was encamped on the banks of a small stream, a cameleopard was killed by a lion whilst in the act of drinking, at no great distance from the wagons. It was a noisy affair; but an inspection of the scene on which it occurred proved that the giant strength of the victim had been paralysed in an instant.

Sometimes the giraffe saves itself from the attacks of its arch-enemy by a timely flight; but when hemmed in, it offers a desperate resistance, and in spite of its naturally gentle and peaceable disposition, gives such desperate kicks with its forefeet as to keep its antagonist at a respectful distance, and finally to compel him to retreat.

The Greeks and Romans were well acquainted with the giraffe; and Aristotle, describing it under the name of hippardion, or panther horse, probably knew it better than Buffon, who never saw more of it than a stuffed skin. Pliny relates that Julius Cæsar (45 B.C.) first exhibited it to the Romans in the amphitheatre, and from that time it often played a conspicuous part in the bloody spectacles with which the military despots of the declining empire used to entertain the rabble of Rome. Even during the middle ages giraffes were sometimes seen in Europe. The sultan of Egypt presented the German emperor, Frederick II., with a cameleopard; and Lorenzo de Medicis was honoured with a similar gift. But since that time three full centuries elapsed before a single giraffe was ever transported across the Mediterranean; and when at length the wily old tyrant Mehemet Ali, who knew how to flatter the French while grinding his poor Fellahs, sent one of them to the Jardin des Plantes in 1827, it raised no less a sensation than if it had been the unicorn itself. Thenceforth, the spell being broken, many giraffes have been imported, so that now there is scarcely a zoological garden of any importance in Europe that has not at least one of them to boast of. How well they bear the change of food and climate is sufficiently proved by the pair originally belonging to the collection in Regent's Park having produced a young family of six, from 1838 to 1853; so that it seems quite possible to acclimatise this denizen of African wilds, which thrives as well upon corn, carrots, and hay, as upon the leaves, shoots, and blossoms of the mokaala, his desert food. There are many analogies between the giraffe and the ostrich; both long-legged, long-necked, fit for cropping the tall mimosas, or scouring rapidly the plain; both, finally, defending themselves by striking their feet forwards, the one against the jackal or hyæna, the other against the assaults of the formidable lion.

As if to make up for the hideous deformity of the rhinoceros and hippopotamus, the African wilds exclusively give birth to the beautifully-striped Zebras, the most gorgeously attired members of the equine race.

Quagga.

The isabelle-coloured Quagga, irregularly banded and marked with dark brown stripes, which, stronger on the head and neck, gradually become fainter, until lost behind the shoulders, has its high crest surmounted by a standing mane, banded alternately brown and white. It used formerly to be found in great numbers within the limits of the Cape Colony, and still roams in vast herds in the open plains farther to the north.

THE ZEBRA

Thus, in the desert of the Meritsane, Major Harris, after crossing a park of magnificent camelthorn trees, soon perceived large herds of quaggas and brindled gnus, which continued to join each other, until the whole plain seemed alive. The clatter of their hoofs was perfectly astounding, and could be compared to nothing but to the din of a tremendous charge of cavalry, or the rushing of a mighty tempest. The accumulated numbers could not be estimated at less than 15,000, a great extent of country being actually chequered black and white with their congregated masses.

The Douw, or Burchell's Zebra, differs little from the common quagga in point of shape or size; but while the latter is faintly striped only on the head and neck, the former is adorned over every part of the body with broad black bands, beautifully contrasting with a pale yellow ground.

Zebra

Major Harris, who had so many opportunities of seeing this fine species in a state of nature, remarks that — "Beautifully clad by the hand of nature, possessing much of the graceful symmetry of the horse, with great bones and muscular power, united to easy and stylish action, thus combining comeliness of figure with solidity of form, this species, if subjugated and domesticated, would assuredly make the best pony in the world. Although it admits of being tamed to a certain extent with the greatest facility — a half-domesticated specimen, with a jockey on its brindled back, being occasionally exposed in Cape Town for sale — it has hitherto contrived to evade the yoke of servitude. The senses of sight, hearing, and smell, are extremely delicate. The slightest noise or motion, no less than the appearance of any object that is unfamiliar, at once rivets their gaze, and causes them to stop and listen with the utmost attention; any taint in the air equally attracting their olfactory organs.

"Instinct having taught these beautiful animals that in union consists their strength, they combine in a compact body when menaced by an attack, either from man or beast; and, if overtaken by the foe, they unite for mutual defence, with their heads together in a close circular band, presenting their heels to the enemy, and dealing out kicks in equal force and abundance. Beset on all sides, or partially crippled, they rear on their hinder legs, fly at their adversary with jaws distended, and use both teeth and heels with the greatest freedom."

The Gnu and the common quagga, delighting in the same situation, not unfrequently herd together; but Burchell's zebra is seldom seen unaccompanied by troops of the brindled gnu, an animal differing materially from its brother of the same genus, from which, though scarcely less ungainly, it is readily distinguishable at a great distance by its black mane and tail, more elevated withers, and clumsier action.

Gnu.

Both the douw and the quagga are more frequently seen in Europe than the real zebra, and might be easily acclimatised, particularly the former, which can bear the cold so well as frequently to be seen lying on the snow in the Jardin des Plantes, exposed to a temperature of three degrees, without the least injury to its health.

Whilst the douw and the quagga roam over the plains, the zebra inhabits mountainous regions only. The beauty of its light symmetrical form is enhanced by the narrow black bands with which the whole of the white-coloured body is covered. Buffon and Daubenton wished to see this elegant creature acclimatised in Europe, which would procure us a beast of burden stronger than the ass, and more beautiful in its nakedness than the horse, even when adorned with the richest trappings. A king of Portugal used frequently to drive about with four zebras; and, about the year 1761, two of these animals that were kept in the park of Versailles had been so far tamed as to allow themselves to be mounted. In spite of the proverbial obstinacy of the zebra, there are thus no insuperable obstacles to its domestication, and a course of training, continued through several generations, would most likely subdue its reluctant nature as completely as that of the original wild horse and ass. The zebra is supposed to be the real *hippotigris*, or tiger-horse of the ancients; and this is the more probable, as he ranges much farther to the north than the quagga or the douw, and approaches the regions of Africa comprised within the Roman empire. Historians inform us that in the year 202 after Christ, Plautius, a governor or prefect of Egypt, sent several centurions to the islands of the Erythræan sea to fetch horses similar to tigers.

The zebra is found in Abyssinia to the present day; and as all the coasts of the Red Sea were open to the Romans, there was nothing to prevent them from becoming acquainted with the zebra.

Travellers through the African wilds have sometimes been startled by piteous wailings, resembling the faint gasps and stifled groanings of a drowning man. On approaching the spot where they supposed some ravenous beast was lacerating an unfortunate native, they were surprised to find a zebra in its last agonies; and well may the dying moans of the animal be sorrowful, when we consider that even its neighings, when heard from a distance, are of a very melancholy sound.

Captain Harris tells us that it seeks the wildest and most sequestered spots, so that it is exceedingly difficult of approach, not only from its watchful habits and very great agility of foot, but also from the inaccessible nature of its abode. The herds graze on the steep hill-side, with a sentinel posted on some adjacent crag, ready to sound the alarm in case of any suspicious approach to their feeding quarters, and no sooner is the alarm given than away they scamper, with pricked ears and whisking their tails aloft, to places where few, if any, would venture to pursue them.

Hippopotamus.

CHAPTER XXXVIII.

THE HIPPOPOTAMUS.

Behemoth—Its Diminishing Numbers and Contracting Empire—Its Ugliness—A Rogue Hippopotamus or Solitaire—Dangerous Meeting—Intelligence and Memory of the Hippopotamus—Methods employed for Killing the Hippopotamus—Hippopotamus Hunting on the Teoge—The Hippopotamus in Regent's Park—A Young Hippo born in Paris.

"BEHOLD now Behemoth, which I made with thee: he eateth grass as an ox; his bones are as strong pieces of brass; his bones are like bars of iron; he lieth under the shady trees in the covert of the reed and fens. The shady trees cover him with their shadow; the willows of the brook compass him about. Behold he drinketh up a river; he trusteth that he can draw up Jordan into his mouth."

Thus, in the book of Job, we find the Hippopotamus portrayed with few words but incomparable power. How tame after this noble picture must any lengthened description appear!

According to the inspired poet, the hippopotamus seems anciently to have inhabited the waters of the Jordan, but now it is nowhere to be found in Asia; and even in Africa the limits of its domain are perpetually contracting before the persecutions of man. It has entirely disappeared from Egypt and the rivers of the Cape Colony, where Le Vaillant found it in numbers during the last century. In many respects a valuable prize;

of easy destruction, in spite, or rather on account of its size, which betrays it to the attacks of its enemies; a dangerous neighbour to plantations, it is condemned to retreat before the waves of advancing civilisation, and would long since have been extirpated in all Africa, if the lakes and rivers of the interior of that vast den of barbarism were as busily ploughed over as ours by boats and ships, or their banks as thickly strewn with towns and villages.

For the hippopotamus is not able, like so many other beasts of the wilderness, to hide itself in the gloom of impenetrable forests, or to plunge into the sandy desert, traversed by the Bedouin on his dromedary; it requires the neighbourhood of the stream, the empire of which it divides with its amphibious neighbour the crocodile.

Hippopotamus.

Occasionally during the day it is to be seen basking on the shore amid ooze and mud, but throughout the night the unwieldly monster may be heard snorting and blowing during its aquatic gambols; it then sallies forth from its reed-grown coverts to graze by the light of the moon, never, however, venturing to any distance from the river, the stronghold to which it betakes itself on the smallest alarm.

In point of ugliness the hippopotamus, or river-horse, as it has also very inappropriately been named, might compete with the rhinoceros itself. Its shapeless carcase rests upon short and disproportioned legs, and, with its vast belly almost trailing upon the ground, it may not inaptly be likened to an overgrown "prize-pig." Its immensely large head has each jaw armed with two formidable tusks, those in the lower, which are always the largest, attaining at times two feet in length; and the inside of the mouth is said to resemble a mass of butcher's meat. The eyes, which are placed in prominences like the garret windows of a Dutch house, the nostrils, and ears, are all on the same plane, on the upper level of the head, so that the unwieldy monster, when immersed in its favourite element, is able to draw breath, and to use three senses at once for hours together, without exposing more than its snout. The hide, which is upwards of an inch and a half in thickness, and of a pinkish brown colour, clouded and freckled with a darker tint, is destitute of covering, excepting a few scattered hairs on the muzzle, the edges of the ears and tail. Though generally mild and inoffensive, it is not to be wondered at that a creature like this, which when full grown attains a length of eleven or twelve feet, and nearly the same colossal girth, affords a truly appalling spectacle when enraged, and that a nervous person may well lose his presence of mind when suddenly brought into contact with the gaping monster. Even Andersson, a man accustomed to all sorts of wild adventure, felt rather discomposed when one night a hippopotamus, without the slightest warning, suddenly protruded its enormous head into his bivouac, so that every man started to his feet with the greatest precipitation, some of the party, in the confusion, rushing into the fire and upsetting the pots containing the evening meal.

As among the sperm-whales, sea bears, elephants, and other animals, elderly males are sometimes expelled the herd, and, for want of company, become soured in their temper, and so misanthropic as to attack every boat that comes near them. The herd is never dangerous except when a canoe passes into the midst of it when all are asleep, and some of them may strike it in terror. To avoid this, it is generally recommended to travel by day near the bank, and by night in the middle of the stream. The "solitaires," or "rogue-hippopotami," frequent certain localities well known to the inhabitants of the banks, and, like the outcast elephants, are extremely dangerous. Dr. Livingstone, passing a canoe which had been smashed to pieces by a blow from the hind foot of one of them, was informed by his men that, in case of a similar assault being made on his boat, the proper way was to dive to the bottom of the river, and hold on there for a few seconds, because the hippopotamus, after breaking a canoe, always looks for the people on the surface, and if he sees none, soon moves off. He saw some frightful gashes made on the legs of the people who, having had the misfortune to be attacked, were unable to dive. One of these "bachelors" one day actually came out of his lair, and, putting his head down, ran with very considerable speed after the missionary and his party; and another time they were nearly overturned by a hippopotamus striking the canoe with its forehead. The butt was so violent as to tilt one of the boatmen out into the river, while the rest sprang to the shore, which was only about ten yards off; the beast looking all the time at the canoe, as if to ascertain what mischief it had done.

In rivers where it is seldom disturbed, such as the Zambesi, the hippopotamus puts up its head openly to blow, and follows the traveller with an inquisitive glance, as if asking him, like the "moping owl" in the elegy, why he came to molest its "ancient solitary reign?" but in other rivers, such as those of Londa, where it is much in danger of being shot, the hippopotamus takes good care to conceal its nose among water plants, and to breathe so quietly that one would not dream of its existence in the river, except by footprints on the banks. Notwithstanding its stupid look, its prominent eyes and naked snout giving it more the appearance of a gigantic boiled calf's head than anything else, the huge creature is by no means deficient in intelligence, knows how to avoid pitfalls, and has so good a memory that, when it has once heard a ball whiz about its ears, it never after ceases to be cautious and "wide awake" at the approach of danger. Being vulnerable only behind the ear, however, or in the eye, it requires the perfection of rifle-practice to be hit; and, when once in the water, is still more difficult to kill, as it dives and swims with all the ease of a walrus or a sea-elephant, its huge body being rendered buoyant by an abundance of fat. Its flesh is said to be delicious, resembling the finest young pork, and is considered as great a delicacy in Africa as a bear's paw or a bison's hump in the prairies of North America. The thick and almost inflexible hide may be dragged from the ribs in strips, like the planks from a ship's side. These serve for the manufacture of a superior description of *sjambok*, the elastic whip with which the Cape boor governs his team of twelve oxen or more, while proceeding on a journey. In Northern Africa it is used to chastise refractory dromedaries or servants; and the ancient Egyptians employed it largely in the manufacture of shields, helmets, and javelins.

But the most valuable part of the hippopotamus is its teeth (canine and incisors), which are considered greatly superior to elephant ivory, and when perfect and weighty, will fetch as much as one guinea per pound, being chiefly used for artificial teeth, since it does not readily turn yellow. All these qualities and uses to which the hippopotamus may be applied are naturally as many prices set upon its head; and the ravages it occasions in the fields are another motive for its destruction. On the White Nile the peasantry burn a number of fires, to keep the huge animal away from their plantations, where every footstep ploughs deep furrows into the marshy ground, to the great injury of the harvest. At the same time, they take care to keep up a prodigious clamour of horns and drums, to scare away the ruinous brute, which, as may well be imagined, is by no means so great a favourite with them as with the visitors of the Zoological Gardens.

They have besides another, and, where it takes effect, far more efficacious method of freeing themselves from the depredations of this animal. They remark the places it most frequents, and there lay a large quantity of pease. When it comes on shore hungry and voracious, it falls to eating what is nearest, and fills its vast stomach with the pease, which soon occasion an insupportable thirst. The river being close at hand, it immediately drinks whole buckets of water, which, by swelling the pease, cause it to blow up, like an overloaded mortar.

The natives on the Teoge, and other rivers that empty themselves into Lake Ngami, kill the hippopotamus with iron harpoons, attached to long lines ending with a float. A huge reed raft, capable of carrying both the hunters and their canoes, with all that is needful for the prosecution of the chase, is pushed from the shore, and afterwards abandoned to the stream, which propels the unwieldy mass gently and noiselessly forward. Long before the hippopotami can be seen, they make known their presence by awful snorts and grunts whilst splashing and blowing in the water. On approaching the herd, for the gregarious animal likes to live in troops of from twenty-five to thirty, the most skilful and intrepid of the hunters stands prepared with the harpoons, whilst the rest make ready to launch the canoes should the attack prove successful. The bustle and noise caused by these preparations gradually subside: at length not even a whisper is heard, and in breathless silence the hunters wait for the decisive conflict. The snorting and plunging become every moment more distinct; a bend in the stream still hides the animals from view; but now the point is passed, and monstrous figures, that might be mistaken for shapeless cliffs, did not ever and anon one or the other of them plunge and reappear, are seen dispersed over the troubled waters. On glides the raft, its crew worked up to the highest pitch of excitement, and at length reaches the herd, which, perfectly unconscious of danger, continue to enjoy their sports. Presently one of the animals is in immediate contact with the raft. Now is the critical moment; the foremost harpooner raises himself to his full height to give the greater force to the blow, and the next instant the iron descends with unerring accuracy, and is buried deep in the body of the bellowing hippopotamus. The wounded animal plunges violently and dives to the bottom, but all its efforts to escape are as ineffectual as those of the seal when pierced with the barbed iron of the Greenlander.

As soon as it is struck, one or more of the men launch a canoe from off the raft, and hastening to the shore with the harpoon line, take a round turn with it about a tree, so that the animal may either be brought up at once, or should there be too great a strain on the line, " played," like a trout or salmon by the fisherman. Sometimes both line and buoy are cast into the water, and all the canoes being launched from off the raft, chase is given to the poor brute, who whenever he comes to the surface is saluted with a shower of javelins. A long trail of blood marks his progress, his flight becomes slower and slower, his breathing more oppressive, until at last, his strength ebbing away through fifty wounds, he floats dead on the surface.

But as the whale will sometimes turn upon his assailants, so also the hippopotamus not seldom makes a dash at his persecutors, and either with his tusks, or with a blow from his head, staves in or capsizes the canoe. Sometimes even, not satisfied with wreaking his vengeance on the craft, he seizes one or other of the crew, and with a single grasp of his jaws, either terribly mutilates the poor wretch or even cuts his body fairly in two.

The natives of Southern Africa also resort to the ingenious plan of destroying the hippopotamus by means of a downfall, consisting of a log of wood with stones attached to it to increase its weight, and a harpoon affixed to its lower end. This formidable weapon depends from the branch of an overhanging tree by means of a line, which is then made to cross horizontally the pathway which the hippopotamus is in the habit of frequenting during the night, at a short distance from the ground. When the animal comes in contact with the line, which is secured on either side of the path by a small peg, it snaps at once, or is disengaged by means of a trigger. The liberated downfall instantly descends with the rapidity of lightning, and the harpoon is driven deep into the back of the monster, who, bellowing with pain, plunges into the river, where he soon after dies in excruciating torments.

Since 1850, the visitors to the Zoological Gardens in the Regent's Park have become familiar with the huge form and the curious manners of the hippopotamus, who, ever since the days of ancient Rome, had not been seen in Europe. The infant animal, a present of the Pacha of Egypt, had been torn from its mother's breast, and was fed during the passage with the milk of two cows and three goats, mixed with maize flour to a porridge. But even this copious allowance hardly sufficed to satisfy the hunger of the interesting stranger, who, on arriving, became at once the lion of the gardens, the observed of all observers, and though the novelties of the aquarium now draw better " houses," the hippopotamus is still one of the chief objects of attraction to strangers. His appetite increasing with his years, one hundred pounds weight of provender— hay, corn, carrots, cabbage — now daily goes down his huge throat, and he thrives so well on this abundant fare, that it is very doubtful whether he would have grown equally fat on the banks of his native Nile.

Unfortunately, his temper seems to have been soured by captivity, for though so gentle when he made his *début* that he could not go to sleep without having his Arab keeper's feet to lay his neck upon, he is now, at times, perfectly furious ; and rushing at the massive oaken door of his enclosure, makes the beams shake and quiver as though they were the lathes of an ordinary paling.

He has a peculiar dislike to the sight of working men, especially if they are employed in doing any jobs about his apartment. The smith of the establishment happening to be passing one day along the iron gallery which runs on one side of his bath, the animal leapt out of the water at least eight feet high (I wish I had been there to see it), and would speedily have demolished the whole construction, had not the terrified man rapidly bolted out of his sight.

It is to be hoped that the young bride who now shares his well-fed captivity may succeed in soothing the spleen of her unruly lord. I know not whether the illustrious strangers have any family to boast of, but the hippopotami of the Jardin des Plantes have added to the list of births, and according to the last accounts, "both mother and child were doing well."

Rhinoceros.

CHAPTER XXXIX.

THE RHINOCEROS.

Brutality of the Rhinoceros—The Borelo—The Keitloa—The Monoho—The Kobaaba—Difference of Food and Disposition between the Black and the White Rhinoceros—Incarnation of Ugliness—Acute Smell and Hearing—Defective Vision—The Buphaga Africana—Paroxysms of Rage—Parental Affection—Nocturnal Habits—Rhinoceros-Hunting—Adventures of the Chase—Narrow Escapes of Messrs. Oswell and Andersson—The Indian Rhinoceros—The Sumatran Rhinoceros—The Javanese Rhinoceros—Its involuntary Suicide.

THE Rhinoceros has about the same range as the elephant, but is found also in the island of Java, where the latter is unknown. Although not possessed of the ferocity of carnivorous animals, the rhinoceros is completely wild and untamable; the image of a gigantic hog, without intelligence, feeling or docility; and if in bodily size and colossal strength it, of all other land animals, most nearly approaches the elephant, it is infinitely his inferior in point of sagacity. The latter, with his beautiful, good-natured, intelligent eye, awakens the sympathy of man; while the rhinoceros might figure as the very symbol of brutal violence and stupidity.

It was formerly supposed that Africa had but one rhinoceros, but the researches of modern travellers have discovered no less than four different species, two white and two black, each of them with two horns.

The black species are the Borelo of the Bechuanas, and the Keitloa, which is longer, with a larger neck and almost equal horns. In both species the upper lip projects over the lower, and is capable of being extended like that of the giraffe, thus enabling the animal to pull down the branches on whose foliage he intends to feast. Both the Borelo and the Keitloa are extremely ill-natured, and, with the exception of the buffalo, the most dangerous of all the wild animals of South Africa.

Rhinoceros.

The white species are the Monoho (*R. simus*), and the Kobaaba (*R. Oswellis*), which is distinguished by one of its horns attaining the prodigious length of four feet.

Although the black and white rhinoceros are members of the same family, their mode of living and disposition are totally different. The food of the former consists almost entirely of roots, which they dig up with their larger horn, or of the branches and sprouts of the thorny acacia, while the latter exclusively live on grasses. Perhaps in consequence of their milder and more succulent food, they are of a timid unsuspecting nature, which renders them an easy prey, so that they are fast melting away before the onward march of the European trader; while the black species, from their greater ferocity and wariness, maintain their place much longer than their more timid relations. The different nature of the black and white rhinoceros shows itself even in their flesh, for while that of the former, living chiefly on arid branches, has a sharp and bitter taste, and but little recommends itself by its meagreness and toughness— the animal, like the generality of ill-natured creatures, being never found with an ounce of fat on its bones—that of the latter is juicy and well-flavoured, a delicacy both for the white man and the negro.

The shape of the rhinoceros is unwieldy and massive; its vast paunch hangs down nearly to the ground; its short legs are of columnar strength, and have three toes on each foot; the misshapen head has long and erect ears, and ludicrously small eyes; the skin, which is completely naked, with the exception of some coarse bristles at the extremity of the tail, and the upper end of the ears, is comparatively smooth in the African species, but extremely rough in the Asiatic, hanging in large folds about the animal like a mantle; so that, summing up all these characters, the rhinoceros has no reason to complain of injustice, if we style it the very incarnation of ugliness. From the snout to the tip of the tail, the African rhinoceros attains a length of from 15 to 16 feet, a girth of from 10 to 12, a weight of from 4,000 to 5,000 pounds; but in spite of its ponderous and clumsy proportions, it is able to speed like lightning, particularly when pursued. It then seeks the nearest wood, and dashes with all its might through the thicket. The trees that are dead or dry are broken down as with a cannon shot, and fall behind it and on its sides in all directions; others that are more pliable, greener, or full of sap are bent back by its weight and the velocity of its motions, and restore themselves like a green branch to their natural position, after the huge animal has passed. They often sweep the incautious pursuer and his horse from the ground, and dash them in pieces against the surrounding trees.

The rhinoceros is endowed with an extraordinary acuteness of smell and hearing; he listens with attention to the sounds of the desert, and is able to scent from a great distance the approach of man; but as the range of his small and deep-set eyes is impeded by his unwieldy horns, he can only see what is immediately before him, so that if one be to leeward of him, it is not difficult to approach within a few paces. The Kobaaba, however, from its horn being projected downwards, so as not to obstruct the line of vision, is able to be much more wary than the other species.

To make up for the imperfection of its sight, the rhinoceros is frequently accompanied by a bird (*Buphaga africana*) which seems to be attached to it like the domestic dog to man, and warns the beast of approaching danger by its cry. It is called "Kala," by the Bechuanas, and when these people address a superior, they call him "my rhinoceros" by way of compliment, as if they were the birds ready to do him service.

The African Buffalo possesses a similar guardian in the *Textor erythrorynchus*. When the beast is quietly feeding, the bird may be seen hopping on the ground, picking up food, or sitting on its back, and ridding it of the insects with which its skin is sometimes infested. The sight of the bird being much more acute than that of the buffalo, it is soon alarmed by the approach of any danger, and when it flies up, the buffaloes instantly raise their heads to discover the cause which has led to the sudden flight of their monitor. The Textor sometimes accompanies the wild ox on the wing, at other times it sits upon its withers, as if enjoying the ride.

The black rhinoceroses are of a gloomy melancholy temper, and not seldom fall into paroxysms of rage without any evident cause. Seeing the creatures in their wild haunts, cropping the bushes, or quietly moving through the plains, you might take them for the most inoffensive good-natured animals of all Africa, but when roused to passion there is nothing more terrific on earth. All the beasts of the wilderness are afraid of the uncouth Borelo. The lion silently retires from its path, and even the elephant is glad to get out of the way. Yet this brutal and stupidly hoggish animal is distinguished by its parental love, and the tenderness which it bestows on its young is returned with equal affection. European hunters have often witnessed that when the mother dies, the calf remains two full days near the body.

Although not gregarious, and most generally solitary or grazing in pairs, yet frequently as many as a dozen rhinoceroses are seen pasturing and browsing together. As is the case with many other inhabitants of the tropical wilderness, the huge beast awakens to a more active life after sunset. It then hastens to the lake or river to slake its thirst or to wallow in the mud, thus covering its hide with a thick coat of clay, against the attacks of flies; or to relieve itself from the itching of their stings, it rubs itself against some tree, and testifies its inward satisfaction by a deep-drawn grunt. During the night, it rambles over a great extent of country, but soon after sunrise seeks repose and shelter against the heat under the shade of a mimosa, or the projecting ledge of a rock, where it spends the greater part of the day in sleep, either stretched at full length or in a standing position. Thus seen from a distance, it might easily be mistaken for a huge block of stone.

The rhinoceros is hunted in various manners. One of the most approved plans is to stalk the animal, either when feeding or reposing. If the sportsman keep well under the wind, and there be the least cover, he has no difficulty in approaching the beast within easy range, when, if the ball be well directed, it is killed on the spot. But by far the most convenient way of destroying the animal is to shoot it from a cover or a screen, when it comes to the pool to slake its thirst. Occasionally it is also taken in pitfalls. Contrary to common belief, a leaden ball (though spelter is preferable) will easily find its way through the hide of the African rhinoceros, but it is necessary to be within thirty or forty paces of the brute, and desirable to have a double charge of powder. The most deadly part to aim at is just behind the shoulder; a ball through the centre of the lobes of the lungs is certain to cause almost instantaneous death. A shot in the head never or rarely proves fatal, as the brain, which, in proportion to the bulk of the animal, does not attain the three hundredth part of the size of the human cerebrum, is protected, besides its smallness, by a prodigious case of bone, hide, and horn. However severely wounded the rhinoceros may be, he seldom bleeds externally. This is attributable in part, no doubt, to the great thickness of the hide and its elasticity, which occasions the hole caused by the bullet nearly to close up, as also from the hide not being firmly attached to the body, but constantly moving. If the animal bleed at all, it is from the mouth and nostrils, which is a pretty sure sign that it is mortally hurt, and will soon drop down dead. It is remarkable that the rhinoceros, when hit by a fatal bullet, does not fall upon one side, but generally sinks on its knees, and thus breathes its last.

From what has been related of the fury of the rhinoceros, its pursuit must evidently be attended with considerable danger, and thus the annals of the wild sports of Southern Africa are full of hair-breadth escapes from its terrific charge. Once Mr. Oswell, having lodged a ball in the body of a huge white rhinoceros, though not with mortal effect, was surprised to see the beast, instead of seeking safety in flight, as is generally the case with this inoffensive species, suddenly stop short, and having eyed him most curiously for a second or two, walk slowly towards him. Though never dreaming of danger, he instinctively turned his horse's head away; but strange to say, this creature, usually so docile and gentle, now absolutely refused to give him his head. When at last he did so, it was too late, for although the rhinoceros had only been walking, the distance was now so small that contact was unavoidable. In another moment the brute bent low his head, and with a thrust upwards, struck his horn into the ribs of the horse with such force as to penetrate to the very saddle on the opposite side, where the rider felt its sharp point against his leg. The violence of the blow was so tremendous as to cause the horse to make a complete somersault in the air, coming heavily down on his back. The rider was, of course, violently precipitated to the ground. While thus prostrated, he actually saw the horn of the monster alongside of him; but without attempting to do any further mischief, the brute started off at a canter from the scene of action. If the rhinoceros imagined it had come off as victor, it was, however, very quickly undeceived; for Mr. Oswell, rushing upon one of his companions, who by this time had come up, and unceremoniously pulling him off his horse, leapt into the saddle, and without a hat, and his face streaming with blood, was quickly in pursuit of the beast, which he soon had the satisfaction to see stretched lifeless at his feet. This adventure teaches us why the rhinoceros is seldom pursued on horseback, as there is no relying on one's steed, and the brute can also be much more easily approached and killed on foot.

Another time, Mr. Oswell was actually gored by a rhinoceros, and tossed in the air. Fortunately he escaped with a wound on the thigh, five inches long; and though it ultimately healed, yet, as may be imagined, it left a deep and indelible scar behind.

Mr. Andersson, another well-known African Nimrod, having one day wounded a black rhinoceros, and being in an unfavourable situation for renewing his shot with deadly effect, the monster, snorting horribly, erecting its tail, keeping its head close to the ground, and raising clouds of dust by its feet, rushed at him furiously. "I had only just time to level my rifle and fire," says this adventurous traveller, "before it was upon me, and the next instant knocked me to the ground. The shock was so violent as to send my rifle, powder-flask, and ball-pouch spinning ten feet high in the air. On the beast charging me, it crossed my mind that, unless gored at once by its horn, its impetus would be such as to carry it beyond me, and I might thus be afforded a chance of escape, and so, indeed, it happened, for having been tumbled over and trampled on with great violence, the fore-quarter of the enraged brute passed over my body. Struggling for life, I seized my opportunity, and as the animal was recovering itself for a renewal of the charge, scrambled out from between its hind legs. But the infuriated rhinoceros had not yet done with me, for scarcely had I regained my feet, before he struck me down a second time, and with his horn ripped up my right thigh (though not very deeply) from near the knee to the hip: with his fore-feet, moreover, he hit me a terrific blow on the left shoulder, near the back of the neck. My ribs bent under the enormous weight and pressure, and for a moment I must, as I believe, have lost consciousness; I have at least very indistinct notions of what afterwards took place. All I remember is, that when I raised my head, I heard a furious snorting and plunging amongst the neighbouring bushes. I now arose, though with great difficulty, and made my way in the best manner I was able towards a large tree near at hand for shelter; but this precaution was needless; the beast, for the time at least, showed no inclination further to molest me. Either in the *mêlée*, or owing to the confusion caused by its wounds, it had lost sight of me, or felt satisfied with the revenge it had taken. Be that as it may, I escaped with life, though sadly wounded and severely bruised, in which disabled state I had great difficulty in getting back to my screen.

"During the greater part of the conflict I preserved my presence of mind, but after the danger was over, and when I had leisure to collect my scattered and confused senses, I was seized with a nervous affection, causing a violent trembling. I have since killed many rhinoceroses, as well for sport as food, but several weeks elapsed before I could again attack these animals with any coolness."

The rhinoceros is hunted for its flesh, its hide (which is manufactured into the best and hardest leather that can be imagined), and its horns, which, being capable of a high polish, fetch at the Cape a higher price than ordinary elephant ivory. It is extensively used in the manufacture of sword-handles, drinking-cups, ramrods for rifles, and a variety of other purposes. Among Oriental princes, goblets made of rhinoceros horn are in high esteem, as they are supposed to have the virtue of detecting poison by causing the deadly liquid to ferment till it flows over the rim, or, as some say, to split the cup.

The number of rhinoceroses destroyed annually in South Africa is very considerable. Captain Harris, who once saw two and twenty together, shot four of them one after the other to clear his way. Messrs. Oswell and Varden killed in one year no less than eighty-nine, and in one journey, Andersson shot, single-handed, nearly two-thirds of this number. It is thus not to be wondered at that the rhinoceros, which formerly ranged as far as the Cape, is now but seldom found to the south of the tropic. The progress of African discovery bodes no good to him, or to the hippopotamus.

The single-horned Indian rhinoceros was already known to the ancients, and not unfrequently doomed to bleed in the Roman amphitheatres. One which was sent to King Emanuel of Portugal in the year 1513, and presented by him to the pope, had the honour to be pictured in a woodcut by no less an artist than Albrecht Dürer himself.

Indian Rhinoceros.

Latterly, rhinoceroses have much more frequently been sent to Europe, particularly the Asiatic species, and all the chief zoological gardens possess specimens of the unwieldy creature.

In its native haunts, the Indian rhinoceros leads a tranquil indolent life, wallowing on the marshy border of lakes and rivers, and occasionally bathing itself in their waters. Its movements are usually slow, and it carries its head low like the hog, ploughing up the ground with its horn, and making its way by sheer force through the jungle. Though naturally of a quiet and inoffensive disposition, it is very furious and dangerous when provoked or attacked, charging with resistless impetuosity, and trampling down or ripping up with its horn any animal which opposes it.

Besides the single-horned species which inhabits the Indian peninsula, Java, and Borneo, Sumatra possesses a rhinoceros with a double horn, which is, however, distinguished from the analogous African species by the large folds of its skin, and its smaller size. It is even asserted that there exists in the same island a hornless species, and another with three horns. There surely can be no better proof of the difficulties which Natural History has to contend with in the wilder regions of the tropical zone, and of the vast field still open to future zoologists, than that, in spite of all investigations and travels, we do not yet even know with certainty all the species of so large a brute as the rhinoceros.

In Java, this huge pachyderm is met with in the jungles of the low country, but its chief haunts are the higher forest-lands, which contain many small lakes and pools, whose banks are covered with high grasses. Here and there, also, the woods are interspersed with dry pasture-grounds, and even in the interior of the forests, numerous species of gramineæ are found increasing in number as they rise above the level of the sea. In these solitudes, which are seldom visited by man, the rhinoceros finds all that it requires for food and enjoyment. As it is uncommonly shy, the traveller rarely meets it, but sometimes, while threading his way through the thicket, he may chance to surprise wild steers and rhinoceroses grazing on the brink of a pool, or quietly lying in the morass.

While the rhinoceros, with his knotty plaited skin, offers a repulsive sight, the black-coated and white-legged Steer, who is nearly equally large, but of much more slender form, is fully entitled to the praise of savage beauty, when, suddenly starting up at sight of the traveller, it rushes loudly snorting into the thicket.

The grooved paths of the rhinoceros, deeply worn into the solid rock, and thus affording proof of their immemorial antiquity, are found even on the summits of mountains above the level of the sea. They are frequently used for the destruction of the animal, for in the steeper places, where, on climbing up or down, it is obliged to stretch out its body, so that the abdomen nearly reaches the ground, the Javanese fix large scythe-like knives into the rock, which they cover with moss and herbage, thus forcing the poor rhinoceros to commit an involuntary suicide, and teaching him, though too late to profit by his experience, how difficult it is to escape the cunning of man, even on the mountain peak.

Elephants.

CHAPTER XL.

THE ELEPHANT.

Love of Solitude and Pusillanimity—Miraculous Escape of an English Officer—Sagacity of the Elephant in ascending Hills—Organisation of the Stomach—The Elephant's Trunk—Use of its Tusks still Problematical—The Rogue-Elephant—Sagacity of the Elephant—The African Elephant—Tamed in Ancient Times—South African Elephant-Hunting—Hair-breadth Escapes—Abyssinian Elephant-Hunters—Importance of the Ivory Trade—The Asiatic Elephant—Vast Numbers destroyed in Ceylon—Major Rogers—Elephant Catchers—Their amazing Dexterity—The Corral—Decoy Elephants—Their astonishing Sagacity—Great Mortality among the captured Elephants—Their Services.

OF a mild and peaceful disposition, the image of strength tempered by good nature, the Elephant loves the shady forest and the secluded lake. Disliking the glare of the midday sun, he spends the day in the thickest woods, devoting the night to excursions and to the luxury of the bath, his great and innocent delight. Though the earth trembles under his strides, yet like the whale, he is timid; but this timidity is accounted for by his small range of vision. Anything unusual strikes him with terror, and the most trivial objects and incidents, from being imperfectly discerned, excite his suspicions. To this peculiarity an English officer, chased and seized by an elephant which he had slightly wounded, owed his almost miraculous escape. The animal had already raised its fore-foot to trample him to death, when, its forehead being caught at the instant by the tendrils of a climbing plant which had suspended itself from the branches above, it suddenly turned and fled.* An instinctive consciousness that his superior bulk exposes him to danger from sources that might be harmless in the case of lighter animals, is probably the reason why the elephant displays a remarkable reluctance to face the slightest artificial obstruction on his passage. Even when enraged by a wound, he will hesitate to charge his assailant across an intervening hedge, suspecting it may conceal a snare or pitfall, but will hurry along it to seek for an opening.

Unlike the horse, he never gets accustomed to the report of fire-arms, and thus he never plays an active part in battle, but serves in a campaign only as a common beast of burden, or for the transport of heavy artillery.

To make up for his restricted vision, his neck being so formed as to render him incapable of directing the range of his eye much above the level of his head, he is endowed with a remarkable power of smell, and a delicate sense of hearing, which serve to apprise him of the approach of danger.

Although, from their huge bulk, the elephants might be supposed to prefer a level country, yet, in Asia at least, the regions where they most abound are all hilly and mountainous. In Ceylon, particularly, there is not a range so high as to be inaccessible to them, and so sure-footed are they, that provided there be solidity to sustain their weight, they will climb rocks, and traverse ledges, where even a mule dare not venture.

Dr. Hooker admired the judicious winding of the elephant's path in the Himalayas, and Sir J. E. Tennent describes the sagacity which he displays in laying out roads, or descending abrupt banks, as almost incredible.

"His first manœuvre is to kneel down close to the edge of the declivity, placing his chest to the ground, one fore-leg is then cautiously passed a short way down the slope, and if there is no natural protection to afford a firm footing, he speedily forms one by stamping into the soil if moist, or kicking out a footing if dry. This point gained, the other fore-leg is brought down in

---

* Sir James Emerson Tennent: Ceylon, vol. ii. p. 288. Fourth Edition.

the same way, and performs the same work, a little in advance of the first, which is thus at liberty to move lower still. Then first one and then the second of the hind-legs is carefully drawn over the side, and the hind-feet in turn occupy the resting-places previously used and left by the fore ones. The course, however, in such precipitous ground is not straight from top to bottom, but slopes along the face of the bank, descending till the animal gains the level below.

"This an elephant has done at an angle of forty-five degrees, carrying a houdah, its occupant, his attendant, and sporting apparatus, and in much less time than it takes to describe the operation." *

The stomach of the elephant, like that of the camel or the llama, is provided with a cavity, serving most probably as a reservoir for water against the emergencies of thirst; but the most remarkable feature in the organisation of the "Leviathan of the land" is his wonderful trunk, which, uniting the flexibility of the serpent with a giant's power, almost rivals the human hand by its manifold uses and exquisite delicacy of touch. "Nearly eight feet in length, and stout in proportion to the massive size of the whole animal, this miracle of nature," as it is well expressed by Mr. Broderip, "at the volition of the elephant will uproot trees or gather grass; raise a piece of artillery or pick up a comfit; kill a man or brush off a fly. It conveys the food to the mouth, and pumps up the enormous draughts of water, which, by its recurvature, are turned into and driven down the capacious throat, or showered over the body. Its length supplies the place of a long neck, which would have been incompatible with the support of the large head and weighty tusks." A glance at the head of the elephant will show the thickness and strength of the trunk at its insertion; and the massy arched bones of the face and thick muscular neck are admirably adapted for supporting and working this incomparable instrument, which is at the same time the elephant's most formidable instrument of defence, for, first prostrating any minor assailant by means of his trunk, he then crushes him by the pressure of his enormous weight.

The use of the elephant's tusks is less clearly defined.

---

* Journal of the Asiatic Society of Bengal, vol. xiii.

Though they are frequently described as warding off the attacks of the tiger and rhinoceros, often securing the victory by one blow, which transfixes the assailant to the earth, it is perfectly obvious, both from their almost vertical position and the difficulty of raising the head above the level of the shoulder, that they were never designed for weapons of attack. No doubt they may prove of great assistance in digging up roots, but that they are far from indispensable, is proved by their being but rarely seen in the females, and by their almost constant absence in the Ceylon elephant, where they are generally found reduced to mere stunted processes.

The elephants live in herds, usually consisting of from ten to twenty individuals, and each herd is a family, not brought together by accident or attachment, but owning a common lineage and relationship. In the forest several herds will browse in close contiguity, and in their expeditions in search of water they may form a body of possibly one or two hundred, but on the slightest disturbance, each distinct herd hastens to reform within its own particular circle, and to take measures on its own behalf for retreat or defence.

Generally the most vigorous and courageous of the herd assumes the leadership: his orders are observed with the most implicit obedience, and the devotion and loyalty evinced by his followers are very remarkable. In Ceylon this is more readily seen in the case of a tusker than any other, because in a herd he is generally the object of the keenest pursuit by the hunters. On such occasions the elephants do their utmost to protect him from danger; when driven to extremity, they place the leader in the centre, and crowd so eagerly in front of him that the sportsmen have to shoot a number which they might otherwise have spared. In one instance, a tusker who was badly wounded by Major Roger was promptly surrounded by his companions, who supported him between their shoulders, and actually succeeded in covering his retreat to the forest.

It is a remarkable fact that among the walruses, the tusked monsters of the Arctic shores, we find similar instances of sagacity and devotion. Thus in the attack on the boats of the Trent by a herd of these animals, so beautifully described by Beechey, "the leader having been mortally wounded, his companions, who immediately desisted from the attack, assembled round him, and in a moment quitted the boat, swimming away as hard as they could with their unfortunate commander, whom they actually bore up with their tusks, and assiduously preserved from sinking." *

When individuals have been expelled from a herd, or by some accident or other have lost their former associates, they are not permitted to attach themselves to any other family, and ever after wander about the woods as outcasts from their kind. Rendered morose and savage from rage and solitude, these *rogue* elephants become vicious and predatory; and so sullen is their disposition, that although two may be in the same vicinity, there is no known instance of their associating, or of a *rogue* being seen in company with another elephant.

These savage solitaires not only commit great injuries in the plantations, trampling down the rice-grounds, and tearing up the trees, but even travellers are exposed to the utmost risk from their unprovoked assaults. Sir J. E. Tennent mentions a "rogue," who in 1847 infested for some months the Rangbodde Pass on the great mountain road leading to the sanatarium at Neuera Ellia. He concealed himself by day in the dense forests on either side of the road, making his way during the darkness to the river below. One morning a poor Caffre pioneer proceeding to his labour came suddenly upon him at a turning in the road, when the spleenish savage, resenting the intrusion, lifted him with his trunk and beat out his brains against the bank.

Among other animals that have been driven from the society of their kind, we find a similar ferocity of disposition. The solitary males of the wild buffalo, when expelled from the herd by stronger competitors for female favour, are apt to wreak their vengeance on whatever they meet; the outcast hippopotamus wantonly strikes the passing canoes; and the *rogue* sperm whale, without waiting for the attack, rushes furiously against the boats sent out against him, and seems to love fighting for its own sake.†

Thus, however soul-improving solitude may prove to the hermit saint, there can be no doubt that it is far from conducing to the amiability of animals; and judging from these facts, we

* The Sea and its Living Wonders Second Edition. Page 108.
† Ibid. Page 90.

may almost doubt whether the anachorets of the Thebaid, far from improving their morals, did not become rather "roguish" from their unnatural seclusion.

As the elephant surpasses all that breathes on earth in strength and weight, his mental faculties also assign to him one of the first places in the animal creation. His docility, his attachment to his master, his ready obedience, are qualities in which he is scarcely inferior to the dog, and it is astonishing how easily he suffers himself to be led by his puny guide.

The dog has been the companion of man through an endless series of generations, a servitude of many centuries has modified his physical and moral type; but the elephant, whom, in spite of his prodigious powers, we train to an equal obedience, is always originally the free-born son of the forest (for he never propagates in a state of captivity), and is often advanced in years before being obliged to change the independence of the woods for the yoke of thraldom. What services might not be expected from an animal like this, were we able to educate the species as we do the individual?

The elephant inhabits both Asia and Africa, but each of these two parts of the world has its peculiar species. The African elephant is distinguished by the lozenge-shaped prominences of

Indian Elephant.

ivory and enamel on the surface of his grinders, which in the Indian elephant are narrow transverse bars of uniform breadth; his skull has a more rounded form, and is deficient in the double lateral bump conspicuous in the former; and he has only fifty-four vertebræ, while the Indian has sixty-one. On the other hand, he possesses twenty-one ribs, while the latter has only nineteen. His tusks are also much larger, and his body is of much greater bulk, as the female attains the stature of the full-grown Indian male. The ear is at least three times the size, being not seldom above four feet long, and broad, so that Dr. Livingstone mentions having seen a negro, who under cover of one of these prodigious flaps effectually screened himself from the rain. All these differences of character appeared so great to M. Cuvier as to induce him to consider the African elephant as a peculiar genus, to which he applied the name of Loxodont.

Ancient medals representing large-eared elephants drawing chariots, are conclusive of the fact that the Romans knew how to catch and tame the African elephant. He was even considered more docile than the Asiatic, and was taught various feats, as walking on ropes, and dancing. The elephants with which Hannibal crossed the Alps, as well as those which Pyrrhus led into Italy, must undoubtedly have been African. At present he is only killed for his ivory, his hide, his flesh, or from the mere wantonness of destruction. The Cape colonists, to whom his services might be of great importance, have never made the attempt to tame him, nor has one of this species ever been exhibited in England.

The African elephant has a very wide range, from Cafraria to Nubia, and from the Zambesi to Cape Verde, and the impenetrable deserts of the Sahara alone prevent him from wandering to the shores of the Mediterranean. Although in South Africa the persecutions of the natives and of his still more formidable enemies — the colonists and English huntsmen, have considerably thinned his numbers, and driven him farther and farther to the north, yet in the interior of the country he is still met with in prodigious numbers. Dr. Barth frequently saw large herds winding through the open plains, and swimming in majestic lines through the rivers with elevated trunks, or bathing in the shallow lakes for coolness or protection against insects.

Dr. Livingstone gives us many interesting accounts of the different modes of South African elephant-hunting.

The Banijai on the south bank of the Zambesi erect stages on high trees overhanging the paths by which the elephants come, and then use a large spear with a handle nearly as thick as a man's wrist, and four or five feet long. When the unfortunate animal comes beneath, they throw the spear, and if it enters between the ribs above, as the blade is at least twenty inches long by two broad, the motion of the handle, as it is aided by knocking against the trees, makes frightful gashes within, and soon causes death. They kill them also by means of a spear inserted in a beam of wood, which being suspended on the branch of a tree by a cord attached to a latch, fastened in the path and intended to be struck by the animal's foot, leads to the fall of the beam, and the spear being poisoned, causes death in a few hours.

The Bushmen select full-moon nights for the chase, on account of the coolness, and choose the moment succeeding a charge when the elephant is out of breath to run in and give him a stab with their long-bladed spears. The huge creature is often bristling with missile weapons like a porcupine, and though singly none of the wounds may be mortal, yet their number overpowers him by loss of blood. On the sloping banks of the Zouga the Bayeiye dig deep pitfalls to entrap the animals as they come to drink, but though these traps are constructed with all the care of savage ingenuity, old elephants have been known to precede the herd and whisk off their coverings all the way down to the water, or, giving proof of a still more astonishing sagacity, to have actually lifted the young out of the pits into which they had incautiously stumbled.

A much more formidable enemy of this noble animal than the spears or pitfalls of the African barbarians is the rifle, particularly in the hands of a European marksman, for while the Griquas, Boers, and Bechuanas generally stand at the distance of a hundred yards or more, and of course spend all the force of their bullets on the air, the English hunters, relying on their steadiness of aim, approach to within thirty yards of the animal, where they are sure not to waste their powder. The consequence is, that when the Griquas kill one elephant, such marksmen as Messrs. Oswell, Varden, Gordon Cumming, and Andersson, will bring at least twenty to the ground, and this difference is the more remarkable as the natives employ dogs to assist them, while the English trust to themselves alone. It requires no little nerve to brave the charge of the elephant, the scream or trumpeting of the brute, when infuriated, being more like what the shriek of a steam-whistle would be to a man standing on the dangerous part of a railroad, than any other earthly sound; a horse unused to it will sometimes stand shivering instead of taking his rider out of danger, or fall paralysed by fear, and thus expose him to be trodden into a mummy, or dashing against a tree, crack his skull against a branch.

Even the most experienced hunters have many dangers to encounter while facing their gigantic adversary. Thus, on the banks of the Zouga in 1850, Mr. Oswell had one of the most extraordinary escapes from a wounded elephant perhaps ever recorded in the annals of the chase. Pursuing the brute into the dense thick thorny bushes met with on the margin of that river, and to which the elephant usually flees for safety, he followed through a narrow pathway by lifting up some of the branches, and forcing his way through the rest; but when he had just got over this difficulty, he saw the elephant, whose tail he had but got glimpses of before, now rushing full speed towards him. There was then no time to lift up branches, so he tried to force the horse through them. He could not effect a passage, and as there was but an instant between the attempt and failure, the hunter tried to dismount, but in doing this one foot was caught by a branch, and the spur drawn along the animal's flank; this made him spring away, and throw the rider on the ground with his face to the elephant, which being in full chase, still went on. Mr. Oswell saw the huge fore-foot about to descend on his legs, parted them, and drew in his breath, as if to resist the pressure of the other foot, which he expected would next descend on his body. His relief may be imagined, when he saw the whole length of the under part of the enormous brute pass over him, leaving him perfectly unhurt.

In Abyssinia the elephant is hunted in an original manner. The men, who make this their chief occupation, dwell constantly in the woods, and live entirely upon the flesh of the animals they kill. They are exceedingly agile and dexterous, both on horseback and on foot; indispensable qualities, partly inherited and partly acquired by constant practice. Completely naked to render their movements more easy, and to prevent their being laid hold of by the trees and bushes; two of these bold huntsmen get on horseback; one of them bestrides the back of the steed, a short stick in one hand, the reins in the other, while behind him sits his companion, armed with a sharp broadsword. As soon as they perceive a grazing elephant, they instantly ride up to him, or cross him in all directions if he flies, uttering at the same time a torrent of abuse, for the purpose, as they fancy, of raising his anger. With outstretched trunk the elephant attempts to seize the noisy intruders, and following the perfectly trained horse, which, springing from side to side, leads him along in vain pursuit, neglects flight into the woods, his sole chance of safety, for while his whole attention is fixed on the rapid movements of the horse, the swordsman, who has sprung unperceived from its back, approaches stealthily from behind, and with one stroke of his weapon, severs the tendon just above the heel. The disabled monster falls shrieking to the ground, and incapable of advancing a step, is soon despatched. The whole flesh is then cut off his bones into thongs, and hung like festoons upon the branches of trees till perfectly dry, when it is taken down and laid by for the rainy season.

African ivory is a not unimportant article of trade. The annual importation into Great Britain alone, for the last few years, has been about 1,000,000 lbs., which, taking the average weight of a tusk at 60 lbs., would require the slaughter of 8,333 male elephants, doomed to destruction in order to provide us with umbrella-stick or knife-handles, card-marks, fancy boxes, or buttons. Above 100,000*l.* worth of Sennaar ivory is annually exported from Alexandria, and Dr. Livingstone informs us that between July 1, 1848 and June 30, 1849, the value of the ivory shipped from St. Paul de Loanda amounted to 48,225*l.* A considerable quantity likewise finds its way to Europe by Zanzibar, the Cape, and the ports on the coast of Guinea; so that the annual produce of Africa probably exceeds 200,000*l.*

As the natives grow more alive to the value of the article, and become better provided with fire-arms, the persecuted giants disappear, and perhaps twenty years hence will become as rare in the countries in which they once abounded, as the sea elephant on Kerguelen's Land.

When Lake Ngami was first discovered (August 1, 1849), a trader, accompanying Messrs. Oswell, Murray, and Livingstone, purchased ivory at the rate of ten good large tusks for a musket worth 13*s.* They were called "bones," and Dr. Livingstone himself saw eight instances in which the tusks had been left to rot, with the remainder of the skeleton, where the elephant fell. The natives never had a chance of a market before, but in less than two years after the discovery, not a man of them could be found who did not fully know the value of ivory, and was of course a more determined elephant-hunter than he had ever been before.

The Asiatic elephant inhabits Hindostan, the Chin-Indian peninsula, Sumatra, Borneo, and Ceylon. In the latter island especially, he was formerly found in incredible numbers, so that thirty years ago, an English sportsman killed no less than 104 elephants in three days. Major Rogers shot upwards of 1,400;

Captain Galloway has the credit of slaying more than half that number; Major Skinner almost as many, and less persevering aspirants follow at humbler distances.

A reward of a few shillings per head offered by the government for taking elephants was claimed for 3,500 destroyed in part of the northern provinces alone, in less than three years prior to 1848, and between 1851 and 1856 a similar reward was paid for 2,000 in the southern provinces. In consequence of this wholesale slaughter, it cannot be wondered at that the Ceylon elephant has entirely disappeared from districts in which he was formerly numerous, and that the peasantry in some parts of the island have even suspended the ancient practice of keeping watchers and fires by night to drive away the elephants from the growing crops. The opening of roads, and the clearing of the mountain-forests of Kandy for the cultivation of coffee, have forced the animals to retire to the low country, where again they have been followed by large parties of European sportsmen, and the Singhalese themselves being more freely provided with arms than in former times, have assisted in the work of extermination.*

The practice in Ceylon is to aim invariably at the head, and, generally speaking, a single ball planted in the forehead ends the existence of the noble creatures instantaneously. Thus, while Prince Waldemar of Prussia, during his visit to the island, was hunting in the forests in company with Major Rogers, they were charged by two elephants, the one furiously trumpeting in their rear, while the other pushed its enormous head through the bushes in front. The major, however, soon put an end to their offensive demonstrations, for springing between them, he instantly lodged one bullet behind the ear of the one, and a second in the temple of the other. As if struck by lightning, they sank to the earth with a deep hollow groan, and the remainder of the herd, terrified by their fall, hurried away into the depth of the woods.

In India and Ceylon, elephants have been caught and tamed from time immemorial, and when we compare their colossal strength with the physical weakness of man, it surely must be considered a signal triumph of his intelligence and courage,

---

* Tennent's Ceylon, vol. ii, p. 273.

that he is able to bend such gigantic creatures to his will. The professional elephant-catchers of Ceylon or Panickeas, as they are called, are particularly remarkable for their daring and adroitness. Their ability in tracing their huge game, rivalling that of the American Indian in following the enemy's trail, has almost the certainty of instinct, and hence their services are eagerly sought by the European sportsmen who go down into their country in search of game. "So keen is their glance, that almost at the top of their speed, like hounds running breast-high, they will follow the course of an elephant over glades covered with stunted grass, where the eye of a stranger would fail to discover a trace of its passage, and on through forests strewn with dry leaves, where it seems impossible to perceive a footstep. Here they are guided by a bent or broken twig, or by a leaf dropped from the animal's mouth on which they can detect the pressure of a tooth. If at fault, they fetch a circuit like a setter, till lighting on some fresh marks, they go a-head again with renewed vigour. So delicate is the sense of smell in the elephant, and so indispensable is it to go against the wind in approaching him, that the Panickeas on those occasions when the wind is so still that its duration cannot be otherwise discerned, will suspend the film of a gossamer to determine it, and shape their course accordingly.

"On overtaking the game, their courage is as conspicuous as their sagacity. If they have confidence in the sportsman for whom they are finding, they will advance to the very heel of the elephant, slap him on the quarter, and then convert his timidity into anger, till he turns upon his tormentor, and exposes his heavy front to receive the bullet which is awaiting him.

"So fearless and confident are they, that two men without aid or attendants will boldly attempt to capture the largest-sized elephant. Their only weapon is a flexible rope made of buffalo's hide, with which it is their object to secure one of the hind-legs. This they effect either by following in his footsteps when in motion, or by stealing close up to him when at rest, and, availing themselves of the propensity of the elephant at such moments to swing his feet backwards and forwards, they contrive to slip a noose over his hind-leg.

"At other times, this is achieved by spreading the noose on

the ground, partially concealed by roots and leaves, beneath a tree on which one of the party is stationed, whose business it is to lift it suddenly by means of a cord, raising it on the elephant's leg at the moment when his companion has succeeded in provoking him to place his foot within the circle, the other end having been previously made fast to the stem of the tree. Should the noosing be effected in open ground, and no tree of sufficient strength at hand round which to wind the rope, one of the Moors, allowing himself to be pursued by the enraged elephant, entices him towards the nearest grove, when his companion, dexterously laying hold of the rope as it trails along the ground, suddenly coils it round a suitable stem, and brings the fugitive to a stand still. On finding himself thus arrested, the natural impulse of the captive is to turn on the man who is engaged in making fast the rope, a movement which it is the duty of his colleague to prevent by running up close to the elephant's head, and provoking him to confront him by irritating gesticulations and incessant shouts of *dah! dah!* a monosyllable, the sound of which the elephant peculiarly dislikes. Meanwhile the first assailant having secured one noose, comes up from behind with another, with which, amidst the vain rage and struggles of the victim, he entraps a fore-leg, the rope being as before secured to another tree in front, and the whole four feet having been thus entangled, the capture is completed.

"A shelter is then run up with branches to protect him from the sun, and the hunters proceed to build a wigwam for themselves in front of their prisoner, kindling their fires for cooking, and making all the necessary arrangements for remaining day and night on the spot, to await the process of subduing and taming his rage.

"Picketed to the ground like Gulliver by the Lilliputians, the elephant soon ceases to struggle, and what with the exhaustion of ineffectual resistance, the constant annoyance of smoke, and the liberal supply of food and water with which he is indulged, a few weeks generally suffice to subdue his spirit, when his keepers at length venture to remove him to their own village, or to the seaside for shipment to India.

"No part of the hunter's performances exhibits greater skill and audacity than this first forced march of the recently captured elephant. As he is still too morose to submit to be ridden, and it would be equally impossible to lead or to drive him by force, the ingenuity of the captors is displayed in alternately irritating and eluding his attacks, but always so attracting his attention, as to allure him along in the direction in which they want him to go.

"In Ceylon, the principal place for exporting these animals to India is Manaar on the western coast, to which the Arabs from the continent resort, bringing horses to be bartered for elephants. In order to reach the sea, open plains must be traversed, across which it requires the utmost courage, agility, and patience of the Panickeas to coax their reluctant charge. At Manaar the elephants are usually detained till any wound on the leg caused by the rope has been healed, when the shipment is effected in the most primitive manner, it being next to impossible to induce the still untamed creature to walk on board, and no mechanical contrivances being provided to ship him. A dhoney, or native boat, of about forty tons burthen, is brought alongside the quay, and being about three parts filled with the strong-ribbed leaves of the Palmyra palm, it is lashed so that the gunwale may be as nearly as possible on a line with the level of the wharf. The elephant, being placed with his back to the water, is forced by goads to retreat till his hind-legs go over the side of the quay; but the main contest commences when it is attempted to disengage his fore-feet from the shore, and force him to entrust himself on board.

"The scene becomes exciting from the screams and trumpetings of the elephants, the shouts of the Arabs, the calls of the Moors, and the rushing of the crowd. Meanwhile the huge creature strains every nerve to regain the land; and the day is often consumed before his efforts are overcome, and he finds himself fairly afloat. The same *dhoney* will take from four to five elephants, who place themselves athwart it, and exhibit amusing adroitness in accommodating their own movements to the rolling of the little vessel, and in this way they are ferried across the narrow strait which separates the continent of India from Ceylon." *

Unfortunately my limits forbid me entering upon a detailed account of the great elephant hunts of India and Ceylon, where

---

* Tennent's Ceylon, vol. ii. pp. 336-340.

whole herds are driven into an enclosure and entrapped in one vast decoy. This may truly be called the sublime of sport, for nowhere is it conducted on a grander scale, or so replete with thrilling emotions.

The *keddah* or *corral*, as the enclosure is called, is constructed in the depth of the forest, several hundred paces long, and half as broad, and of a strength commensurate to the power of the animals it is intended to secure. Slowly and cautiously the doomed herds are driven onwards from a vast circuit by thousands of beaters in narrowing circles to the fatal gate, which is instantly closed behind them, and then the hunters, rushing with wild clamour and blazing torches to the stockade, complete the terror of the bewildered animals. Trumpeting and screaming with rage and fear, they rush round the corral at a rapid pace, but all their attempts to force the powerful fence are vain, for wherever they assail the palisade, they are met with glaring flambeaux and bristling spears, and on whichever side they approach, they are repulsed with shouts and discharges of musketry. For upwards of an hour their frantic efforts are continued with unabated energy, till at length, stupified, exhausted, and subdued by apprehension and amazement, they form themselves into a circle, and stand motionless under the dark shade of the trees in the middle of the corral.

To secure the entrapped animals, the assistance of tame elephants or decoys is necessary, who, by occupying their attention and masking the movements of the nooser, give him an opportunity of slipping one by one a rope round their feet until their capture is completed.

The quickness of eye displayed by the men in watching the slightest movement of an elephant, and their expertness in flinging the noose over its foot, and attaching it firmly before the animal can tear it off with its trunk, are less admirable than the rare sagacity of the decoys, who display the most perfect conception of the object to be attained, and the means of accomplishing it. Thus Sir Emerson Tennent saw more than once, during a great elephant hunt which he witnessed in 1847, that when one of the wild elephants was extending his trunk, and would have intercepted the rope about to be placed over his leg, the decoy, by a sudden motion of her own trunk, pushed his aside and prevented him; and on one occasion, when successive efforts had failed to put the noose over the leg of an elephant who was already secured by one foot, but who wisely put the other to the ground as often as it was attempted to pass the noose under it, he saw the decoy watch her opportunity, and when his foot was again raised, suddenly push in her own leg beneath it, and hold it up till the noose was attached and drawn tight. Apart from the services which from their prodigious strength the tame elephants are alone capable of rendering, in dragging out and securing the captives, it is perfectly obvious that, without their sagacious cooperation, the utmost prowess and dexterity of the hunters would not avail them to enter the enclosure unsupported, or to ensnare and lead out a single captive.

It may easily be imagined that the passage from a life of unfettered liberty in the cool and sequestered forest to one of obedience and labour, must necessarily put the health of the captured animals to a severe trial. Many perish in consequence of the fearful wounds on the legs occasioned by their struggling against the ropes, and it has frequently happened that a valuable animal has lain down and died the first time it was tried in harness from what the natives designate a "*broken heart.*" Official records prove that more than half of the elephants employed in the public departments of the Ceylon government die in one year's servitude, and even when fully trained and inured to captivity, the working elephant is always a delicate animal, subject to a great variety of diseases, and consequently often incapacitated from labour. Thus, in spite of his colossal strength, which cannot even be employed to its full extent, as it is difficult to pack him without chafing the skin, and wagons of corresponding dimension to his muscular powers would utterly ruin the best constructed roads, it is very doubtful whether his services are in proportion to his cost, and Sir J. E. Tennent is of opinion that two vigorous dray horses would, at less expense, do more effectual work than any elephant. Most likely from a comparative calculation of this kind, the strength of the elephant-establishments in Ceylon has been gradually diminished of late years, so that the government stud, which formerly consisted of upwards of sixty elephants, is at present reduced to less than one quarter of that number.

In no kind of labour does the elephant display a greater

ingenuity than in dragging and piling felled timber, going on for hours disposing of log after log, almost without a hint or a direction of his attendant. In this manner two elephants, employed in piling ebony and satin wood in the yards attached to the Commissariat stores at Colombo, were so accustomed to the work that they were enabled to accomplish it with equal precision and with greater rapidity than if it had been done by dock-labourers. When the pile attained a certain height which baffled their conjoint efforts to raise one of the heavy logs of ebony to the summit, they had been taught to lean two pieces against the heap, up the inclined plane of which they gently rolled the remaining logs, and placed them trimly on the top.

Such is the earnestness and perseverance displayed by the sagacious creatures while accomplishing their task, that supervision might almost be thought superfluous, but as soon as the eye of the keeper is withdrawn, their innate love of ease displays itself, and away they stroll lazily to browse, or to enjoy the luxury of fanning themselves and blowing dust over their backs.

Leopard and Cheetah.

CHAPTER XLI.

THE FELIDÆ OF THE OLD WORLD.

The Lion—Conflicts with Travellers on Mount Atlas—The Lion and the Hottentot—A Lion taken in—Narrow Escapes of Andersson and Dr. Livingstone—Lion-Hunting by the Arabs of the Atlas—By the Bushmen—The Asiatic Lion—The Lion and the Dog—The Tiger—The Javanese Jungle—The Peacock—Wide Northern Range of the Tiger—Tiger-Hunting in India—Miraculous Escape of an English Sportsman—Animals announcing the Tiger's Presence—Turtle-Hunting of the Tiger on the Coasts of Java—The Panther and the Leopard—The Leopard attracted by the Smell of Small-Pox—The Cheetah—The Hyæna—Fables told of these abject Animals—The Striped Hyæna—The Spotted Hyæna—The Brown Hyæna.

THE majestic form, the noble bearing, the stately stride, the fine proportions, the piercing eye, and the dreadful roar of the Lion, striking terror into the heart of every other animal, all combine to mark him with the stamp of royalty. All nerve, all muscle, his enormous strength shows itself in the tremendous

bound with which he rushes upon his prey, in the rapid motions of his tail, one stroke of which is able to fell the strongest man to the ground, and in the expressive wrinkling of his ' row.

No wonder that, ever inclined to judge from outward appearances, and to attribute to external beauty analogous qualities of mind, man has endowed the lion with a nobility of character which he in reality does not possess. For modern travellers, who have had occasion to observe him in his native wilds, far from awarding him the praise of chivalrous generosity and noble daring, rather describe him as a mean-spirited robber, prowling about at night-time in order to surprise a weaker prey.

Lion.

The lion is distinguished from all other members of the feline tribe by the uniform colour of his tawny skin, by the black tuft at the end of his tail, and particularly by the long and sometimes blackish mane, which he is able to bristle when under the influence of passion, and which contributes so much to the beauty of the male, while it is wanting in the lioness, who, as everyone knows, is very inferior in size and comeliness to her stately mate.

His chief food consists of the flesh of the larger herbivorous animals, very few of which he is unable to master, and the swift-footed antelope has no greater enemy than he. Concealed in the high rushes on the river's bank, he lies in ambush for the timorous herd, which at night-fall approaches the water to quench its thirst. Slowly and cautiously the children of the waste advance; they listen with ears erect, they strain their eyes to penetrate the thicket's gloom, but nothing suspicious appears or moves along the bank. Long and deeply they quaff the delicious draught, but suddenly with a giant spring, like lightning bursting from a cloud, the lion bounds upon the unsuspecting revellers, and the leader of the herd lies prostrate at his feet, while his companions fly into the desert.

During the daytime the lion seldom attacks man, and sometimes even when meeting a traveller he is said to pass him by unnoticed; but when the shades of evening descend, his mood undergoes a change. After sunset it is dangerous to venture into the woody and wild regions of Mount Atlas, for there the lion lies in wait, and there one finds him stretched across the narrow path. It is then that dramatic scenes of absorbing interest not unfrequently take place. When, so say the Bedouins, a single man thus meeting with a lion is possessed of an undaunted heart, he advances towards the monster brandishing his sword or flourishing his rifle high in the air, and, taking good care not to strike or to shoot, contents himself with pouring forth a torrent of abuse:—"Oh, thou mean-spirited thief! thou pitiful waylayer! thou son of *one* that never ventured to say *no!* think'st thou I fear thee? Knowest thou whose son I am? Arise, and let me pass!" The lion waits till the man approaches quite near to him; then he retires, but soon stretches himself once more across the path; and thus by many a repeated trial puts the courage of the wanderer to the test. All the time the movements of the lion are attended with a dreadful noise, he breaks numberless branches with his tail, he roars, he growls; like the cat with the mouse, he plays with the object of his repeated and singular attacks, keeping him perpetually suspended between hope and fear. If the man engaged in this combat keeps up his courage,—if, as the Arabs express themselves, "he holds fast his soul," then the brute at last quits him and seeks some other prey. But if the lion perceives that he has to do with an opponent whose courage falters, whose voice trembles, who does not venture to utter a menace, then to terrify him still more he redoubles the described manœuvres. He approaches his victim, pushes him from the path, then leaves him and approaches again, and enjoys the agony of the wretch, until at last he tears him to pieces.

The lion is said to have a particular liking for the flesh of the Hottentots, and it is surprising with what obstinacy he will follow one of these unfortunate savages. Thus Mr. Barrow relates the adventure of a Namaqua Hottentot, who, endeavouring to drive his master's cattle into a pool of water enclosed between two ridges of rocks, espied a huge lion couching in the midst of the pool. Terrified at the unexpected sight of such a beast, that seemed to have his eyes fixed upon him, he instantly took to his heels. In doing this he had presence of mind enough to run through the herd, concluding that if the lion should pursue he would take up with the first beast that presented itself. In this, however, he was mistaken. The lion broke through the herd, making directly after the Hottentot, who, on turning round and perceiving that the monster had singled him out, breathless and half dead with fear, scrambled up one of the tree-aloes, in the trunk of which a few steps had luckily been cut out to come at some birds' nests that the branches contained. At the same moment the lion made a spring at him, but missing his aim, fell upon the ground. In surly silence he walked round the tree, casting at times a dreadful look towards the poor Hottentot, who screened himself from his sight behind the branches. Having remained silent and motionless for a length of time, he at length ventured to peep, hoping that the lion had taken his departure, when to his great terror and astonishment, his eyes met those of the animal, which, as the poor fellow afterwards expressed himself, flashed fire at him. In short, the lion laid himself down at the foot of the tree, and did not remove from the place for twenty-four hours. At the end of this time, becoming parched with thirst, he went to a spring at some distance in order to drink. The Hottentot now, with trepidation, ventured to descend, and scampered off home, which was not more than a mile distant, as fast as his feet could carry him.

Another time an elderly Hottentot observed a lion following him at a great distance for two hours together. He thence naturally concluded that the lion only waited the approach of darkness in order to make him his prey; and in the meantime expected nothing else than to serve for this fierce animal's supper, as he had no other weapon of defence than a staff. But as he was well acquainted with the nature of the lion and the manner of his seizing upon his prey, and at the same time had leisure at intervals to ponder on the ways and means in which it was most probable that his existence would be put an end to, he at length hit upon an expedient for saving his life. For this end, instead of making his way home, he looked out for a *klip-krans*, or a rocky place, level at top, and having a perpendicular precipice on one side of it; and sitting down on the edge of one of these precipices, he found to his great satisfaction that the lion also made a halt, and kept the same distance as before. As soon as it grew dark, the Hottentot sliding a little forwards, let himself down below the upper edge of the precipice upon a projecting part of the rock, where he could barely keep himself from falling. But in order to deceive the lion still more, he set his hat and cloak on the stick, making with it at the same time a gentle motion just over his head, and a little way from the edge of the mountain. This crafty expedient had the desired effect, for he did not remain long in that situation, before the lion came creeping softly towards him like a cat, and mistaking the skin-cloak for the Hottentot himself, took his leap with such exactness and precision as to fall headlong down the precipice, where he dashed out his brains upon the rock.

On account as well of the devastations which he causes among the herds as of the pleasure of the chase, the lion is pursued and killed in North and in South Africa wherever he appears, a state of war which, as may well be supposed, is not without danger for the aggressive party.

Thus Andersson, the well-known Swedish Nimrod, once fired upon a black-maned lion, one of the largest he ever encountered in Africa. Roused to fury by the slight wound he had received, the brute rapidly wheeled, rushed upon him with a dreadful roar, and at the distance of a few paces, couched as if about to spring, having his head imbedded, so to say, between his fore paws. Drawing a large hunting knife, and slipping it over the wrist of his right hand, Andersson dropped on one knee, and thus prepared, awaited the onset of the lion. It was an awful moment of suspense, and his situation was critical in the extreme. Still his presence of mind (a most indispensable quality in a South African hunter) never for a moment forsook him; indeed, he felt that nothing but the most perfect coolness and absolute self-command would be of any avail. He would now have become the assailant; but as, owing to the intervening bushes and clouds of dust raised by the lion's lashing his tail against the ground, he was unable to see his head, while to aim at any other part would have been madness, he refrained from firing. Whilst intently watching every motion of the lion, the animal suddenly made a prodigious bound; but whether it was owing to his not perceiving his intended victim, who was partially concealed in the long grass, and instinctively threw his body on one side, or to miscalculating the distance, he went clear over him, and alighted on the ground three or four paces beyond. Quick as thought Andersson now seized his advantage, and wheeling round on his knee, discharged his second barrel;

and as the lion's broadside was then towards him, lodged a ball in his shoulder, which it completely smashed. The infuriated animal now made a second and more determined rush; but owing to his disabled state was happily avoided, though only within a hair's breadth, and giving up the contest, he retreated into a neighbouring wood, where his carcase was found a few days after.

Dr. Livingstone once had a still more narrow escape, for he was actually under the paws of a lion, whose fury he had roused by firing two bullets into him. " I was upon a little height; he caught my shoulder as he sprang, and we both came to the ground below together. Growling horribly close to my ear, he shook me as a terrier-dog does a rat. The shock produced a stupor, similar to that which seems to be felt by a mouse after the first shake of the cat. It caused a sort of drowsiness in which there was no sense of pain nor feeling of terror, though quite conscious of all that was happening. It was like what patients partially under the influence of chloroform describe, who see all the operation, but feel not the knife. This singular condition was not the result of any mental process; the shake annihilated fear, and allowed no sense of horror in looking round at the beast. This peculiar state" (a fine remark) "is probably produced in all animals killed by the carnivora; and if so, is a merciful provision by our benevolent Creator for lessening the pain of death. Turning round to relieve myself of the weight, as he had one paw on the back of my head, I saw his eyes directed to Mebalwé, who was trying to shoot him at a distance of ten or fifteen yards. His gun, a flint one, missed fire in both barrels; the lion immediately left me, and attacking Mebalwé, bit his thigh. Another man attempted to spear the lion while he was biting Mebalwé. He left Mebalwé and caught this man by the shoulder, but at that moment the bullets he had received took effect and he fell down dead. The whole was the work of a few moments, and must have been his paroxysm of dying rage. A wound from this animal's tooth resembles a gun-shot wound; it is generally followed by a great deal of sloughing and discharge, and pains are felt in the part periodically ever afterwards. I had on a tartan jacket on the occasion, and I believe that it wiped off all the virus from the teeth that pierced the flesh, for my two companions in this affray have both suffered from the peculiar pains, while I have escaped with only the inconvenience of a false joint in my limb.

The man whose shoulder was wounded showed me his wound actually burst forth afresh on the same month of the following year. This curious point deserves the attention of inquirers."

In the Atlas, the lion is hunted in various ways. When he prowls about the neighbourhood of a Bedouin encampment, his presence is announced by various signs: at night, his dreadful roar resounds; now an ox, now a foal is missing from the herd; at length even a member of the tribe disappears. Terror spreads among all the tents, the women tremble for their children, everywhere complaints are heard. The warriors decree the death of the obnoxious neighbour, and congregate on horse and on foot at the appointed hour and place. The thicket in which the lion conceals himself during the daytime has already been discovered, and the troop advances, the horsemen bringing up the rear. About fifty paces from the bush they halt, and draw up in three rows, the second ready to assist the first in case of need, the third an invincible reserve of excellent marksmen. Then commences a strange and animated scene. The first row abusing the lion, and at the same time sending a few balls into his covert to induce him to come out, utters loud exclamations of defiance. "Where is he who fancies himself so brave, and ventures not to show himself before men? Surely it is not the lion, but a cowardly thief, a son of Scheitan, on whom may Allah's curse rest!"

At length, the roused lion breaks forth. A momentary silence ensues. The lion roars, rolls flaming eyes, retreats a few paces, stretches himself upon the ground, rises, smashes the branches with his tail. The front row gives fire, the lion springs forward, if untouched, and generally falls under the balls of the second row, which immediately advances towards him. This moment, so critical for the lion whose fury is fully excited, does not end the combat till he is hit in the head or in the heart. Often his hide has been pierced by a dozen balls before the mortal wound is given, so that sometimes in case of a prolonged contest several of the hunters are either killed or wounded. The horsemen remain as passive spectators of the fray so long as the lion keeps upon hilly ground, but when driven into the plain, their part begins, and a new combat of a no less original and dramatic character commences; as every rider, according to his zeal or courage, spurs his horse upon the monster, fires upon him at a short distance, then rapidly wheels as soon as the shot is made, and reloads again, to prepare for a new onset. The lion attacked on all sides and covered with wounds, fronts everywhere the enemy, springs forward, retreats, returns, and only falls after a glorious resistance, which must necessarily end in his defeat and death, as he is no match for a troop of well-mounted Arabs. After he has spent his power on a few monstrous springs, even an ordinary horse easily overtakes him. One must have been the witness of such a fight, says Dumas, to form an idea of its liveliness. Every rider utters loud imprecations, the white mantles that give so spectral an appearance to their dusky owners, fly in the air like "streamers long and gay," the carbines glisten, the shots resound, the lion roars; pursuit and flight alternate in rapid succession. Yet in spite of the tumult, accidents are rare, and the horsemen have generally nothing to fear but a fall from their steed, which might bring them under the claws of their enemy, or, what is oftener the case, the ball of an incautious comrade.

The Arabs have noticed that the day after the lion has carried away a piece of cattle, he generally remains in a state of drowsy inactivity, incapable of moving from his lair. When the neighbourhood, which usually resounds with his evening roar, remains quiet, there is every reason to believe that the animal is gorged with his gluttonous repast. Then some huntsman, more courageous than his comrades, follows his trail into the thicket, levels his gun at the lethargic monster, and sends a ball into his head. Sometimes even, a hunter, relying on the deadly certainty of his aim, and desirous of acquiring fame by a display of chivalrous courage, rides forth alone into the thicket, on a moonlight night, challenges the lion with repeated shouts and imprecations, and lays him prostrate before he can make his fatal bound.

Dr. Livingstone informs us that the Bushmen likewise avail themselves of the torpidity consequent upon a full meal, to surprise the lion in his slumbers: but their mode of attack is very different from that which I have described as practised by the fiery Arabs of Northern Africa. One discharges a poisoned arrow from the distance of only a few feet, while his companion simultaneously throws his skin-cloak over the beast's head. The sudden surprise makes him lose his presence of mind, and he bounds away in the greatest confusion and terror. The poison which they use is the entrails of a caterpillar named N'gwa, half an inch long. They squeeze out these, and place them all around the bottom of the barb, and allow the poison to dry in the sun. "They are very careful in cleaning their nails after working with it, as a small portion introduced into a scratch acts like morbid matter in dissection wounds. The agony is so great that the person cuts himself, calls for his mother's breast, as if he were returned in idea to his childhood again, or flies from human habitations a raging maniac. The effects on the lion are equally terrible. He is heard moaning in distress, and becomes furious, biting the trees and ground in rage."

The Arabs of the Atlas consider it much less dangerous to hunt the lion himself than to rob him of his young. Daily about three or four o'clock in the afternoon, the parent lions roam about, most likely to espy some future prey. They are seen upon a rising ground surveying the encampment, the smoke arising from the tents, the places where the cattle is preserved, and soon after retire with a deep growl.

During this absence from their den, the Bedouins cautiously approach to seize the young, taking good care to gag them, as their cries would infallibly attract the parent lion. After a razzia like this, the whole neighbourhood increases its vigilance, as for the next seven or eight days the fury of the lion knows no bounds, and it would then not be well to meet his eye.

In ancient times, the lion was an inhabitant of south-eastern Europe. Herodotus relates that troops of lions came down the Macedonian mountains, to seize upon the baggage camels of Xerxes' army, and even under Alexander the Great, the animal, though rare, was not yet completely extirpated.

In Asia also, where the lion is at present confined to Mesopotamia, the northern coast of the Persian Gulf, and the north-western part of Hindostan, he formerly roamed over far more extensive domains. The Asiatic lion differs from the African, by a more compressed form of body, a shorter mane, which sometimes is almost entirely wanting, and a much larger tuft of hair at the end of the tail.

Africa is the chief seat of the lion, the part of the world where he appears to perfection with all the attributes of his peculiar strength and beauty. There he is found in the wilds of the Atlas as in the high mountain-lands of Abyssinia, from the Cape to Senegal, and from Mozambique to Congo, and more than one species of the royal animal, not yet accurately distinguished by the naturalists, roams over this vast expanse. The lion is frequently brought to Europe, and forms one of the chief objects of attraction in zoological gardens. When taken young, he easily accustoms himself to captivity, and even propagates within his prison bounds, but the cubs born in our climate generally die young. He is not difficult to tame, shows himself thankful towards his keepers, and lives with them upon so friendly a footing, that they stroke and play with him.

There are many examples of lions having spared the life of dogs that had been thrown into their dens for food, and of the strongest affection having been formed between them. One of these couples existed a few years ago, in the Zoological Gardens at Antwerp, and it was most interesting to see the mighty African throwing himself on his back, and playfully tossing his tiny friend between his enormous paws. The dog — a regular spoilt child — frequently plagued his mighty comrade, though without ever making him impatient or angry.

The lion reigns in Africa, but the Tiger is lord and master of the Indian jungles. A splendid animal — elegantly striped with black on a white and golden ground; graceful in every movement, but of a most sanguinary and cruel nature. The lengthened body resting on short legs wants the proud bearing of the lion,

Tiger.

while the naked head, the wildly rolling eye, the scarlet tongue constantly lolling from the jaws, and the whole expression of the tiger's physiognomy indicate an insatiable thirst of blood, a pitiless ferocity, which he wreaks indiscriminately on every living thing that comes within his grasp. In the bamboo jungle on the banks of pools and rivers, he waits for the approaching herd; there he seeks his prey, or rather multiplies his

Axis.

murders, for he often leaves the carcase of the axis or the nylghau still writhing in the agony of death to throw himself upon new victims, whose bodies he rends with his claws, and then plunges his head into the gaping wound to absorb with deep and luxurious draughts the blood whose fountains he has just laid open.

Nothing can be more delightful than the aspect of a Javanese savannah, to which clumps of noble trees, planted by Nature's hand, impart a park-like character; yet even during the daytime, the traveller rarely ventures to cross these beautiful wilds without being accompanied by a numerous retinue. In Italy armed guards are necessary to scare the bandit; here the tiger calls for similar precautions. The horses frequently stand still, trembling all over, when their road leads them along some denser patch of the jungle, rising like an island from the grassy plain, for their acute scent informs them that a tiger lies concealed in the thicket, but a few paces from their path.

Nyighau.

It is a remarkable fact that the peacock and the tiger are so frequently seen together. The voice of the bird is seldom heard during the daytime, but as soon as the shades of evening begin to veil the landscape, his loud and disagreeable screams awaken the echoes, announcing, as the Javanese say, that the tiger is setting forth on his murderous excursions.

Then the traveller carefully bolts the door of his hut, and the solitary Javanese retreats to his palisadoed dwelling, for the tyrant of the wilderness is abroad. At night his dreadful roar is heard, sometimes accompanied by the peacock's discordant voice. Even in the villages, thinly scattered among the grass or alang-wilds of Java, there is no security against his attacks, in spite of the strong fences with which they are enclosed, and the watch fires carefully kept burning between these and the huts. During Junghuhn's sojourn in Tjurug Negteg (August, 1851) more than one family deplored the loss of a member. The monotonous, dreary neighbourhood and the depressed spirits of the wretched inhabitants, equally persecuted by tigers and poverty, made a melancholy impression on the naturalist, who hastened to leave the gloomy spot.

At night when he had been delayed on his journey, he often saw the eyes of the tiger glistening like balls of fire through the darkness. Then the horses trembled and "shook like the aspen leaves in wind," but loud screams constantly repelled the brute, who was perhaps no less afraid of the superiority of man.

India, South China, Sumatra, and Java, are the chief seats of the tiger, who is unknown both in Ceylon and Borneo, while to the north he ranges as far as Mandschuria and the Upper Obi, and Jennisei (55°—56° N. lat.). A species of tiger identical with that of Bengal is common in the neighbourhood of Lake Aral, near Sussac (45° N. lat.), and Tennant mentions that he is found among the snows of Mount Ararat in Armenia. As Hindostan is separated from these northern tiger haunts by the great mountain chains of Kuen-Lun (35° N.), and of Mouztagh (42° N. lat.), each covered with perpetual snows, mere summer excursions are quite out of the question, and it is evident that the animal is able to live in a much more rigorous climate than is commonly imagined. Even in India the tiger is by no means confined to the sultry jungle, for we learn from Mr. Hodgson's account of the mammalia of Nepaul, that in the Himalaya he is sometimes found at the very edge of perpetual snow.

If the French boast of their lion-killer Gerard, England possesses in Colonel Rice a no less bold and dexterous tiger-slayer, who, in the space of five campaigns (1850—1854) destroyed no less than sixty-eight of these formidable brutes and wounded a great many more. The gallant sportsman's work * contains, besides an account of his personal adventures, many excellent remarks on the habits of the animal, whom he so often encountered in his native wilds.

Tiger-hunting is a chief pleasure of the Indian rajahs and zemindars, who, anxious that their favourite amusement may suffer no diminution, forbid anyone else to chase on their domains, however much their poor vassals may have to suffer in consequence. But the delight they take in tiger-shooting never leads these cautious Nimrods so far as to endanger their precious persons. On some trees of the jungle a scaffolding is prepared, at a ludicrous height for his Highness, who, at the appointed hour makes his appearance with all the pomp of a petty Asiatic despot. The beating now begins, and is executed by a troop of miserable peasants, who most unwillingly submit to this forced and unpaid labour, which is the more dangerous for them as

* Tiger Shooting in India.

they are dispersed on a long line, instead of forming a troop, the only way to secure them against the attacks of the tiger. Thus they advance with a dreadful noise of drums, horns and pistol-firing, driving the wild beasts of the jungle towards the scaffolding of their lord and master. At first the tigers, startled from their slumbers, retreat before them, but generally on approaching the scaffolding they guess the danger that awaits them and turn with a formidable growl upon the drivers. Sometimes, however, they summon resolution to rush with a few tremendous bounds through the perilous pass, and their flight is but rarely impeded by the ill-aimed shots of the ambuscade. Nevertheless, great compliments are paid to the noble sportsman for his ability and courage, and nobody says a word about the poor low-born wretches, that may have been killed and mutilated by the infuriated brutes.

Our gallant countryman managed his tiger-shooting excursions on a very different plan. Provided with excellent double-barrelled rifles, and accompanied by a troop of well-armed, well-paid drivers, and a number of courageous dogs, he boldly entered the jungle to rouse the tiger from his lair. In front of the party generally marched the shikarree or chief driver, who attentively reconnoitring the traces of the animal, pointed out the direction that was to be followed. On his right and left hand walked the English sportsmen, fully prepared for action, and behind them the most trustworthy of their followers, with loaded rifles ready for an exchange with those that had been discharged. Then followed the music, consisting of four or five tambourins, a great drum, cymbals, horns, a bell, and the repeated firing of pistols, and convoyed by men armed with swords and long halberds. A few slingsmen made up the rear, who were constantly throwing stones into the jungle over the heads of the foremost of the party, and even more effectually than the noise of the music drove the tiger from his retreat. From time to time one of the men climbed upon the summit of a tree to observe the movements of the grass. The whole troop constantly formed a close body. The tiger in cold blood is never able to attack a company that announces itself in so turbulent a manner. If he ventures it is only with half a heart; he hesitates, stops at a short distance, and gives the hunter time to salute him with a bullet.

While strictly following an order of march like the one described, the drivers run little risk, even in the thickest jungle; but the difficulty is to keep them together, as the least success immediately tempts them to disperse.

On one of these hunting expeditions Ensign Elliott, a friend of the Colonel's, had an almost miraculous escape. Accompanied by about forty drivers they had entered a jungle, which did not seem to promise much sport, and had mounted with their rifles upon some small trees to await the issue of the explorations, when suddenly their people roused a beautiful tiger, who advanced slowly towards them. They remained perfectly silent, but one of their followers posted upon another tree, and fearing they might be surprised by the animal, called out to them to be upon their guard. This was enough to make the tiger change his direction, so that they had scarcely time to send a bullet after him. His loud roar announced that he was wounded, but the distance was already too great to admit of his being effectually hit a second time, so that the impatient sportsmen now pursued him with more eagerness than caution. At the head of their troop, they marched through the jungle, following the bloody trail of the animal, until at length they emerged into an open country, when all further traces were lost. In vain some of their people climbed upon the highest trees, nothing was to be perceived either in the bushes or in the high grass. Meanwhile the Englishmen slowly walked on, about twenty paces in advance, attentively gazing upon the ground, when suddenly with a terrific roar the tiger bounded upon Colonel Rice from a hollow, concealed beneath the herbage. The gallant sportsman had scarcely time to fire both his barrels at the head of the monster, who diverted from the attack by this warm reception, now made an enormous spring at Ensign Elliott before he had time to aim. All this was the work of a moment, for on turning towards the tiger, the Colonel saw his unfortunate friend prostrate under the paws of the furious brute.

Immediately the shikarree with admirable coolness handed him a freshly loaded rifle; he discharged one of the barrels without effect, but was then obliged to pause, as the tiger had seized his friend by the arm and was dragging him towards the hole from whence he had sprung forth. Thus it was absolutely necessary that the next shot should hit the animal in the brain, as any other wound not immediately fatal would only have increased its fury.

Closely following the tiger, and watching all his movements with the most intense attention, the Colonel, after having aimed several times, at length fired and hit the temple of the tiger, who fell over his victim a lifeless corpse.

Fortunately, the Ensign was not mortally wounded, the stroke of the tiger's paw, which had been aimed at his head, having been parried by his rifle. The blow, however, had been so furious, as to flatten the trigger, and thus he escaped with a terrible wound in the arm.

The tiger is particularly fond of dense willow or bamboo bushes on swampy ground, as he there finds the cool shades he requires for his rest during the heat of the day, after his nocturnal excursions. It is then very difficult to detect him, but the other inhabitants of the jungle, particularly the peacock and the monkey, betray his presence. The scream of the former is an infallible sign that the tiger is rising from his lair; and the monkeys, who during the night are so frequently surprised by the panther or the boa, never allow their watchfulness to be at fault during the day. They are never deceived in the animal, which slinks into the thicket. If it is a deer or a wild boar, they remain perfectly quiet, but if it is a tiger or a panther, they utter a cry, destined to warn their comrades of the approach of danger. When, on examining a jungle, the traveller sees a monkey quietly seated on the branches, he may be perfectly sure that no dangerous animal is lurking in the thicket.

Jackal.

During the night the cry of the jackal frequently announces the tiger's presence. When one of these vile animals is no longer able to hunt from age, or when he has been expelled from his troop, he is said to become the provider of the tiger, who, after having satiated himself on the spoil, leaves the remains to his famished scout.

Though the male-tiger ventures to attack the buffalo, an animal equally distinguished by courage and strength, the female, even when pressed by hunger, refrains from the perilous contest. A single buffalo may find it difficult to withstand the sharp claws and furious onset of the formidable brute; but when he belongs to a herd, his companions immediately come to his assistance and put the tiger to flight.

The Indian herdsmen, riding on a buffalo of their herd, are therefore not in the least afraid of entering the jungles that are infested by tigers. Colonel Rice once saw a troop of buffaloes, excited by the blood of a tiger he had wounded, throw themselves furiously into the thicket where the beast had sought refuge, beat about the bushes, and tear up the ground with their horns. Their rage became at length so ungovernable that they began to fight among each other, to the great despair of the herdsmen.

The she-buffalo is more timid, and knows not how to defend herself. Thus the tiger loves to steal upon her when alone, springs upon her, knocks her down with a stroke of his fore-paws, tears open her throat, and rejoices in the gushing streams of her blood.

As soon as the tiger has strangled his prey, the carrion vultures come flying in troops from a great distance. The jackals also assemble, and when the satiated monster retires, begin to fight with the vultures for the bloody remains. The birds are soon driven away, but they ultimately enjoy the picking of the bones, on which the jackals, who are not such expert dissectors, have left them many a delicate morsel.

The tiger, who on the declivities of the Himalaya tears to pieces the swift footed antelope, lacerates on the desert sand coasts of Java the tardy tortoise, when at nightfall it leaves the sea to lay its eggs in the drift-sand at the foot of the dunes. "Hundreds of tortoise skeletons lie scattered about the strand, many of them five feet long and three feet broad; some bleached by time, others still fresh and bleeding. High in the air a number of birds of prey wheel about, scared by the traveller's approach. Here is the place where the turtles are attacked by the wild dogs. In packs of from twenty to fifty, the growling rabble assails the poor sea animal at every accessible point, gnaws and tugs at the feet and at the head, and succeeds by united efforts in turning the huge creature upon its back. Then the abdominal scales are torn off, and the ravenous dogs hold a bloody meal on the flesh, intestines and eggs of their defenceless prey. Sometimes, however, the turtle escapes their rage, and dragging its lacerating tormentors along with it, succeeds in regaining the friendly sea. Nor do the dogs always enjoy an undisturbed repast, often during the night, the 'lord of the wilderness,' the royal tiger, bursts out of the forest, pauses for a moment, casts a glance over the strand, approaches slowly, and then with one bound, accompanied by a terrific roar, springs among the dogs, scattering the howling band like chaff before the wind. And now it is the tiger's turn to feast; but even he, though rarely, is sometimes disturbed by man. Thus on this lonely, melancholy coast, wild dogs and tigers wage an unequal war with the inhabitants of the ocean." *

After the tiger and the lion, the Panther and the Leopard are the mightiest felidæ of the Old World. Although differently spotted, the ocelli or rounded marks on the panther being larger and more distinctly formed, they are probably only varieties of one and the same species, as many intermediate individuals have been observed.

Both animals are widely diffused through the tropical regions of the Old World, being natives of Africa, Persia, China, India, and many of the Indian islands; so that they have a much more extensive range than either the tiger or the lion. The manner in which they seize their prey, lurking near the sides of woods, and darting forward with a sudden spring, resembles that of the tiger, and the chase of the panther is said to be more dangerous than that of the lion, as it easily climbs the trees and pursues its enemy upon the branches.

According to Sir J. E. Tennent, the panthers or leopards of Ceylon are strongly attracted by the peculiar odour which accompanies small-pox. The reluctance of the natives to expose themselves or their children to vaccination exposes the island to frightful visitations of this disease; and in the villages in the interior it is usual on such occasions to erect huts in the jungle to serve as temporary hospitals. Towards these the leopards are certain to be allured; and the medical officers are obliged to resort to increased precautions in consequence.

The Cheetah, or hunting leopard (*Gueparda jubata, guttata*), which inhabits the greater part both of Asia and Africa, exhibits in its form and habits a mixture of the feline and canine tribes.

---

* The Sea and its Living Wonders, p. 154.

Resembling the panther by its spotted skin, it is more elevated on its legs and less flattened on the fore part of its head. Its brain is more ample, and its claws touch the ground while walking like those of the dog, which it resembles still further by its mild and docile nature. In India and Persia, where the Cheetahs are employed in the chase, they are carried chained and hoodwinked to the field in low cars. When the hunters come within view of a herd of antelopes, the Cheetah is liberated, and the game is pointed out to him: he does not, however, immediately dash forward in pursuit, but steals along cautiously till he has nearly approached the herd unseen, when with a few rapid and vigorous bounds, he darts on the timid game and strangles it almost instantaneously. Should he, however, fail in his first efforts and miss his prey, he attempts no pursuit, but returns to the call of his master, evidently disappointed, and generally almost breathless.

The Cheetah.

While the sanguinary felidæ may justly be called the eagles, the carrion-feeding Hyænas are the vultures, among the four-footed animals. Averse to the light of day, like the owl and the bat, they conceal themselves in dark caverns, ruins, or burrows, as long as the sun stands above the horizon, but at night-fall they come forth from their gloomy retreats with a lamentable howl or a satanic laugh, to seek their disgusting food on the fields, in churchyards, or on the borders of the sea. From the prodigious strength of their jaws and their teeth, they are not only able to masticate tendons, but to crush cartilages and bones; so that carcases almost entirely deprived of flesh still provide them with a plentiful banquet.

Though their nocturnal habits and savage aspect have rendered them an object of hatred and disgust to man, they seem destined to fill up an important station in the economy of nature, by cleansing the earth of the remains of dead animals, which might otherwise infect the atmosphere with pestilential effluvia.

Among other fabulous qualities, a courage has been attributed to the hyæna which is completely alien to his base and grovelling nature. Far from venturing to attack the panther, or putting even the lion to flight, as Kämpfer pretended to have seen, he is in reality a most pusillanimous creature, and cautiously avoids a contest with animals much weaker than himself. Although his jaws are strong, he has not the sharp retractile claws of the felidœ, nor their formidable spring, his hind legs being comparatively feeble, and thus he can hardly become dangerous to the herds, though Bruce assures us that the hyœnas destroyed many of his mules and asses.

In Barbary, the Arabs pursue the hyænas on horseback, and run them down with their greyhounds, never thinking of wasting their powder on so abject a game. They are held in such contempt, that huntsmen will fearlessly penetrate into the caverns where they are known to sojourn, first carefully stopping the opening with their burnus, to keep out the light of day. They then advance towards the snarling brute, address it in menacing language, seize and gag it, without its venturing upon the least resistance, and cudgel the animal out of the den. The rough and ugly hide of the hyæna is but of little value, and in many tents its sight is not even tolerated, as so unworthy a spoil could only bring misfortune to its owner.

The intractability of the hyæna is as fabulous as his courage or his cruelty. On the contrary, he is very easily tamed, and may be rendered as docile as the dog himself.

The striped hyæna is a native of Asiatic Turkey, Syria, and North Africa, as far as the Senegal, while the spotted hyæna ranges over South Africa, from the Cape to Abyssinia. Both species attain the size of the wolf, and have similar habits. As the shark follows the ship, or the crow the caravan, they are said to hover about the march of armies, as if taught by instinct that they have to expect the richest feast from the insanity of man.

Striped Hyæna.

The moonlight falling on the dark cypresses and snow-white tombs of the Oriental churchyards not seldom shines upon hungry hyænas, busily employed in tearing the newly-buried corpses from their graves.

A remarkable peculiarity of the spotted hyæna is that when he first begins to run he appears lame, so that one might almost fancy one of his legs was broken; but after a time this halting disappears, and he proceeds on his course very swiftly.

"One night, in Maitsha," says Bruce, "being very intent on observation, I heard something pass behind me towards the bed, but upon looking round could perceive nothing. Having finished what I was then about, I went out of my tent, intending directly to return, which I immediately did, when I perceived large blue eyes glaring at me in the dark. I called upon my servant for a light, and there was a hyæna standing nigh the head of the bed, with two or three large bunches of candles in his mouth. To have fired at him, I was in danger of breaking my quadrant or other furniture; and he seemed, by keeping the candles steadily in his mouth, to wish for no other prey at that time. As his mouth was full, and he had no claws to tear with, I was not afraid of him, but with a pike struck him as near the heart as I could judge. It was not till then he showed any sign of fierceness, but upon feeling his wound he let drop the candles and endeavoured to run up the shaft of the spear to arrive at me, so that in self-defence I was obliged to draw a pistol from my girdle and shoot him, and nearly at the same time my servant cleft his skull with a battle-axe."

The *brown* hyæna, which is found in South Africa, from the Cape to Mozambique and Senegambia, and has a more shaggy fur than the preceding species, has very different habits. He is particularly fond of the crustaceæ which the ebbing flood leaves behind upon the beach, or which the storm casts ashore in great quantities, and exclusively inhabits the coasts, where he is known under the name of the sea-shore wolf. His traces are everywhere to be met with on the strand, and night after night he prowls along the margin of the water, carefully examining the refuse of the retreating ocean.

CHAPTER XLII.

THE FELIDÆ OF THE NEW WORLD.

The Jaguar — His Boldness — Jaguar Hunting — Heroic Conflict of Three Brazilian Herdsmen with a Jaguar — The Couguar, or the Puma — His Cowardice — The Ocelot.

THE same radical differences which draw so wide a line of demarcation between the simiæ of the Old and the New World are found also to distinguish the feline races of both hemispheres, so that it would be as vain to search in the American forests and savannas for the Numidian lion, or the striped tiger, as on the banks of the Ganges or the Senegal for the tawny puma, or the spotted jaguar. While in the African plains the swift-footed spring-bok, or the koodoo, unrivalled among the antelopes for his bold and widely-spreading horns, falls under the impetuous bound of the panther — or while the tiger and the buffalo engage in mortal combat in the Indian jungle — the blood-thirsty Jaguar, concealed in the high grass of the American llanos, lies in wait for the wild horse or the passing steer.

Jaguar.

The arrival of the Spaniards in the New World, so destructive to most of the Indian tribes with whom they came into contact, was beneficial at least to the large felidæ of tropical America, for they first introduced the horse and the ox into the western hemisphere, where these useful animals, finding a new and congenial home in the boundless savannas and pampas, which extend almost uninterruptedly from the Apure to Patagonia, have multiplied to an incredible extent. Since then the jaguar no longer considers the deer of the woods, the graceful agouti, or the slow capybara as his chief prey, but rejoices in the blood of the steed or ox, and is much more commonly met with in the herd-teeming savannas than in the comparatively meagre hunting-grounds of the forest.

Of all the carnivora of the New World, perhaps with the sole exception of the grisly and the polar bears, the tyrants of the North American solitudes, the jaguar is the most formidable, resembling the panther by his spotted skin, but almost equalling the Bengal tiger in size and power. He roams about at all times of the day, swims over broad rivers, and even in the water proves a most dangerous foe, for when driven to extremities he frequently turns against the boat, and forces his assailants to seek their safety by jumping overboard. Many an Indian, while wandering through thinly populated districts, where swampy thickets alternate with open grass plains, has been torn to pieces by the jaguar, and in many a lonely plantation the inhabitants hardly venture to leave their enclosures after sunset, for fear of his attacks. During Tschudi's sojourn in Northern Peru, a jaguar penetrated into the hut of an Englishman who had settled in those parts, and dragging a boy of ten years out of his hammock, tore him to pieces and devoured him. Far from being afraid of man, this ferocious animal springs upon him when alone, and when pressed by hunger will even venture during the day time into the mountain villages to seek its prey. The distinguished traveller whom I have just quoted mentions the case of an Indian in the province of Vito, who hearing during the night his only pig most piteously squeaking, rose to see what was the matter, and found that a jaguar had seized it by the head and was about to carry it away. Eager to rescue his property, he sprang forward, and seizing the pig by the hind legs, disputed its possession with the beast of prey, that with eyes gleaming through the darkness, and a ferocious growl, kept tugging at its head. This strange struggle between the undaunted Indian and the jaguar lasted for some time, until the women coming out of the hut with lighted torches put to flight the monster, which slowly retreated into the forest.

The same traveller relates that in some parts the jaguars had increased to such a degree, and proved so destructive to the inhabitants, that the latter were obliged to emigrate, and settle in less dangerous districts. Thus, the village of Mayunmarca, near the road from Huanta to Anco, had been long since abandoned, and the neighbourhood was still considered so dangerous that few Indians ventured to travel through it alone.

The chase of these formidable animals requires great caution, yet keen sportsmen will venture, single-handed, to seek the jaguar in his lair, armed with a blow-pipe and poisoned arrows, or merely with a long and powerful lance. The praise which is due to the bold adventurers for their courage is, however, too often tarnished by their cruelty. Thus, a famous jaguar hunter once showed Pöppig a large cavity under the tangled roots of a giant bombax-tree, where he had some time back discovered a female jaguar with her young. Dexterously rolling down a large stone, he closed the entrance, and then with fiendish delight slowly smoked the animals to death, by applying fire from time to time to their dungeon. Having lost one half of his scalp in a previous conflict with a jaguar, he pleaded his sufferings as an excuse for his barbarity.

To attack these creatures with a lance, a sure arm, a cool determined courage, and great bodily strength and dexterity are required, but even these qualities do not always ensure success if the hunter is unacquainted with the artifices of the animal. The jaguar generally waits for the attack in a sitting posture, turning one side towards the assailant, and, as if unconcerned, moves his long tail to and fro. The hunter, carefully observing the eye of his adversary, repeatedly menaces him with slight thrusts of his lance, which a gentle stroke of the paw playfully wards off; then seizing a favourable moment, he suddenly steps forward and plunges his weapon into his side. If the thrust be well aimed, a second is not necessary, for pressing with his full weight on the lance, the huntsman enlarges and deepens the mortal wound. But if the stroke is parried or glances off, the jaguar, roused to fury, bounds on his aggressor, whose only hope now lies in the short knife which he carries in his girdle.

All those that have escaped from one of these death-struggles affirm that the breath of the enraged animal is of a suffocating heat, with a smell like that of burning capsicum, and that its pestilential contact produces an inflammation of the throat, which lasts for several days. Those who are less inclined to desperate conflicts destroy the jaguar by poisoned pieces of meat, or else they lay pitfalls for him, when they kill him without running any personal risk. Like the cayman, the jaguar, after having once tasted the flesh of man, is said to prefer it to anything else. During his first solitary journeys through the American wilds, the traveller's sensations, on meeting with the fresh footmarks of the monster, are like those of Robinson Crusoe when he discovered the vestiges of the savage on the beach of his lonely island; but as the animal itself very rarely crosses the wanderer's path, he at length becomes completely indifferent, and roams about the wilderness as unconcernedly as if no beasts of prey existed under the forest shade, or among the high grasses of the savannah. During his long residence in Yuurmangua Pöppig met but one jaguar, who, not deeming it advisable to engage in hostilities, slowly retreated into the woods.

In the Brazilian campos great devastations are caused among the herds by the jaguar, who has strength enough to drag an ox to some distance. He frequently kills several bullocks in one night, and sucks their blood, leaving their flesh for a future repast. When, after having satiated himself, he retires to a neighbouring thicket, the vaqueros or herdsmen follow his bloody trail with their hounds, and as soon as the jaguar sees the pack approach, he seeks to climb the inclined trunk of a tree, and is then shot down from his insecure station. But the chase does not always terminate without accident or loss of life, as very strong jaguars will face the dogs, kill several of them, and frequently carry them away and devour them.

While Prince Maximilian of Neu Wied was travelling through the campos, he heard of the heroic conflict of three vaqueros with a monstrous jaguar that had never been known to retreat. One day, while following their herds through the woods, their dogs discovered the fresh foot-prints of the beast, and following the scent, soon brought it to a stand. Armed merely with their long lance-like *varas*, the bold men did not long deliberate, but resolutely advanced towards the jaguar, who stood confronting the dogs, and immediately bounding upon his new antagonists, wounded them one after the other, though not without receiving repeated thrusts of their lances and knives. The least determined of the three, appalled by his wounds, at first retreated, but seeing the boldest of his companions lying prostrate under the paws of the monster, his courage revived, and the attack being vigorously renewed, the jaguar was at length killed. The bleeding and exhausted heroes were hardly able to crawl home in the evening. They pointed out the spot where they had fought, and where the jaguar was found swimming in his blood, surrounded by the dogs which he had torn to pieces.

There is a black variety of the jaguar, on whose dark skin the ring-formed spots are still visible, and which is said to surpass the common species in size and ferocity.

The Couguar, or the Puma, as he is called by the Indians, is far

Puma.

inferior to the jaguar in courage, and consequently far less dangerous to man. On account of his brownish-red colour and great size, being the largest felis of the new world, he has also been named the American lion, but he has neither the mane nor the noble bearing of the "king of animals." In spite of his strength he is of so cowardly a disposition that he invariably takes to flight at the approach of man, and consequently inspires no fear on being met with in the wilderness; while even the boldest hunter instinctively starts back, when, winding through the forest, he suddenly sees the sparkling eye of the jaguar intently fixed upon him.

The puma has a much wider range than the jaguar, for while the latter reaches in South America only to the forty-fifth degree of latitude, and does not rove northwards beyond Sonora and New Mexico, the former roams from the Straits of Magellan to the Canadian lakes. The jaguar seldom ascends the mountains to a greater height than 3,000 feet, while in the warmer lateral valleys of the Andes the puma frequently lies in ambush for the vicunas at an elevation of 10,000 feet above the level of the sea. He can climb trees with great facility, ascending even vertical trunks, and, like the lynx, will watch the opportunity of springing on such animals as happen to pass beneath. No less cruel than cowardly, he will destroy without necessity forty or fifty sheep when the occasion offers, and content himself with licking the blood of his victims. When caught young, he is easily tamed, and, like the common cat, shows his fondness at being caressed by the same kind of gentle purrings.

Tschudi informs us that the Indians of the northern provinces frequently bring pumas to Lima, to show them for money. They either lead them by a rope, or carry them in a sack upon their back, until the sight-seers have assembled in sufficient number.

Besides the puma or the jaguar, tropical America possesses the beautifully variegated Ocelot (*Felis pardalis*); the Oscollo (*F. celidogaster*); the spotless, black-grey Jaguarundi (*F. jaguarundi*), which is not much larger than the European wild cat; the long-tailed, striped, and spotted Margay or Tiger-cat, and several other felidæ.

Ocelot.

All these smaller species hardly ever become dangerous to man, but they cause the death of many an acouchi and cavy; and, with prodigious leaps, the affrighted monkey flies from their approach into the deepest recesses of the forest.

CHAPTER XLIII.

THE SLOTH.

Miserable Aspect of the Sloth—His Beautiful Organisation for his Peculiar Mode of Life—His Rapid Movements in the Trees—His Means of Defence—His Tenacity of Life—Fable about the Sloth refuted—The Ai—The Unau—The Mylodon Robustus.

"THE piteous aspect, the sorrowful gestures, the lamentable cry of the Sloth, all combine to excite commiseration.

The Sloth.

While other animals assemble in herds, or roam in pairs through the boundless forest, the sloth leads a lonely life in those immeasurable solitudes, where the slowness of his movements exposes him to every attack. Harmless and frugal, like a pious anchorite, a few coarse leaves are all he asks for his support. On comparing him with other animals, you would say that his deformed organisation was a strange mixture of deficiency and superabundance. He has no cutting teeth, and though possessed of four stomachs, he still wants the long intestines of ruminating animals. His feet are without soles, nor can he move his toes separately. His hair is coarse and wiry, and its dull colour reminds one of grass withered by the blasts of surly winter. His legs appear deformed by the manner in which they are attached to the body, and his claws seem disproportionably long. Surely a creature so wretched and ill-formed stands last on the list of all the four-footed animals, and may justly accuse Nature of step-motherly neglect!"

When seeing a captured sloth painfully creeping along on even ground, sighing and moaning, and scarcely advancing a few steps after hours of awkward toil, the observer might well be disposed to acquiesce in the foregoing remarks, and to fancy he had discovered a flaw among the general beauty of the Creator's works; but let him view the animal in the situation for which it was ordained, and he will soon retract his hasty judgment, and discover it to be no less perfect in its kind, and no less admirably fitted for its sphere of existence, than the most highly organised of the mammalian tribes.

For the sloth, in his wild state, spends his whole life in the trees, and never once touches the earth but through force or by accident. Like the monkey, he has been formed for an exclusively sylvan life, high above the ground, in the green canopy of the woods; but while the nimble simiæ constantly live *upon* the branches, the sloth is doomed to spend his whole life *under* them. He moves, he rests, he sleeps suspended from the boughs of trees, a wonderfully strange way of life, for which no other four-footed animal of the Old or the New World has been destined.

And now examine his organisation with reference to this peculiar mode of existence, and all his seeming deficiencies and deformities will appear most admirably adapted to his wants, for these strong, muscular, preposterously long fore-feet, while the hinder extremities are comparatively short and weak, these slender toes armed with enormous claws, are evidently as well suited for clasping the rugged branch as the enormous hind legs of the kangaroo for bounding over the arid plain. Indeed, in every case, we shall find the fundamental type or idea of the four extremities belonging to the vertebrated animals most admirably modified according to their wants: here shortened, there prolonged; here armed with claws, there terminating in a hoof; here coalescing to a tail, there assuming the shape of a fin; here clothed with feathers to cleave the air, there raised to the perfection of the human hand, the wonderful instrument of a still more wonderful intelligence; and who, seeing all this, can possibly believe that the world is ruled by chance, and not by an all-pervading and almighty power.

Thus the sloth, so helpless when removed from his native haunts, is far from exhibiting the same torpidity in his movements when seen in the place for which Nature fitted him.

"One day, as we were crossing the Essequibo," says Mr. Waterton, "I saw a large sloth on the ground upon the bank; how he had got there nobody could tell; the Indian said he had never surprised a sloth in such a situation before: he would hardly have come there to drink, for both above and below the place the branches of the trees touched the water, and afforded him an easy and safe access to it. Be this as it may, though the trees were not above twenty yards from him, he could not make his way through the sand time enough to escape before we landed. As soon as we came up to him, he threw himself upon his back, and defended himself in gallant style with his fore-legs. 'Come, poor fellow!' said I to him, 'if thou hast got into a hobble to-day, thou shalt not suffer for it; I'll take no advantage of thee in misfortune; the forest is large enough both for thee and me to rove in. Go thy ways up above, and enjoy thyself in these endless wilds; it is more than probable thou wilt never have another interview with man. So fare thee well.' On saying this I took up a long stick which was lying there, held it for him to hook on, and then conveyed him to a high and stately mora. He ascended with wonderful rapidity, and in about a minute he was almost at the top of the tree. He now went off in a side direction, and caught hold of the branch of a neighbouring tree; he then proceeded towards the heart of the forest. I stood looking on, lost in amazement at his singular mode of progress. I followed him with my eye till the intervening branches closed in betwixt us, and then lost sight for ever of the sloth. I was going to add that I never saw a sloth take to his heels in such earnest, but the expression will not do, for the sloth has no heels."

The Indians, to whom no one will deny the credit of being acute observers of animal life, say that the sloth wanders principally when the wind blows. In calm weather he remains still, probably not liking to cling to the brittle extremity of the branches, lest they should break under his weight in passing from one tree to another; but as soon as the breeze rises, the branches of the neighbouring trees become interwoven, and then he seizes hold of them and pursues his journey in safety. There is seldom an entire day of calm in the forests of Guiana. The trade-wind generally sets in about ten o'clock in the morning, and since the sloth, as we have just seen, is able to travel at a good-round pace when he has branches to cling to, there is nothing to prevent him making a considerable way before the sun sinks, and the wind goes down.

During night, and while reposing in the day-time, the sloth

constantly remains suspended by his feet, for his anatomy is such that he can feel comfortable in no other position. In this manner he will rest for hours together, expressing his satisfaction by a kind of purring, and from time to time his dismal voice may be heard resounding through the forest, and awakening at a distance a similar melancholy cry.

The colour of the sloth's hair so strongly resembles the hue of the moss which grows on the trees, that the European finds it very difficult to make him out when he is at rest, and even the falcon-eyed Indian, accustomed from his earliest infancy to note the slightest signs of forest life, is hardly able to distinguish him from the branches to which he clings. This no doubt serves him as a protection against the attacks of many enemies; but, far from being helpless, his powerful claws and the peculiarly enduring strength of his long arms, make very efficient weapons of defence against the large tree snakes that may be tempted to make a meal of him.

Among other strange stories related of the sloth, it has been asserted that he lives but on the leaves of one particular tree, the cecropia, and that when he has entirely stripped it, and feels the torments of hunger, he drops upon the ground, and then with the greatest difficulty, and many a piteous moan, strives to attain the nearest cecropia that has not yet been plundered of its foliage. It is hardly necessary to point out how very improbable it is that, among the infinite variety of leaves in the primeval forest, the choice of the sloth should thus be limited to one species, and that with his great facility of locomotion in the trees he should unnecessarily subject himself to a painful fall and a most irksome journey. During his extensive wanderings in the forests of Guiana, Schomburgk never once saw a tree that had been robbed of its verdure, though sometimes ten or twelve sloths might be seen clinging to its branches; and Waterton even hazards a conjecture that by the time the animal had finished the last of the old leaves, there would be a new crop on the part of the tree he had stripped first, ready for him to begin again, so quick is the process of vegetation in these damp and sultry wilds.

The sloth possesses a remarkable tenacity of life, and withstands the dreadful effects of the wourali poison of the Macushi Indians longer than any other animal. Schomburgk slightly scratched a sloth in the upper lip, and rubbed a minimum of the venom in the wound, which did not even emit a drop of blood; he then carried the animal to a tree, which it began to climb, but after having reached a height of about twelve feet, it suddenly stopped, and swinging its head about from side to side, as if uncertain which way to go, tried to continue its ascent, which, however, it was unable to accomplish. First it let go one of its fore-feet, then the other, and remained attached with its hind legs to the tree until, these also losing their power, it fell to the ground, where, without any of the convulsive motions or the oppressive breathing which generally mark the effect of the wourali, it expired in the thirteenth minute after the poison had been administered.

The sloths attain a length of about two feet and a half, and form two genera—the Unaus, with two-toed fore-feet and three-toed hinder extremities, and the Aïs, with three toes on each foot. The former have forty-eight ribs, the latter only thirty-two. Their way of living is the same, and their range is limited to the forests of Guiana and the Brazils. They bring forth and suckle their young like ordinary quadrupeds, and the young sloth, from the moment of its birth, adheres to the body of its parent till it acquires sufficient size and strength to shift for itself.

Sloth-like animals of colossal dimensions — Megatheriums, Mylodons—extinct long before man appeared upon the scene, inhabited the forests of South America during the tertiary ages of the world.

From the dentition of the mylodon, it may be concluded that, like the sloth of the present day, this monstrous animal fed on the leaves or slender terminal twigs of trees, but while the former, from the comparatively light weight of his body, is enabled to run along the under side of the boughs till he has reached a commodious feeding-place, the elephantine bulk of the mylodon evidently rendered all climbing utterly impossible.

First scratching away the soil from the roots of the tree on whose foliage he intended to feast, he next grasped it with his long fore-legs, and rocking it to and fro, to right and left, soon brought it to the ground, for "extraordinary must have been the strength and proportions of that tree," says Professor Owen, "which in such an embrace could long withstand the efforts of its ponderous assailant."

CHAPTER XLIV.

THE ANT-EATERS OF THE NEW AND THE OLD WORLD.

The Great Ant-Bear — His Way of Licking up Termites — His Formidable Weapons — A Perfect Forest-Vagabond — His Peculiar Manner of Walking — The Smaller Ant-Eaters — The Manides — The African Orycteropi — The Armadillos — The Glyptodon — The Porcupine Ant-Eater of Australia — The Myrmecobius Fasciatus.

THE great Ant-bear is undoubtedly one of the most extraordinary denizens of the wilds of South America, for that a powerful animal, measuring above six feet from the snout to the end of the tail, should live exclusively on ants, seems scarcely less remarkable than that the whale nourishes his enormous body with minute pteropods and medusæ.

Ant-bear.

The vast mouth of the leviathan of the seas has been most admirably adapted to his peculiar food, and it was not in vain that Nature gave such colossal dimensions to his head, as it was necessary to find room for a gigantic straining apparatus, in which, on rejecting the engulphed water, thousands upon thousands of his tiny prey might remain entangled; but the ant-bear has been no less wonderfully armed for the capture of the minute animals on which he feeds, and if, on considering the use for which it was ordained, we become reconciled to the seeming disproportion of the whale's jaws, the small and elongated, snout-like head of the ant-bear will also appear less uncouthly formed when we reflect that it is in exact accordance with the wants of the animal. For here no deep cavity was required for the reception of two rows of powerful teeth, as in most other

K K

quadrupeds, but a convenient furrow for a long and extensile tongue — the use of which will immediately become apparent on following the animal into the Brazilian campos, where, as we have seen in a former chapter, the wonderful cities of the white ant are dispersed over the plains in such incalculable numbers. Approaching one of these structures, the ant-bear strikes a hole through its wall of clay, with his powerful crooked claws, and as the ants issue forth by thousands to resent the insult, stretches out his tongue for their reception. Their furious legions, eager for revenge, immediately rush upon it, and, vainly endeavouring to pierce its thick skin with their mandibles, remain sticking in the glutinous liquid with which it is lubricated from two very large glands situated below its root. When sufficiently charged with prey, the ant-bear suddenly withdraws his tongue and swallows all the insects.

Without swiftness to enable him to escape from his enemies, for man is superior to him in speed; without teeth, the possession of which would assist him in self-defence; without the power of burrowing in the ground, by which he might conceal himself from his pursuers, without a cave to retire to, the ant-bear still ranges through the wilderness in perfect safety, and fears neither the boa nor the jaguar, for he has full reliance on his powerful fore-legs and their tremendous claws. Richard Schomburgk had an opportunity of witnessing a young ant-bear make use of these formidable weapons.

On the enemy's approach it assumed the defensive, but in such a manner as to make the boldest aggressor pause, for, resting on its left fore-foot, it struck out so desperately with its right paw as would undoubtedly have torn off the flesh of any one that came in contact with its claws. Attacked from behind, it turned round with the rapidity of lightning, and on being assailed from several quarters at once, threw itself on its back, and, desperately fighting with both its fore-legs, uttered at the same time an angry growl of defiance. In fact, the ant-bear is so formidable an opponent that he is said not unfrequently to vanquish even the jaguar, the lord of the American forests, for the latter is often found swimming in his blood, with ripped-up bowels, a wound which, of all the beasts of the wilderness, the claws of the ant-bear are alone able to inflict.

On seizing an animal with these powerful weapons, he hugs it close to his body, and keeps it there till it dies through pressure or hunger. Nor does the ant-bear, in the meantime, suffer much from want of aliment, as it is a well-known fact that he can remain longer without food than perhaps any other quadruped, so that there is very little chance indeed of a weaker animal's escaping from his clutches.

Peaceable and harmless, the ant-bear when unprovoked never thinks of attacking any other creature; and as his interests and pursuits do not interfere with those of the more formidable denizens of the wilderness, he would, without doubt, attain a good old age, and be allowed to die in peace, if, unfortunately for him, his delicate flesh did not provoke the attacks of the large carnivora and man. To be sure, the Indian fears his claws, and never ventures to approach the wounded ant-bear until he has breathed his last; nor can he be hunted with dogs, as his skin is of a texture that perfectly resists a bite, and his hinder parts are effectually protected by thick and shaggy hair; yet, armed with the dreadful wourali poison, the Indian knows how to paralyse in a few minutes his muscular powers, and to stretch him dead upon the earth.

A perfect forest vagabond, the ant-bear has no den to retire to, nor any fixed abode; his immense tail is large enough to cover his whole body, and serves him as a tent during the night, or as a waterproof mantle against the rains of the wet season, so that he might boast, like Diogenes, of carrying all he required about him.

The peculiar position of his paws, when he walks or stands, is worthy of notice. He goes entirely on the outer side of his fore-feet, which are quite bent inwards, the claws collected into a point and going under the foot. In this position he is quite at ease, while his long claws are disposed of in a manner to render them harmless to him, and are prevented from becoming dull and worn, which would inevitably be the case did their points come in actual contact with the ground, for they have not that retractile power which is given to animals of the feline race, enabling them to preserve the sharpness of their claws on the most flinty path. In consequence of its resting perpetually on the ground, the whole outer side of the foot is not only deprived of hair, but hard and callous, while, on the contrary, the inner side of the bottom of the foot is soft and hairy.

The intellectual faculties of the ant-bear are said to be very feeble, and in fact a large tongue seems much more necessary to the creature than a large brain. His small cerebral developement is another reason, besides his want of teeth, for the smallness of his head, as the casket was naturally made to conform to its diminutive contents.

Besides the great ant-bear, there are two other species of American ant-eaters, one nearly the size of a fox, and the smallest not much larger than a rat. Being provided with prehensile tails, they are essentially arboreal, while the great ant-bear, incapable of climbing, always remains on the ground, where, thanks to the abundance of his prey, he is always sure of obtaining a sufficient supply of food with very little trouble.

The Manides, Pangolins, or scaly Ant-eaters of South Africa and Asia, resemble the myrmecophagi of America in having a very long extensile tongue, furnished with a glutinous mucus for securing their insect food, and in being destitute of teeth, but differ wholly from them in having the body, limbs, and tail covered with a panoply of large imbricated scales, overlapping each other like those of the lizard tribes, and also in being able to roll themselves up when in danger, by which their trenchant scales become erect, and present a formidable defensive armour, so that even the tiger would vainly attempt to overcome the Indian pangolin (*Manis pentadactyla*).

The manides are inoffensive animals, living wholly on ants and termites, and chiefly inhabit the most obscure parts of the forest, burrowing in the ground to a great depth, for which purpose, as also for extracting their food from ant-hills and decaying wood, their feet are armed with powerful claws, which they double up in walking, like the ant-bear of Brazil.

Orycteropus Capensis.

Besides several species of manides, Africa possesses a peculiar class of ant-eaters in the Orycteropi, which are found from the Cape to Senegambia and Abyssinia, all over the sultry plains where their food abounds. Their legs are short, and provided with claws fit for burrowing in the earth, which they can do with great rapidity; and when once the head and fore-feet have penetrated into the ground, their hold is so tenacious that even the strongest man is incapable of dragging them from their hole.

The Orycteropi, or earth-hogs (*Aard-varks*) as they are called by the boors, from their habit of burrowing and their fancied resemblance to small short-legged pigs, have an elongated head, though less tapering than that of the American myrmecophagi, and are provided with peculiarly formed teeth, with a flat crown and undivided root, which is pierced with a multitude of little holes, like those of a ratan-cane when cut transversely, while the ant-bears have no teeth at all. Their way of feeding is the same, and to enable them to retain their nimble-footed prey, their tongue is likewise lubricated with a glutinous liquid. Their flesh is considered very wholesome and palatable, and at the Cape they are frequently hunted both by the colonists and the Hottentots. There are several species, all very much resembling each other: their stout body measures about five feet from the tip of the snout to the end of the tail, the latter being nearly half the length of the body.

The American Armadillos have many points in common with the myrmecophagi, manides, and orycteropi. They have neither fore nor canine teeth, but a number of conical grinders, and are distinguished by having the upper part of their bodies defended by a complete suit of armour, divided into joints or bands, folding one over the other like the parts of a lobster's tail, so as to accommodate themselves to all the motions of the animal.

Poyou Armadillo.

In life this shell is very limber, so that the armadillo is able to go at full stretch, or to roll himself up into a ball as occasion may require. These animals are very common both in the forests and in the open plains of South America, where they burrow in the sand-holes like rabbits. The armadillo is seldom seen abroad during the day, and when surprised he is sure to be near the mouth of his hole; but after sunset he sallies forth in search of roots, grain, worms, insects, and other small animals, and when disturbed, coils himself up in his invulnerable armour like the hedge-hog, or squats close to the ground, or, if he has time enough, escapes by digging into the earth, a work which he performs with masterly dexterity. " As it often takes a considerable time to dig him out of his hole," says Mr. Waterton, " it would be a long and laborious business to attack each hole indiscriminately, without knowing whether the animal were there or not. To prevent disappointment, the Indians carefully examine the mouth of the hole, and put a short stick down it. Now if, on introducing the stick, a number of mosquitos come out, the Indians know to a certainty that the armadillo is in it; whenever there are no mosquitos in the hole, there is no armadillo. The Indian having satisfied himself that the armadillo is there by the mosquitos which come out, he immediately cuts a long and slender stick, and introduces it into the hole; he carefully observes the line the stick takes, and then sinks a pit in the sand to catch the end of it; this done, he puts it further into the hole, and digs another pit, and so on, till at last he comes up with the armadillo, which had been making itself a passage in the sand till it had exhausted all its strength through pure exertion. I have been sometimes three quarters of a day in digging out one armadillo, and obliged to sink half a dozen pits, seven feet deep, before I got up to it. The Indians and negroes are very fond of the flesh, but I considered it strong and rank.

"On laying hold of the armadillo, you must be cautious not to come in contact with his feet; they are armed with sharp claws, and with them he will inflict a severe wound in self-defence: when not molested, he is very harmless and innocent; he would put you in mind of the hare in Gay's fable:—

> 'Whose care was never to offend,
> And every creature was her friend.'"

The family of the armadillos has been subdivided into no less than six genera and fifteen species, which are chiefly distinguished from each other by the number of their shelly bands, their teeth, and their toes. They might also be conveniently divided into two tribes, the one with a long and conical tail, the other with a short caudal appendage, formed like a club. They differ greatly in size, for while the giant armadillo (*Pri*-

odontes gigas) is at least four feet long from the tip of the snout to the tip of the tail, the *Chlamyphorus truncatus*, which inhabits the province of Mendoza in the Andes, and is remarkable for its mole-like propensities, passing the greatest part of its life under ground, scarcely measures six inches in length.

Giant Armadillo.

But even the giant armadillo is a pigmy when compared to the extinct mail-clad animals, which at times of unknown antiquity, peopled the plains of South America. Mr. Darwin saw, in the possession of a clergyman near Monte Video, the fragment of a tail of one of these monsters of the past, from which he conjectured that it must have been from six to ten feet long; and the glyptodon, of which the College of Surgeons possesses an admirable specimen, and which, like the armadillos of the present day, was covered with a tesselated bony armour, was equal in size to the rhinoceros! How formidable must have been the enemies which made it necessary for an animal like this to move about with harness on its back!

The curious Echidna, or Porcupine Ant-eater (*Echidna hystrix*) of Australia, is a striking instance of those beautiful gradations so frequently observed in the animal kingdom, by which creatures of various tribes or genera are blended, as it were, or linked together, and of the

Porcupine Echidna.

wonderful diversity which Nature has introduced into the forms of creatures destined to a similar mode of life. It has the general appearance and external coating of the porcupine, with the mouth and peculiar generic characters of the ant-eaters. It is about a foot in length, and burrows with wonderful facility by means of its short muscular fore-feet and its sharp-pointed claws. When attacked, it rolls itself into a ball like the hedgehog, erecting the short, strong, and very sharp spines with which the upper parts of the body and tail are thickly coated.

Australia is likewise the native country of another ant-eating animal, the marsupial *Myrmecobius fasciatus*. It is formed like a squirrel, and is of the size of the rat; its brown red fur, with six or seven light yellow transverse bands over the back, gives it an elegant appearance. It was discovered about thirty years ago in the neighbourhood of Swan River.

Indian Pangolin.

Flying Foxes.

CHAPTER XLV.

TROPICAL BATS.

Wonderful Organisation of the Bats—The Fox-Bat—The Vampire—Its Blood-Sucking Propensities—The Vampire and the Scotchman—The Horse-Shoe Bats—The Nycteribia—The Flying Squirrel—The Galeopithecus—The Anomalurus.

WHEN the sun has disappeared below the horizon, and night falls on the landscape, which a little while ago was bathed in light, then from hollow trees, and creviced rocks, and ruined buildings, a strange and dismal race comes forth.

Silently hovering through the glades of the woods, or skimming along the surface of the streams, it catches the crepuscular or nocturnal moths, and serves like the swallow by day to check the exuberant multiplication of the insect tribes. But while man loves the swallow, and suffers him to build his nest under the eaves of his dwelling, he abhors the bat, which like an evil spirit avoids the light of day, and seems to feel happy only in darkness. The painter gives to his angels the white pinions of the swan, while his demons are made to bear the black wings of the bat. And yet the bat, in Europe at least, is a most inoffensive creature, which, may well claim the gratitude of the farmer, from the vast number of cockchafers and other noxious insects which it destroys; while a closer inspection of its wonderful organisation proves it to be far more deserving of admiration than of repugnance. Can anything be better adapted to its wants than the delicate membrane, which, extending over the long slim fingers, can be spread and folded like an umbrella, so as to form a wing when the animal wishes to fly, and to collapse into a small space when it is at rest? How slight the bones, how light the body, how beautifully formed for flight! Admire also the tiny unwebbed thumb, which serves the bat to hook itself fast while resting, or to clip off the wings of the flies or moths, which it never devours with the rest of the body.

But the exquisite acuteness of the senses of smell, feeling, and hearing in the bat is still more wonderful than its delicate flying apparatus. Naturalists, more curious than humane, have blinded bats, and seen, to their astonishment, that they continued to fly about, as if still possessed of the power of vision. They always knew how to avoid branches suspended in the room in which they were flitting, and even flew betwixt threads hung perpendicularly from the ceiling, though these were so near each other, that they were obliged to contract their wings in order to pass through them.

To explain these wonderful phenomena, Spallanzani and other naturalists of the last century believed the bats to be endowed with a sixth sense; but Carlyle found that, on closing the ears of the blinded creatures, they lose their wonderful power, and hit against the sides of the room, without being at all aware of their situation.

How they are able to distinguish night from day, when shut up in a dark box, is a fact still unexplained. As long as the sun stands above the horizon, they will remain perfectly quiet.

but as soon as twilight begins to darken the earth, a strange piping and chirping and scratching is heard within the lightless dungeon, and scarcely has the lid been raised, when the prisoners rapidly escape.

Though temperate Europe possesses many bats, yet they are most numerous and various in the woody regions of the tropical zone, where the vast numbers of the insect tribes and forest fruits afford them a never-failing supply of food. There also they attain a size unknown in our latitudes, so that both from their dimensions and their physiognomy, many of the larger species have obtained the name of flying-dogs or flying-foxes.

On approaching a Javanese village, you will sometimes see a stately tree, from whose branches hundreds of large black fruits seem to be suspended.

A strong smell of ammonia and a piping noise soon, however, convince you of your mistake, and a closer inspection proves them to be a large troop of Kalongs or Fox-bats (*Pteropus*) attached head downwards to the tree, where they rest or sleep during the day time, and which they generally quit at sunset, though some of them differ so much from the usual habits of the family as to fly about in the broad light of day.

Many species of pteropi are found all over the torrid zone in the Old World, but they abound particularly in the East Indian archipelago. They belong to the rare quadrupeds indigenous in some of the South Sea Islands, such as Tonga or Samoa, and extend northwards as far as Japan, and southwards to Van Diemen's Land. They occasion incalculable mischief in the plantations, devouring indiscriminately every kind of fruit, from the cocoa-nut or the banana to the matchless mangosteen; but on the other hand, the gigantic kalong of Java (*Pteropus edulis*), whose body attains a length of a foot and a half, and whose outstretched wings measure no less than four feet and a half from tip to tip, is eaten and esteemed as a delicacy by the natives.

The same essential differences which we observe between the monkeys of both hemispheres, are also found to exist between the large bats of the Old and the New World. Not a single pteropus is to be found in all America, while the Phyllostomidæ, distinguished by the orifices of their nostrils being placed in a kind of membranous scutcheon, surmounted by a leaf-like expansion, like the head of a lance, and supposed to extend in an extraordinary degree the sense of smelling, are exclusively confined to the western continent. These phyllostomidæ are remarkable for their blood-sucking propensities, and under the name of vampires have brought the whole race of the large tropical bats into evil repute.

Prince Maximilian of Neu Wied * often saw by moonshine, or in the twilight, the Guandiru (*Phyllostoma hastatum*), a bat five inches long, and measuring twenty-three inches with outstretched wings, hover about his horses and mules while grazing after their day's journey. The animals did not seem incommoded by its presence, but on the following morning, he generally found them covered with blood from the shoulders to the hoofs. The muscular under-lip of the phyllostoma can be completely folded together in the shape of a sucking-tube, which, after the sharp canine teeth have penetrated the skin, continues to pump forth the blood. Even man himself is liable to the attacks of the larger phyllostomidæ.

"Some years ago," says Mr. Waterton, " I was in Demarara with a Scotch gentleman, by name Tarbet. We hung our hammocks in the thatched loft of a planter's house. Next morning I heard this gentleman muttering in his hammock, and now and then letting fall an imprecation or two, just about the time he ought to have been saying his morning prayers. 'What is the matter, Sir?' said I, softly: 'is anything amiss?' 'What's the matter?' answered he surlily; 'why, the vampires have been sucking me to death.' As soon as there was light enough, I went to his hammock, and saw it much stained with blood. 'There,' said he, thrusting his foot out of the hammock, 'see how these infernal imps have been drawing my life's blood.' On examining his foot, I found the vampire had tapped his great toe: there was a wound somewhat less than that made by a leech; the blood was still oozing from it. I conjectured he might have lost from ten to twelve ounces of blood. Whilst examining it, I think I put him into a worse humour by remarking that an European surgeon would not have been

---

* Travels in the Brazils.

so generous as to have blooded him without making a charge. He looked up in my face, but did not say a word; I saw he was of opinion that I had better have spared this piece of ill-timed levity."

Captain Stedman, while in Surinam, was attacked in a similar way. "On waking about four o'clock one morning in my hammock, I was extremely alarmed at finding myself weltering in congealed blood, yet without feeling any pain whatever. Having started up, I ran for the surgeon, with a firebrand in one hand, and all over besmeared with gore, to which, if added my pale face, short hair, and tattered apparel, he might well ask the question:

> 'Be thou a spirit of health or goblin damn'd,
> Bring with thee airs from heaven or blasts from hell?'

The mystery, however, was soon solved, for I then found I had been bitten by the vampire or spectre of Guiana."

Other instances of the same kind are mentioned by Tschudi, Schomburgk, Azara (who was phlebotomised no less than four times by the vampire), and other naturalists of equal repute, so that there is no reason to doubt the fact, although Prince Maximilian of Neu Wied was unable to ascertain its truth during the course of his travels.

The general food of the phyllostomidæ consists, however, in vespertine and nocturnal moths, and Waterton is of opinion that they also partake of vegetable food. As there was a free entrance and exit to the vampire in the loft where he slept at Mibiri Creek in Demerara, he had many a fine opportunity of paying attention to the nocturnal surgeon. When the moon shone bright, and the fruit of the banana tree was ripe, he could see him approach and eat it. He would also bring into the loft from the forest a green round fruit, something like the wild guava, and about the size of a nutmeg. There was something also in the blossom of the Sawarri nut tree which was grateful to him, for on coming up Waratella Creek in a moonlight night, he saw several vampires fluttering round the top of the Sawarri tree, and every now and then the blossoms which they had broken off, fell into the water. They certainly did not drop off naturally, for on examining several of them, they appeared quite fresh and blooming; and so Waterton concluded the vampires pulled them from the tree, either to get at the young fruit or to catch the insects which often take up their abode in flowers.

The Vampire (*Phyllostoma spectrum*), in general, measures about twenty-six inches from wing to wing extended, so that his dimensions are not equal to those of the oriental kalong. Like the flying foxes, he may sometimes be seen in the forest hanging in clusters, head downwards, from the branch of a tree, a circumstance of which Goldsmith seems to have been aware, for in the "Deserted Village," speaking of America, he says —

> "And matted woods, where birds forget to sing,
> But silent bats in drowsy clusters cling."

Some of the phyllostomidæ have a tongue once as long again as the head, and armed at the extremity with recurved bristles, like that of the wood-pecker, no doubt a very serviceable instrument for extracting insects from the narrow hollows and crevices of trees and rocks.

The Rhinolophi or Horse-shoe Bats of the old continent, have also a more or less complicated nasal appendage, or foliaceous membrane at the end of the nose, but differing in its conformation, from that of the phyllostomidæ. They are insectivorous, like most of their order, and none of them seem to indulge in the blood-sucking propensities of the large American vampires. They chiefly inhabit the tropical regions of Africa and Asia, and more particularly the Indian archipelago, but the *Rhinolophus unihastatus* ranges in Europe as far as England.

Rhinolophus.

Numerous genera and species of tropical bats, distinguished

from each other by the formation of their teeth, lips, nostrils, heads, wings, and tails, have already been classified by naturalists, but many, no doubt, still live unknown in their gloomy retreats, for who is able to follow them into the obscure nooks of the forest, or in intricate caverns, and accurately to observe them during their nocturnal rambles? It may give an idea of their vast numbers throughout the torrid zone, when we hear that in Ceylon alone about sixteen species have been identified, and of these two varieties are peculiar to the island. Unlike the sombre bats of the northern climates, the colours of some of them are as brilliant as the plumage of a bird, bright yellow, deep orange, or of a rich ferruginous brown, thus contradicting the general belief which attires nocturnal animals in vestures as dark as their pursuits.

The torrid zone, which produces the largest bats, also gives birth to the tiniest representatives of the order, such as the minute Singhalese variety of *Scotophilus Coromandelicus*, which is not much larger than the humble bee, and of a glossy black colour. "It is so familiar and gentle," says Sir J. E. Tennent, "that it will alight on the cloth during dinner, and manifests so little alarm, that it seldom makes any effort to escape before a wine-glass can be inverted to secure it."

The fur of this pretty little creature, like that of many other bats, is frequently found infested with a most singular insect. Unlike most parasites, which are either extremely sluggish in their movements, or even condemned to utter immobility, the velocity of the nycteribia is truly marvellous, and as its joints are so flexible as to yield in every direction, it tumbles through the fur of the bat, rotating like a wheel on the extremities of its spokes, or like the clown, in a pantomime, hurling himself forward on hands and feet alternately. To assist its mountebank movements, each foot is armed with two sharp hooks, with elastic pads opposed to them, so that the hair can not only be rapidly seized and firmly clasped, but as quickly disengaged as the creature whirls away in its headlong career. But the strangest peculiarity of the nycteribia is the faculty which it possesses of throwing back or inverting its head so completely, that the underside becoming uppermost, the mouth, the eyes, and the antennæ are completely hid between its shoulders, and then again projecting it forward by a sudden jerk of its long flexible neck. By means of this wonderful organisation, the nimble parasite feels completely at home in the furry coat which has been assigned to it as a pasture ground, and whisks along as easily through the hairy thicket as the monkey through the bush-ropes of the forest.

Though incapable of a prolonged flight like the bats, several other tropical quadrupeds have been provided with extensions of the skin, which give them the power of supporting themselves for some time in the air, and of making prodigious leaps. Thus, by means of an expansile furry membrane reaching from the fore-feet to the hind, the Flying Squirrels (*Pteromys*) bound, or rather swiftly sail, to the distance of twenty fathoms or more, and thus pass from one tree to another, always directing their flight obliquely downwards. They very rarely descend to the ground, and when taken or placed on it, run or spring somewhat awkwardly with their tail elevated, beginning to climb with great activity as soon as they reach a tree.

The Galeopitheci are in like manner enabled to take long sweeping leaps from tree to tree, by means of an extension of their skin between the anterior and posterior limbs on each side, and between the posterior limbs, including also the tail. These

Galeopithecus volans.

extraordinary animals are natives of the islands of the Indian archipelago. They inhabit lofty trees in dark woods, to which they cling with all four extremities. During the day time, they suspend themselves like bats from the branches, with their heads downwards, but at night they rouse themselves and make an active search for food, which consists of fruits, insects, eggs, and birds. They are inoffensive, but on attempting to seize them, they inflict a sharp scratch with their trenchant nails.

The Anomaluri of the west coast of Africa, which have only been known to the world since 1842, and possess a most remarkable tail, covered on the lower surface of its base with imbricated horny scales, resemble the galeopitheci by the wing-

THE ANOMALURI

like expansion of their skin, and no doubt the investigations of travellers will bring to light other animals endowed with similar parachutes, for when we consider the large number of quadrupeds that have been discovered since the beginning of the century, we have every reason to believe that many still remain unknown.

Flying Foxes at rest.

Monkeys of the Old World.

CHAPTER XLVI.

THE SIMIÆ OF THE OLD WORLD.

Forest-Life — Excellent Climbers — Bad Pedestrians — Imperfectly known to the Ancients — Similitude and Difference between the Human Race and the Apes — The Chimpanzee — Chim in Paris — The Gorilla — The Uran — The Gibbons — A Siamang on Board — The Proboscis Monkey — The Huniman — The Wanderoo — The Cercopitheci — The Maimon — "Happy Jerry" — The Pig-Faced Baboon — The Derryas — The Loris and Makis.

IN the midst of tropical vegetation, the Simiæ lead a free forest-life, for which they might well be envied. The green canopy of the woods protects them at every season of the year from the burning rays of a vertical sun, flowers of the most delicious fragrance embalm the air they breathe, and an endless supply of fruits and nuts never allows them to know want, for should the stores near at hand be exhausted, an easy migration to some other district soon restores them to abundance. With an agility far surpassing that with which the sailor ascends the rigging, and climbs even to the giddy top of the highest mast, they leap from bush-rope to bush-rope, and from bough to bough, mocking the tiger-cat and the boa, which are unable to follow them in their rapid evolutions.

Formed to live on trees, and not upon the ground, they are as excellent climbers as they are bad pedestrians. Both their fore and hind-feet are shaped as hands, generally with four fingers and a thumb, so that they can seize or grasp a bough with all alike.

Buffon erroneously remarks of the chimpanzee, that he always walks erect, even when carrying a weight; but this ape, as well as the other anthropomorphous simiæ, proves by the slowness and awkwardness of his movements, when by chance he walks upon even ground, that this position is by no means natural to him, or congenial to his organisation. Man alone, of all creatures, possesses an upright walk; the ape, on the contrary, always stoops, and not to lose his equilibrium when walking, is obliged to place his hands upon the back of his head, or on his loins. Thus, in his native wilds, he rarely has recourse to this inconvenient mode of progression, and when forced by some chance or other to quit the trees, he leans while walking upon the finger-knuckles of his anterior extremities, a position which in fact very much resembles walking on all-fours.

It is, indeed, only necessary to compare the long, robust, and muscular arms of the chimpanzee with his weaker and shorter hind-feet, to be at once convinced that he was never intended for walking. But see with what rapidity, with what power and grace, he moves from branch to branch, his hind-legs serving him only as holdfasts, while his chief strength is in his arms. The tree is, without all doubt, for him what the earth is for us, the air for the bird, or the water for the fish.

We cannot wonder at the ancients having known but few species of simiæ, as these animals chiefly belong to the torrid zone, with which the Greeks and Romans were so imperfectly acquainted. It is only since a wide extent of the tropical regions has been opened by trade or conquest to European research, that many of the mysteries of monkey-existence have been brought to light from the darkness of the primeval forest, and particularly within the last twenty years, naturalists and travellers have devoted so much attention to this interesting family, that while in the year 1840 only 128 species were known, their numbers had increased in 1852 to no less than 210, a stately list, to which, since then, many more have been added, and which even now is far from being closed.

The simiæ of the Old World are all distinguished by the common character of a narrow septum or partition of the nose like that of man, and by the same number of teeth, each jaw being provided with ten grinders, two canine teeth, and four incisors, as in the human race.

The large apes, or tailless monkeys, resemble us besides in many other respects, as well in their external appearance as in their anatomical structure; and form, as it were, the caricature of man, both by their gestures and by glimpses of a higher intelligence.

Creatures so remarkably endowed have naturally at all times attracted a great share of attention, for if even the lowest links in the chain of animated beings lay claim to our interest, how much more must this not be the case with beings whose faculties seem almost to raise them to the rank of our relations. The question how far this similarity extends has naturally given rise to many acute investigations and been differently answered, according as naturalists were more or less inclined to depress man to the level of the ape, or to widen the gulph between them. The former, pointing to the brutality of the lowest savages, would willingly make us believe that we are nothing but an improved edition of the Uran, while the latter complacently cite in favour of their opinion, the incommensurable distance which exists between even the most degraded specimens of humanity and the most perfect quadrumana. Man alone is capable of continually progressive improvement; in him alone each generation inherits the acquirements of its fathers and transmits the growing treasure to its sons, while the ape, like all other animals, constantly remains at the same point. The lowest savage knows how to make fire; the ape, though he may have seen the operation performed a thousand times, and have enjoyed the genial warmth of the glowing embers, will never learn the simple art. His hairy skin is a sufficient proof of his low intellect, an infallible sign that as he never would be able to provide himself with an artificial clothing, Nature was obliged to protect him against the inclemencies of the cold nights and the pouring rain. As man advances in age, his mind acquires a greater depth and a wider range. In the ape, on the contrary, signs of a livelier intelligence are only exhibited during youth, and as the animal waxes in years, its physiognomy acquires a more brutal expression; its forehead recedes, its jaws project, and instead of expanding to a higher perfection, its mental faculties are evidently clouded by a premature decline.

Both in Africa and Asia, we find large anthropomorphous apes, but while the chimpanzee and the gorilla exclusively belong to the African wilds, the uran and the gibbons are confined to the torrid regions of South Asia.

Chimpanzee.

The Chimpanzee (*Simia troglodytes*) attains a height of about five feet, but seems much smaller from his stooping attitude. He inhabits the dense forests on the west coast of Africa, particularly near the river Gaboon, and as his travels are facilitated by his fatherland not being too far distant from Europe, there is hardly a Zoological Garden of any note that does not exhibit a chimpanzee among its lions. One of the finest specimens ever seen was kept a few years since in the Jardin des Plantes in Paris, where the mild climate, agreeable diet (he drank his pint of Bordeaux daily), and lively society of the French maintained him in wonderful health and spirits.

"The last time I saw him" (May, 1854), says an accomplished naturalist,[*] "he came out to inhale the morning air in the large circular inclosure in front of the monkey palace, which was built for our poor relations by M. Thiers. Here Chim began his day by a leisurely promenade, casting pleased and thankful glances towards the sun, the beautiful sun of early summer.

"He had three satellites, Coatimondis, either by chance or to amuse him, and while making all manner of eyes at a young lady, who supplied the *Singerie* with pastry and cakes, one of the coatimondis came up stealthily behind, and dealt him a

---

[*] Quarterly Review, 1855, p. 22.

small but malicious bite. Chim looked round with astonishment at this audacious outrage on his person, and put his hand hastily upon the wound, but without losing his temper in the least.

Rufous Coatimondi.

He walked deliberately to the other side of the circle, and fetched a cane which he had dropped in his promenade. He returned with majestic wrath upon his brow, mingled, I thought, with contempt, and taking coati by the tail, commenced punishment with his cane, administering such blows as his victim could bear without permanent injury, and applied with equal justice on the ribs at either side. When he thought enough had been done, he disposed of coati, without moving a muscle of his countenance, by a left-handed jerk, which threw the delinquent high in air, head over heels.

"He came down a sadder and a better coati, and retired with shame and fear to a distant corner. Having executed this act of justice, Chim betook himself to a tree. A large baboon, who had in the meantime made his appearance in the circle, thought this was a good opportunity of doing a civil thing, and accordingly mounted the tree, and sat down smilingly, as baboons smile, upon the next fork. Chim slowly turned his head at this attempt at familiarity, measured the distance, raised his hind foot, and as composedly as he had caned the coati, kicked the big baboon off his perch into the arena below. This abasement seemed to do the baboon good, for he also retired like the coati, and took up his station on the other side. To what perfection of manners and developement of thought the last year and a half may have brought him, I can scarcely guess; but one day, doubtless, some one will say of him, as an Oriental prince once said to me after looking at the uran 'Peter'—'Does he speak English yet?'"

The body of the chimpanzee is covered with long hair on the head, shoulders, and back, but much thinner on the breast and belly. The arms and legs are not so disproportionate as those of the uran, the fore-fingers not quite touching the knees when the animal stands upright. The upper part of the head is very flat, with a retiring forehead, and a prominent bony ridge over the eye-brows, the mouth is wide, the ears large, the nose flat, and the face of a blackish-brown colour.

From this short notice it will be seen at once that friend Chim has not the least claim to beauty, but yet he is far from equalling the hideous deformity of the Gorilla, whom M. Du Chaillu has so prominently introduced to public notice. This savage animal, which is covered with black hair like the chimpanzee, and resembles it in the proportion of its body and limbs, though its form is much more robust, unites a most ferocious and undaunted temper with an herculean bodily strength, and is said to hold undisputed dominion of the hill-forests in the interior of Lower Guinea, forcing even the panther to ignominious flight.

To kill a gorilla is considered by the negroes as a most courageous exploit; and Dr. Savage, an American missionary on the coast of Guinea, who, in a memoir published at Boston in the year 1847, was the first to point out the generic differences between this formidable ape and the chimpanzee, tells us that a slave having shot a male and female gorilla, whose skeletons afterwards came into his possession, was immediately set at liberty and proclaimed the prince of hunters.

M. Du Chaillu's description of his first encounter with an adult gorilla, which entirely agrees with the accounts given to Dr. Savage by the natives of the mode of attack of this monstrous creature, shows that this distinction was by no means unmerited, and that it requires all the coolness and determination of an accomplished sportsman to face an animal of such appalling ferocity and power.

"The under-bush swayed rapidly just a-head, and presently before us stood an immense male gorilla. He had gone through the jungle on his all-fours, but when he saw our party he erected himself, and looked us boldly in the face. He stood about a dozen yards from us, and was a sight I think I shall never forget. Nearly six feet high (he proved four inches shorter), with immense body, huge chest, and great muscular arms, with fiercely glaring, large deep-grey eyes, and a hellish expression of face, which seemed to me like some night-mare vision; thus stood before us the king of the African forest. He was not afraid of us. He stood there and beat his breast with his huge fists, till it resounded like an immense bass-drum, which is their mode of offering defiance, meantime giving vent to roar after roar. The roar of the gorilla is the most singular and awful noise heard in these African woods. It begins with a sharp bark like an angry dog, then glides into a deep bass roll which literally and closely resembles the roll of distant thunder along the sky, for which I have been sometimes tempted to take it when I did not see the animal. So deep is it that it seems to proceed less from the mouth and throat than from the deep chest and vast paunch. His eye began to flash deeper fire as we stood motionless on the defensive, and the crest of short hair which stands on his forehead began to twitch rapidly up and down, while his powerful fangs were shown as he again sent forth a thunderous roar.

"And now truly he reminded me of nothing but some hellish dream-creature; a being of that hideous order, half-man, half-beast, which we find pictured by old artists in some representations of the infernal regions. He advanced a few steps, then stopped to utter that hideous roar again, advanced again, and finally stopped when at a distance of about six yards from us. And here, just as he began another of his roars, beating his breast in rage, we fired and killed him. With a groan which had something terribly human in it, and yet was full of brutishness, he fell forward on his face. The body shook convulsively for a few minutes, the limbs moved about in a struggling way, and then all was quiet — death had done its work, and I had leisure to examine the huge body. It proved to be five feet eight inches high, and the muscular developement of the arms and breast showed what immense strength he had possessed."

Uran.

Deep in the swampy forests of Sumatra and Borneo, lives the famous Uran, or "wild man of the woods" as he is called by the Malays. He is less human in his shape than the chimpanzee, as his hind-legs are shorter and his arms so long that they reach to his ankles, but in intelligence he is supposed to be his superior. The jaws are more projecting, and the thick pouting lips add to the brutal expression of his physiognomy. While in a well-proportioned human face the distance from the chin to the nose forms but a third of the total length, it amounts to one half in the uran. Even in his native wilds this ape is rare, and as he is but little able to withstand the change of climate and the fatigues of a long sea-journey, he generally falls a speedy prey to consumption when brought to Europe. But little of the restlessness of the monkey race is to be seen in him. He loves an indolent repose, even while still enjoying his native freedom, and the necessity for procuring food seems alone capable of rousing him from his laziness. When satiated, he immediately resumes his favourite position, sitting for hours together upon a branch, with bent back, with eyes immovably staring upon the ground, and uttering from time to time a melancholy growl. When pursued by the Dayaks or Malays, who highly esteem his flesh, and kill him with poisoned arrows, he conceals himself among the dense foliage of the highest trees, and remains quiet until the danger is past. He is very difficult to catch, as his ear is sharp, and his suspicious temper keeps him perpetually on the alert; the most stealthy footstep, the least rustling of the leaves, suffices to warn him of his danger, and to make him seek his safety by a speedy retreat. He generally spends the night on the crown of a nibong-palm or of a screw pine; he often also seeks a refuge against the wind and cold among the orchids and ferns which cover the branches of the giant trees. There he spreads his couch of small twigs and leaves, for he distinguishes himself from all other apes by his not sleeping in a sitting position, but on the back or on one side, and in inclement weather he is even said to cover his body with a layer of foliage.

The series of the large anthropomorphous apes closes with the Gibbons. Their arms, which reach to the ankle joints when the animal is standing erect, are longer than those of the uran; their brain, and consequently their intelligence, is less developed; and moreover, like all the following simiæ of the Old World, they possess callosities on each side of the tail. Their size is inferior to that of the uran, and their body is covered with thicker hair, grey, brown, black, or white — according to the species — but never party-coloured, as is the case with many of the long-tailed monkeys.

To the gibbons belong the black Siamang of Sumatra—who, assembled in large troops, hails the first blush of early morn, and bids farewell to the setting sun with dreadful clamours—the black, white-bearded Lar of Siam and Malacca, and the Wou-Wou (*Hylobates leuciscus*) who, hanging suspended by his long arms,

Gibbon, or Long-armed Ape.

and swinging to and fro in the air, allows one to approach within fifty yards, and then, suddenly dropping upon a lower branch, climbs again leisurely to the top of the tree. He is a quiet, solitary creature of a melancholy peaceful nature, pursuing a harmless life, feeding upon fruits in the vast untrodden recesses of the forest; and his peculiar noise is in harmony with the sombre stillness of these dim regions, commencing like the gurgling of water when a bottle is being filled, and ending with a long loud wailing cry, which resounds throughout the leafy solitude to a great distance, and is sometimes responded to from the depths of the forest by another note as wild and melancholy.

Mr. G. Bennett has given a very interesting account of a siamang, that accompanied him on board during his homeward voyage. "This ape was two feet four inches high, and walked tolerably erect when on a level surface; his arms either hanging down or uplifted, with hands pendant. He preferred a vegetable diet — plantains, rice — but was so ravenously fond of carrots that he lost all decorum when they appeared at dinner. A piece of carrot would draw him from one end of the table to the other, over which he would walk without disturbing a single article, preserving his balance in an admirable manner although the ship might be strongly rolling and pitching from side to side. He would drink tea, coffee, and chocolate, but neither wine nor spirits. He was particularly fond of sweetmeats, and would not unfrequently enter the cabin in which some Manilla cakes were kept, and endeavour to lift up the cover of the jar.

"His attachment to liberty was excessive, and being so docile when free, and so very much irritated at being confined, he was permitted to range about the deck or rigging, which he did with such agility as to excite the astonishment and admiration of the crew. He usually slept on the main-top, coming on deck regularly at daylight. But arriving off the Cape, he expressed an eager desire to pass the night in my cabin, and on re-entering the warmer latitudes, would not again return to the main-top. Always desirous of retiring to rest at sunset, he would approach me with a peculiar begging chirping noise, and beg to be put to bed. When once taken into the arms of a friend, he was as adhesive as Sinbad's old man of the sea, and any attempt to remove him was at once resisted with violent screams. When he came at sunset to be taken into my arms and was refused, he exhibited all the freaks of temper of a spoilt child, lying on deck, rolling about, and dashing everything aside; but finding his paroxysms of rage unattended to, would mount the rigging, and hanging over that part of the deck on which I was walking, would suddenly drop into my arms. His look was grave, his manner mild, and he evinced a degree of intelligence beyond common instinct.

"Once or twice I lectured him on taking away my soap continually from the washing-stand. One morning I was writing, the ape being present in the cabin, when, casting my eyes upon him, I saw the little fellow taking the soap. I watched him without his perceiving me, occasionally giving him a furtive glance. I pretended to write, and seeing me occupied, he took the soap, and moved away with it in his paw. When he had walked half the length of the cabin, I spoke quietly without frightening him, when he instantly walked back and deposited the soap with the evident consciousness of having done wrong. I never observed him lap with his tongue when drinking, but when tea or coffee was given him he protruded it to ascertain the temperature, and if too hot, shaking his sapient head violently, but still undeterred, would wait patiently till it cooled. His mildness of disposition and playfulness of manner rendered him the universal favourite, but he became particularly attached to a little Papuan child (Elau, a native of Erromango), whom he probably considered as having affinity. They might often be seen, the animal with his long arm round her neck, eating biscuit together. He would often roll on the deck with her, as if in a mock combat, but on her attempting to play with him when he had no inclination, he would make a slight impression with his teeth on her arm, just enough for a warning or sharp hint.

"There were several small monkeys on board, who repelled the kind approaches of the little man in black by chattering. He was, however, determined to punish them for their impudence; so the next time they united as before in a body, he watched his opportunity, and seizing a rope, and swinging towards them, caught one of them by the tail, and dragged him up the rigging. If, in the ascent, he required both hands, he would pass the tail into his foot. It was most grotesque to see his perfect gravity of countenance, whilst the poor suffering monkey grinned, chattered, and twisted about, making the most strenuous and fruitless endeavours to escape, during which his countenance, at all times funny, had now terror added to its usual beauty. After having dragged the culprit some distance, the siamang let him go, and had he not seized a rope in falling, he would hardly have escaped a compound fracture.

"At dinner his station was on a corner of the table, between the captain and myself. When, from any of his ludicrous actions, we all burst out in loud laughter, he would vent his indignation by a barking noise, and by regarding the persons laughing with a most serious look, until they ceased their indecorous behaviour.

"When we spoke a ship at sea, his curiosity seemed to be much excited, and he would invariably mount the rigging until it was out of sight, wistfully gazing after it. When strangers came on board he approached them with caution.

"When the poor animal lay on a bed of sickness from dysentery, produced by cold, as much inquiry was made after his health as if he had been of human form divine, and his death excited great regret. Even when ill, he preferred going on deck in the cold air with persons to whom he was attached, to remaining in the warm cabin with those for whom he had no regard."

We shall see in the next chapter that the American monkeys are totally different from those of the Old World; but also in the eastern hemisphere, each part of the world has its peculiar families and genera of simiæ. Thus, besides the uran and the gibbon, Asia exclusively possesses the semnopitheci and the macaques, while Africa, besides the chimpanzee and the gorilla, enjoys the undivided honour of giving birth to the families of the cercopitheci, mangabeys, colobi, magots, and baboons.

The Semnopitheci are characterised by a short face, rounded ears, a slender body, short thumbs, and a strong muscular tail, terminated by a close tuft of hair, and surpassing in length that of all the other quadrumana of the Old World. To this genus belongs the celebrated Proboscis Monkey (*Semnopithecus nasicus*) of Borneo, who is distinguished from all other simiæ by the possession of a prominent nasal organ, which lends a highly ludicrous expression to the melancholy aspect of his physiognomy. "When excited and angry," says Mr. Adams, who had many opportunities of examining this singular creature in its native woods, "the female resembles some tanned and peevish hag, snarling and shrewish. They progress on all-fours, and sometimes, while on the ground, raise themselves upright and look about them. When they sleep, they squat on their hams, and bow their heads upon the breast. When disturbed, they utter a short impatient cry, between a sneeze and a scream, like that of a spoilt and passionate child; and in the selection of their food they appear very dainty, frequently destroying a fruit, and hardly tasting it. When they emit their peculiar wheezing or hissing sound, they avert and wrinkle the nose, and open the mouth wide. In the male, the nose is a curved, tubular trunk, large pendulous, and fleshy; but in the female it is smaller, recurved, and not caruncular."

Under the ugly form of the Huniman (*Semnopithecus entellus*), the Hindoos venerate the transformed hero who abstracted the sweet fruit of the mango from the garden of a giant in Ceylon, and enriched India with the costly gift. As a punishment for this offence he was condemned to the stake, and ever since his hands and face have remained black. Out of gratitude for his past services, the Hindoos allow him the free use of their gardens, and take great care to protect him from sacrilegious Europeans. While the French naturalist Duvaucel was at Chandernagor, a guard of pious Bramins was busy scaring away the sacred animals with cymbals and drums, lest the stranger, to whom they very justly attributed evil intentions, might be tempted to add their skins to his collection.

The semnopitheci are scattered over Asia in so great a multiplicity of forms, that Ceylon alone possesses four different species, each of which has appropriated to itself a different district of the wooded country, and seldom encroaches on the domain of its neighbours. "When observed in their native wilds," says Sir J. E. Tennent, " a party of twenty or thirty of the Wanderoos of the low country, the species best known in Europe (*Presbytes cephalopterus*), is generally busily engaged in the search for berries and buds. They are seldom to be seen on the ground, and then only when they have descended to recover seeds or fruit that have fallen at the foot of their favourite trees. In their alarm, when disturbed, their leaps are prodigious, but generally speaking their progress is made not so much by *leaping* as by swinging from branch to branch, using their powerful arms alternately, and when baffled by distance, flinging themselves obliquely so as to catch the lower bough of an opposite tree; the momentum acquired by their descent being sufficient to cause a rebound, that carries them again upwards till they can grasp a higher branch, and thus continue their headlong flight. In these perilous achievements wonder is excited less by the surpassing agility of these little creatures, frequently encumbered as they are by their young, which cling to them in their career, than by the quickness of their eye and the unerring accuracy with which they seem to calculate almost the angle at which a descent would enable them to cover a given distance, and the recoil to elevate themselves again to a higher altitude."

The African Colobi greatly resemble the Asiatic semnopitheci, but differ by the remarkable circumstance of having no thumb on the hands of their anterior extremities.

The Cercopitheci likewise possess a large tail, which is, however, not more or less pendulous, as in the semnopitheci, but generally carried erect over the back. They have also a longer face, and their cheeks are furnished with pouches, in which, like the pelican or the hamster, they are capable of stowing part of their food; an organisation which seems to denote that they are inhabitants of a country where the forests are less extensive.

Diana Monkey.

They are not devoid of intelligence, but extremely restless and noisy. Many that were mild and amiable while young, undergo at a later period a complete change of character. The only way, according to M. Isidore Geoffroy, to curb the temper of one of these full-grown monkeys is to extract the sharp and formidable canine teeth, with which it is capable of inflicting the most dangerous wounds. When disarmed, it immediately alters its manners, as it now feels its impotence. Several of the monkeys belonging to this group are distinguished by the lively colours of their fur; that of the Diana Monkey (*Cercopithecus diana*) among others, which is a native of Congo and Guinea, sells for a considerable price.

The tribes of the mangabeys, macaques, magots, and cynopitheci form the links between the cercopitheci and the baboons. Their shape is less slender than that of the former, their frontal bone is more developed, particularly above the eye-brows, and their face is longer. They are all of them provided with cheek-pouches. Several of the macaques have a very short tail, and the magots, or Barbary apes, and the cynopithecus of the Philippine Islands, have none, thus resembling the large anthropomorphous apes, but widely differing from them in other respects.

The Magot is the only European species, and seems exclusively confined in our part of the world to the rock of Gibraltar, though some authors affirm that it is found in other parts of Andalusia, and even in the province of Grenada. It would no doubt long since have been extirpated, if the British Government had not taken it under its especial protection, and imposed the penalty of a heavy fine upon its wanton destruction.

The Cynocephali (Baboons and Mandrills) show at once by their Greek name that a dog-like snout gives them a more bestial expression than belongs to the rest of the monkey tribes, and that of all the simiæ of the Old World they are most widely distant from man. In size they are only surpassed by the gorilla and the uran, and if in the latter the physiognomy becomes more brutal in its expression with advancing age, this degradation is much greater in the baboons.

Their canine teeth in particular acquire a greater sharpness than those of almost every other carnivorous animal, so that these malignant and cruel animals, armed with such powerful weapons, may well be reckoned among the most formidable of the wild beasts of Africa. As if to render them complete pictures of depravity, their manners also are so shamelessly filthy, that the curiosity they excite soon changes into horror and disgust.

The short-tailed mandrills inhabit the west coast of Africa. The Maimon is the most remarkable of the whole genus for brilliancy and variety of colour; its furrowed cheeks are magnificently striped with violet, blue, purple, and scarlet, so as more to resemble an artificial tattooing than a natural carnation. As the creature increases in age, the nose also becomes blood-red. On the loins the skin is almost bare, and of a violet-blue colour, gradually altering into a bright blood-red, which is more conspicuous on the hinder parts, where it surrounds the tail, which is generally carried erect.

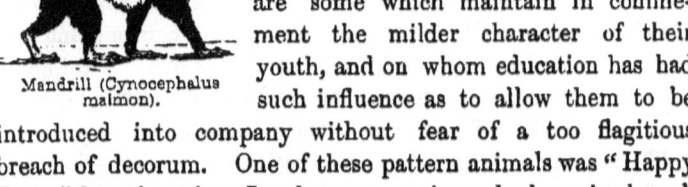

Mandrill (Cynocephalus maimon).

Even among the base mandrills there are some which maintain in confinement the milder character of their youth, and on whom education has had such influence as to allow them to be introduced into company without fear of a too flagitious breach of decorum. One of these pattern animals was "Happy Jerry," long kept in a London menagerie, and who gained such fame by his good manners, as to be honoured by a special invitation to Windsor. Jerry knew how to sit upon a chair, and worthily to fill it, as he was nearly five feet long. He relished his pot of porter, which he used to drink out of a pewter can, and smoked his pipe with all the gravity of a German philosopher. But even Jerry was not to be trusted out of the sight of his keepers.

The real baboons are distinguished from the mandrills by a long tail, terminated by a tuft of hair. The great baboon of Senegal (*Cynocephalus Sphinx*) is by no means devoid of intelligence, and learns many tricks when taught from early youth. His

Pig-faced Baboon.

temper, however, is brutal and choleric, though less so than that of the Chacma (*Cynocephalus porcarius*), or pig-faced baboon, which is found in the vicinity of Cape Town, among others on the celebrated Table Mountain. It frequently commits great devastations in the fields. Young chacmas are often kept as domestic animals, performing the offices of a mastiff, whom they greatly surpass in strength. Thus they immediately announce by their growling the approach of a stranger, and are even employed for a variety of useful purposes which no dog would be able to perform. Here one is trained to blow the bellows of a smith; there another to guide a team of oxen. When a stream is to be crossed, the chacma immediately jumps upon the back of one of the oxen, and remains sitting till he has no longer to fear the wet, which he loves as little as the cat.

In Abyssinia, Nubia, and South Arabia we find the Derryas (*C. Hamadryas*), which enjoyed divine honours among the ancient Egyptians. The general colour of the hair is a mixture of light-grey and cinnamon, and in the male that of the head and neck forms a long mane, falling back over the shoulders. The face is extremely long, naked, and of a dirty flesh-colour. This ugly monkey was revered as the symbol of Thoth, the divine father of literature and the judge of man after death. Formerly temples were erected to his honour, and numerous priests ministered to his wants, but now, by a sad change of baboon-fortune, he is shot without ceremony, and his skin pulled over his ears to be stuffed and exhibited in profane museums.

In the forests of tropical Africa and Asia we find a remarkable group of animals, which, though quadrumanous like the monkeys, essentially differs from them by possessing long curved claws on the index, or also on the middle finger of the hinder extremities; by a sharp, projecting muzzle, and by a different dentition. The Loris, remarkable for the slowness of their gait and their large glaring eyes, are exclusively natives of the East Indies; the Galagos, which unite the organisation of the monkeys with the graceful sprightliness of the squirrels, are solely confined to Africa, where they are chiefly found in the gum-forests of Senegal; the Tarsii, thus named from their elongated tarsi, giving to their hinder limbs a disproportionate length, are restricted to part of the Indian archipelago; but the large island of Madagascar, where, strange to say, not a single monkey is found, is the chief seat of the family, being the exclusive dwelling-place of the short-tailed Indri, (whom, from his black thick fur and anthropomorphous shape, one would be inclined to reckon among the gibbons), and of the long-tailed Lemurs or Makis. All these gentle and harmless animals are arboreal in their habits, avoid the glaring light of day under the dense covert of the forest, and awaken to a more active existence as soon

Slow-paced Lemur.

as night descends upon the earth. Then the loris, who during the day have slept clinging to a branch, prowl among the forest-boughs in quest of food. Nothing can escape the scrutiny of their large glaring eyes; and when they have marked their victim, they cautiously and noiselessly approach till it is within their grasp. The Galagos have at night all the activity of birds, hopping from bough to bough on their hind limbs only. They watch the insects flitting among the leaves, listen to the fluttering of the moth as it darts through the air, lie in wait for it, and spring with the rapidity of an arrow, seldom missing their prize, which is caught by the hands. They make nests in the branches of trees, and cover a bed with grass and leaves for their little ones. The tarsii leap about two feet at a spring, and feed chiefly on lizards, holding their prey in their fore-hands, while they rest on their haunches.

Handed Lemur.

CHAPTER XLVII.

THE SIMIÆ OF THE NEW WORLD.

Wide Difference between the Monkeys of both Hemispheres—The Prehensile Tail—The Wourali Poison—Mildness of the American Monkeys—The Stentor Monkey—The Spider Monkeys—The Sajous—The Fox-tailed Monkeys—The Saïmiris—Friendships between various kinds of Monkeys—Nocturnal Monkeys—Squirrel Monkeys—Their Lively Intelligence.

THE monkeys of the New World differ still more widely from those of the Old than the copper-coloured Indian from the woolly negro. One sees at once on comparing them that whole oceans roll between them, that they have not migrated from one hemisphere to another, but belong to two different phases of creation. While the nasal partition of the Old World simiæ is narrow as in man, it is broad without exception in all the American monkeys, so that the nostrils are widely separated and open sideways. The dental apparatus is also different, for while the monkeys of our hemisphere have thirty-two teeth, those of the western world generally possess thirty-six.

The tailless monkeys or apes, and the short-tailed baboons, with a dog-like projecting snout and formidable fangs, are peculiar to our hemisphere, and it is only here that we find almost voiceless simiæ, while the American quadrumana are all of them tailed, short-snouted, and generally endowed with stentorian powers. Finally, it would be as useless to look among the western monkeys for cheek-pouches and sessile callosities, as among those of the Old World for prehensile tails.

In the boundless forests of tropical South America, the monkeys form by far the greater part of the mammalian inhabitants, for each species, though often confined within narrow limits, generally consists of a large number of individuals. The various arboreal fruits which the savage population of these immeasurable wilds is unable to turn to advantage, fall chiefly to their share; many of them also live upon insects. They are never seen in the open campos and savannas, as they never touch the ground unless compelled by the greatest necessity. The trees of the forest furnish them with all the food they require in inexhaustible abundance; it is only in the woods that they feel "at home" and secure against the attacks of mightier animals; why then should they quit them for less congenial haunts? For their perpetual wanderings from branch to branch, Nature has bountifully endowed many of them not only with robust and muscular limbs, and large hands, whose moist palms facilitate the seizure of a bough, but in many cases also with a prehensile tail, which may deservedly be called a fifth hand, and is hardly less wonderful in its structure than the proboscis of the elephant. Covered with short hair, and completely bare underneath towards the end, this admirable organ rolls round the boughs as though it were a supple finger, and is at the same time so muscular, that the monkey frequently swings with it from a branch like the pendulum of a clock.

Scarce has he grasped a bough with his long arms, when immediately coiling his fifth hand round the branch, he springs on to the next, and secure from a fall, hurries so rapidly through the crowns of the highest trees that the sportsman's ball has scarce time to reach him in his flight. When the Miriki (*Ateles hypoxanthus*), the largest of the Brazilian monkeys, sitting or stretched out at full length, suns himself on a high branch, his tail suffices to support him in his aërial resting-place, and even when mortally wounded, he remains a long time suspended by it, until life being quite extinct, his heavy body, whizzing through the air, and breaking many a bough as it descends, falls with a loud crash to the ground.

The famous wourali poison is alone capable of instantly annihilating his muscular powers, and of sparing the wounded animal a long and painful agony. Slow and with noiseless step, so as scarcely to disturb the fallen leaves beneath his feet, the wily Indian approaches. His weapons are strange and peculiar, and of so slight an appearance as to form a wondrous contrast to their terrific power. A colossal species of bamboo (*Arundinaria Schomburgkii*), whose perfectly cylindrical culm often rises to the height of fifteen feet from the root before it forms its first knot, furnishes him with his blow-pipe, and the slender arrows which he sends forth with unerring certainty of aim are made of the leaf-stalks of a species of palm-tree (*Maximiliana regia*), hard and brittle, and sharp-pointed as a needle. You would hardly suppose these fragile missiles capable of inflicting the slightest wound at any distance, and yet they strike more surely and effectually than the rifleman's bullet, for their point is dipped in the deadly juice of the Strychnos Urari, whose venomous powers are not inferior to those of the dreaded bushmaster or the fatal cobra.

It is chiefly on the Camuku mountains in Guiana that this formidable creeping-plant is found, whose sombre-coloured, brown-haired leaves and rind seem by their sinister appearance to betray its dreadful qualities. The savage tribes of the South American woods know how to poison their arrows with the juices of various plants, but none equals this in virulence and certainty of execution, and yearly the Indians of the Orinoco, the Rio Negro, and even of the Amazons, wander to the Camuku mountains to purchase by barter the renowned Urari or Wourali poison of the Macusis. Nature has vouchsafed to these sons of the wilderness an inestimable gift in the venomous juice of the strychnos, for by no other means would they be able to kill the birds of the forest and the monkeys on whose flesh they chiefly subsist. How they made the discovery of its powers is unknown; at all events the combination of so many means for the attainment of the end in view — the preparation of the poison, the blow-pipe, the arrows—denotes a high degree of ingenuity, and shows at once the infinite superiority of the savage over the monkey. Fatal to every animal it touches, the wourali poison sometimes proves destructive to the archer. "One day," says Waterton, "while we were eating a red monkey, erroneously called the baboon in Demarara, an Arawack Indian told an affecting story of what happened to a comrade of his. He was present at his death, and as it did not interest this Indian in any point to tell a falsehood, it is very probable that his account was a true one. If so, it appears that there is no certain antidote, or at least no antidote that could be resorted to in a case of urgent need, for the Indian gave up all thoughts of life as soon as he was wounded. The Arawack Indian said it was but four years ago, that he and his companion were ranging in the forest in quest of game. His companion took a poisoned arrow and sent it at a red monkey in a tree above him. It was nearly a perpendicular shot. The arrow missed the monkey, and in its descent struck him in the arm, a little below the elbow. He was convinced it was all over with him. 'I shall never,' said he to his companion in a faltering voice, and looking at his bow as he said it, 'I shall never,' said he 'bend this bow again.' And having said that, he took off his little bamboo poison-box, which hung across his shoulder, and putting it together with his bow and arrows on the ground, he laid himself down close by them, bid his companion farewell, and never spoke more."

In a less concentrated or diluted form the wourali poison merely benumbs or stuns the faculties without killing, and is thus made use of by the Indians when they wish to catch an old monkey alive and tame him for sale. On his falling to the ground they immediately suck the wound, and wrapping him up in a strait-jacket of palm leaves, dose him for a few days with sugar-cane juice, or a strong solution of saltpetre. This method generally answers the purpose, but should his stubborn temper not yet be subdued, they hang him up in smoke. Then after a short time his rage gives way, and his wild eye, assuming a plaintive expression, humbly sues for deliverance. His bonds are now loosened, and even the most unmanageable monkey seems henceforward totally to forget that he ever roamed at liberty in the boundless woods.

In general, however, the American simiæ are distinguished by a much milder disposition than those of the eastern hemisphere, and retain at an advanced age the playful manners of their youth. They are commonly more easy to tame, and learn many little tricks which are taught with much greater difficulty to their restless Asiatic or African cousins. Their weakness, their short canine teeth, their good temper, render them harmless play-fellows, and thus they are generally preferred in Europe to the Old World monkeys, though they are not so lively, and constantly have a more or less dejected mien, as if they still regretted the primitive freedom of the forest.

The American monkeys may be conveniently divided into two large groups; with or without a prehensile tail. To the first great subdivision belong the Howling Monkeys or Aluates

(*Mycetes*), the Spider Monkeys (*Ateles*), the Sajous, and several other intermediate genera.

The aluates are chiefly remarkable for their stentorian powers, which no other animal can equal or approach.

When the nocturnal howl of the Large Red Aluate (*Mycetes ursinus*) bursts forth from the woods, you would suppose that all the beasts of the forest were collecting for the work of carnage. Now it is the tremendous roar of the jaguar as he springs on his prey; now it changes to his terrible and deep-toned growlings as he is pressed on all sides by superior force; and now you hear his last dying moan, beneath a mortal wound. Some naturalists have supposed that these awful sounds can only proceed from a number of the red monkeys howling in concert, but one of them alone is equal to the task. In dark and cloudy weather, and just before a squall of rain, the aluate often howls in the day-time; and on advancing cautiously to the high and tufted tree where he is sitting, one may then have a wonderful opportunity of seeing the large lump in his throat, the sounding-board which gives such volume to his voice, move up and down as he exerts his stentorian lungs.

Howling Monkey.

Pöppig compares the howling of the aluate to the noise of ungreased cart wheels, but very much stronger, and affirms that it may be heard at the distance of a league. In spite of his loud harsh voice, this monkey is of a very mild disposition, and easily familiarises himself with man. His flesh is good food, but when skinned, his appearance is so like that of a young one of our own species, that a delicate stomach and lively fancy might possibly revolt at the idea of putting a knife and fork into him. Waterton, however, affirms from experience, that after a long and dreary march through the remote forests of Guiana, his flesh is not to be despised, when boiled in Cayenne pepper, or roasted on a stick over a good fire. A young one tastes not unlike kid, and the old ones have somewhat the flavour of the goat.

The howling monkeys are the most robust of the American simiæ, and in spite of their long tail have a certain analogy with

the urans, whom they may be said to represent in the New
World. Their shaggy hair is of a brown or blackish colour, and
they chiefly subsist on fruits and foliage.

Their various species range from Paraguay to Honduras, while
the Ateles or Spider Monkeys, thus named from their long
slender limbs and sprawling movements, extend over the whole
surface of tropical America. The Marimonda (*Ateles Belzebub*)
is even found on the eastern slopes of the Andes at a height of
10,000 feet above the level of the sea, an elevation attained by
him alone of all the quadrumanous tribes.

Like the African Colobi, the spider monkeys have no thumb
on their fore-hands; their voice is a soft and flute-like whistling,
resembling the piping of a bird. It is
said that when a mother burthened
with her young hesitates to take too
wide a leap, *paterfamilias* seizes
the branch she intends to reach, and
swings himself to and fro with it,
until his companion is able to attain
it by a spring. But when a young
monkey that is already sufficiently
strong, is fearful, the mother, to give him courage, repeats
the manœuvre several times before him.

Black Spider Monkey.

The spider monkeys live in more or less numerous troops, and
chiefly subsist on insects, though when near the sea they will
also come down upon the beach and feed on mollusks, particularly on oysters, whose shells they are said to crack with a
stone.

Though the sajous have a prehensile tail, it is not denuded on the under surface of its extremity, as in the above-
mentioned tribes, and consequently not so delicate an organ
of touch. They are lively, active, mild, and intelligent creatures.

The second group of American monkeys, consisting of those
with a non-prehensile tail, comprises the sakis, the saïmiris, the
ouistitis, &c.

The Sakis, or Fox-tailed Monkeys, are distinguished by their
bushy tail, which, however, in some species, is very short. They
usually live in the outskirts of forests, in small societies of
ten or twelve. Upon the slightest provocation, they display a

morose and savage temper, and like the howling-monkeys utter loud cries before sunrise and after sunset.

The elegant ease of their movements, their soft fur, the large size of their brilliant eyes, and their little round face, entitle the Saïmiris to be called the most graceful of monkeys. Humboldt, who frequently observed them in tropical America, tells us that they are extremely affectionate, and that when offended, their eyes immediately swim in tears. On speaking to them for some time, they listen with great attention, and soon lay their tiny hand upon the speaker's mouth, as if to catch the words as they pass through his lips. They recognise the objects represented in an engraving even when not coloured, and endeavour to seize the pictured fruits or insects. The latter, and particularly spiders, which they catch most dexterously with their lips or hands, seem to be their favourite food. The weak little creatures are very fond of being carried about by larger monkeys, and cling fast to their back.

At first the animal to which they thus attach themselves endeavours to get rid of its burden, but finding it impossible, it soon becomes reconciled to its fate, and after a short time an intimate affection arises between them, so that when the saïmiri is busy chasing insects, his friend, before leaving the spot, first gives him notice by a gentle cry. A similar dependent and affectionate intercourse is not rare among other species of monkeys.

"When an Indian," says Pöppig, "meets with a large troop of these animals, he takes care not to disturb the small forerunners, as he knows full well that the larger ones will soon follow. Thus the Frailecito (*Callithrix Sciurus*) generally forms the vanguard of the machini; the pinchecito likes to stroll about with the tocon, and the choro is hand and glove with the maquisapa. It need hardly be mentioned that the saïmiris are in great request as pet animals, but unfortunately they are very rare.

The habits of the Nyctopitheci, or nocturnal monkeys, bear a great resemblance to those of the bats or flying foxes. The shy and quiet little animals sleep by day concealed in the dense thickets of the forest. Their eye and motions are completely feline. Those which Von Martius observed in his collection, crept by day into a corner of the cage, but after sunset their agility made up for their diurnal torpor.

In Guiana, Schomburgk met with the *Nyctipithecus trivirgatus* as a domestic animal. "A very neat little monkey, shy of light as the owl or the bat. A small round head, extremely large yellow eyes, shining in the dark stronger than those of the cat, and tiny short ears, give it a peculiarly comical appearance. When disturbed in its diurnal sleep and dragged forth to the light, its helpless movements excite compassion; it gropes about as if blind, and lays hold of the first object that comes within its reach, often pressing its face against it to escape the intolerable glare. The darkest corner of the hut is its seat of predilection, where it lies during day in a perfect asphyxia, from which it can only be roused by blows. But soon after sunset it leaves its retreat, and then it is impossible to see a more lively, active, and merry creature. From hammock it springs to hammock, generally licking the faces of the sleepers, and from the floor to the rafters of the roof, overturning all that is not sufficiently fastened to resist its curiosity."

Its hair, which is grey on the back and orange-coloured on the belly, is much thicker than that of the other monkeys, and somewhat woolly, thus being admirably suited to the colder temperature of its nocturnal rambles. It seems to range over a great part of South America, but on account of its retirement during the day is very rarely caught. Its voice is remarkably strong, and according to Humboldt is said to resemble the jaguar's roar, for which reason it is called the Tiger Monkey in the missions along the Orinoco. It lives chiefly on nocturnal insects, thinning their ranks like the bat, but is also said to prey upon small birds like the owl.

In the Andes of New Granada, in the large forests of Quindiu, the *N. lemurinus* lives at an elevation of from four to five thousand feet above the level of the sea, and makes the woods resound during the night with his clamorous cry of "dūrūcŭli."

The Ouistitis, or Squirrel Monkeys, are distinguished from all the other American quadrumana by the claws with which all their fingers except the thumbs of their hands are provided, and which render them excellent service in climbing. They have a very soft fur, and are extremely light and graceful in their movements, as well as elegant in their forms. The young are often not bigger than a mouse, and even a full-grown ouistiti is hardly larger than a squirrel, whom it resembles both in its mode of life, and by its restless activity, as its little head is never quiet. They use their tail, which in many species is handsomely marked by transverse bars, as a protection against the cold, to which they are acutely sensitive. Their numerous species are dispersed over all the forests of tropical America, where they live as well upon fruits and nuts as upon insects and eggs; and when they can catch a little bird, they suck its brain with all the satisfaction of an epicure. They are easily tamed, but very suspicious and irritable. The learned French naturalist Audouin made some interesting observations on a pair of tame ouistitis, which prove their intelligence to be far superior to that of the squirrels, to whom they are so often compared.

One of them having one day, while regaling on a bunch of grapes, squirted some of the juice into its eye, never failed from that time to close its eyes while eating of the fruit. In a drawing they recognised not only their own likeness, but that of other animals. Thus the sight of a cat, and what is still more remarkable, that of a wasp, frightened them very much, while at the aspect of any other insect, such as a cricket or a cockchafer, they at once rushed upon the engraving, as if anxious to make a meal of the object that deluded them with the semblance of life.

INDEX.

AAR

AARD-VARKS, 501
— Abies Brunoniana of the slopes of Sikkim, 90
— Webbiana of the slopes of Sikkim, 90
Abyssinia, the tsalt-salya or zimb of, 253
— mode of hunting the elephant in, 457
Acacia latronum and tomentosa, thorns of the, 126
Acclimatisation of plants, 64
Adams, Mr., his fight with a water-lizard, 334
Adanson, attacked by termites, at Goree, 282
Adansonia digitata, 102
Aden, coffee first introduced into, 190
Adjutant, the, 322, 381
— his destruction of reptiles, 322
Æolus, Greek worship of, 4
Africa, influence of the heated plains of, in deflecting the trade-winds, 10
— gigantic trees of, 104, et seq.
— the coffee tree first planted by nature in, 189
— cultivation of cotton in, 213
— swarms of locusts in, 255
— ivory of, 458
— Eastern, gigantic forests of, 61
— South, long-continued droughts of, 6
— constant winds of, 7
— giant timber of the east coast and arid wastes of the interior, causes of the difference, 8
— reason why droughts are prevalent in, 61
— fruits of the deserts of, 62, 63
Agades, tower in, 66
Agave Americana, 89, 116
— its uses, 117
Agouti, the, 84
Agriculture in the Puna, or high table-lands of Peru, 26
Air-currents, their effects in the equatorial regions, 4
— the trade-winds, 4
— polar and equatorial air-currents, 4
Aïs, the, 496

AMA

Akyab, quantities of rice exported from, 159
Alango, the night-blowing, or moon-flower, 225
Albatross, the, compared with the condor, 389
Alexander the Great, said to have introduced the peacock into Europe, 376
Algeria, domestication of the ostrich in, 405
Alligators, torpor of, during the dry season in the savannahs of South America, 14
— their return to life, 19
— of the lagunes of the Amazons, 49
— their delight in the sunshine, 83
— the Caymen, of the New World, 351
— mode of seizing their prey, 351
— their voice, 352
— their preference for human flesh, 352
— their tenacity of life, 353
— their tenderness for their young, 354
— their friends and enemies, 355
Allspice, 233
Aloes, the, of torrid Africa, 116
Alpaca, value of its wool, 28
— question of its acclimatisation in Europe and Australia, 29
— herds of, in the high table-lands of Peru and Bolivia, 29
Alps, origin of the Föhn-wind of the, 73
Aluate, or howling monkey, 535
Amache house-ant, 277
Amaryllis toxicaria, 400
Amazons river, 44
— — its length, width, and course, 44
— — its tributaries, 45
— — called the Solimoens from the Brazilian frontier to the influx of the Rio Negro, 45
— — its unfathomable depth at the Strait of Obydos, 46
— — its tide-waves, 46, 47
— — its width below Gurupa, 47
— — and when it reaches the ocean, 47
— — extent of territory drained by the Amazons, 47
— — its colossal rise, 48

INDEX

AMA

Amazons river, lagunes of the, and their beautiful scenery, 49
— — fish-catching on a grand scale, 49
— — different character of the forests beyond and within the verge of the inundation of the river, 50
— — a sail on the river, and a night's encampment, 52
— — the voracious pirangas, 54
— — storms on the river, 55
— — rapids and whirlpools, 55
— — the Amazons regarded as the stream of the future, 55
— — discovery of the Amazons by Vincent Yañez Pinson, 57
— — adventures of Pizarro and Madame Godin on the, 57–60
— — primitive forests of the banks of the Amazons, 78
— — the mosquito plagues of, 246
— — manufacture of tortoise-oil on the banks of the, 342
Amblyrhynchus of the Galapagos Islands, 339
Amboyna, taken from the Portuguese by the Dutch, 227
— clove trade of, 229
America, growth of cotton in, 209, 210
— insect plagues of, 248, 250
— snakes of the United States of, 316
— South, influence of the Peruvian stream on the climate of the, 5
— — gigantic scale of the mountains, rivers, &c., of, 14
— — savannahs of, 14
— — a savannah on fire, 17
— — cultivation of maize in, 162
— — primitive forests of, 79
— — vegetation of, 112
— — effects of the civil war in, on the cotton trade, 213
Ammophila arundinacea of the coasts of the North Sea, 64
Amsterdam, a spice fire in, 229
Anaconda, or water-boa (Eunectes murinus), 320
Andersson, Mr., his adventure with a rhinoceros, 445
— and with a lion, 470
Andes, the, considered as colossal refrigerators, 5
— causes of the perpetual verdure of their eastern and aridity of their western slopes, 7
— mosquitoes of the eastern slopes of the, 246
Angola, ferocious mosquitoes of, 246
— the red ant of, 274
Angostura, alligators in the streets of, 353
Animal life in the Llanos in the dry season, 15–17
— — and in the rainy season, 19

APH

Animal life, decomposition arrested by the dry sand of the rainless regions and the winds of the Puna, 25
— — of the Puna, 26–34
— — of the Peruvian sand-coast, 40
— — of the arid plains of South Africa, 64
— — of the Sahara, 75
— — of the Sikkim mountains, 92
— — of the mangrove forests, 95
— — insect plagues of the tropics, 245
— — in the tropical ocean, 303
— — of the Galapagos Islands, 339
— — tropical birds, 358
Anolis, the, 330
— battles of the, 330
— faculty of changing colour, 331
Anomaluri, the, 512
Anona tripetala, of Peru, 179
Ant-eaters, 497
— the great ant-bear, 497
— his mode of licking up termites, 497
— his characteristics, 498
— the manides, or pangolins, 500
— the orycteropi, 500
— the porcupine ant-eater, 503
— the Myrmecobius fasciatus, 503
Ant-hills, various forms and kinds of, 279
Antilles, climate of the, 5
Antilope melampus, the, always found near water, 65
Ants, their ravages in sugar plantations, 187
— vast numbers of, in the tropical zone, 272
— excruciating pain caused by the bite of the Ponera clavata, 273
— the ferocious ant of the Triplaris Americana, 273
— the black fire-ant of Guiana, 274
— the great red ant (Dimiya) of Ceylon, 274
— the red ant of Angola, 274
— the coffee and sugar ants, 275
— the Atta cephalotes, or umbrella ant, 276
— house-ants, 277
— ranger-ants, 278
— societies of ants, 278
— ant-hills, 279
— sagacity of ants, 279
— slave-making expeditions of some kinds of ants, 280
— the cow-keeper ants, 280
— food of the ants of tropical America, 280
— the honey-ant of Mexico, 281
— termites, or white ants, 281. *See* Termites.
— wars between black and white ants, 288
Apes, anthropomorphous, compared and contrasted with man, 516
Aphides " milked " by ants, 280

INDEX

APU

Apure river in the rainy season, 19
Arabia, coffee first introduced into, 189
— mode of cultivating coffee in, 194
Arabs, their mode of hunting the lion, 472
Arachnotheræ, 295
Arachuna river during the rainy season, 19
Arandi (Bombyx Cynthia), soft threads spun by the, 259
Araneæ of the tropics, 291
Aras of America, the, 411, 415
Arauca river in the rainy season, 19
Archipelago, the Eastern, bamboos of, 114
— — screw pine of the, 118
— — climate of the, 5
Archipelago, the Mulgrave, importance of the screw-pine to the inhabitants of, 118
Arctic day and night, 302
Areca palm (Areca catechu), the, 137, 153
— — Malay and Singalese habit of chewing the nuts with lime and betel-pepper leaves, 137
Areca sapida of New Zealand, 151
Arenaria rupifraga, great elevation at which it grows, 91
Argala, or adjutant-bird, of Africa and India, 381
Argus pheasant, 376
Aristolochia, gigantic, of South America, 102
Arkwright, his invention of the cotton-spinning frame, 208
Armadillos, the, 501
— on the sand-coast of Peru, 40
— genera of the Armadillos, 502
Arnatto (Bixa orellana), used as a dye, 241
Arracan, rice trade of, 158, 159
Arrowroot, from what obtained, 177
— mode of obtaining it, 178
Arrows, poisoned, of the bushmen of South Africa, 400
Artocarpus incisa, or bread-fruit tree, 166
Ascension, turtles of the island of, 345, 346
Asp of ancient authors, 319
Ateles hypoxanthus, 532
— — the, or spider monkeys, 536
Athene cunicularia of the Peruvian sand-coast, 41
Atlantic, limits of the trade-winds in the Northern, 6, 7
Atlas mountains, ephemeral streams of the, 70
— — the lions of, 467
— — a lion-hunt in, 472
Atlas-moth, cinnamon-eating, of Ceylon, 276
Atolls, or coral islands, 309
Atta cephalotes, its ravages amongst the banana and cassava fields, 276
Attalia funifera, 149
Aura vulture, 390
Australia, North, long-continued droughts of, 6
— — no monkey or tortoise indigenous in, 342
— birds of, 378

BAY

Avaca, or Manilla hemp, 175
Avicennia tomentosa, its magnificence, 98
Axis, the, 475

BABOONS, 527
Baboon, the great, of Senegal, 528
Bacha, the (Falco bacha), 394
Bactrian camel, 420
Bahama Islands, mode of catching turtle on the, 346
— — the guana lizard of the, 333
Bahia, two rainy seasons of, 9
— toad of, 337
Bakalahari, the, of the Kalahari, 65
— their love for agriculture and domestic animals, 65
Balize, mahogany trees of, 113
Balsam-bog of the Falkland Islands, 91
Baltimore bird (Icterus Baltimore), 367
Bamboos, the (Bambusaceæ), of the tropics, 114
— variety of uses to which they are applied, 115
Bambusaceæ, the, of the tropics, 114
— rapidity of their growth, 114
Banana (Musa sapientum), its importance as food, 154, 173
— the wild, of the lower slopes of Sikkim, 90
— ravages of the Atta cephalotes in the banana fields, 276
Banana-bird (Icterus xanthornus), of America, 383
Banda, nutmeg trees of, 228, 230
Banijai, their mode of hunting the elephant, 455
Banyan tree (Ficus indica), 106
— — fondness of the Hindoos for it, 107
— — its historical celebrity, 107
Baobab, African, or monkey-bread tree (Adansonia digitata), 102
— — immense specimens of, 103
— — used as a vegetable cistern, 103
— — its age, 104
— — family to which it belongs, 104
Barima river, the Upper, gigantic trees of, 114
Basilisk, the, 336
Batavia, the coffee-tree introduced into, 190
Batocera rubrus, 133
Bats, huge, of tropical forests, 84
— organisation of, 505
— the kalongs, or fox-bats, of Java, 507
— the vampire, 507, 509
— the Rhinolophi, or horse-shoe bats, 510
— the Scotophilus Coromandelicus of Ceylon, 511
— flying squirrels, 511
— the Galeopitheci, 512
— the Anomaluri, 513
Batticaloa, forests of satinwood at, 111
Baya birds of Hindostan, their nests, 383

INDEX

BAY

Bayeiye, their mode of hunting the elephant, 456
Bear, the cocoa-nut (Ursus malayanus), 134
Bechuana country, alleged poisonous spider of, 297
Bechuanas, their love for agriculture and domestic animals, 65
— their mode of drawing water, 66
— their umbrellas, 405
Beer brewed from maize, 165
— — from millet, 166
Beetles of the tropical forests, 83, 84
— used as food and ornaments, 263
— numbers of, in various countries from the poles to the tropics, 265
— the carrion-feeding beetles of the temperate zone, 266
— peculiarity of beetle-life in the torrid zone, 266
— the Hercules beetle (Megasomina Hercules), 266
— habits of tropical beetles, 267
— the Goliaths of the coast of Guinea, 267
— the South American incas, 267
— luminous beetles, 270
Behemoth of the Bible, 432
Bela dates, 143
Bell-bird, or campanero, 365
Bengal, indigo of, 234, 236
Beni river, a tributary of the Madeira, 46
Benin, palm-oil trade of, 146
Berber tribes of the oases of the Sahara, 71
Berbice river, the Victoria Regia of the, 123
Bête rouge, the, of Guiana and the West Indies, 250
Betel-pepper leaves chewed, 137
Biledulgerid, or oases south of the Atlas, toddy drunk in, 143
Birds of the Puna, or high table-lands of tropical America, 33
— of the tropical forests, 82
— of the Mexican plateaus, 87
— of the tropical seas, 303
— singular birds of the Galapagos Islands, 339
— their strange voices, 369
— breeding season, 374
— of prey of the tropics, 387–396
Birds'-nests, edible, 305
Blast, a sugar-cane disease, 187
Blatta gigantea, or the drummer, 256
Bligh, Lieutenant, his voyage in the "Bounty," and adventures subsequently, 168
Boa constrictor, 319
— — his habitat, 320
Boaquira (Crotalus horridus), 316
Bogota, perennial rainy seasons of, 9
Bolas, or hand stones of the Indians of Peru, 31

BRE

Bolivia, immense consumption of cocoa in, 202
— the Indian's mode of preparing and chewing cocoa in, 203
Bombax ceiba of the forests of Yucatan, 112
Bombyx cynthia, 259
— niori, 259
— mylitta, 259
Bonito, the, 306
Bonny, palm-oil trade of, 146
— divine honours paid to water-lizards at, 335
Boomerang, or kiley, of the Australian savage, 414
Borassus flabelliformis, 137
Borelo rhinoceros, 441
Boselephus oreas of South Africa, 65
Bo tree, the sacred, of India (Ficus religiosa), 108
— — antiquity of one at Anarajapoora, in Ceylon, 108
— — veneration of the Buddhists for it, 110
Bottle tree of tropical Australia, 124
"Bounty," mutiny of the, 168
— fate of the mutineers, 172
Bouradi, the, of Guiana, 360
Bourbon, nutmeg trade of, 229
Bower-birds of Australia, 378
— — — — their singular nests, 379
Brachypterus aurantius, 382
Branco, Rio, 56
Brass country, palm oil of, 147
Brazil, constant winds of, 7
— weak and jealous government of, 56
— impenetrable forests of, 79
— the Porliera bygrometrica of, 120
— the bushropes or lianas of, 121
— quantity and quality of coffee exported from, 191, 193
— cotton cultivation of, 214
— immense number of beetles found in, 265
— comparative rareness of venomous snakes in, 312
— the bush-master of, 315
— the giant-toad of, 338
— tree-frog of, 338
— colossal turtles of the coast, 344
— birds of Brazil do not migrate, 373
— jaguars of Brazil, 489
— wood (Cæsalpinia crista), description of the tree producing, 240
Brodin, Mr., his steam cocoa-nut oil mill on Foa Island, 130
Bromelia ananas, 117
— sagenaria, 117
— variegata, 117
Bromelias of tropical forests, 123
Bread-fruit tree (Artocarpus incisa) of Polynesia, 166
— — — the harvest, 167
— — — the mahei, or sour paste, 167

INDEX

BRE

Bread-fruit tree, its uses besides those of food, 168
— first mentioned by Dampier, 168
— history of its transplantation to the West Indies, 168–171
Breviceps gibbosus, of Guinea, 338
Buceros Malabaricus, 386
— rhinoceros, 374
Buddhists, their veneration for the sacred Bo tree at Anarajapoora, 110
Buffalo always found near water, 65
— the African, his guardian bird, 442
Buffalo-thorn (Acacia latronum), thorns of the, 126
Buffaloes, ferocity of the male solitaires of the, 453
— attacked by the tiger, 480
Bufo gigas, agua, of the Brazilian campos, 338
Bull-frog of Virginia, 338
Bulls, wild, of the Puna mountain valleys, 34
Buphaga Africana, the, 442
Buprestis gigas, the elytra of the, worn as an ornament, 264
Burgomaster bird (Larus glauca), 391
Burgundy, former woollen cloth trade of, 207
Bushmen, African, 61, 65
— their mode of catching ostriches, 399, 400
— their mode of hunting the elephant, 456
— their mode of hunting the lion, 473
Bush-master snake (Lachesis rhombeata), 315
Bushropes, or lianas, of tropical vegetation, 120
Butterflies of the Sikkim mountains, 93

CABBAGE-PALM of the Antilles (Oreodoxa oleracea), its magnificence, 148
— — grub eaten, 149
Cabeza di Negro (Phytelephas), hard white nuts of the, 149
Cacao tree (Cacao theobroma), 199
— — indigenous, in Mexico, 199
— — Humboldt's description of a cacao plantation, 199
— — mode of cultivation, 200
— — management of the beans, 200
Cacatua cristata, 413
— bauksii, 414
Cachalot, the, 303
Cactus Opuntia of the shores of the Mediterranean, 119
Cactus cochinellifer, 261
Cactuses, description of the, 118
— their usefulness to man, 119
— did not exist in the Old World previous to the discovery of America, 119
— range of their growth, 119
Cæsalpinia crista, 240
— echinata, 241
— brasiliensis, 241

CAS

Cæsalpinia Sappan, 241
— bimas, 241
Caffa and Enarea, the original home of the coffee plant, 189
Calabar, New and Old, palm-oil trade of, 146
Caladium esculentum of the Sandwich Islanders, 178
Calami, the, 152
Calao, or rhinoceros horn-bill (Buceros rhinoceros), 374
Caledonia, New, spiders eaten by the natives of, 297
Caliatour wood, 241
Calms, zone of, 8
— intense heat of the, 8
— heavy afternoon rains of the, 8
Camel, its resemblance to the ostrich, 403
— the dromedary the ship of the desert, 417
— adaptation of its organisation to its mode of life, 418
— Bedouin mode of training it, 420
— the Bactrian camel, 420
— immemorial slavery of the camel, 421
— its unamiable character, 422
Cameleopard. See Giraffe
Camelides of tropical America, 26–32
Campanero, or bell-bird, 365
Canary Islands, gigantic dragon-trees of the, 105
— — cochineal trade of the, 261
Canis azaræ of the high table-land of Peru and Bolivia, 33
— — of the Peruvian sand-coast, 40
— ingæ, of the Puna, 34
Caoutchouc tree (Siphonia elastica), 216
— — description of the tree, 217
— — introduction of caoutchouc into Europe, 217
— — mode of collecting the resin, 217
— — other trees yielding caoutchouc, 217
— — various uses of India rubber, 218
— — vulcanisation, 218
— — supply of India rubber, 221
Caouana, or loggerhead turtle, (Chelonia caouana), 349
Capybara, or water-pig, 351
Carabi, comparative numbers of, in the tropical and temperate zones, 266
Cardinal bird of Mexico, 87
Carinaria vitrea, the, 308
Carnauba palm (Corypha cerifera), wax obtained from the, 148
— — other uses of the tree, 148
Caroa (Bromelia variegata), fishing-nets made from the fibres of the, 117
Carolina, South, introduction of rice into, 157
Cartwright, his invention of the power-loom, 208
Cassava, or Mandioca, root (Jatropha manihot), how prepared as food, 175
— the sweet cassava (Jatropha janipha), 176
— ravages of the Atta cephalotes in the cassava fields, 276

CAS

Cassia, 224
Cassicus cristatus, 368
— uber, 368
— persicus, 368
Cassiques, the, 368
— their pendulous nests, 368
Cassiquiare river, 56
Cassowary, the galeated (Casuarius galeatus), 407
Caterpillars eaten by man in Africa, 262
Catoblepas gnu, always found near water, 65
Cats unable to live at a great height above the sea, 25
— wild, their ravages in coffee plantations, 197
Cayenne, two rainy seasons of, 9
Cayman. *See* Alligator
Ceiba (Bombax ceiba), the, of the forests of Yucatan, 112
Centipede, its virulent bite, 301
— its phosphorescence, 301
Cephalopoda, gigantic, 308
Cephalopterus ornatus, 369
Cephalopus mergens of South Africa, 65
Cerastes of the Sahara, 75
— or horned viper, of the Egyptian jugglers, 319
Cercopitheci, their characteristics, 526
Cereals, traditional birthplace of the, 174
Cerei, or torch-cactuses, 118
Cerro de Pasco, silver mines of, 45
Cervus antisiensis, an animal peculiar to the Punas, 32
Ceroxylon andicola, wax obtained from the, 148
— height at which it will grow, 151
Ceylon, timber-trees of, 111
— abundance of the cocoa-nut tree in, 129
— cocoa-nut oil trade of, 130
— palmyra toddy of, 138
— cultivation of rice in, 155
— the ruined tanks of, 156
— Singalese mode of clearing the mountain forests, 195
— the coffee cultivation of, 192, 195
— cinnamon gardens of, 222
— taken by the Dutch, 227
— land-leeches of, 252
— immense number of ants in, 274
— the formidable dimiya, or great red ant, 274
— the kaddiya, 274
— snakes of, 317
— comparative rareness of venomous snakes in, 311
— the rat-snake and cobra domesticated in, 326
— barbarous mode of selling turtle-flesh in, 348
— birds of, 382
— elephants of, 450, 452, 458, 459
— elephant-catchers of, 460
— species of bats in, 510

CLI

Chacma, or pig-faced baboon (Cynocephalus porcarius), 528
Chamœrops humilis, 150, 152
— palmetto, 150
Chameleon, the, 331
— its habitat, 331
— its manner of hunting for its food, 331
— peculiarities of its organisation, 331, 332
Champsa vallifrons, 353
Chancay, sand-hills of, 42
Cheetah, or hunting leopard, 466, 482
Chegoe, Pique, or Jigger, of the West Indies (Pulex penetrans), 248
— its mode of working, 248
— native method of extirpating it, 248
Chelonia imbricata, 347, 349
— midas, 347
— caouana, 349
Chelonians, 339
Chenopodium quinoa, 177
Chicha, or maize beer, 165
Chilicabra cockroach of Peru, 258
Chimpanzee, the (Simia troglodytes), 515, 517
China, bamboo of, 114
— — care with which it is cultivated, 115
— probably the original home of the sugar-cane, 183
— cotton cultivation of, 214
Chinese allow nothing edible to go to waste, 263
Chinese sea, north-east monsoon of the, 9
— — violence of the storms of the, 11
Chinchilla lanigera, the, of the high tablelands of Peru, 33
— — its appearance and habits, 33
Chirimoya (Anona tripetala), a Peruvian fruit, 179
Chloëphaga melanoptera, the, of the Puna, 33
Chloroxylon Swietenia, forests of, in Ceylon, 111
Chocolate, 201
Chorisia ventricosa of the Brazilian forests, 124
Christian, his mutiny on board the "Bounty," 168
— his abode in Pitcairn Island, 172
Chunu, or chaps, caused by the biting winds of the Puna, 24
Cicada Anglica, 263
Cicadæ, or frog-hoppers, eaten by man, 263
Cilgero bird of Cuba, his song, 370
Cinnamon plant, 222
— gardens of Ceylon, 223
— immense profits of the Dutch, 223
— decline of the trade, 224
— export of, from Java, 224
— mode of cultivating the plant and procuring the rind, 225, 226
— the Ceylon chalias, 226
— oil, 227
Cleopatra, her death, 319
Climates, diversity of, within the tropics, 3

INDEX

CLI

Climates, causes by which this diversity is produced, 3
— climate of the Llanos of Venezuela and New Grenada, 14
— of the Puna or high table-lands of Peru and Bolivia, 23
— vegetation always suited to the peculiar climate of a country, 62
— variety of the climate of Mexico, 87
Climbing trees of the tropics, 121
Cloves, substitute for the essential oil of, 227
— history of the cruel monopoly of the Dutch in, 227
— clove-tree groves, 229
— essential oil, 229
Coary river, a tributary of the Amazons, 45
Coatimondi, the, 517, 518
Cobra di Capello, the, 317
— tamed by the Indian jugglers, 317
— its habitat, 318
Coca (Erythroxylon coca), 202
— its immense consumption in Peru and Bolivia, 202
— its wonderfully strengthening effects, 203
— fatal consequences of its abuse, 204
— divine honours paid to the shrub by the Peruvians, 205
— its use interdicted by the Spanish conquerors, but finally allowed and encouraged, 206
— analysis of, by Professor Wöhler of Göttingen, 206
Cocaïn, Professor Wöhler's discovery of, 206
Cocci of Europe, blights caused by the, 260
— the cochineal coccus of Mexico, 260
Coccus hesperidum of Mexico, 260
— lacca, or lac-insect, 262
Cochineal insect, food of the, 119
— — exportation of, forbidden by the Spaniards in Mexico, 261
— — introduced into the Canary Islands, Spain, and other places, 261
— — cultivation of the, 260
— — history of cochineal, 261
Cock of the rock (Rupicola aurantia), 366
Cockatoo, the, 413
— the great white, 413
— the black, of Australia, 414
— cockatoo killing in Australia, 414
Cockroaches (Blattæ), tropical plague of, introduced into England, 256
— the giant cockroach of the tropics (Blatta gigantea), 256
— encounter between a spider and a cockroach, 298
Cocoa-nut tree (Cocos nucifera), the, 128
— — extent of its domain, 129
— — its abundance in Ceylon, 129
— — its many uses to man, 129, 131, 132
— — cocoa-nut oil and the oil trade, 130

COR

Cocoa-nut tree, toddy made from the, 131
— — timber of the, 132
— — cultivation of the, 132
— — enemies of the, 133
Cocoa de mer (Ladoicea Sechellarum), 140
— — its former and present value, 141
Cocos nucifera, the, 128. *See* Cocoa-nut tree
— butyracea, or oil palm-tree of West Africa, 144
Cocujus beetle of South America, its luminous qualities, 270
Coffee, original home of the plant, 189
— the use of, introduced into Arabia, 189
— history of coffee-drinking, 190
— the first coffee-houses in London and Paris, 190
— present state of coffee production throughout the world, 191
— Brazil and Java, 191
— Ceylon, Hayti, and Venezuela, 192
— Mocha coffee, its quality, 193
— mode of cultivation of the coffee-tree, 193
— profits of coffee-plantations, 196, 197
— enemies of the coffee tree, 197
— ravages of the Viviagua ant, 275
Coffee bug (Lecanium coffeæ), of Ceylon, 197, 198
Coir, or cocoa-nut fibre, uses to which it is applied, 131
Colobi of Africa, 526
Colocalia esculenta, 305
Colombo, cinnamon gardens of, 224
Columbus, Christopher, the first to import maize from America into Europe, 160
Combretias, the, of the banks of the rivers of Guiana, 82
Condamine, M. La, his voyage from Brancamoros to Para, 59
— introduces caoutchouc into Europe, 217
Condor, the, of the high table-lands of tropical America, 33, 387
— his marvellous flight, 387
— his cowardice, 388
— modes of capturing him, 388, 389
— compared with the albatross, 389
Condors of the sand-coast of Peru, 43
Congo, baobab trees of, 103
Coniferæ of the slopes of the Sikkim mountains, 90
Convolvulus batatas, or sweet potato, 177
Cooroominya beetle (Batocera rubrus), its attacks on the young cocoa-nut tree, 133
— — eaten by the Singalese, 133
Coot, the gigantic (Fulica gigantea), of tropical America, 33
Coppersmith bird of Ceylon (Megalasara Indica), 382
Coquero, the, or habitual coca chewer, 205
Coral islands, 302
— — formation of, 309

COR

Coral-snake (Elaps corallinus), domesticated in Brazil, 326
Coriaceous turtle (Sphargis coriacea), 348
Corozo palm (Elæis oleifera), oil of the, 149
Corypha australis, 150
Coryphaena, the, 306
Coryphodon Blumenbachii, 326
Cotingas, the, 365
Cotton, 207
— amazing rise of the cotton manufacture, 208
— known in ancient times, 208
— invention of the spinning-jenny, 208
— and of the spinning-frame, the mule-jenny, and the power-loom, 208
— statistics of the cotton trade, 209
— different species of the cotton plant, 210
— cultivation of the plant, 210
— the cotton harvest, 211
— the cotton trade of India, present and prospective, 211, et seq.
— effects of the civil war in America on the cotton trade, 213
— cotton production of Africa, 213
— efforts of the English to introduce the cultivation of cotton and abolish the slave trade in Africa, 213
— Brazilian, Egyptian, and Chinese cotton, 214
— cotton seed oil, 214
Cougar, or puma, the, 490
Counacutchi, or bush-master snake (Lachesis rhombeata), 315
Crab, cocoa-nut (Birgus latro), 134
— its mode of operation, 135
Crab, land, 308
— injuries done by, to the sugar-cane, 187
Crabs of the tropical seas, 308
Cranata de rede (Bromelia sagenaria), cordage made from the, 117
Cray-fish, 308
Creeping-plants, their importance in the deserts of South Africa, 64
Crocodiles, 14
— their return to life, 19
— their torpidity, 356
— their power of fascinating their prey, 356
— their wanderings, 356
— anecdote of one in Ceylon, 357
— their habitat, 351
Crompton, Mr., his invention of the mule-jenny, 208
Crotalus horridus, 316
— durissus, 316
Crustaceans of the tropics, 307
— decapod, 308
Cuba, cultivation of coffee in, 193
Cubagua, Island of, 58
Cubbeer-burr banyan tree, the famous, 107
Cucharacha cockroach of Peru, 258

DIM

Cuculi pigeons of the Peruvian sand-coast 41
Cucumis cuffer, 63
Curcuma longa, 242
Cyclones, 11
Cynocephali, 527
Cynocephalus porcarius, 528
— sphinx, 528
Cyphorinus cantans, 371
Cypraea aurora, 308

DAMARA LAND, reason why droughts are prevalent in, 62
— — umbrellas used in, 405
Dampier, the bread-fruit first mentioned by, 168
— his description of a logwood-cutter, 238
— attacked by a Guinea worm, 250
Darwin, Mr., his account of the giant tortoise of the Galapagos Islands, 340
— his ride on a tortoise 342
Dasypus tatuny of the sand-coast of Peru, 40
Date-palm (Phœnix dactylifera), 142
— — range of cultivation, 142
— — mode of propagation, 143
— — tamr and bela dates, 143
— — varieties of dates, 143
— — trees of Nubia, 144
Decomposition arrested by sand and the winds of the Punas, 25
Deer of the Punas, 32
Degleh dates, 143
Delabechea, or bottle-tree, of tropical Australia, 124
Deleb palms of Kordofan, 144
Delphinium glaciale, great elevation at which it grows, 91, 92
Demerara, the goatsucker of, 369
Derryas, the (Cynocephalus hamadryas), formerly regarded with divine honours, 529
Doscleux, Captain, his introduction of the coffee-plant into the West Indies, 191
Desert, the ship of the, 417
— horrors and beauties of the, 419
— the Bedouins of the, and their camels, 420
Devil-bird, or gualama, of Ceylon, 382
Dew, abundance and distribution of, within the torrid zone, 6
Dhourra, or millet beer, of Nubia, 166
Diactor bilineatus, 269
Diadem spider, 299
Diamond-beetle (Entimus nobilis, used as an ornament, 264
Diana monkey, 526
Diodon, the, 307
Dilolo, Lake, sagacity of the ants near, 279
Dimiya, or great red ant of Ceylon, 274

INDEX

DIS

Diseases to which the traveller is liable in the Punas, or high table-lands of Peru and Bolivia, 24
Dogs, half wild (Canis Ingæ), of the Punas, 34
Dolphins, 307
Donkiah mountain, 89
Dorjiling, sanatorium of, in the Sikkim mountains, 89
Doum-palm (Hyphæne Thebaica), 144
Douw, or Burchell's zebra, 420
Dracænas, or dragon-trees, 105
— gigantic ones of the Canary Islands, Madeira, and Porto Santo, 105
— celebrated specimen at Orotava, in Teneriffe, 105
Dragon-flies of the tropics, 267
Dragons, flying, 336
Dragon-trees. *See* Dracænas.
Drill, the, 528
Dromalus Novæ Hollandiæ, 407
Dromedary. *See* Camel.
Droughts, reason why they are prevalent in South Africa, 61
Drummer cockroach (Blatta gigantea), 256, 257
Du Chaillu, M., his description of the gorilla, 519
Duiker (Cephalopus mergens), the, of South Africa, 65
Durissus (Crotalus durissus), 316
Duruculi monkey, the, 538
Dutch, their progress in the Indian Ocean and cruel monopolies, 227
Dyes, tropical vegetable, 234
— indigo, 234
— logwood, 238
— Brazil wood, 240
— red saunders wood, 241
— arnatto, 241
— fustic, 242
— turmeric, 242

EAGLE, the harpy, 392
— his habitat, 393
— the fishing, of Africa (Haliætus vocifer), 395
— the sea, of the north, its method of procuring food, 402
Earth-hogs of the Cape, 50
Eastern coasts or eastern slopes better watered than western, why, 7
Echidna, the, or porcupine ant-eater, 503
— ocellata, its fatal bite, 315
Echinocacti, the, 118
Echinocactus nana, or dwarf-cactus, 119
— visnaga, its immense size, 119
Egypt, sycamore groves of, 105
— cotton cultivation of, 214
Eider-down, mode of obtaining, 402
Elæis guineensis, or oil-palm tree of West Africa, 144

FEA

Elæis oleifera, oil of the, 149
Elands (Boselaphus oreas) of South Africa, 65
Elaps corallinus, 326
Electric eel (Gymnotus electricus), 20
— — Indian mode of capturing them, 20
Elephant-thorn of Ceylon, 126
Elephant, his love of solitude, 449
— his senses of smell and of hearing, 450
— his mode of ascending and descending abrupt banks, 450
— his stomach, 451
— his trunk, 451
— uses of his tusks, 451
— his discipline, 452
— his sagacity and devotion, 452
— rogues, 453
— value of the elephant to man, 454
— species of the, 454
— wide range of the African, 455
— mode of hunting him in various countries, 455
— ivory of the African elephant, 458
— the Asiatic, 458
— catchers, of Ceylon, 460
— corrals, 462
Elliott, Ensign, his escape from a tiger, 479
Emu or nandu (Rhea Americana), its habitat, 406
— of Australia (Dromaius Novæ Hollandiæ), 407
Emydæ, or marsh-tortoises, of America and the Indian Archipelago, 342
Emys picta, 342
Enarea and Caffa, the original home of the coffee plant, 189
England, annual imports of wool into, 207
— former wool trade of, 207
— cotton trade of, 209
— number of beetles found in, 265
Entimus nobilis, 264
Epeiras, beautiful colouring of the, 294
Epiphytes of the forests of the tropics, 123
Essequibo river, 56
Eunectes murinus, 320
Euphorbia arborescens of Africa, 105
— tree, the juice of the, used as a poison in South Africa, 400
Everest, Mount, 89
Eyes, acute inflammation of the, in the Puna, 24
— of the chameleon, 331

FALCO BACHA, 394
Falcon (Falco sparverius) of the Peruvian sand-coast, 41
Fayal, destruction of the orange trees of, 260
Feathers of the ostrich, 405

INDEX

FEE

Feejee Islands, barbarous mode of treating turtles in the, 347
— — visit from a crocodile to the, 357
— — sandal tree of, 112
Felidæ of Peru, 40
— of the tropical forests, 84
— of the Old World, 466
— of the New World, 486
Felis dogaster, 491
— jaguarundi, 491
— pardalis, 491
Ferns, arborescent, of tropical forests, 80, 123
Ficus elastica, singular formation of the roots of the, 125
— — caoutchouc of the, 217, 220
— indica, astonishing size of the, 107
— religiosa, of India, 108
— sycomorus, of Africa, 105
Fig, the Indian, the fruit of the melocacti, 119
Fig trees, climbing, of Polanarrua, 122
— — marriage of the fig tree and palm, 123
Filaria medinensis, or Guinea worm, 249
— — its mode of working, 249
— — method of extracting it, 249, 250
Fir, the silver (Abies Webbiana), of the slopes of Sikkim, 90
Fire-ant, the black, of Guiana, 274
Fishes, tropical, 306
Fishing-eagle of Africa (Haliætus vocifer), 395
Flamingo (Phœnicopterus ruber), 371
— its habits, 371
— its nests, 371
— the rose-coloured (P. antiquorum), 371
— of the Puna, 33
Flowers of the primitive forest of tropical America, 81
Flute-bird (Cyphorinus cantans), 371
Fly-catcher, crowned (Myoarchus coronatus), of the Peruvian sand-coast, 41
Flying-fishes (Exocœtus volitans, Pterois volitans), 306
Foa, one of the Tonga Islands, steam cocoa-nut oil-mill on, 130
Föhn-wind of the Alps; origin of the, 73
Forbes, Mr., his narrow escape from a cobra di capello, 318
Forests inundated by the Amazons river, 48
Forest, primitive tropical, 78
— its peculiar charms and terrors, 78
— troubles of the botanist in the, 79
— endless varieties of trees in tropical forests, 79
— and of their sites, 80
— lowland forests during the rainy season, 81
— a hurricane in, 81
— beauty of the forests after the rainy season, 82

GIR

Forests, birds of the tropical, 82, 83
— morning, noon, and night, in the forests, 83-85
— the sylvan wonders of Sikkim, 89
Formica analis and saccharivora, ravages of the, in sugar plantations, 276
— rubescens, 279
— sanguinea, 280
Fox (Canis azaræ), the, of the high table-lands of Peru and Bolivia, 33
— of the Peruvian sand-coast, 40
Foxes, flying, 505, 507
France, cotton manufacture of, 209
— number of beetles found in, 265
Frigate-bird, 303
— — its mode of operation, 304
Frog, the Brazilian and Surinam tree, 338
Frog-fish, the, 307
Frog-hoppers (Cicadæ), edible, 263
Fruit trees of the tropics, 179
— — the chirimoya of Peru, 179
— — the litchi, 180
— — the mangosteen, 180
— — the mango, 181
Fulgora lanternaria, 271
Fulica gigantea of tropical America, 33
Fustic, from what obtained, 242

GAD-FLY of South America (Œstrus hominis), ulcers produced by the, 247
Galapagos, or Tortoise Islands, 339
— singular animal and vegetable life of the, 339
Galagos, the, 530
Galeodes vorax, its fatal bite, 301
Galeopitheci, the, 512
Gallinazos, or turkey-buzzards, 390
Garapats (Ixodes sanguisuga), a kind of tick, 250
Garua, or drizzling mists, of the Peruvian sand-coasts, 39
Gasteracantha arcuata, 292
Gavials of the Ganges, 351
— their attack of the tiger, 351
Gazelle, chase of the, in the Sahara, 75
Gecko, the, 328
— its usefulness to man, 329
— anatomy of its feet, 329
— different species of, 329
Geese, wild, of the heights of the Sikkim mountains, 92
Gemaledie, the mufti of Aden, introduces the use of coffee to the dervises of Aden, 189
Gemsbucks of South Africa, 65
Geography, physical, of the tropics, 3
Germany, cotton manufacture of, 209
Gibbon, the, 521
Gibraltar, monkeys of, 527
Gilolo, island of, birds of paradise of, 378
Ginger, its commercial importance, 233
— its growth and cultivation, 233
Giraffe, or camelopard, its beauty, 423

INDEX

GIR

Giraffe, its wide range of vision, 424
— use of its horns, 424
— its gregarious habits, 425
— hunting, 425
— his enemies in the forest, 427
— known to the ancients, 427
— his probable acclimatisation, 428
— always found near water, 65
Glagah, or tall grass of the Indian jungle, 156
Glow-worms (Lampyrides) of Europe, 270
— — of Sarawak, 271
— — worn as ornaments, 271
Glue, marine, 219
Gnu (Catoblepas gnu), always found near water, 65
— the, of South Africa, 425, 426
Goatsucker, singular voice of the, 369
— his usefulness, 370
— his food, 370
Godin des Odonnais, M., accompanies La Condamine on his voyage, 59
Godin, Madame, her adventures, 59
Goliath beetles of the coast of Guinea, 267
— — eaten, 263
Golunda coffee-rat, the, 197
Gomuti palm, wine of the, 136
Gomutus vulgaris, 136
Good Hope, Cape of, discovery of the road round the, 227
Goodyear, Mr., his mode of hardening India-rubber, 218
Goose, the white Huachua, of the Puna, 33
Gordonia Wallichii, the, of the Sikkim slopes, 90
Goree, ravages of termites at, 282
— giant spiders of, 292
Gorilla, the, 519
Greenland, number of beetles found in, 265
Green turtle (Chelonia midas), 347
Grenada, New, the Llanos of, 14
— — importance of the guadua bamboo of, 114
Grosbeak, the, 384
Gua Gede, cavern of, 305
Gua Rongkop, cave of, and its esculent swallows' nests, 305
Guadeloupe, tornado in, 11
Guadua bamboo, its importance in New Grenada and Quito, 114
Guama, Rio, singular vegetation on the banks of the, 123
Guanas of the Bahama Islands, 333
Guandiru (Phyllostoma hastatum), 508
Guano or Chincha Islands, 36, 42, 43
Guapore river, a tributary of the Madeira, 46, 57
Guarana Indians, importance of the Mauritia palm to the, 21
— — their singular habitations, 21
Guatemala, indigo of, 236
Guayaquil, perennial rainy season of, 9
Gueparda jubata, guttata, 482

HIM

Guiana, beauty of the vegetation of the banks of the rivers of, after the rainy season, 82
— birds of, 82, 366, 367
— magnificence of the forests of, 113
— insect plagues of, 250
— the black fire-ant of, 274
— the bush-master snake of, 315
— the toucan of, 360
Guinea, deflections from the ordinary course of the trade-winds on the coast of, 10
Guinea worm (Filaria medinensis), 249
Gulf stream, influence of the, on our climate, 5
Gulielma speciosa, its uses to man, 149
Gumatty, or fibres of the saguer palm, 136
Gumer, in Fassokl, immense baobab tree near, 103
Gutta percha, or gutta tuban (Icosandra gutta), its native country, 219
— — its introduction into Europe, 219
— — Malay mode of collecting the gum, 219
— — properties of gutta percha, 220
— — uses of gutta percha, 220
— — supply of gutta percha, 221
Gymnotus electricus, 20
Gyrophora of the Sikkim heights, 92

HACHA, RIO, red dye-wood from, 241
Haje (Naja Haje), of Egypt, 319
Haliætus vocifer, 395
Halig dates, 143
Hancock and Broding, Messrs., their India-rubber manufactures, 218
Hare, Mr., and the cocoa-nut oil of the Keeling islands, 130
Hares of the Sahara, 75
Hargraves, Mr., his invention of the spinning-jenny, 208
Harpy eagle (Thrasaëtes harpya), 392
Hayti, coffee trade of, 192
— attempts of M. Thierry de Meronville to introduce the cochineal insect into, 261
Hawaii, sandal tree of, 112
Hawk, the sparrow, of Africa (Melierca musicus), 395
Hawk-moths, giant, of the tropical forests, 84
Hawksbill turtle (Chelonia imbricata), 347, 349
Hebrides, New, sandal tree of the, 112
Hedgehog of the Sahara, 75
Heliconias of tropical forests, 81
Hemp, Manilla, or avaca, 175
Herbivorous animals of South Africa, 64
Herbs of the upper slopes of the Sikkim mountains, 91
Hercules beetle (Megasomina Hercules), 266
Himalaya mountains, 89
— — considered as colossal refrigerators, 5
— — land-leeches of the southern slopes of the, 252

HIN

Hindoos, their fondness for the banyan tree, 107
Hippopotamus, the Behemoth of the Book of Job, 432
— its diminishing numbers, 433
— "rogue hippopotami," or "bachelors," 434
— intelligence and memory of the hippopotamus, 435
— uses of its skin and teeth, 435
— modes of scaring it away from the fields on the White Nile, 436
— methods of killing it, 436
— the specimen in the zoological gardens, 438
Hippotigris, the, of the ancients, 430
Hog, the chief enemy of the rattlesnake, 316
Hogs, wild, of Peru, 40
Honduras, two rainy seasons of, 9
— mahogany trees of, 113
— turkey of (Meleagris ocellata), 376
Honey-ants of Mexico (Myrmecocystus Mexicanus), their singular habits, 281
Honey-eaters (Melithreptes), 375
Hornbill, red-beaked, 385, 386
— large, of Malabar, 386
— rhinoceros (Buceros rhinoceros), 374
Horses, their amphibious life during the rainy season in the savannahs, 19
— their enemies in all parts of the world, 19, 20
— effects of the rarefaction of the air on horses in the high table-lands of Peru and Bolivia, 24
Hottentots, fondness of the lion for the flesh of, 468
House-ants in the river Amazon, 277
Huallaga river, a tributary of the Amazons, 45
Huanacu, the, of Peru, 28
Humboldt, M. von, on tropical vegetation, 102
Humming-birds, 361
— — their wide range over the New World, 362
— — their habits, 362
— — their courage, 363
— — their enemies, 363
— — impossibility of bringing them alive to England, 363
— — mode of catching them, 364
— — various species, 364
Hunger, how long a man can withstand, 38
Huniman, the (Semnopithecus entellus), 525
Hyæna, the, 483
— hunting, 484
— varieties of the, 484
Hyænas of the Sahara, 75
Hydrocanthari, comparative numbers of, in the tropical and temperate zones, 266
Hyla crepitans, 338
— micans, 338

INS

Hyphæne coriacea, southern limits of the, 151
Hyphæne Thebaica, or doum palm, 144

IBIS, the, of the Puna, 33
— of America, 372
— of Egypt, 372
Ibo, palm oil of, 147
Iça river, a tributary of the Amazons, 45
Icebergs, wanderings of, 302
Ichneumon, or mongoos, his destruction of venomous serpents, 322, 355
Icterus Baltimore, 367
— xanthornus, 383
Iguana tuberculata, 352
Iguanas of the Peruvian sand-coast, 41
Inca beetles of South America, 267
India, teak forests of, 111
— bamboos of, 114
— cultivation of cotton in, present and prospective, 211
— hopes from the railways in progress, 212
— irrigation in India in ancient times, 212
— the indigo trade of, 237
— comparative rareness of venomous snakes in, 311
India-rubber tree (Ficus elastica), singular formation of the roots of the, 125. See Caoutchouc tree
Indian Ocean, north-east monsoon of the, 9
— — violence of the storms of the, 11
Indies, West, annual fall of rain in the, 6
— — insect plagues of, 250
— — turtles of the, 347
Indigo plant (Indigofera tinctoria), Bengalese cutting the plant, 234
— — mode of cultivation, 235
— — and of preparing the colour, 236
— — the indigo trade of the East Indies, 237
Indri, the, of Madagascar, 529
Inga, the, of the banks of the rivers of Guiana, 82
Insects of the Sikkim mountains, 93
— their aversion for sandal wood, 112
— of the tropical world, 245
— insect plagues, 245
— the universal dominion of, 245
— mosquitoes, 246
— the Œstrus hominis, 247
— the chegoe or jigger, 248
— Filaria medinensis, 249
— the bête rouge, 250
— blood-sucking ticks, 250
— land-leeches of Ceylon, 251
— the tsetse-fly, 252, 260
— the Soudan fly, 254
— the locust, 254
— the cucharacha and chilicabra cockroaches of Peru, 258
— tropical insects directly useful to man, 259
— silk-worms, 259

INDEX 553

Insects, cochineal, 260
— the gum-lac insect, 262
— eaten by man, 262
— worn as ornaments, 263
— increasing numbers in various countries advancing from the poles to the tropics, 265
— vast numbers of Coleoptera in Brazil, 267
— dragon-flies of the tropics, 267
— similarity of some, to the soil or object on which they are found: the walking-leaf and walking-stick insects, 268
— luminous, 270
— ants and termites, 272
— spiders and scorpions, 291
Island of Ascension, 345
— Banda, 228
— Ceylon, 282
— Cubagua, 58
— Goree, 282
— Lonshoir, 228
— Madeira, 184
— Melville, 265
— Pitcairn, 172
— Pulo Aij, 228
— Tahiti, 184
Islands:—
— Bahamas, 333, 346
— Carolinas, 130
— Coral, 309
— Feejee, 347
— Friendly, 169
— Galapagos or Tortoise, 339
— Keeling, 130, 134
— Kingsmill, 130
— Moluccas, 347
— Sandwich, 178, 376
— Society, 130
— South Sea, 375
— Tonga, 130
— Tortoise, or Galapagos, 339
Italy, origin of the sirocco of, 73
Ivory, African, 458
Ixodes sanguisuga, 250

JABIRU of America, 372
Jacana, the, or surgeon-bird, 372
Jackal, the, of the Sahara, 75, 480
Jaffna, forests of Palmyra palms in the peninsula of, 137
Jaggery, or Singalese sugar of the palmyra palm tree, 139
Jagua palm, elegance of the, 151
Jaguar, the, of the sand-coast of Peru, 40, 486
— his habits in the impenetrable forests of South America, 78
— his boldness, 487
— hunting, 488
Jaguarundi (Felis jaguarundi), the, 491

Jamaica, two rainy seasons of, 9
— pimento of, 232
Jatropha manihot, 175
— janipha, 176
Java, teak forests of, 111
— cocoa-nut oil trade of, 131
— ratans of, 141
— sparrow, or rice-bird (Loxia oryzivora), 155
— state of the coffee culture in, 191
— Javanese mode of cultivating coffee, 195
— cultivation of vanilla in, 201
— giant spiders of, 292
— the crocodile considered a sacred animal in, 354
— rhinoceros of, 448
— a Javanese jungle and its terrors, 476
— the kalongs, or fox-bats of, 507
Jelly-fish, 309
Jiboya, or boa constrictor, 319, 320
Jigger of the West Indies (Pulex penetrans), 248
Job, his description of Behemoth, 432
Jungle-fires in Sikkim, 115, 156
Jungle-fowl (Megapodius tumulus), mound-like nest of the, 381
Jungle-nail of Ceylon, 126
Jutay river, a tributary of the Amazons, 45

KADDIYA ANT of Ceylon, 274
Kala bird, the (Buphaga Africana), 442
Kalahari, causes of drought in the, 62
— abundance of vegetation in the, 62
— singular and useful plants of the, 62
— animal life of the, 64
— Bushmen and Bakalahari of the, 65
— skins of animals of the, 66
Kalongs, or fox-bats, of Java, 507
Karabamite, the largest of the humming-birds, 364
Kaross, or skin dress of the deserts of South Africa, 66
Keeling Island, method of catching turtle on, 346
— — cocoa-nut oil mills of, 130
— — cocoa-nut crab of, 134
Keitloa rhinoceros, 441
Kengwe, or Kēme, plant (Cucumis Caffer), of the Kalahari, 63
Khaidge pheasants, 92
Khamsin, or sand-storm of the Sahara, 72
Khaya Senegalensis, or mahogany tree of Africa, 113
Kiley, or boomerang, of the Australian savage, 414
Kinchinghow mountain, 89
Kinchinginga mountain, 89
Kingsmill Islands, perennial rainy season of the, 9
— — cocoa-nut oil trade of the, 130

KLI

Klipdachs, the, 394
Kobaaba rhinoceros (R. Oswellis), 441
Koodoo (Strepsiceros capensis), of South Africa, 65
Korakorum mountain, 89
Kordofan, baobab trees of, 103
— deleb palms of, 144
Korwê (Tockus erythrorynchus), 385
— its singular nest, 385
Kuisip river, vegetation of the banks of the, 63
Kunthia montana, height at which it will grow, 151

LABARRI snake, 312
— — the Prince of Neu Wied and one, 313
Lac, or gum-lac, 262
Lac, insect, the (Chermes lacca), 262
Lachesis rhombeata, 315
Ladang, or mountain rice, cultivation of, 156
Lagidium peruanum and pallipes of Peru, 32
Lagostomus viscacha of the Pampas, 32
Lagunes of the Marañon, 49
— fish catching on a grand scale, 49
Lalo, the, of the Africans, 104
Land, growth of, promoted by mangrove vegetation, 95
Land-crabs, 308
Land-leeches of Ceylon, 251
Lantern-fly (Fulgora lanternaria), its fabled phosphorescence, 271
Lapland, number of beetles found in, 265
Lar, the, of Siam and Malacca, 522
Larus serranus of tropical America, 33
Lathami talegalla, 380
Laurels of tropical forests, 81
Lecanium coffeæ, or coffee bug, 197
Lecanora miniata of the heights of the Sikkim mountains, 92
Leeches of the streams of the Sikkim mountains, 93
— of the southern slopes of the Himalaya, 252
— land, of Ceylon, the plague of, 251
Leguminosas of tropical forests, 81
Lemur, slow-paced, 529
— handed, 530
Leopard, the, 466
— the, attracted by the smell of the smallpox, 482
— the hunting leopard, or cheetah, 482
Lepidoptera, comparative numbers of, in Siberia and Brazil, 266
Leroshúa plant, its value to the inhabitants of the Kalahari, 62
Liana, or bushrope, of the primitive tropical forest, 79, 121
Libellula lucretia dragon-fly, 267
Lichens of the heights of the Sikkim mountains, 92
Licli, the, a bird of the Puna, 33
Lima wood, 241

LOC

Limulus, the, 307
Lion, similarity of the voice of the ostrich to that of the, 403
— on the borders of the Sahara, 75
— not a noble animal, 467
— his conflicts with travellers on Mount Atlas, 467
— his fondness for the flesh of the Hottentot, 468
— hunting, 470
— different species of the, 474
— his affection for the dog, 475
Liquidambar tree of Mexico, 88
Litchi (Nephelium litchi), of China and Cochin China, 180
Lithophytes, or stone polyps, 309
Liverpool, its monopoly of the palm-oil trade, 145
— trade of, 209
Livingstone, Dr., his efforts to promote the cultivation of cotton in Africa, 214
— attacked by red ants, 274
— his adventure with a lion, 471
Lizards of the Peruvian sand-coast, 4
— of the Sahara, 76
— their delight in the sunshine, 83
— their vast numbers in the tropics, 328
— the gecko, 328
— the anolis, 330
— chameleons, 331
— iguanas, 332
— guanas, 333
— monitor-lizard, 333
— water-lizards, 334
— flying-dragons, 335
— the basilisk, 336
— peculiar, of the Galapagos Islands, 339
Llama, its use to the ancient Peruvians, 26
— the only animal domesticated by the aboriginal Americans, 26
— its similarity to the dromedary of the Old World, 27
— its appearance, its ordinary load, and price, 27
— herds journeying described by Tschudi, 27
— kindness of the Indians towards them, 28
Llanos, the, of Venezuela and New Grenada, their extent, 14
— their aspect in the dry season, 14
— and in the rainy season, 18
— their appearance at the end of the rainy period, 21
— almost uninhabited when discovered by the Spaniards, 22
Lobsters of the tropical seas, 308
Locust (Gryllus migratorius), description of the, 254
— vast numbers of them, 254
— superstition of the Moslems respecting them, 255
— their devastation of the Empire of Morocco, 255

Locusts, Mr. Barrow's description of a swarm of, in South Africa, 255
— army of them at Poonah, 255
— larvæ of the, or "Voetgangers," 255
— Southey's description of them, 256
— eaten by man in the Sahara and South Africa, 262
Loggerhead turtle, 349
Logwood, 238
— a native of America, 238
— logwood-cutters, their mode of life, 238, 239
— disputes with the Spaniards, 238, 239
Lomas, or chains of hills, which bound the east of the sand-coast of Peru, 39
— the pasture-grounds of the Lomeros, 39
— beasts of prey in the Lomas, 40
Lonthoir, nutmeg trees of, 228
Loris, the, 529
Lory, the, 413
Loxia oryzivora, 155
Lucan, sugar alluded to by, 183
Luminous beetles, 270
Lum tree of Usian, singular formation of the roots of the, 125
Lyre-bird, 376, 377

MACA, a tuberous root, cultivated by the Indians in the high table-lands of Peru and Bolivia, 26
Mucauba palm trees, encased by parasitic fig trees, 123
Macaw (Macrocercus macao), 415
Mace, 230
Mackintosh, Mr., his India-rubber manufactures, 218
Macrocercus macao, 415
Madagascar, traveller-tree of (Ravenala speciosa), 175
Madeira river, a tributary of the Amazons, 46
— — its own tributaries, 46
— gigantic dragon trees of, 105
— the sugar-cane introduced into, 184
Magot, the, of the rock of Gibraltar, 527
Mahogany tree (Swietenia mahagoni), of British Honduras and Balize, 113
— of Africa (Khaya Senegalensis), 113
Maimon monkey, 527
Mainas, on the Amazons, house-ants in, 277
Maize, cultivation of, 159, 162
— imported from America by Columbus, 160
— its present cultivation in the eastern hemisphere, 160
— its magnificent growth, 160
— its enormous productiveness, 161
— production of, in the United States in 1853, 161, *note*.

Maize, the least subject to disease of all the cereals, 162
— the harvest of, 163
— its wide zone of cultivation, 163
— Dr. Franklin on the various uses of the plant and grain, 164
— beer, or chicha, 165
— nutrition of, 166
Makis, or murel, of Madagascar, 529
Makololo, spotted spider of, 292
— poisonous spider of, 296
Malabar, annual fall of rain on the coast of, 6
Malacca, peninsula of, its climate, 5
Malaria of the Mexican tierra caliente, 87
Mammee tree (Mammæa Americana), 266
Mammillariæ, the, 118
Mamore river, a tributary of the Madeira, 46
Manaar, in Ceylon, baobab trees at, 104
Manakins (Pipra), the, 366
Manchester, trade of, 209
Mandrill, the, 528
Mango (Mangifera indica), fruit of the, 181
— varieties grown at Kew Gardens, 181
Mangosteen (Garcinia mangostana), 180
— its delicious fruit, 181
Mangrove tree (Rhizophora gymnorrhiza, R. Mangle), 94
— — of the delta of the Amazons, 51
— — its peculiarities of growth and adaptation to its site, 94
— — its importance in furthering the growth of land, 95
— — animal life in the mangrove forests, 95
Manilla hemp, or Avaca, 175
Manis pentadactyla, 500
Mantis, or soothsayer, its habits, 268, 269
Mantis religiosa, 269
Marajo Island, size of the, 47
Marañon river. *See* Amazons.
Marantha arundinacea, arrowroot made from the, 177, 178
Margay, or tiger-cat, 491
Marimonda, the (Ateles Belzebub), 536
Marine glue, 219
Marsh forests of the torrid zone, 98
Marsh-tortoises of America and the Indian Archipelago, 342
Martha, Saint, red dye wood so called, 241
Mauritia palm, 15, 18, 21, 22
— — its importance to the South American Indian, 21
Mauritius, tornado in, 11
— the coffee tree introduced into, 190
Maynas, in Peru, mosquitoes of, 247
Medanos, or sand-hillocks, of the coast of Peru, 38
Megalasara Indica, 382
Megapodius tumulus, 381
— Duperreyii, 381
Megasomina Hercules, 266

Melierca musicus, 395
Melithreptes, or honey-eater birds of the South Sea Islands, 375
Melocacti, the pulp of the, 119
Melon-cactus of the llanos, the, 15
Melon, water, of the deserts of Africa, 63
Melville Island, scanty insect life on, 265
Membracidæ used as food by ants, 280
Menura, or lyre-bird, 376, 377
Mesembryanthemum, its admirable adaptation to the deserts of Africa, 64
— various kinds of, 64
Mexico, Gulf of, causes of the distribution of rain in the, 10
— New, influence of the heated plains of, in deflecting the trade-winds, 10
Mexico, geological formation of, 86
— the *tierra caliente*, or lowlands of, 87
— vegetable and animal life of, 87
— the *tierra templada*, 88
— the *tierra fria*, 89
— the Agave Americana of, 116
— the pulque of, 117
— cultivation of vanilla in, 201
— indigo cultivation in, 234
— giant spiders of, 292
Micrometers, spiders' webs used for, 299
Mikania huaco, an antidote against snake-bites, 314
Millet (Sorghum vulgare), cultivation of, 166
— the beer, or dhourra, of the Nubians, 166
Mimosas of the tropics, their beauty, 120
— the Porliera hygrometrica, 120
Minelle, a kind of cobra di capello, 318
Mirage in the llanos in the dry season, 16
— on the Amazons river, 55
Miriki monkey (Ateles hypoxanthus), 532
Misantla, cultivation of vanilla at, 201
Mists of the Peruvian sand-coast, 39
Mocha, quality of its coffee, 193
Mocking-bird (Cassicus persicus), 87, 88, 368
Mokuri plant, its importance to the inhabitants of the Kalahari, 63
Molasses, or uncrystallised sugar, 186
Moluccas, the, in the hands of the Dutch, 227
— turtles of the, 347
Mollusks, 308
Monakhir dates, 143
Mongoos, or ichneumon, 322
Monitor-lizard (Tejus monitor), 76, 333
Monkey-bread tree. *See* Baobab
Monkeys of the primitive forests, 83
— of the Sikkim mountains, 92
— their destruction of the sugar-cane, 186
— their fondness for coffee berries, 197
— not indigenous in Australia, 342
— of the Old World, 514
— their climbing powers, 515
— bad pedestrians, 515

Monkeys, contrasted and compared with man, 516
— the chimpanzee, 517
— the gorilla, 519
— the uran, or wild man of the woods, 520
— gibbons, 521
— the semnopitheci, 524
— the proboscis monkey, 525
— the huniman, 525
— the wanderoos of Ceylon, 525
— the colobi and cercopitheci, 526
— the magots of Gibraltar, 527
— the baboon, 527
— the maimon, 527
— the mandrill and drill, 528
— wide difference between the monkeys of the New and Old World, 531
— Indian mode of taming a monkey, 534
— the aluate, or howling monkey, 535
— the spider monkey, 536
— sakis, or fox-tailed monkeys, 536
— the nyctopitheci, or nocturnal, 537
— the duruculi and ouistitis, 538
— its points of resemblance to the parrot, 408
Monoho rhinoceros (R. simus), 441
Monsoon, the north-east, 9
— the south-west, 10
Montgomery, Mr, his introduction of gutta percha into Europe, 219
Moon-flower (Alango), the, 225
Mora excelsa of the forests of Guiana, Waterton's description of the, 113
Murmolyce, the Javanese, 270
Morning in the tropical forest, 82
Morocco, Empire of, laid waste by locusts, 255
Morus tinctoria, 242
Mosquitoes, 246
— ferocious, of the river Seuza, 246
— and of tropical America, 246
Mygale avicularia, effects of its bite, 296
Mosses of the Sikkim heights, 92
Moth, Atlas, of Ceylon, 267
"Mother of the Waters," Indian superstition respecting the, 53
Mountain: Donkiah, 89
— Everest, 89
— Kinchinghow, 89
— Kinchinginga, 89
— Korakorum, 89
Mountains: Andes, 89
— Cordilleras, 89
— Sikkim, 89
Mozambique, mode of catching turtles at, 347
Mule, the "ship of the desert" in Peru, 36
Mule-jenny, Crompton's invention of the, 208
Mulgrave Archipelago, importance of the screw pine of the, 118
Mummy-coffins, wood of which they are made, 106
Musa paradisiaca, 173

INDEX

MUS

Musa sapientum, 173
— textilis, 175
Mushrooms, enormous, 50
Muslin, antiquity of the manufacture of, in Hindostan, 208
Musk-deer on the slopes of the Sikkim mountains, 92
Mycetes ursinus, 535
Mygales, or trap-door spiders, 284
Mylodon, the, 496
Myrmecocystus Mexicanus, 281
Myrmecophagi of America, 500
Myrmica erythrothorax, 280
— paleata, 280
Myrtus pimenta, 233
Mzab, the Beni, and the inundation from the Atlas, 70

NADLER, Mr., his India-rubber manufactures, 218
Naja Haje of Egypt, 319
Namaqua country, reason why droughts are prevalent in the, 62
Napo river, 57
Naras, a fruit of the west coast of Africa, 63
Nefta, date-palms of the oasis of, 143
Negro, Rio, meaning of its name, 45
— — a tributary of the Amazons, 46
Negroes, great mortality of, in cultivating rice in South Carolina, 158
Nelumbias of the tropics, 123
Nepenthes, the, of the East Indian forests, 15
Nests of Australian birds, 378–381
— of the Baltimore bird, 367
— of the baya of Hindostan, 383
— of the bower-bird of Australia, 378, 379
— of the cassique, 368
— of the flamingo, 371
— of the grosbeak, 384
— of the humming-birds, 362
— of the korwé, 385
— of the lyre-bird, 377
— of the talegalla, or brush-turkey, 380
— of the variegated troopial, 367
Neu Wied, the Prince of, anecdote of, 313
Nicaragua wood, 241
Night in the tropical forests, 84
Nile, palm trees of the banks of the, 144
Nipa fruticans, 98
Noddy bird (Sterna stolida), its attacks on the cocoa-nut tree, 134
Noon in the tropical forest, 82
Nopal (Cactus opuntia), the, of the shores of the Mediterranean, 119
— (Cactus cochinellifer), 261
Nowaja Semlja warmed by the Gulf Stream, 5
— — nearly total absence of insect life at, 265

OST

Nubia, date-palms of, 144
Nutmegs, cultivation of, confined by the Dutch to Banda, Lonthoir, and Pulo Aij, 228
— their present extended range, 229
— description of the tree, 229
— mode of cultivation, 230
— nutmeg oils, 230
Nyctopitheci, or nocturnal monkeys, 537
Nycteribia, the, 511
Nylghau, the, 475, 476
Nymphæas of the tropics, 123

OAK, Indian, or teak tree (Tectona grandis), 110
Oases of the Sahara, character of the, 70
— — — — tribes inhabiting the, 71
Obydos, Strait of, 46
Ocelot (Felis pardalis), the, 491
Œnocarpus disticha, oil of the, 149
Œstrus hominis, 247
Oil made from cotton seed, 214
— cocoa-nut, 130
— essential, of cloves, 229
— palm trees of West Africa, 144
— of cinnamon bark, 227
— of the Corozo palm, 149
— of the Œnocarpus disticha, 149
Olas, or strips of talipot paper of Ceylon, 140
Opossum of the tropical forests, 84
— on the sand-coast of Peru, 40
Opuntias, the, 118
Orange trees in the Azores destroyed by a plant bug, 260
Orchids, flowering, of tropical forests, 80
— of the slopes of Sikkim, 90, 91
— of tropical forests, 123
Orellana, Francis, his voyage on the Amazons, 56
Oreodoxa oleracea, 148
— regia, 150
Organist bird (Troglodytes leucophrys), 370
Oricou, or sociable vulture (Vultur auricularis), 393
Orinoco river, 56, 78
— — manufacture of tortoise-oil on the banks of the, 343
Oriolus varius, 367
Orotava, in Teneriffe, gigantic dragon tree near, 105
Orycteropi, the, 500
Oryctes nasicornis, 266
Oryza sativa, 152, et seq.
Oscollo (Felis dogaster), the, 491
Ostrich, its endurance of thirst, 75
— its favourite haunts, 76
— the American, his attacks on snakes, 321
— swiftness of the ostrich of the Old World, 397

OST

Ostrich, mode of hunting it, 399
— its stratagem for protecting its young, 400
— its enemies, 402
— its young, 403
— its resemblance to the camel, 403
— similarity of its voice to that of the lion, 403
— its voracity, 404
— its feathers, 405
— domesticated in Algeria, 405
Oswell, Mr., his adventure with a rhinoceros, 444
—— his escape from a wounded elephant, 456
Otaheite, geckoes of, 329
Ouistitis, or squirrel monkeys, 538
Owl, burrowing (Athene cunicularia), of the Peruvian sand-coast, 41
— the pearl, of the same region, 41
Oxen, herds of, in the mountain valleys of the Puna, 34
— grunting, or yacks, 91

PACIFIC OCEAN, limits of the trade-winds in the, 6
—— causes of the distribution of rain on the Pacific off Central America, 10
—— violent tropical storms of, 12
Paddy, or uncleaned rice, 159
Pajara river during the rainy season, 19
Pallah (Antilope melampus), always found near water, 65
Palma real (Oreodoxa regia), the, 150
Palmetto of Azufral, height at which it will grow, 151
Palm-martin (Paradoxus typus or Pougouni), its fondness for cocoa-nuts, 134
Palm oil, 145
—— vast trade in, 145, *et seq.*
—— mode of trading in, at Bonny, 146
—— quantity exported from West Africa, 147
Palms of the tropical forest, 81
Palm-squirrel (Sciurus palmarum), its fondness for cocoa-nuts, 134
Palm trees of the lower slopes of Sikkim, 90
Palm trees, 128
— the cocoa-nut tree, 128
— the sago-palm, 135
— the saguer or gomuti, 136
— the areca palm, 137
— the palmyra palm, 137
— the talipot or talipot palm, 140
— cocoa de mer, 140
— date-palms, 142
— doum-palms, 144
— oil-palms, 145
— the carnuaba, 148

PER

Palm trees, the Ceroxylon andicola, 148
— the cabbage-palm, 148
— the corozo, 149
— the pirijao and piaçava palms, 149
— cabeza di negro, 149
— difficulties in ascertaining the different species of palms, 149
— their wide geographical range, 150
— different physiognomy of palms according to their heights, 151
— position and form of their fronds, 151
Palmyra palm (Borassus flabelliformis), extent of its range, 137
—— its uses to man, 137, 138
Pangolin, the Indian (Manis pentadactyla), 500
Panther, the, 481
Pao Barrigudo (Chorisia ventricosa), singular shape of the, 124
Papantla, cultivation of vanilla at, 201
Paper, Chinese, material of which it is made, 115
— made from the talipot tree of Ceylon, 140
Para, perennial rainy season of, 9
— population of, 56
Paradise, great bird of, 377
Paradoxus typus or Pougouni, 134
Paraguay, constant east winds of, 7
— river, 56
Parima river, 56
Paroquets, or parakeets, 415
— ring and green, 416
Parrots of the primitive forests, 83, 84
— their peculiar manner of climbing, 408
— their resemblance to monkeys, 408
— their food, 409
— their sociability, 409
— their connubial love, 410
— their powers of mimicry, 410
— American Indian mode of catching them, 412
— various species of them, 412
— the colours of, artificially changed, 413
Parsley, a deadly poison to parrots, 416
Peacock, the, 376
— frequently seen in company with the tiger, 476
Pearls, oriental, 309
Pegu, rice trade of, 158
Peireskia, the leafy cactus, of the Lake of Titicaca, 119
Pepper, 231
— description of the vine, 231
— mode of cultivation, 231
— its habitat, 231
Pernambuco, two rainy seasons of, 9
Peru, constant drought of the sand-coast of, 6
— the Puna, or high table-lands of, 23
— Puna chases in the times of the Incas, 32
— the sand-coast of, 36

INDEX 559

PER

Peru, desolation of, 36, 37
— sand-spouts and sand-hillocks of, 38
— summer season in, 38
— spring and the damp season, 38
— extreme dryness of the soil in the northern coast districts, 39
— animal world of the coast, 40
— the Guano or Chincha Islands, 42
— the chirimoya fruit of, 179
— cockroach pests of, 258
— the small brown viper of, 315
— the condor of, 388
— immense consumption of coca in, 202
— the Peruvian Indian's mode of chewing and preparing the coca acullico, 203
— divine honours paid to the shrub in, 205
Peruvian stream, influence of the, on climate, 5
Phasma dragon-fly, 267
Phasmas, the herbivorous, 268
Pheasants, Khaidge, 92
Phocæ of the Peruvian coasts, 42
Phœnix paludosa, on the islands of the Sunderbunds, 98
— dactylifera, or date palm, 142
Phœnicopterus ruber, 371
— antiquorum, 371
Phosphorescence of the centipede, 301
Phylliums, the herbivorous, 268
Phyllosomas, 307
Phyllostomidæ, the, 507
Physalia, or "Portuguese man-of-war," 309
Phytelephas (Cabeza di Negro), hard white nuts of the, 149
Piaçava palm (Attalia funifera), uses of the nuts and fibres of the, 149
Pigeons of the Peruvian sand-coast, 41
Pig-faced baboon, 528
Pimento, or allspice (Myrtus pimenta), 232
— cultivation of the plant, 232
Pimpla arachnitor, its mode of proceeding, 296
Pine-apple (Bromelia ananas), its abundance in Brazil, 117
Pines, the screw, of the East Indian and South Sea Isles, 118
— their importance to the inhabitants of the Mulgrave Archipelago, 118
Pinson, Vincent Yañez, discovers the Amazons river, 57
Pipa Surinamensis, 337
Pippul tree of India. See Bo tree.
Pipra, the, 366
Pique, or Jigger, of the West Indies (Pulex penetrans), 248
Piranga, voracity of the, 54
Pirijao palm (Gulielma speciosa), 149
Pisang, the, of the South Sea Islands, 174
Pitcairn Island, storm and famine in, 11
— — and the mutineers of the "Bounty," 172
Pitcher plant, 15

PUN

Pizarro, Gonzalo, his expedition from Quito over the Andes, to the Marañon river, 57
Plantain (Musa paradisiaca), its importance as food, 154, 173
Plants, tropical, 101, et seq.
— number of known, in the world, 102
Platalea Ajaja, 372
Pliny, sugar spoken of by, 183
Poison used by the Bushman of South Africa, 400
Polanarrua, climbing fig trees of, 122
Polyps, stone, or lithophytes, 309
Ponera clavata, excruciating pain caused by the bite of the, 273
Pongo de Manseriche, defile of, 44
Poonah, army of locusts, of a red species, at, 255
Porcupine, the, of the Sahara, 75
— ant-eater, 503
— wood, 132
Porliera hygrometrica, the, of Brazil, 120
Pororocca, or spring-tide wave of the Amazons, 47
"Portuguese man-of-war," 309
Potato, the sweet (Convolvulus batatas), 177
— when introduced into Europe, 177
Porto Santo, gigantic dragon trees of, 105
Pothos family of epiphytes of the tropical forests, 123
— beauty of the leaves, 90
Power-loom, Cartwright's invention of the, 208
Praong bamboo, of Sikkim, 115
Prêcheur insect, 269
Presbytes cephalopterus, 525
Prie Dieu, Le, insect, 269
Priodontes gigas, 501
Pristis, saw-snouted, 306
Proboscis monkey, the (Semnopithecus nasicus), 525
"Providence," voyage of Lieutenant Bligh in the, 169
Psittacus Amazonicus, 413
— erithacus, 411
— infuscatus, 409
— passerinus, 410
— pullarius, 410
Pterocarpus santalinus, 241
Pteropi, or fox-bats, 507
Pucaticse, house-ant, 277
Pulex penetrans of the West Indies, 248
Pulla, or palm oil, 145
Pulo Aij, nutmeg trees of, 228
Pulque, or Mexican agave wine, 117
Puma, or American lion, in the high table-lands of tropical America, 33, 490
— and on the Peruvian sand-coast, 40
Puna, or "Uninhabited" high table-lands of Peru and Bolivia, 23
— their contrast with the Llanos, 23
— violent changes in their temperature, 24
— plagues of the Puna, 24
— vegetable life of the, 25

PUN

Puna, animal life, 26 et seq.
— chases in the times of the Incas, 32
— beasts of prey of the, 33
— birds of the, 33
— flocks and herds of the Puna valleys, 34
— the mountain valleys, 34, 35
— the cacti of the, 120
Purus river, a tributary of the Amazons, 45
Puynipet Island, cocoa-nut oil trade of, 130
Pythons, 319
— their mode of attack, 319
— their long fasts, 319

QUAGGA, the, of South Africa, 428
Quinoa (Chenopodium quinoa), used as food in the Peruvian and Bolivian Andes, 177
Quinquina trees, region of the, 80
Quito, perennial rainy season of, 9
— the capital of Pizarro's government, 57
— importance of the guadua bamboo of, 114
Quiulla gull (Larus serranus), of tropical America, 33

RABBITS of the prickly shrubs of the Sahara, 75
Rain, abundance and distribution of, within the torrid zone, 6
— causes which produce an abundance or want of, 6
— annual fall of, on the coast of Malabar and in the West Indies, 6
— heavy afternoon showers of the zone of calms, 8
— zone of two distinct rainy seasons, 9
— and of one rainy season, 9
— immense quantity of, in the tropics, 10
— rainy season in the savannahs of South America, 18
— no rain in the northern coast-districts of Peru, 39
— enormous fall of, in the equatorial plains of the New World, 49
— reasons why so little rain falls in South Africa, 61
—*showers of liquid mud, 74
— periodical rains of the Sahara, 77
Ranger-ants, 278
— their venomous bite, 278
Rarotonga Island, devastation of, by a tropical storm, 12
Rat, its attacks on the cocoa-nut tree, 134
— his destructive ravages in sugar plantations, 187
— the Golunda, or coffee rat, 189, 197
Ratans, their immense length, 141
— uses of, 141
Rat-snake of Ceylon (Coryphodon Blumenbachii), domesticated, 323, 326
— its agility in seizing its prey, 327

RIV

Rattlesnakes, 315
— their rattle, 315
— different species, 316
— their chief enemy, 316
— divine honours paid to them, 319
Ravenala speciosa, or traveller-tree of Madagascar, 175
Rehoboth, larvæ of locusts in myriads at, 255
Reptiles of the Peruvian sand-coast, 41
Rhamphastidæ, 360
Rhea Americana, 406
— Darwinii, 406
Rhinoceros, the, always near water, 65
— its brutality and stupidity, 440
— different species of, 440, 441
— food and dispositions of the black and white kinds, 441
— their ugliness, 441
— their size, 442
— their acuteness of smell and hearing, 442
— defective vision, 442
— their friend the Buphaga Africana, 442
— their paroxysms of rage, 443
— their parental affection, 443
— their nocturnal habits, 443
— rhinoceros hunting and its perils, 444
— the Indian rhinoceros, 447
— the Sumatran kind, 447
— the Javanese rhinoceros, 448
— mode of killing it, 448
Rhinoceros Oswellsii, 441
— simus, 441
Rhinoceros beetle (Oryctes nasicornis), 266
Rhizophora gymnorrhiza and R. Mangle, 94
Rhododendron nivale, great elevation at which it grows, 91
Rhododendrons, arborescent, of the slopes of Sikkim, 90
— region of the Alpine, in the Sikkim mountains, 91
Rice (Oryza sativa), 152
— original seat of its cultivation, 154
— various aspects of the rice-fields at different seasons, 155
— ladang, or mountain rice, 156
— Malay "sawah," or artificially irrigated rice-fields, 157
— introduction of rice into South Carolina, 157
— great mortality of the negroes employed in cultivating it, 158
— rice trade of Arracan and Pegu, 158
— paddy, or uncleaned rice, 159
— mountain rice, 179
Rice, Colonel, his tiger-hunts, 477
Rice-bird, or Java sparrow (Loxia oryzivora), 155
Rivers of the tropics :—
— Amazons, 44 et seq.
— Apure, 19
— Arachuna, 19
— Aranca, 19
— Burima, Upper, 114

RIV

Rivers of the tropics, *continued*:—
— Beni, 46
— Berbice, 123
— Branco, 56
— Cassiquiare, 56
— Coary, 45
— Guama, 123
— Guapore, 46, 56
— Hacha, 241
— Huallaga, 45
— Iça, 45
— Jurua, 45
— Jutay, 45
— Kuisip, 63
— La Plata, 56
— Madeira, 46
— Mamore, 46
— Marañon, 44, *et seq.*
— Negro, 45, 56, 247
— Orinoco, 56, 78
— Parima, 56
— Pajara, 19
— Paraguay, 56
— Purus, 45
— Senza, 246
— Tapajos, 46
— Teffee, 45
— Tooge, 436
— Tocantines, 47
— Tunguragua, 44
— Ucayale, 45
— Xanari, 45
— Xingu, 47, 246
— Yapura, 45
— Zambesi, 435, 455
— Zouga, 436
Rodentia of the sand-coast of Peru, 40
Roots of trees, singular formation of the, 125
Ropes made from the bromelias, 117
Rupicola aurantia, 366
— Peruviana, 369

SAGO-PALM (Sagus fariniferus), the, of the Indian Archipelago, 135
Saguer, or Gomuti palm (Gomutus vulgaris), uses to which it is put, 136
Sahara, the, 6, 7, 68
— constant drought of the, 6
— north-easterly winds of, 7
— its uncertain limits, 68
— caravan routes, 69
— its desolate appearance, 69
— chasms and mountain streams, 70
— deposits of salt, 70
— the oases of the wilderness, 70
— tribes of the Sahara, 71
— contrast between the sterile desert and the oases, 71
— grandeur of the desert scene, 71
— its fascination for the traveller, 72
— the khamsin or simoom, 72
— sand-spouts, 74

SEC

Sahara, animal life of the, 75
— periodical rains of the, 77
Saïmiris monkey, the, 537
Sakis, or fox-tailed monkeys, 536
Salamanders, their delight in the sunshine, 83
Salt, deposits of, in the basins of the Sahara, 70
Sancudos, or mosquitoes, of tropical America, 247
Sand-coast of Peru, 36
Sandal tree of China, 112
— smaller variety of Hawaii, Feejee, and the New Hebrides, 112
Sand-hills of Chancay, on the Peruvian coast, 42
Sand-reed (Ammophila arundinacea), of the coasts of the North Sea, 64
Sand-spouts in the llanos in the dry season, 16
— — of the coast of Peru, 38
— — description of one at Chiggre, 74
Sand-storms of the Sahara, 72
— — at Schiebun on the Nile, 73
Sandwich Islanders, food of the, 178
— — their indolence, 179
Sarcoramphus papa, 390
Sarsaparilla, habitat of the, 50
Satinwood tree (Chloroxylon swietenia), forests of, in Ceylon, 111
Saul tree (Shorea robusta), excellent timber of the, 111
Saunders wood, red (Pterocarpus santalinus), 241
Saurians, 350
Savannahs of South America during the dry season, 14
— a savannah on fire, 17
— their aspect during the rainy season, 18
— and at the end of the rainy period, 21
Sawah, or artificially irrigated rice-fields of the Malays, 157
Saw-fishes, 306
Scalaria pretiosa, 308
Sciurus palmarum, 134
Scheidbun on the Nile, sand-storm at, 73
Schomburgk, Robert, his discovery of the Victoria Regia, 123
— bitten by a Ponera clavata, 273
— anecdote of him, 315
Scorpions, immense size of, in the torrid zone, 299
— fatal effects of their bite, 300
— their habitat, 300
— poison lodged in their tails, 301
Scotophilus Coromandelicus, the, 511
Sea-birds, tropical, 303
— on the coast of Peru, 42
Sea-eagle of the north, its devices for procuring food, 402
Seals of the Peruvian sand-coast, 42
Sea-otters of the Peruvian coast, 42
Secretary-bird, his destruction of snakes, 321

Secretary-eagle (Serpentarius cristatus), 396
Semnopitheci, the, 524
Senegal, ostriches of, 398
Seringueros, or caoutchouc-gatherers, 217
Serpentarius cristatus, 396
Serpents. *See* Snakes
Shark, the white, his ferocity, 307
Sheep, flocks of, in the Puna mountain valleys, 34
Shells, rare and beautiful, 308
Sherbet, preparation of, 144
Ship of the desert, 417
Shoes made of India-rubber, 218
Shorea robusta of India, 111
Siamang of Sumatra, the, 512–524
Sikkim mountains, slopes of the, 89
— — sylvan wonders of the, 89
— — changes of the forests on ascending, 90
— — the torrid zone of vegetation, 90
— — the temperate zone, 90
— — the coniferous belt, 90
— — limits of arboreal vegetation, 91
— — animal life, 92
— — the Praong bamboo of, 115
— — firing the jungle in, 115
Silk-worm (Bombyx mori), its importance to man, 259
— antiquity of silk in China, 259
Silk-spiders, 298
Simiæ of the Old World, 514
Simoom, the, of the Sahara, 72
Singalese, their mode of avoiding snake-bites, 311
Singapore, sago trade of, 136
Siphonia elastica, 216
Sirocco, origin of the, of Italy, 73
Skink of the Sahara, 76
Slavery amongst ants, 279
Sloth, the, 83
— his miserable appearance, 492
— adaptation of his organisation to his peculiar mode of life, 493
— his means of defence, 495
— his tenacity of life, 496
— genera of the sloth, 496
Small-pox, panthers or leopards attracted by the smell of the, 482
Snake-tree, the, 125
Snakes of the Sahara, 75
— of the tropical forests, 83, 310
— comparative rareness of venomous, 311
— habits of venomous, and their external characteristics, 312
— bite of the trigonocephalus, 313
— antidotes, 313
— fangs of venomous serpents, 314
— the enormous bush-master, 315
— the rattlesnake, 316
— the cobra di capello, 317
— the asp and viper, 319
— boas and pythons, 319

Snakes, enemies of, 319
— sometimes feed on one another, 323
— their means of locomotion, 323
— anatomy of their jaws, 324
— feeding-time at the Zoological Gardens, 325
— useful to man, 326
— adaptability of their colour to their pursuits, 327
— water, 327
Society Islands, cocoa-nut oil trade of the, 130
Solaneas of tropical forests, 81
Sonneratias, the, of the rivers of India, the Moluccas, and New Guinea, 98
Soudan, cotton of the, 213
— destructive fly of, 254
Southey, Robert, his description of the locust, 256
South Sea Islands; screw pine of the, 118
Sparrow-hawk of Africa (Melierca musicus), 395
Sparrow, Java, or rice-bird (Loxia oryzivora), 155
Sperm whales, 303
— — attacked by sword-fish, 306
Sphargis coriacea, 348
Spices of the tropics, 222
— cinnamon, 222
— nutmegs and cloves, 227–230
— mace, 230
— pepper, 231
— importations of spices into Great Britain during part of 1861, 232, *note*
— pimento, 232
— ginger, 233
Spiders, tropical, formation of, 291
— spotted spider of Makololo, 292
— giant webs of several tropical species, 292
— harmony of colour between the Araneæ and their usual haunts, 293
— beautiful colouring of the epeiras, 294
— the mygales, or trap-door, 294
— maternal instincts of, 295
— enemies of, 295
— venom of the, 296
— services rendered by spiders to man, 297
— eaten by several savage nations, 297
— encounter between a spider and a cock-roach, 298
— silk-spiders of America, 298
— threads of this spider used for micrometers, 299
Spider monkeys, 536
Spines and thorns of tropical plants, 126
Spinning-frame, invention of the, 208
Spinning-jenny, invention of the, 208
Spitzbergen, warmed by the Gulf Stream, 5
Spondylus, the royal, 308
Spoonbill of America (Platalea ajaja), 372

SQU

Squirrels, their fondness for coffee berries, 197
— flying, 511
Steer, the black-legged and white-coated, 448
Storms, tropical, violence of, 11
— tornados and cyclones, 11
— on the Amazons river, 55
— in the Sahara, 72
— a hurricane in the primitive forest, 81
Strepsiceros capensis of South Africa, 65
Struthio camelus, 397, *et seq.*
Sucking-fish (Remora), used for turtle-catching, 347
Sucuriaba, or water-boa (Eunectes murinus), 320
Sugar, or jaggery, of the palmyra palm, of Ceylon, 139
— commercial importance of sugar, 182
— original home of the sugar-cane, 183
— progress of its cultivation throughout the tropical zone, 183, 184
— mentioned by several classical authors, 183
— transplanted from the Canary Islands to Hispaniola, 184
— the Chinese species supplanted by the Tahitian kind, 184
— description of the cane, 185
— manufacture of sugar, 185
— molasses, or uncrystallised sugar, 186
— the enemies of the sugar-cane, 186
— nutritive qualities of its juice, 187
Sugar-ants, ravages of the, 275
Sumatra, crocodiles of, 356
— rhinoceros of, 447
Sumpits, Malay, materials used for forming the shafts of, 136
Sun, unequal influence of the, on the sea and land, 4, 5
Sun-birds, or suimangas (Cinnyris), 375
Sunderbunds, the Phœnix paludosa of the islands of the, 98
Sun-fish, the, 307
Superstition, an Indian, 53
Surgeon-bird, 372
Surinam toad, 337
— tree-frog, 338
Surumpe, or acute inflammation of the eyes in the Puna, 24
Swallow, the esculent (Colocalia esculenta), 305
— mode of getting the nests, 306
Sweden, number of beetles found in, 265
Sword-fishes, 306
Sword-tail fishes, 307
Sycamore tree (Ficus sycomorus), gigantic specimens of the, in Africa, 105
Sylvia sutoria, 383

TACCA PINNATIFIDA, arrowroot made from the, 177, 178

TER

Tahiti, sugar-cane of, 184
Tailor-bird of Hindostan (Sylvia sutoria), 383
Talegalla, or brush-turkey, 380
Talgaha tree the sacred, of the Brahminical Tamils of Ceylon, 137
Talpot, or talipot tree of Ceylon, uses to which it is applied, 140
Tamehameha, the Great, his royal mantle, 375
Tamr dates, 143
Tangaras, the, 365
Tapajos river, a tributary of the Amazons, 46, 47
Tapioca, from what prepared, 176
Tapir, the, of the tropical forests, 84
Tarantula spider of Sicily, stories of its bite, 296
Taro roots (Caladium esculentum) of the Sandwich Islanders, 178
— — its abundant growth, 178
— — mode of cooking it, 178
— — mountain taro (Caladium cristatum), 179
Tarush, an animal peculiar to the Punas, 32
Teak tree, or Indian oak (Tectona grandis), 110
— — its excellent timber, 110
Tectona grandis, or Indian oak, 110
Teffe river, a tributary of the Amazons, 4'
Teju, or monitory lizard (Tejus monitor), of South America, 333
Temperature of a country, decrease of, in proportion to its elevation, 5
Teneriffe, cochineal trade of, 261
Teoge river, hippopotamus hunting on the, 436
Terebinthinaceas of tropical forests, 81
Termites, or white ants, 281
— their devastations, 281
— their introduction into Europe, 283
— their services and uses, 283
— their communities and astonishing buildings, 283
— formation of a termite colony, 285
— wonderful fecundity of the queen, 285
— courage and obstinacy of the termite soldier, 286
— foes of the termites, 133, 287
— their wars with the black ants, 288
— termites used as food, 288
— marching termites, 289
— mysteries of termite life, 290
— their deserted nests inhabited by the cobra di capello, 318
Termes bellicosus, their clay-built citadels or domes, 284
— destructor arborum, their dwellings in trees, 283
— flavicollis introduced from Northern Africa into Marseilles, 283
— flavipes, introduced into Portugal and Vienna, 283

TER

Termes lucifuga of France, 283
— vinrum, or marching white ants, 285
Testudo Indica, elephantina, 339
— — its habitat and habits, 340
Texas, influence of the heated plains of, in deflecting the trade-winds, 10
— the maize harvest of, 163
Textor erythrorhynchus, 442
Theridion, its maternal instincts, 295
Thierry de Meronville, his attempts to introduce cochineal into San Domingo, 261
Thirst, how long a man can withstand, 38
Thorinsia, maternal instincts of the, 295
Thorns and spines of tropical plants, 126
Thrasaëtus harpya, 392
Tibbo tribe of the Great Sahara, 71
Ticks, the blood-sucking, of the tropical regions, 250
— the garapata (Ixodes sanguisuga), 250
Tierra caliente, the, of Mexico, 87
— templada, 88
— fria, 89
Tiger, the time for his bloodthirsty excursions, 476
— his chief seats, 477
— tiger-hunting, 477
— destroyed by the gavial of the ganges, 351
— his mode of attack, 481
— his destruction of the tortoise, 481
Tiger-cat, or margay, 491
Tillandsias of the tropical forests, 123
Timber, valuable, of the teak or Indian oak, 110
— of the Saul tree (Shorea robusta), 111
— of the satinwood tree of Ceylon, 111
Titicaca, Lake, cacti of the, 120
— maize of the islands of, 163
Tlaouili, a gigantic kind of Mexican maize, 163
Toads of the tropics, 337
— the Pipa Surinamensis, 337
— the Bahia toad, 337
— the giant toad, 338
— the musical toad of Guinea, 338
Tocantines river, 47
Toddalia aculeata, thorns of the, 126
Toddy-bird of Ceylon (Artamus fuscus), 138
Toddy made from the Mauritia palm, 21
— extracted from the cocoa-nut tree, 131
— and from the palmyra palm, 138
— and from the date-palm, 143
Tofoa, mutiny of the "Bounty" at the island of, 169
Tornados, 11
Tortoise-oil, manufacture of, 342
Tortoises of the tropics, 339
— the gigantic land-tortoise, 339
— Mr. Darwin's ride on one, 342
— tortoises not indigenous in Australia, 342

TUR

Tortoises, marsh, of America and the Indian Archipelago, 342
— river, 343
— attacked by wild dogs and tigers, 481
Toropishu bird (Cephalopterus ornatus), 369
Touat, oases of the, annual pilgrimage from the, 69
Toucans (Ramphastidæ), 360
— their quarrelsome habits, 360
Tozer, date-palms of the oasis of, 143
Trade-winds, the, 4, 6
— their limits in the Northern Atlantic, 6
— and in the Pacific, 6
— moisture carried by them from the coldest to the warmer regions, 7
— deflections from the ordinary course of the trade-winds, 10
Trap-door spiders, 294, 297
Traveller-tree of Madagascar (Ravenala speciosa), 175
Tridacna, the colossal, 308
Trigonocephalus, fatal effect of the bite of the, 313
Trincomalee, forests of satinwood at, 111
Tripe de roche of the Sikkim heights, 92
Trochilus minimus, the smallest of the humming-birds, 364
Troglodytes audax of Peru, 356
— leucophrys, 370
Troopials (Icterus, Xanthornus) of Guiana, 367
— the variegated troopial, 367
Troughton, Mr., his use of spiders' threads for micrometers, 299
Trunk-fish, the, 307
Tryothorus plateus, 383
Tsalt-salya, or zimb, of Abyssinia, 253
Tsetsé-fly of South Africa (Glossitans mositans), 252, 260
— its destruction to cattle and horses, 252, 253
— range of its pestiferous influence, 253
Tuaryk tribe of the Sahara, 71
— chieftain of the, 77
Tunguragua river, 44
Tunqui bird (Rupicola Peruviana), 369
Tunuhy, the Sierra, 45
Tupinambaranas, Island of, 46
Turmeric or Indian saffron, 242
Turkey of Honduras (Meleagris ocellata), 376
— the brush, or talegalla, 380
Turkey-buzzards, 390
Turtles of the tropics, 344
— colossal, of the Brazilian coast, 344
— foes of the turtle tribe, 345
— of the island of Ascension, 345
— mode of taking them at Ascension, the Bahamas, and at Keeling Island, 346
— and at Mozambique, 347

INDEX

TUR

Turtles, barbarous treatment of, at Feejee and Ceylon, 347, 348
— food of, 349
Tusseh-worm (Bombyx mylitta), silk filaments of the, 259

U

UJALAN, island of, singular roots of a tree on the, 125
Ucayale river, a tributary of the Amazons, 45
Umbrella ants, 276
Umbrellas of ostrich feathers, 405
Unaus, the, 496
Uran, or wild man of the woods, 520
Urceola elastica, caoutchouc of the, 217, 220
Uropeltis Philippinus, 310
Ursus malayanus, its fondness for cocoanuts, 134
Urticeae, dendritic, of tropical forests, 80
Urubu, the, of America, 390
— his habits, 391
Utah, influence of the heated plains of, in deflecting the trade-winds, 10

V

VAMPIRES, 507, 509
Vanikoro, geckoes of, 329
Vanilla (Vanilla aromatica), 201
— cultivation of, in Mexico and Java, 201
Vargas, Sanchez, his fate, 58
Vegetable life in the savannahs of South America in the dry season, 16
— — in the rainy season, 18
— — vegetation of the Puna or high table-lands of Peru, 25
— — vegetation of the sand-coast of Peru, 36
— — of the banks of the Amazons, 50, 56
— — of the Kalahari, 62
— — of the oases of the Sahara, 70
— — of the primitive forests of the tropics, 78
— — of the Mexican plateaus, 87, 88
— — of the slopes of the Sikkim mountains, 90
— — of the mangrove forests, 94
— — giant trees and characteristic forms of tropical vegetation, 101
— — gigantic vegetation of tropical America, 112
— — singular formation of the trunks and roots of trees, 124, 125
— — thorns and spines of tropical plants, 126
— — palms, 128
— — chief nutritive plants of the torrid zone, 154
— — fruit trees of the tropics, 179
— — vegetable life of the Galapagos Islands, 339

WAX

Vejuco de huaco (Mikania Huaco), an antidote against snake-bites, 314
Velella, the, 309
Venado, a species of deer, of the sand-coast of Peru, 40
Venezuela, the llanos of, 14
— cultivation of coffee in, 192
Veta, a disease caused by the rarefaction of the air in the high table-lands of Peru and Bolivia, 24
Victoria regia, the, 123
Vicuña, its solitary habits, 29
— value of its wool, 29
— its appearance, 29
— Indian mode of hunting it, 30
— mode of preparing its flesh, 31
— its enemies, 32
Viper, small brown (Echidna ocellata), of Peru, its fatal bite, 315
— the horned (Cerastes) of the Egyptian jugglers, 319
— idolum, divine honours paid to the, 319
Virginia, bull-frog of, 338
Viscachas, the, of Peru, 32
— of the Pampas, 32
Viviagua ant of the West Indies, its ravages in the coffee plantations, 275
" Voet-gangers," or pedestrians, the larvæ of locusts so called, 255
Vomito, or malaria, of the tierra caliente of Mexico, 87
Vulcanisation of India-rubber, 218
Vultures, white, their destruction of ostrich eggs, 402
— of the sand-coast of Peru, 40
— of America, 390
— king of the, 390
— of the Old World, 393

W

WALKING-LEAF insect, 268
Walking-stick insect, 268
— — — of Van Diemen's Land, 269
Wanderoos of Ceylon (Presbytes cephalopterus), 525
Water, Bechuana mode of drawing, in the desert, 66
— temperature at which it boils in the Puna, 25
Water-boa (Eunectes murinus), 320
Water-lizards (Hydrosauri), 334
— — Mr. Adams's contest with one, 334
— — their habitat, 335
— — worshipped at Bonny, 335
Water-plants of the tropics, 123
Water-snakes, 327
— buried during the dry season, in the savannahs, 16
— their return to life in the rainy season, 19
Wax obtained from the Carnauba palm, 148
— and from the Ceroxylon andicola, 148

WES

West Indies, introduction of the coffee-plant into the, 191
— present state of its culture, 192
— Martinique, introduction of the coffee-plant into, 191
Widah, divine honours paid to the viper in, 319
Winds, the system of, and its importance, 4
— trade-winds, and polar and equatorial air-currents, 4, 6, 10
— effects of the Puna-winds in arresting decomposition, 25
— origin of the sirocco of Italy and Fön-wind of the Alps, 73
Wine of the Agave Americana, 117
— of the saguer or gomuti palm, 136
Wöhler, Professor, of Göttingen, his analysis of coca, 206
Woodpecker, the ivory-billed, 83
— orange-coloured (Brachypterus aurantius), 382
Wood-pigeon, cries of the, 83
Wool, annual imports of, into England, 207
Woollen cloth, manufacture of, in former times in Flanders and England, 207

ZOU

XAVARI river, a tributary of the Amazons, 45
Xingu river, a tributary of the Amazons, 47

YACKS, or grunting-oxen, 91
Yams (Dioscorea sativa and alata), 176
Yapura river, a tributary of the Amazons, 45
Yarriba country, cultivation of cotton in the, 213
Yriartea exorrhiza, 152
— ventricosa, 152
Yuccas, the American, 116

ZAMBESI river, hippopotami of the, 435
— — — elephant-hunting on the, 485
Zebra, Burchell's, or douw, 429
— always found near water, 65
— the hippotigris of the ancients, 430
— its piteous wailings,
— its inaccessible

www.ingramcontent.com/pod-product-compliance
Lightning Source LLC
Chambersburg PA
CBHW021229300426
44111CB00007B/477